# DYNAMICS OF CLASSICAL AND QUANTUM FIELDS

## An Introduction

# DYNAMICS OF CLASSICAL AND QUANTUM FIELDS

## An Introduction

**Girish S. Setlur**

**Indian Institute of Technology**
**Guwahati, India**

CRC Press
Taylor & Francis Group
Boca Raton London New York

CRC Press is an imprint of the
Taylor & Francis Group, an **informa** business

Taylor & Francis
Taylor & Francis Group
6000 Broken Sound Parkway NW, Suite 300
Boca Raton, FL 33487-2742



ISBN 13: 978-1-4665-5628-7 (hbk)



**Visit the Taylor & Francis Web site at**
**http://www.taylorandfrancis.com**

**and the CRC Press Web site at**
**http://www.crcpress.com**

To my readers

# Contents

# List of Figures

# Preface

In the hyper-connected world we live in—where information in all its many avatars not only travels at the speed of light but is also assimilated equally fast—the traditional book is struggling to justify its existence. This struggle is especially acute for a book that attempts to revisit topics on which weighty tomes have been penned by Nobel Laureates, no less. Another introductory book on classical and quantum field theory had better have a good reason to be. A good reason there is. Firstly, most books are either on quantum fields or classical fields, never both, barring perhaps a brief introduction to classical fields just enough to find application to quantum fields. Very few introductory texts mention dynamical symmetries which lead to conservation of the Runge Lenz vector in discussions of Noether's theorem. The words 'tensor', 'Lorentz covariance', 'gauge invariance' and so on continue to instill confusion and trepidation in the minds of the novice. Topics such as elasticity theory have fallen into disfavor through no fault of their own. The equations of fluid mechanics are seldom shown to be derivable from point-particle mechanics as the continuum limit of these traditional ideas. Stokes drag, which every high school student knows but whose derivation is kept a closely guarded secret, is typically found only in specialized texts in fluid mechanics that are rarely consulted by instructors teaching general field theory. In quantum mechanics, discussion of fermions is limited to wavefunctions of a few particles. Few attempts are made in introductory texts to study the behavior of fermions in the presence of a filled Fermi sea taking into account Pauli blocking—an exercise that naturally leads to the important notions of right and left movers. While path integrals are typically taught well in traditional texts on field theory, even they do not address the question of how to use a path integral in the presence of a filled Fermi sea. How Wick's theorem may be derived from path integrals is usually left implicit. The core concepts of Fock space of fermions and bosons, definitions of creation and annihilation operators of a fermion or boson at a point in space and related notions, so important in many-body physics, rarely find explicit mention in available texts. Likewise, lattice models such as the Hubbard model, the t-J model, and so on are rarely shown to be derivable from more familiar continuum pictures. The non-equilibrium many-body Green function method, originally due to Schwinger, is rarely covered in general field theory texts. Saving the best for last, the final two chapters on non-local operators in many-body physics—a subject of ongoing research—take the reader to the place where she should be—at the frontiers. It is my claim that this book does all of this and deserves to be on the shelf of whoever is reading this preface now.

The main motivation for writing this book came from the feedback given by our Ph.D. scholars who felt that a comprehensive book on topics not typically covered well in their M.Sc. coursework would go a long way in speeding up progress in their research careers. Departmental meetings, where course curricula are debated ad nauseam over countless cups of tea, with decisions on the length of the Ph.D. coursework reversed after every alternate semester, helped, too.

An author attempting to provide an original perspective on standard and core topics runs the very real risk of getting the basics wrong, for in physics, there are only a few ways of doing a problem right and countless ways of getting it wrong. It is my hope that such mistakes are absent. I have purposely made some of the exercises open ended. A few questions are deliberately vaguely posed because being able to pose a proper question in research is a job half done. These questions challenge the mind to think like a researcher. By and large, I have made an effort to make the book's treatment complementary to what may typically be found in standard treatises. The common thread running through the entire book is the notion of a field—an infinite collection of dynamical variables where the infinity is of the continuous kind. The prose, namely, wordy explanations are given only when the equations are not self-explanatory. They say a picture is worth a thousand words. A wise physicist once observed that an equation is worth a thousand pictures.

I don't necessarily endorse the contents of the references listed in the bibliography. They are meant to point the reader to literature relevant to the topics discussed in this book. I apologize to authors who are left out and urge them to inform me about their works.

Now I have to be upfront and point out all the shortcomings I already know about. Gross injustice has been done to our high-energy friends. Their favorite topics, such as QED and so forth, have been completely left out—Dirac, Klein Gordon and QCD Lagrangian are just mentioned. The connection between spin and statistics viz. that fermions have half integer spin and bosons have integer spin does not even find a mention, let alone an explanation. No sophisticated theoretical tools such as renormalization group, dynamical mean field theory, etc. other than bosonization are discussed. I make a distinction between tools and models on which the tools are applied. This book has plenty of the latter but very few of the former—the implication being that development of effective tools is still a subject of ongoing research.

All these shortcomings have their reasons. I don't feel qualified to write about any of these topics as of now. Pauli's proof of the spin statistics connection has been criticized by E.C.G. Sudarshan and I. M. Duck. Thus, I don't feel comfortable including either Pauli's proof or its criticism, and instead refer the reader to the extensive bibliography. I have not studied any tool in depth other than bosonization, which is what I wish to emphasize. This book therefore has a hidden agenda of wanting to 'convert' readers into making the vision of the last few chapters widely successful.

This book has to be read sequentially. Unlike in most Indian movies where one could walk in anytime and still would have not missed anything, the uninitiated reader cannot simply walk into some chapter and follow the logic as use is made of earlier chapters. The point particle picture eases into the classical continuum. The classical picture merges into the quantum one and so forth. That said, an instructor could easily pick and choose the topics he wants to cover, provided he is willing to bridge the gaps that inevitably result.

No book is written in isolation. This one is no exception. A large number of resources, both human and cybernetic (as in the Internet), have been employed in writing the book. Human contributors include many good friends and former and present students. In no particular order they are, Anurag Anshu (currently a final-year undergraduate in IIT Guwahati) who critically read the manuscript and suggested improvements. He also provided some illustrations and helped with LaTeX. Dr. V. Meera, a former Ph.D. student at IIT Guwahati, provided many of the references in the bibliography. She also helped with LaTeX and illustrations. Upendra Kumar (current Ph.D. scholar at IITG) typed some sections from my lecture notes for which only pdf files were available. Colleagues here in the Physics Department, 'coffee mates' as it were, helped a lot too by allowing me to bounce ideas off their welcoming chests. Naming some will run the risk of forgetting others, but still, again in no particular order, Dr. Amarendra Sarma and Dr. Tarak Dey for sharing their expertise and a vast collection of actual books and e-books, and Dr. Charudutt Kadolkar for being a reliable 'stress tester' for novel ideas and also for sharing his considerable expertise and e-resources. All faculty members here have been really friendly and cooperative.

Non-human contributors include the IIT Guwahati Physics Department (thank you, third floor and department office printers), Department of Science and Technology Project #(DST-SERC) SR/S2/CMP/46 2009, various websites of Physics professors all over the world, Google search engine, Mathematica software ... the list is endless.

On a personal front, my parents Veda and Sampath have been a source of great support and inspiration. They have passed on their dedication to their fields of expertise to me. My wife Uma has been a source of delicious four-course meals daily for years. She has also been ever ready with a pair of blinkers whenever I felt distracted. My brother Sudarshan (the Wall Street tycoon, not the physicist) and his lovely family always put a smile on my face.

# About the Author

Girish S. Setlur, Ph.D. is an associate professor of physics at the Indian Institute of Technology Guwahati, Guwahati, India. He received his B.Tech. in engineering physics from the Indian Institute of Technology Bombay, India. He earned his M.S. and Ph.D. degrees in physics from the University of Illinois at Urbana-Champaign, USA. He has nearly two decades of teaching and research experience both at the undergraduate and postgraduate levels. He is currently a mentor of several bachelor's, master's and Ph.D. students.

# Chapter 1

# The Countable and the Uncountable

This chapter deals with the analogy between the classical mechanics of discrete particles and that of a continuum. The reader is expected to be familiar with Lagrangian mechanics at the level of *Classical Mechanics* by Goldstein. Nevertheless, we provide a brief introduction to, and a derivation of Lagrange equations starting from Newton's equations of motion. This is followed by a discussion that motivates the notion of a dynamical field as the continuum analog of a dynamical generalized coordinate (of a system with a finite number of such coordinates). The chapter also includes worked-out examples of dynamical fields encountered in everyday situations. The aim is to introduce readers to the subject through quasi-realistic examples that will enable them to formulate and solve problems.

## 1.1 Review of Lagrangian Mechanics

The notion of a countable set is intuitive and easy to grasp even for children. We learn to count with our fingers at an early age: add, subtract, multiply, and divide integers. What is less intuitive is the notion of a fraction and the concept of an irrational number, even more so. One may suspect that the need for such a notion stems from a desire to abstract the properties of space and time that we perceive as continuous rather than discrete. From this emerges all of the mathematical subject of calculus. The present book deals with the subject of mechanics (be it classical or quantum) of systems with the number of degrees of freedom being an infinity of the continuous kind. This chapter starts with some background material on classical mechanics of systems with finite degrees of freedom that readers are expected to be familiar with. But we show that the same ideas may be generalized quite naturally to deal with systems with infinitely many degrees of freedom. We use examples to drive home this point.

In classical mechanics we deal with generalized coordinates. We identify a finite number of degrees of freedom $\{q_1(t), q_2(t), ..., q_s(t)\}$ that is sufficient to specify the configuration $\{\mathbf{r}_i(q_1, q_2, ..., q_s) : i = 1, 2, ...., N\}$ of the system (collection of particles subject to various constraints) at any given time. If we imagine that the system contains a countable (and finite) number of particles with positions $\{\mathbf{r}_1, \mathbf{r}_2, ..., \mathbf{r}_{N-1}, \mathbf{r}_N\}$, then we may think of the position of each particle $\mathbf{r}_i(q_1, q_2, ..., q_s)$ as depending on the generalized coordinates, which are going to be fewer in number depending on the number of constraints the system obeys. For example, a particle in three dimensions would normally require three coordinates $x, y, z$ or $r, \theta, \phi$ for fixing its location. But when it is constrained to only move on a sphere of radius $a$, the number of coordinates needed are fewer viz. the angles, $\theta$ and $\phi$. These are precisely the generalized coordinates that we are talking about. We may imagine that each particle is acted on by a predetermined force $\mathbf{F}_i(\mathbf{r}, \mathbf{v})$, which could in a general case depend on the locations and velocities of all the particles. In this case, the force law reads as follows.

$$m_i \frac{d^2}{dt^2} \mathbf{r}_i = \mathbf{F}_i(\mathbf{r}, \mathbf{v}). \tag{1.1}$$

The number of unknown functions that these equations purport to solve for is the number of generalized coordinates, or degrees of freedom consistent with the constraints. However, it is evident that the number of such force equations is the number of particles times the dimension of space (3N in three dimensions) which is more than the number of unknown functions. This seeming paradox is resolved when one realizes that the force on each particle is never fully known, as the forces of constraints are not given explicitly; it is just stated that the particles move in such a way that the change in their positions is accomplished by an appropriate change in the generalized coordinates and not in any other way. Thus one may suspect that there are going to be $3N - s$ number of such unknowns in the force components that constrain the system. Thus we have to reduce the system of 3N equations to a system of 's' equations in the unknown functions q and their (time) derivatives. As the reader will discover in the exercises, this is usefully accomplished by taking the dot product of the two sides of Eq. (1.1) with $\frac{\partial \mathbf{r}_i}{\partial q_k}$ and summing over the index 'i', which denotes the particle number. Since the position in terms of generalized coordinates is

$$\mathbf{r}_i = \mathbf{r}_i(q_1, q_2, ...., q_s), \tag{1.2}$$

the velocity of the i-th particle is

$$\frac{d}{dt} \mathbf{r}_i = \sum_{j=1}^{s} \frac{\partial \mathbf{r}_i}{\partial q_j} \dot{q}_j, \tag{1.3}$$

and the acceleration is

$$\frac{d^2}{dt^2} \mathbf{r}_i = \sum_{j=1}^{s} \frac{\partial \mathbf{r}_i}{\partial q_j} \ddot{q}_j + \sum_{j,k=1}^{s} \frac{\partial^2 \mathbf{r}_i}{\partial q_k \partial q_j} \dot{q}_k \dot{q}_j. \tag{1.4}$$

Mass times this acceleration is the force acting. Hence,

$$m_i\left(\sum_{j=1}^{s}\frac{\partial \mathbf{r}_i}{\partial q_j}\ddot{q}_j+\sum_{j,k=1}^{s}\frac{\partial^2 \mathbf{r}_i}{\partial q_k \partial q_j}\dot{q}_k\dot{q}_j\right)=\mathbf{F}_i(\mathbf{r},\mathbf{v}). \tag{1.5}$$

Figure 1.1: Sir Isaac Newton PRS MP (25 December 1642 to 20 March 1727) was an English physicist and mathematician who is widely regarded as one of the most influential scientists of all time and as a key figure in the scientific revolution. His book *Philosophiae Naturalis Principia Mathematica* (Mathematical Principles of Natural Philosophy), first published in 1687, laid the foundations for most of classical mechanics. Newton also made seminal contributions to optics (source: Wikipedia).

As we have pointed out, these equations are not directly useful since all the components of $\mathbf{F}_i$ are not known beforehand. To make it useful and reduce the number of equations to $s$—the number of unknown functions, we take dot product with $m_i\frac{\partial \mathbf{r}_i}{\partial q_k}$ and sum over 'i'.

$$\sum_{i=1}^{N}\frac{\partial \mathbf{r}_i}{\partial q_k}\cdot m_i\left(\sum_{j=1}^{s}\frac{\partial \mathbf{r}_i}{\partial q_j}\ddot{q}_j+\sum_{j,l=1}^{s}\frac{\partial^2 \mathbf{r}_i}{\partial q_l \partial q_j}\dot{q}_l\dot{q}_j\right)=\sum_{i=1}^{N}\frac{\partial \mathbf{r}_i}{\partial q_k}\cdot\mathbf{F}_i(\mathbf{r},\mathbf{v}) \tag{1.6}$$

This is really the force equation in generalized coordinates. But it is written in a not so elegant form. To write it more compactly, we consider the kinetic energy,

$$T = \frac{1}{2}\sum_{i=1}^{N} m_i \ \mathbf{v}_i \cdot \mathbf{v}_i \tag{1.7}$$

$$T = \frac{1}{2}\sum_{i=1}^{N} m_i \ \mathbf{v}_i \cdot \mathbf{v}_i = \frac{1}{2}\sum_{i=1}^{N} m_i \sum_{j,k=1}^{s} \left( \frac{\partial \mathbf{r}_i}{\partial q_j} \cdot \frac{\partial \mathbf{r}_i}{\partial q_k} \right) \ \dot{q}_j \dot{q}_k. \tag{1.8}$$

Now consider (all these steps are left as an exercise to the reader),

$$\frac{\partial T}{\partial \dot{q}_k} = \sum_{i=1}^{N} m_i \sum_{j}^{s} \left( \frac{\partial \mathbf{r}_i}{\partial q_j} \cdot \frac{\partial \mathbf{r}_i}{\partial q_k} \right) \ \dot{q}_j, \tag{1.9}$$

and

$$\frac{d}{dt}\frac{\partial T}{\partial \dot{q}_k} = \sum_{i=1}^{N} m_i \sum_{j}^{s} \left( \frac{\partial \mathbf{r}_i}{\partial q_j} \cdot \frac{\partial \mathbf{r}_i}{\partial q_k} \right) \ \ddot{q}_j + \sum_{i=1}^{N} m_i \sum_{j,l}^{s} \left( \frac{\partial^2 \mathbf{r}_i}{\partial q_j \partial q_l} \cdot \frac{\partial \mathbf{r}_i}{\partial q_k} \right) \ \dot{q}_l \ \dot{q}_j$$

$$+ \sum_{i=1}^{N} m_i \sum_{j,l}^{s} \left( \frac{\partial \mathbf{r}_i}{\partial q_j} \cdot \frac{\partial^2 \mathbf{r}_i}{\partial q_k \partial q_l} \right) \ \dot{q}_l \ \dot{q}_j. \tag{1.10}$$

Using Eq.(1.6) in the above equation we get,

$$\frac{d}{dt}\frac{\partial T}{\partial \dot{q}_k} = \sum_{i=1}^{N} \frac{\partial \mathbf{r}_i}{\partial q_k} \cdot \mathbf{F}_i(\mathbf{r},\mathbf{v}) + \sum_{i=1}^{N} m_i \sum_{j,l}^{s} \left( \frac{\partial \mathbf{r}_i}{\partial q_j} \cdot \frac{\partial^2 \mathbf{r}_i}{\partial q_k \partial q_l} \right) \ \dot{q}_l \ \dot{q}_j. \tag{1.11}$$

Also,

$$\frac{\partial T}{\partial q_k} = \sum_{i=1}^{N} m_i \sum_{j,l=1}^{s} \left( \frac{\partial \mathbf{r}_i}{\partial q_j} \cdot \frac{\partial^2 \mathbf{r}_i}{\partial q_k \partial q_l} \right) \ \dot{q}_j \dot{q}_k. \tag{1.12}$$

Subtracting one from the other we obtain,

$$\frac{d}{dt}\frac{\partial T}{\partial \dot{q}_k} - \frac{\partial T}{\partial q_k} = \sum_{i=1}^{N} \frac{\partial \mathbf{r}_i}{\partial q_k} \cdot \mathbf{F}_i(\mathbf{r},\mathbf{v}) = Q_k(q,\dot{q}) \tag{1.13}$$

These are the most general version of the Euler-Lagrange equations of the system in generalized coordinates. The quantity $Q_k(q,\dot{q})$ is known as the generalized force. It is typically specified and does not include components that are due to constraints. These equations may be further simplified if the force is velocity independent and derivable from a potential. Thus, if

$$\mathbf{F}_i(\mathbf{r},\mathbf{v}) = -\nabla_i V(\mathbf{r}), \tag{1.14}$$

then,

$$\frac{d}{dt}\frac{\partial T}{\partial \dot{q}_k} - \frac{\partial T}{\partial q_k} = -\sum_{i=1}^{N} \frac{\partial \mathbf{r}_i}{\partial q_k} \cdot \nabla_i V(\mathbf{r}) = -\frac{\partial}{\partial q_k} V(\mathbf{r}). \tag{1.15}$$

Call $L = T - V$; then since $\frac{\partial}{\partial \dot{q}_l} V(\mathbf{r}) \equiv 0$,

$$\frac{d}{dt}\frac{\partial L}{\partial \dot{q}_k} - \frac{\partial L}{\partial q_k} = 0. \tag{1.16}$$

Figure 1.2: Joseph-Louis Lagrange (25 January 1736 to 10 April 1813), was an Italian mathematician and astronomer born in Turin, Piedmont, who lived part of his life in Prussia and part in France. He made significant contributions to all fields of analysis, number theory, and classical and celestial mechanics (source: Wikipedia).

These are the celebrated Lagrange equations of a (conservative) system. These equations may also be derived from an extremum principle. Define the action $S_q = \int_{t_i}^{t_f} dt\ L(q, \dot{q})$ where $L$ is the Lagrangian above. This is defined for each trajectory $q(t) = \{q_1(t), q_2(t), ..., q_N(t)\}$. We consider that the end points $q(t_i) = q_i$ and $q(t_f) = q_f$ of any trajectory are given fixed quantities. Thus a new trajectory, such as $\tilde{q} = q + q^{'}$, will have the same property, making the deviation $q^{'}(t_i) = q^{'}(t_f) = 0$. We look at small deviations from the trajectory $q$ determined by Eq. (1.16).

$$S_{q+q^{'}} = \int_{t_i}^{t_f} dt\ L(q + q^{'}, \dot{q} + \dot{q}^{'}) \tag{1.17}$$

Let us Taylor expand both sides in powers of $q^{'}$ and retain only up to linear terms. Thus,

$$S_{q+q^{'}} = \int_{t_i}^{t_f} dt\ L(q, \dot{q}) + \int_{t_i}^{t_f} dt \sum_{k=1,...,s} q_k^{'} \frac{\partial L(q, \dot{q})}{\partial q_k}$$

$$+\int_{t_i}^{t_f} dt \sum_{k=1,\dots,s} \dot{q}_k^{'} \frac{\partial L(q,\dot{q})}{\partial \dot{q}_k}. \tag{1.18}$$

We now write,

$$\dot{q}_k^{'} \frac{\partial L(q,\dot{q})}{\partial \dot{q}_k} = \frac{d}{dt}\left(q_k^{'} \frac{\partial L(q,\dot{q})}{\partial \dot{q}_k}\right) - q_k^{'} \frac{d}{dt}\frac{\partial L(q,\dot{q})}{\partial \dot{q}_k}. \tag{1.19}$$

Substituting in the earlier equation and performing the integration from $t_i$ to $t_f$, keeping in mind that $q^{'}(t_i) = q^{'}(t_f) = 0$, we get,

$$S_{q+q'} = \int_{t_i}^{t_f} dt\ L(q,\dot{q}) + \int_{t_i}^{t_f} dt \sum_{k=1,\dots,s} q_k^{'}(t) \left(\frac{\partial L(q,\dot{q})}{\partial q_k} - \frac{d}{dt}\frac{\partial L(q,\dot{q})}{\partial \dot{q}_k}\right). \tag{1.20}$$

Thus, by combining with Eq. (1.16) we may conclude,

$$S_{q+q'} = S_q + O(q^{'2}). \tag{1.21}$$

Since the above equation has to be true for every choice of $q^{'}(t)$ at each time, it must be that the coefficients vanish independently.

$$\frac{\delta S_q}{\delta q_k(t)} = 0 \tag{1.22}$$

Thus the path that the system takes is one that makes the action an extremum. We will have occasion to consider systems where the generalized coordinates are further constrained by functions of the form $F[q] = 0$. The action has to be minimized subject to these additional constraints. This is accomplished by introducing Lagrange multipliers (see box description). If the constraints are $F_j[q] = 0$ where $j = 1,\dots,p$, then there are Lagrange multipliers $m_1, m_2, \dots, m_p$ so that the system takes the path given by the solution to

$$\frac{\delta(S_q - \sum_{j=1}^{p} m_j F_j[q])}{\delta q_k(t)} = 0 \tag{1.23}$$

where in the above evaluation we treat all the $q$'s as independent and later on enforce $F_j[q] = 0$.

**Lagrange multipliers**: A typical problem in physics or mathematics is that of optimization. Given a set of constraints, one is supposed to minimize or maximize a function. Let us take a simple example, from statistical mechanics. There are $N$ particles and they are to be distributed in 3 states $S_1, S_2, S_3$. For each particle, energy of $S_1$ is $\varepsilon_1$, of $S_2$ is $\varepsilon_2$ and of $S_3$ is $\varepsilon_3$. We want to find a distribution of particles in these energy levels which maximizes the number of possible configurations and keeps the number of particles and total energy constant. Let $n_1$ be number of particles in $S_1$; similarly for $n_2$ and $n_3$. So we are constrained by $n_1\varepsilon_1 + n_2\varepsilon_2 + n_3\varepsilon_3 = E$ and $n_1 + n_2 + n_3 = N$. With this constraint, we are required to maximize $N!/(n_1!n_2!n_3!)$. Using Stirling's approximation, this is same as maximizing $N^N/((n_1)^{n_1}\,(n_2)^{n_2}\,(n_3)^{n_3})$. Conventionally, we take the log of this, and ignore $N\,log(N)$ (since it is constant already) and now we want to maximize: $F(n_1,n_2,n_3) = -n_1\,log(n_1) - n_2\,log(n_2) - n_3\,log(n_3)$ given the constraints: $n_1\,\varepsilon_1 + n_2\,\varepsilon_2 + n_3\,\varepsilon_3 = E$ and $n_1 + n_2 + n_3 = N$. A trivial way of solving is to find $n_1$ and $n_2$ in terms of $n_3$ and put in $F$. Then differentiate $F$ to get the desired answer. Let's try a different approach, as this is difficult. Let's directly differentiate all equations and expressions with respect to $n_1$, $n_2$, and $n_3$ to get:

$$log(n_1)\,dn_1 + log(n_2)\,dn_2 + log(n_3)\,dn_3 = 0$$

$$dn_1 + dn_2 + dn_3 = 0$$

$$\varepsilon_1\,dn_1 + \varepsilon_2\,dn_2 + \varepsilon_3\,dn_3 = 0. \tag{1.24}$$

This means that the matrix

$$\begin{pmatrix} log(n_1) & log(n_2) & log(n_3) \\ 1 & 1 & 1 \\ \varepsilon_1 & \varepsilon_2 & \varepsilon_3 \end{pmatrix} \tag{1.25}$$

must have at least one null vector. In order to avoid a trivial solution, this implies one row must be a linear combination of remaining two rows. This implies, $log(n_i) + a + b\,\varepsilon_i = 0$, for some $a, b$ and all $i$. This gives $n_i = e^a e^{b\,\varepsilon_i}$. We put $n_1, n_2$, and $n_3$ in the constraints $n_1\,\varepsilon_1 + n_2\,\varepsilon_2 + n_3\,\varepsilon_3 = E$ and $n_1 + n_2 + n_3 = N$ to obtain the values of a and b. The numbers a and b are known as Lagrange multipliers. Their significance emerges from the fact that they give a very convenient way to solve the problem. The algorithm for this method is therefore as follows. Extremize $F(n_1,n_2,n_3...)$ given the constraints $C_1(n_1,n_2...), C_2(n_1,n_2...), ...$ etc. We introduce Lagrange multipliers $l_1, l_2...$ and change the problem to this: find variables $(n_1,n_2,n_3...,l_1,l_2...)$ which extremize the function

$$F(n_1,n_2...) + l_1\,C_1(n_1,n_2...) + l_2\,C_2(n_1,n_2...) + ... \tag{1.26}$$

So there are no constraints here in the beginning and differentiating with respect to $l_1, l_2$, etc. gives us the required constraints. Differentiating with respect to $n_1, n_2$, etc. gives more equations to be solved. We can appreciate the usefulness of this approach from the fact that it gives a very neat way to approach problems in statistical mechanics. They are also invaluable in understanding behavior of particles under constraints.

## 1.2 The Hamiltonian Formulation

It is possible to have a different formulation of a dynamical system where instead of the equation of motion being second order in time, like it was in the previous section, Lagrange equations involve two time derivatives, and there could be two coupled first-order equations. This is known as the Hamiltonian formulation. To understand how this comes about, one has to study the mathematical transformation known as the Legendre transformation. Consider a convex function $L(v)$ of some single variable called $v$ (convex function means $L^{''}(v) > 0$, generalization to many variables is left to the exercises). The Legendre transformation of this function is defined as $H(p) = max_{v\in\mathbb{R}}(p\,v - L(v))$. This means that we vary $v$ until the expression $G(v) \equiv (p\,v - L(v))$ reaches a maximum value. This value is denoted by $H(p)$ since it is clearly a function of $p$ alone. The maximum value of $G(v)$ is reached at $v = v_*$ where $G^{'}(v_*) = 0$ and $G^{''}(v_*) < 0$ is guaranteed by the convex property of $L(v)$. Thus $H(p) = pv_* - L(v_*)$ where $v_*$ is defined as the solution to the equation $L^{'}(v_*) = p$.

Figure 1.3: Sir William Rowan Hamilton (4 August 1805 to 2 September 1865) was an Irish physicist, astronomer, and mathematician, who made important contributions to classical mechanics, optics, and algebra. His studies of mechanical and optical systems led him to discover new mathematical concepts and techniques. His greatest contribution is perhaps the reformulation of Newtonian mechanics, now called Hamiltonian mechanics.

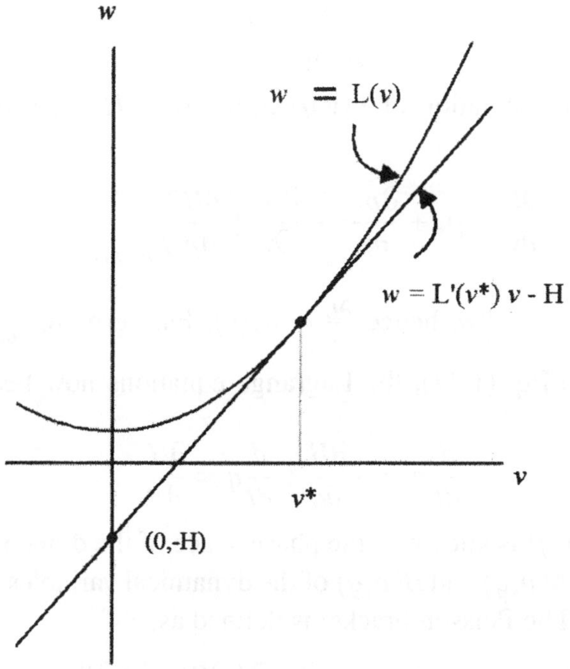


Figure 1.4: This is a pictorial description of the Legendre transformation.

In other words, mathematically we wish to claim that any convex function $L(v)$ may be described by providing the value of $L(v)$ at each $v$, or equivalently, by providing the value of this function's slope $p$ and a corresponding function $H(p)$ of this slope ($H(p)$ is the vertical intercept of the tangent). This equivalence is possible only if the transformation from the $L(v)$ description to the $H(p)$ description is invertible. To verify this, we first show that $H(p)$ is also convex. The first derivative is $H^{'}(p) = v_*(p) + p\frac{\partial v_*(p)}{\partial p} - L^{'}(v_*(p))\frac{\partial v_*(p)}{\partial p} = v_*(p)$. The last result follows since $L^{'}(v_*(p)) = p$. Thus $H^{''}(p) = \frac{\partial v_*(p)}{\partial p}$. To show that this is non-negative, we simply differentiate the defining equation for $v_*$ to get $L^{''}(v_*(p))\frac{\partial v_*(p)}{\partial p} = 1$. Since $L^{''}(v_*) > 0$, it follows that $\frac{\partial v_*(p)}{\partial p} \equiv H^{''}(p) > 0$. Now we may define the inverse Lagrange transform as $L_*(v) = max_{p\in\mathbb{R}}(pv - H(p)) = (p_*v - H(p_*))$ where $v = \left(\frac{\partial H(p)}{\partial p}\right)_{p=p_*} = v_*(p_*)$, from the earlier result. Thus $H(p_*) = p_*v_* - L(v_*) = p_*v - L(v)$ since $v_* = v$. Thus $L_*(v) = p_*v - H(p_*) = L(v)$, thereby proving that the Legendre transformation procedure is invertible. Thus the information contained in the $L(v)$ or Lagrangian description is the same as the $H(p)$ or the Hamiltonian description.

Furthermore, it is possible to derive the respective equations of motion if we realize that in addition to dependence of $v$ and $p$ there is also dependence on a generalized coordinate $q$. Thus we write $L(v,q)$ and $H(p,q)$. The Lagrange equations then read as follows.

$$\frac{d}{dt}\frac{\partial L}{\partial v} = \frac{\partial L}{\partial q} \tag{1.27}$$

We now use the representation of $L(v,q)$ in terms of $H(p,q)$ to write, $L(v,q) = p_*(v)v - H(p_*(v),q)$

$$\frac{\partial L}{\partial v} = p_* + \frac{\partial p_*}{\partial v}v - \frac{\partial p_*}{\partial v}\left(\frac{\partial H}{\partial p}\right)_{p=p_*(v)}. \tag{1.28}$$

But $\left(\frac{\partial H}{\partial p}\right)_{p=p_*(v)} = v = \frac{d}{dt}q$, hence $\frac{\partial L}{\partial v} = p_*(v)$. Furthermore, $\frac{\partial L}{\partial q} = -\frac{\partial H}{\partial q}$. Inserting $\frac{\partial L}{\partial v} = p_*(v)$ into Eq. (1.27), the Lagrange equations now become Hamilton's equations,

$$\frac{d}{dt}p = -\frac{\partial H}{\partial q}; \ \frac{d}{dt}q = \frac{\partial H}{\partial p}. \tag{1.29}$$

The set of points $(p,q)$ is known as the phase space of the dynamical system. Consider two functions $A(p,q)$ and $B(p,q)$ of the dynamical variables $p,q$ in the Hamiltonian description. The Poisson bracket is defined as,

$$\{A(p,q),B(p,q)\} = \frac{\partial A}{\partial q}\frac{\partial B}{\partial p} - \frac{\partial A}{\partial p}\frac{\partial B}{\partial q}. \tag{1.30}$$

Consider now a general function $F(p,q,t)$. We wish to determine its rate of change with respect to time. This function changes with time due to two possible reasons. First, it may be explicitly time dependent. Second, because it depends on the dynamical variables which themselves change with time according to Hamilton's equations. Thus we may write,

$$\frac{d}{dt}F(p,q,t) = \frac{\partial F}{\partial t} + \frac{dp}{dt}\frac{\partial F}{\partial p} + \frac{dq}{dt}\frac{\partial F}{\partial q}$$

$$= \frac{\partial F}{\partial t} - \frac{\partial H}{\partial q}\frac{\partial F}{\partial p} + \frac{\partial H}{\partial p}\frac{\partial F}{\partial q} = \frac{\partial F}{\partial t} + \{F,H\}. \tag{1.31}$$

Now it is easy to see the condition for a dynamical variable to be a constant of the motion. If a variable does not depend on time explicitly or implicitly, then it follows that its Poisson bracket with the Hamiltonian should vanish.

$$\{F,H\} = 0 \tag{1.32}$$

Since $\{H,H\} = 0$ it follows that if the Hamiltonian is explicitly time independent then it is also implicitly time independent, or it is a constant of the motion. Two variables $A$ and $B$ are said to be conjugates of each other if $\{A,B\} = 1$. The simplest example is $q,p$—they are conjugates of one another for, $\{q,p\} = 1$, if $q$ is an angular displacement, then $p$ would be the corresponding angular momentum $l$ so that $\{\theta,l\} = 1$.

Figure 1.5: A French mathematician and physicist, Simon Denis Poisson (21 June 1781 to 25 April 1840) was a master of a wide range of topics such as classical mechanics, analysis, probability theory, electromagnetism, and differential equations. His study of stability of planetary systems stands next to that of Laplace and his work on classical mechanics influenced the work of William Hamilton. He derived the Navier-Stokes equation independent of Claude Navier and was the last great supporter of corpuscle theory of light, which was shown to be flawed and replaced by the wave theory.

## 1.3 Flows and Symmetries

We may think of Hamilton's equations for the coordinates as a kind of flow with respect to time. The answer to the question of how the coordinates and momenta change with time is determined by integrating Hamilton's equations. We could equally well be interested in how the coordinates and momenta change with respect to some other variable. To be concrete, let us consider a system in two dimensions so that $x, y, p_x, p_y$ are the coordinates in phase space. These change with time in a manner determined by integrating Hamilton's equations, as we have pointed out earlier. But now, imagine instead that we rotate the coordinate system by an angle $\alpha$, then the coordinates and momenta change with $\alpha$. If we denote the new coordinates by $x(\alpha), y(\alpha), p_x(\alpha), p_y(\alpha)$,

$$Z_x(\alpha) = cos(\alpha)\ Z_x + sin(\alpha)\ Z_y \tag{1.33}$$

$$Z_y(\alpha) = -sin(\alpha)\, Z_x + cos(\alpha)\, Z_y \tag{1.34}$$

where $Z_x$ is one of $x, p_x$ and $Z_y$ is one of $y, p_y$. We wish to see these as the consequence of integrating a pair of equations analogous to Hamilton's equations where the flow parameter, namely time $t$, is replaced with this angle $\alpha$. In Eq. (1.29) we may see that the left side contains time as the parameter of differentiation. The right-hand side involves the Hamiltonian. Just as we may think of linear momentum and linear coordinates as canonically conjugate variables and angular momentum and angular coordinate in a similar manner, we may intuitively suspect that the same must be true for time and Hamiltonian. However, this cannot be made rigorous since time is a parameter and not a dynamical variable. We shall merely be content at using this as a cue to write the analog of Hamilton's equation in case of general flows. If $\alpha$ is some parameter with respect to which we need the flow, we may suspect that there exists a canonically conjugate variable $G$ such that,

$$\frac{d}{d\alpha}p = -\frac{\partial G}{\partial q}\ ;\ \frac{d}{d\alpha}q = \frac{\partial G}{\partial p}. \tag{1.35}$$

Here $\{\alpha, G\} = 1$ and $p = p_x, p_y$ and $q = q_x, q_y$. There are some important properties that are worth noting. First, we show that $G$ is independent of $\alpha$. To do this consider,

$$\frac{d}{d\alpha}G(p,q) = \frac{dp}{d\alpha}\frac{\partial G(p,q)}{\partial p} + \frac{dq}{d\alpha}\frac{\partial G(p,q)}{\partial q}. \tag{1.36}$$

If $\alpha$ is the angle alluded to in Eq. (1.34), it is easy to see that since $q_{x\ (y)} \equiv x\ (y)$,

$$\frac{d}{d\alpha}x(\alpha) = y(\alpha) = \frac{\partial G}{\partial p_x}\ ;\ \frac{d}{d\alpha}y(\alpha) = -x(\alpha) = \frac{\partial G}{\partial p_y}. \tag{1.37}$$

Similarly,

$$\frac{d}{d\alpha}p_x(\alpha) = p_y(\alpha) = -\frac{\partial G}{\partial x}\ ;\ \frac{d}{d\alpha}p_y(\alpha) = -p_x(\alpha) = -\frac{\partial G}{\partial y}. \tag{1.38}$$

Thus,

$$\begin{aligned}\frac{d}{d\alpha}G(p,q) &= \frac{dp_x}{d\alpha}\frac{\partial G(p,q)}{\partial p_x} + \frac{dp_y}{d\alpha}\frac{\partial G(p,q)}{\partial p_y} \\ &\quad + \frac{dx}{d\alpha}\frac{\partial G(p,q)}{\partial x} + \frac{dy}{d\alpha}\frac{\partial G(p,q)}{\partial y} \\ &= p_y(\alpha)\, y(\alpha) + p_x(\alpha)\, x(\alpha) - y(\alpha)\, p_y(\alpha) - x(\alpha)\, p_x(\alpha) = 0.\end{aligned} \tag{1.39}$$

Thus the conjugate does not flow relative to the flow parameter. This is more easily seen by the general statement,

$$\frac{d}{d\alpha}Q = \frac{\partial}{\partial\alpha}Q + \{Q, G\} \tag{1.40}$$

when $Q \equiv G$, $G$ is seen to be independent of the flow parameter $\alpha$ if $G$ depends on $\alpha$ only through the phase space variables. We may see that Eq. (1.37) and Eq. (1.38) are consistent with the statement,

$$G(q,p) = p_x y - p_y x. \tag{1.41}$$

This is nothing but the component of the angular momentum along the $-z$ axis. Thus, angular momentum is conjugate to (also known as a generator of) simple rotations in coordinate space. Later when we discuss Noether's theorem we will see that there is a different kind of rotation that mixes coordinates and momenta that also has a generator distinct from angular momentum.

## 1.4 Dynamics of a Continuous System

We now turn to a discussion of continuous systems. The problem at hand is to generalize the earlier sections to accommodate situations when the number of degrees of freedom are an infinity of the continuous kind. Specifically, this involves reinterpreting the list of coordinates $\{q_i; i = 1,2,...,N\}$ by a function of a parameter $s$ such that the list is now written as $\{q_s; s \in [a,b]\}$ where $s$ is a continuous variable belonging to an interval $[a,b]$, for fields in one dimension, $\{q_{s_x,s_y}; s_x \in [a,b], s_y \in [c,d]\}$ for fields in two dimensions, and so on, which now replaces the index $i$ that was used while describing a system with a finite number of degrees of freedom. In the process of making this generalization, we will have occasion to reinterpret various definitions related to the discrete index that characterize the number of degrees of freedom. For instance, the summation $\sum_{i=1}^{N} F(q_i)$ will have to be reinterpreted as,

$$\sum_{i=1}^{N} F(q_i) \to \int_a^b ds\, F(q_s). \tag{1.42}$$

This would be the case if the fields were described in one dimension. In, say, three dimensions they would be,

$$\sum_{i=1}^{N} F(q_i) \to \int_{s\in\Omega} d^3s\, F(q_{s_x,s_y,s_z}). \tag{1.43}$$

A term such as a difference between successive values would be a derivative.

$$F(q_{i+1}) - F(q_i) = \frac{d}{ds} F(q_s) \tag{1.44}$$

Now, just as we may differentiate and integrate with respect to the discrete number of degrees of freedom, we may do so even when there are a continuum of them.

Consider a function of all the N degrees of freedom : $R(q_1, q_2, ...., q_N)$. We may contemplate differentiating with respect to one of these coordinates - $\frac{\partial}{\partial q_j}R(q)$ (here $q$ refers to the list of all possible coordinates). Then,

$$\frac{\partial}{\partial q_j}R(q) \rightarrow \frac{\partial}{\partial q_s}R(q). \tag{1.45}$$

Using this we may derive some basic identities in functional differential calculus. The first of these is,

$$\frac{\partial}{\partial q_s}q_{s'} = \delta(s - s') \tag{1.46}$$

where $s$ could mean a single continuous variable or two or three continuous variables depending upon the dimension of the field. Similarly, $\delta(s - s')$ could refer to the Dirac delta function in one, two, or three dimensions.

> Readers unfamiliar with the concept of the Dirac delta function may find this description useful. First consider the distinction between rational and irrational numbers. Specifically, consider the sequence $\{x_1 = 1, x_2 = 1.5, x_3 = 1.4, ....., x_{n-1}, x_n = 1 + \frac{1}{1+x_{n-1}}, ...\}$. It is easy to see that the limit $x_\infty = Lim_{n\to\infty} x_n = \sqrt{2}$. Thus, while each element of the sequence is rational, the limit of the sequence is an irrational quantity. Similarly, we define the Dirac delta function as the limit of a sequence of perfectly well-behaved functions namely $f_N(x) = \frac{N}{\pi\,(x^2N^2+1)}$. We identify $\delta(x) \equiv Lim_{N\to\infty} f_N(x)$. Each $f_N(x)$ is an 'entire function' (differentiable infinitely many times) of its dependent variable $x$. Further, for each $N$, $\int_{-\infty}^{\infty} dx\ f_N(x) = 1$ and yet $\delta(x) \equiv Lim_{N\to\infty} f_N(x)$ is anything but regular. This 'function' is zero everywhere except at $x = 0$ where it becomes infinite. However, $\int_{-\infty}^{\infty} \delta(x)\ dx = 1$. With this concept, one may differentiate discontinuous functions such as Heaviside's unit step function $\theta(x)$ where we write $\frac{d}{dx}\theta(x) = \delta(x)$. Both the left- and right-hand sides are zero when $x \neq 0$. Both are infinite when $x = 0$. When integrated over a region containing the origin, the result in both cases is unity.

To derive this we write, $R(q) = \sum_{i=1}^{N} q_i$ in the first instance, to get,

$$\frac{\partial}{\partial q_j}R(q) = 1, \tag{1.47}$$

whereas the same result in the continuum language would be, $R(q) = \int_a^b ds\ q_s$ (for fields in one dimension) and (setting $j \rightarrow s'$),

$$\frac{\partial}{\partial q_{s'}}\int_a^b ds\ q_s = 1. \tag{1.48}$$

This means (since obviously, $\frac{\partial}{\partial q_{s'}} q_s = 0$ if $s \neq s'$),

$$\frac{\partial}{\partial q_{s'}} q_s = \delta(s - s'). \tag{1.49}$$

Furthermore, we may later have occasion to use quantities such as,

$$\frac{\partial}{\partial q_{s'}} \frac{dq_s}{ds} = \frac{d}{ds}\delta(s - s') = \delta'(s - s'). \tag{1.50}$$

More generally, we define functional differentiation as,

$$\frac{\delta}{\delta J(x)} F[\{J(y)\}] = Lim_{\varepsilon \to 0} \frac{F[\{J(y) + \varepsilon\,\delta(x-y)\}] - F[\{J(y)\}]}{\varepsilon}. \tag{1.51}$$

Imagine a functional $F[g]$ defined as,

$$F[g] \equiv \frac{1}{2}\int_a^b ds\, g^2(s). \tag{1.52}$$

Then we see that,

$$\frac{\delta}{\delta g(x)} F[g] = \int_a^b ds\, g(s)\, \frac{\delta g(s)}{\delta g(x)}. \tag{1.53}$$

We may derive an expression for the quantity $\frac{\delta g(s)}{\delta g(x)}$ using the formal definition for functional differentiation.

$$\frac{\delta g(s)}{\delta g(x)} = Lim_{\varepsilon \to 0} \frac{(g(s) + \varepsilon\delta(s-x)) - g(s)}{\varepsilon} = \delta(s-x) \tag{1.54}$$

Now we wish to evaluate the functional derivative of a slightly different functional. Set,

$$H[g] \equiv \frac{1}{2}\int_a^b ds\, g'^2(s) \tag{1.55}$$

where $g'(s) = \frac{dg(s)}{ds}$. Consider a variable $x$ such that $a < x < b$. It is clear that,

$$\frac{\delta}{\delta g(x)} H[g] \equiv \int_a^b ds\, \frac{dg(s)}{ds} \frac{\delta}{\delta g(x)} \frac{dg(s)}{ds} = \int_a^b ds\, \frac{dg(s)}{ds} \frac{d}{ds} \frac{\delta}{\delta g(x)} g(s)$$

$$= \int_a^b ds\, \frac{dg(s)}{ds} \frac{d}{ds} \delta(s-x) = -\int_a^b ds\, g''(s)\delta(s-x) = -g''(x) \tag{1.56}$$

where the last few steps follow from integration by parts.

We provide a quick tutorial for integration by parts. Consider the integral

$$\int_a^b dx\, F(x)G'(x) = \int_a^b dx\, \frac{d}{dx}(F(x)G(x)) - \int_a^b dx\, G(x)F'(x). \tag{1.57}$$

The first term on the right-hand side is called the boundary term. Since we are integrating over all space, the boundary is at infinity. If $F(x)$ and $G(x)$ vanish at the boundaries $a$ and $b$ (as we shall assume always), then this term is zero. Thus the above equation may be rewritten as,

$$\int_a^b dx\, F(x)G'(x) = -\int_a^b dx\, G(x)F'(x). \tag{1.58}$$

Repeated application of this rule yields,

$$\int_a^b dx\, F(x)G''(x) = -\int_a^b dx\, F'(x)G'(x) = \int_a^b dx\, F''(x)G(x). \tag{1.59}$$

Now we provide specific examples of dynamical systems with a finite number of degrees of freedom and a prescription on how to generate a system with infinitely many degrees of freedom by taking the continuum limit.

■ Imagine a collection of $N$ identical masses $m$ confined to the circumference of a circle of radius $\frac{L}{2\pi}$. Imagine also that each mass is tied to its adjacent mass on either side by two identical springs with stiffness $k$ and the motion is along the circumference. We wish to write down the Lagrangian of this system. In order to do this, we choose the generalized coordinate of the $n$-th mass to be the distance $S_n$ ($n = 1,2,...,N$ and $S_{N+1} = S_1$) along the circumference from a point designated as the origin (say the north pole at $n = 1$). The kinetic energy of the system of masses is then $T = \sum_{n=1}^{N} \frac{1}{2} m\dot{S}_n^2$. The potential energy is given by $V = \frac{1}{2}k\sum_{n=1}^{N-1}(S_{n+1} - S_n - l)^2$. At equilibrium, both the potential energy and kinetic energy vanish identically. This means that at equilibrium $S_n^0 = (n-1)l$ since we choose $n = 1$ to be the pole that remains fixed at equilibrium. We choose to measure the displacement relative to this equilibrium position. Thus we write $S_n = (n-1)l + s_n$. The Lagrangian becomes,

$$L = \sum_{n=1}^{N} \frac{1}{2} m\dot{s}_n^2 - \frac{1}{2}k\sum_{n=1}^{N-1}(s_{n+1} - s_n)^2. \tag{1.60}$$

In order to make the transition to the continuum limit, we write $x = (n-1)l$, $dx = l$,

$m = \rho dx$, $kl^2 = \kappa dx$ and $\sum_n \equiv \int_0^L$, $s_{n+1} - s_n = s(x+l) - s(x) = l\frac{\partial s(x,t)}{\partial x}$.

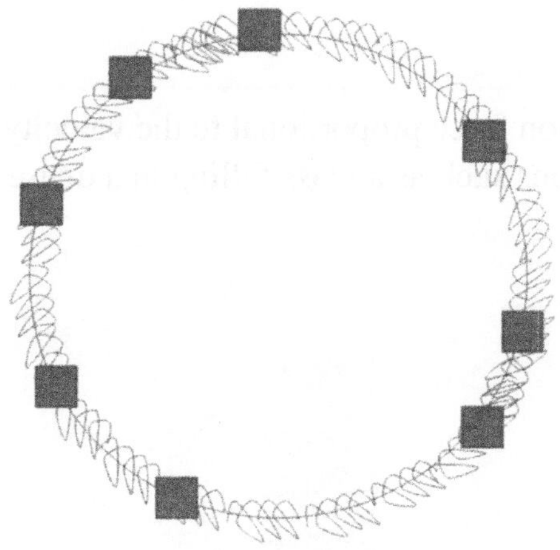

Figure 1.6: This is an illustration of the masses connected to adjacent ones by springs, constrained to move on a circle.

With these substitutions the Lagrangian becomes,

$$L(s,\dot{s}) = \int_0^L \frac{1}{2}\rho dx \left(\frac{\partial s(x,t)}{\partial t}\right)^2 - \frac{1}{2}\kappa \int_0^L dx \left(\frac{\partial s(x,t)}{\partial x}\right)^2. \tag{1.61}$$

Here we may see that the role of the n-th degree of freedom is taken up by the symbol $x$. We now derive the Lagrange equations of this Lagrangian using,

$$\frac{d}{dt}\frac{\delta L}{\delta \dot{s}(y,t)} = \frac{\delta L}{\delta s(y,t)}. \tag{1.62}$$

We evaluate the generalized momentum as,

$$\frac{\delta L}{\delta \dot{s}(y,t)} = \rho \left(\frac{\partial s(x,t)}{\partial t}\right) \tag{1.63}$$

Using the tutorials on functional differentiation we may write,

$$\frac{\delta L}{\delta s(y,t)} = -\frac{\delta}{\delta s(y,t)}\frac{1}{2}\kappa \int_0^L dx \left(\frac{\partial s(x,t)}{\partial x}\right)^2$$

$$= -\kappa \int_0^L dx \left(\frac{\partial s(x,t)}{\partial x}\right)\left(\frac{\partial \delta(x-y)}{\partial x}\right) = \kappa \left(\frac{\partial^2 s(y,t)}{\partial y^2}\right). \tag{1.64}$$

Therefore, the Lagrange equation in this case is nothing but the wave equation.

$$\rho \frac{\partial^2 s(x,t)}{\partial t^2} = \kappa \frac{\partial^2 s(y,t)}{\partial y^2} \tag{1.65}$$

■ Next, imagine a friction force proportional to the velocity of the mass $F = -kv$. This is typical of situations such as a mass falling in a dense fluid.

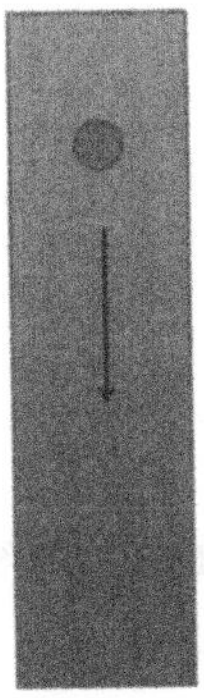

Figure 1.7: Mass falling in a dense fluid. In this problem, we have to imagine several masses experiencing a drag force of the type shown only when there is relative motion between adjacent masses.

Imagine now two masses 1 and 2 that are connected by a 'spring' such that the force on 1 is proportional to the velocity of 1 relative to 2. Thus $F_{12} = -k(v_1 - v_2)$. Now imagine a chain of such masses also subject to a constant external force. In this case, the force equations read as follows:

$$m\frac{dv_n}{dt} = -k(v_n - v_{n-1}) - k(v_n - v_{n+1}) + f. \tag{1.66}$$

In the continuum limit we write $x = (n-1)l$ and $l = dx$. Therefore, $v_n - v_{n-1} = v(x) - v(x-l) \approx l\,\frac{\partial v(x)}{\partial x} - \frac{l^2}{2}\frac{\partial^2 v(x)}{\partial x^2} + \ldots$ and $v_{n+1} - v_n \approx l\frac{\partial v(x)}{\partial x} + \frac{l^2}{2}\frac{\partial^2 v(x)}{\partial x^2} + \ldots$.

$$m\frac{\partial v(x,t)}{\partial t} = kl^2\frac{\partial^2 v(x)}{\partial x^2} + f. \tag{1.67}$$

Further $m = \rho\, l$ where $\rho$ is the linear mass density and $f = ma$ would be the external force on each mass, which would suffer an acceleration of $a$ in the absence of friction. Just as in the earlier example, the friction constant $k$ diverges in the continuum limit so that $0 < \frac{kl}{\rho} = \eta < \infty$. The continuum description would be,

$$\frac{\partial v(x,t)}{\partial t} = \eta \frac{\partial^2}{\partial x^2} v(x,t) + a. \tag{1.68}$$

This is nothing but the driven diffusion equation.

■ As a next example, let us consider a special kind of system—a slack (but inextensible) rope of length $L$ with ends tied at the same level to two tree trunks separated by a distance $d < L$. We wish to describe this system using the methods just described but taking into account that this system contains an infinite number of closely spaced particles. The natural generalized coordinates are $(x_s(t), y_s(t))$ where the parameter $s$—the distance from one end along the contours of the rope—plays the role of the index $i$ we used when the number of particles were finite. The summation over the number of particles is replaced by an integration $\sum_i m_i G_i = \int dm_s\, G_s = \rho \int ds\ \ G_s$, for a system with uniform density $\rho = \frac{dm_s}{ds}$. The kinetic energy is,

$$T = \frac{1}{2}\sum_i m_i\, (\dot{x}_i^2 + \dot{y}_i^2) = \frac{1}{2}\rho \int_0^L ds\, (\dot{x}_s^2 + \dot{y}_s^2). \tag{1.69}$$

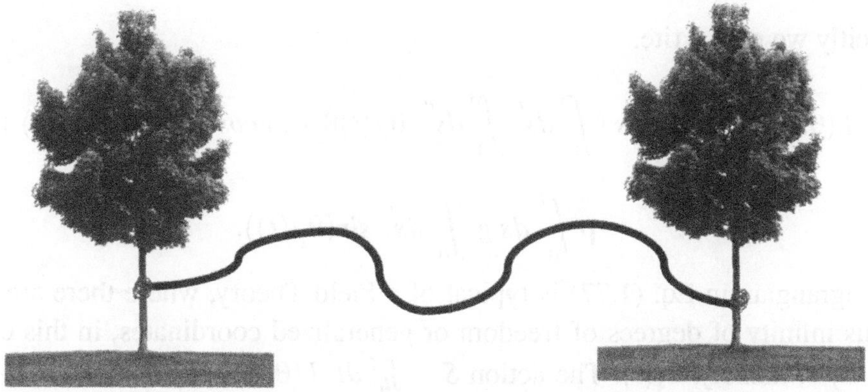

Figure 1.8: This is an illustration of the rope.

Since we assume that the rope is inextensible, the distance (along the rope) between any two points is time independent, hence $ds^2 = dx_s^2 + dy_s^2$ or,

$$\left(\frac{\partial x_s}{\partial s}\right)^2 + \left(\frac{\partial y_s}{\partial s}\right)^2 = 1. \tag{1.70}$$

Thus we may parameterize using a single angle $\theta_s(t)$.

$$\frac{\partial x_s}{\partial s} = cos(\theta_s(t)) \; ; \; \frac{\partial y_s}{\partial s} = sin(\theta_s(t)) \tag{1.71}$$

Taking into account the boundary conditions: $x_{s=0} = 0$, $x_{s=L} = d$, and $y_{s=0} = y_{s=L} = 0$, we get,

$$x_s(t) = \int_0^s ds' \; cos(\theta_{s'}(t)) \; ; \; y_s(t) = \int_0^s ds' \; sin(\theta_{s'}(t)). \tag{1.72}$$

Further,

$$d = \int_0^L ds' \; cos(\theta_{s'}(t)) \; ; \; 0 = \int_0^L ds' \; sin(\theta_{s'}(t)). \tag{1.73}$$

These constraints have to be incorporated while minimizing the action. Considering that the potential energy is given by $V = \int dm_s \; g \; y_s(t) = \rho \int_0^L ds \; g \; y_s(t)$,

$$\dot{x}_s(t) = -\int_0^s ds' \; sin(\theta_{s'}(t)) \; \dot{\theta}_{s'}(t) \tag{1.74}$$

$$\dot{y}_s(t) = \int_0^s ds' \; cos(\theta_{s'}(t)) \; \dot{\theta}_{s'}(t). \tag{1.75}$$

The Lagrangian is,

$$L(\theta,\dot{\theta}) = \frac{1}{2}\rho \int_0^L ds \; (\dot{x}_s^2 + \dot{y}_s^2) - \rho \int_0^L ds \; g \; y_s(t). \tag{1.76}$$

Explicitly we may write,

$$L(\theta,\dot{\theta}) = \frac{1}{2}\rho \int_0^L ds \; (\int_0^s ds' \int_0^s ds'' \; \dot{\theta}_{s'}(t)\dot{\theta}_{s''}(t) \; cos(\theta_{s'}(t) - \theta_{s''}(t)) \; )$$

$$-\rho \int_0^L ds \; g \int_0^s ds' \; sin(\theta_{s'}(t)). \tag{1.77}$$

The Lagrangian in Eq. (1.77) is typical of a Field Theory, where there are a continuous infinity of degrees of freedom or generalized coordinates, in this case labeled by $s$, namely, $\theta_s(t)$. The action $S = \int_{t_i}^{t_f} dt \; L(\theta,\dot{\theta})$ subject to $\theta(t_i) = \theta_i$ and $\theta(t_f) = \theta_f$ has to be minimized subject to additional constraints namely Eq. (1.73). This may be done by the method of Lagrange multipliers, where the unconstrained Lagrangian may be written as,

$$L(\theta,\dot{\theta}) = \frac{1}{2}\rho \int_0^L ds \; (\int_0^s ds' \int_0^s ds'' \; \dot{\theta}_{s'}(t)\dot{\theta}_{s''}(t) \; cos(\theta_{s'}(t) - \theta_{s''}(t)) \; )$$

$$-\rho \int_0^L ds \; g \int_0^s ds' \; sin(\theta_{s'}(t))$$

$$-m_1(d-\int_0^L ds' \; cos(\theta_{s'}(t))) - m_2\int_0^L ds' \; sin(\theta_{s'}(t)). \tag{1.78}$$

As can be seen, the general equations for $\theta_s(t)$ are quite formidable indeed. In order gain further insight into the physics, it is always useful to focus on some simplifying situation. We consider small deviations from equilibrium. We may see easily from the Lagrange equations or otherwise, that a configuration that minimizes the total potential energy is the equilibrium one. Thus we start with

$$\delta(-\rho\int_0^L ds\, g\int_0^s ds' \; sin(\theta_{s'}) - m_1(d-\int_0^L ds \; cos(\theta_s)) - m_2\int_0^L ds \; sin(\theta_s)) = 0. \tag{1.79}$$

In other words the equilibrium is achieved at $\theta_s = \theta_s^0$, where

$$\left(-\rho\int_0^L ds' \, g\,\theta(s'-s)\, cos(\theta_s^0) - m_1 \, sin(\theta_s^0) - m_2\, cos(\theta_s^0)\right) = 0 \tag{1.80}$$

or

$$(-\rho(L-s)\, g - m_2) = m_1 \, tan(\theta_s^0). \tag{1.81}$$

Substitute the formula for $\theta_s$ above into Eq. (1.73) to obtain formulas for the Lagrange multipliers,

$$m_2 = -\frac{gL}{2}\rho \tag{1.82}$$

$$d = \frac{m_1}{g\rho} Log\left(\frac{\frac{gL}{2}\rho + \sqrt{m_1^2 + \frac{(gL\rho)^2}{4}}}{-\frac{gL}{2}\rho + \sqrt{m_1^2 + \frac{(gL\rho)^2}{4}}}\right). \tag{1.83}$$

In order to study the dynamics we examine small deviations from the equilibrium configuration: $\theta_s(t) = \theta_s^0 + \tilde{\theta}_s(t)$. We may expand the Lagrangian with the Lagrange multipliers and retain terms up to the second order in $\tilde{\theta}$,

$$\begin{aligned} L(\theta,\dot{\theta}) = {} & \frac{1}{2}\rho\int_0^L ds \, (\int_0^s ds' \int_0^s ds'' \; \dot{\tilde{\theta}}_{s'}(t)\dot{\tilde{\theta}}_{s''}(t) \; cos(\theta_{s'}^0 - \theta_{s''}^0)\,) \\ & +\rho\int_0^L ds\, g\int_0^s ds' \; \frac{1}{2} sin(\theta_{s'}^0)\; \tilde{\theta}_{s'}^2(t) \\ & -\frac{1}{2}m_1\int_0^L ds' \; cos(\theta_{s'}^0)\tilde{\theta}_{s'}^2(t) - \frac{gL}{2}\rho\int_0^L ds' \; sin(\theta_{s'}^0)\tilde{\theta}_{s'}^2(t) \end{aligned} \tag{1.84}$$

where,

$$\rho g(s-\frac{L}{2}) = m_1 \, tan(\theta_s^0). \tag{1.85}$$

We further make the approximation that $m_1 \gg \rho g \frac{L}{2}$ making $\theta_s^0$ small. In this limit we may solve for $m_1$ explicitly.

$$d = \frac{m_1}{g\rho}\left(\frac{2a}{m_1} - \frac{a^3}{3m_1^3}\right) \tag{1.86}$$

$a = \frac{gL}{2}\rho$,

$$m_1^2 \approx \frac{(gL)^3}{8}\frac{\rho^2}{3(L-d)g} \tag{1.87}$$

$m_1$ being large then implies that $L$ is only slightly larger than $d$.

$$L(\theta,\dot{\theta}) = \frac{1}{2}\rho\int_0^L ds\,(\int_0^s ds' \int_0^s ds''\ \dot{\tilde{\theta}}_{s'}(t)\dot{\tilde{\theta}}_{s''}(t)) - \frac{1}{2}m_1\int_0^L ds'\ \tilde{\theta}^2_{s'}(t) \tag{1.88}$$

There is no loss of generality in assuming that $\theta_s(t)$ is periodic in $s$ since we are only interested in solutions in the region $[0,L]$. Hence we may write,

$$\tilde{\theta}_s(t) = \sum_n e^{i\frac{2\pi n}{L}s}\ \omega_n(t). \tag{1.89}$$

One more simplification is desirable. We assert that the average deviation over the length of the rope is zero. This means $\omega_{n=0} = 0$; further, we also assert that $\omega_{-n} = \omega_n = \omega_n^*$ without loss of generality.

$$L(\theta,\dot{\theta}) = \frac{1}{2}\rho\sum_{n\neq 0}\frac{L^3}{(2\pi n)^2}\ \dot{\omega}_n(t)\ \dot{\omega}_{-n}(t) - \frac{1}{2}m_1 L\ \sum_n \omega_n(t)\ \omega_{-n}(t) \tag{1.90}$$

The Lagrange equations read as follows.

$$\rho\frac{L^3}{(2\pi n)^2}\ \ddot{\omega}_n(t) = -m_1 L\ \omega_n(t) \tag{1.91}$$

The solution may be written as follows:

$$\omega_n(t) = e^{\pm i\sqrt{m_1\ \frac{(2\pi n)^2}{\rho L^2}}\ t}. \tag{1.92}$$

Thus the general solution is,

$$\tilde{\theta}_s(t) = \sum_n a_n\ e^{i\frac{2\pi n}{L}(s\pm\sqrt{\frac{m_1}{\rho}}\ t)}. \tag{1.93}$$

We can see that this implies that the general motion of the rope is a superposition of waves traveling along the rope with speed $\sqrt{\frac{m_1}{\rho}}$ in either direction. This moderately simple example shows that it is possible to study systems with infinite degrees of freedom using the methods of Larangians just as well as systems with finitely many degrees of freedom. In the exercises, the reader is invited to explore more possibilities.

■ Lest the reader go away with the impression that only systems that are related to everyday tangible objects such as masses, springs, ropes, and pulleys are the basis for writing down continuum classical field theories, we give a slightly unusual example. Consider the Lagrangian with $q_1(t) \to \psi(x,t)$ and $\dot{q}_1(t) \to \partial_t \psi(x,t) \equiv \dot{\psi} \equiv \frac{\partial}{\partial t}\psi(x,t)$ and $q_2(t) \to \psi^*(x,t)$ and $\dot{q}_2(t) \to \partial_t \psi^*(x,t) \equiv \dot{\psi}^* \equiv \frac{\partial}{\partial t}\psi^*(x,t)$,

$$L[\psi, \dot{\psi}; \psi^*, \dot{\psi}^*] = \frac{1}{2}\int_{-\infty}^{\infty} dx\, \psi^*(x,t)(i\frac{\partial}{\partial t} + \frac{\hbar^2}{2m}\frac{\partial^2}{\partial x^2} - V(x,t))\psi(x,t)$$

$$+\frac{1}{2}\int_{-\infty}^{\infty} dx\, \psi(x,t)(-i\frac{\partial}{\partial t} + \frac{\hbar^2}{2m}\frac{\partial^2}{\partial x^2} - V(x,t))\psi^*(x,t) \qquad (1.94)$$

The classical Lagrange equation of this Lagrangian is nothing but the Schrodinger equation of quantum mechanics! (Prove this.) What does this mean ? What then is the quantum version of this theory, if the classical equations already reproduce quantum mechanics? The answer is that $\psi(x,t)$ here represents not the wave function of a particle but the 'coordinate' of a matter field just as $\mathbf{E}(\mathbf{x},t)$ represents the strength of the electric field, which we shall see later is the 'coordinate' in a Lagrangian whose Lagrange equations are the Maxwell equations of electrodynamics. These matter fields then have excitations that may be identified with particles, just as the excitations of the electromagnetic field are identified with photons. The main advantage of such a point of view is that both material particles and quanta of force carriers such as photons may be treated on an equal footing. This allows for the possibility seen in nature that material particles may be created from the vacuum by conversion of energy to matter. Clearly, this requires in addition the introduction of special relativity, but for the purposes of this problem it suffices to state that since $\psi(x,t)$ does not have the interpretation of a wave function, it is not required to obey $\int_{-\infty}^{\infty} dx\, \psi^*(x,t)\psi(x,t) = 1$. This allows for the possibility that the number of particles in the system is not fixed.

## 1.5 Variational Methods

Variational methods are perhaps historically the first application of what may be termed functional calculus to physics. The variational method refers to an approach where the desired solution of a problem is expressed as a function which minimizes a certain functional (most interesting situations involve minimizing, but there may be situations where we can only claim that the solution is extremal). The Brachistochrone problem (see illustration) was an attempt to find the path that a particle should take in order to reach from an elevated point in a gravitational field to another point at a lower level in the shortest possible time. Fermat's principle in optics states that the path taken by light is the one that has the shortest optical path length, which allows for the derivation of Snell's law. The Schrödinger equation of quantum mechanics is expressible as a variational principle. Now we describe in detail the Brachistochrone problem.

### 1.5.1 The Brachistochrone Problem

A Brachistochrone curve (Gr. *brachistos* 'the shortest,' *chronos* 'time'), or path of quickest descent, is the path between two points that is covered in the least time by a point-like body that starts at the first point from rest and moves along the curve frictionlessly to a second point at a lower level, under the action of constant gravity. Mathematically, this problem is stated as follows. The rate of change of velocity along the x and y directions are (the force equations),

$$m\frac{dv_x}{dt} = N\,cos(\theta)\ ;\ m\frac{dv_y}{dt} = N\,sin(\theta) - mg \tag{1.95}$$

where $N$ is the normal reaction exerted by the surface on which the mass is sliding (see fig 1.9). We may eliminate $N$ to write,

$$\frac{m\frac{dv_y}{dt} + mg}{m\frac{dv_x}{dt}} = tan(\theta) \tag{1.96}$$

But, $\frac{dx}{dy} = -tan(\theta)$. In other words, $v_x = \frac{dx}{dt} = -tan(\theta)\ \frac{dy}{dt} = -tan(\theta)\ v_y$. From simple trigonometry we may write $v_y = v\,cos(\theta)$; $v_x = -v\,sin(\theta)$ where $v = \sqrt{v_x^2 + v_y^2}$ is the instantaneous speed of the mass. Thus the earlier equation becomes,

$$\frac{dv_y + g\,dt}{dv_x} = -\frac{v_x}{v_y}; \tag{1.97}$$

rewriting this we get,

$$v_y\,dv_y + g\,v_y\,dt = -v_x\,dv_x. \tag{1.98}$$

But $v_y dt = dy$, thus after integration we obtain,

$$\frac{1}{2}v^2 + gy = const. \tag{1.99}$$

This is nothing but conservation of energy. Of course, we could have started with this but we just managed to show that the normal reaction does no work and therefore has no contribution to the total mechanical energy. Since the mass starts from rest at a height $H$ we may write,

$$v = \frac{ds}{dt} = \sqrt{2g(H-y)}. \tag{1.100}$$

Customarily, the path is defined parametrically to be $(x(s), y(s))$ where $s$ is the distance along the curve. We may now integrate the above equation to get,

$$\int_{t_i}^{t_f} dt = \int_{s_i}^{s_f} \frac{ds}{\sqrt{2g(H - y(s))}}. \tag{1.101}$$

The problem with this is that the length of the curve is not fixed, whereas the problem statement tells us that $(x_i, y_i) = (0, H)$ and $(x_f, y_f) = (L, 0)$. Thus we have to rewrite $ds = \sqrt{dx^2 + dy^2} = dx\,\sqrt{1 + y'^2(x)}$. The time of flight is,

$$T = \int_0^L dx\,\frac{\sqrt{1 + y'^2(x)}}{\sqrt{2g(H - y(x))}}. \tag{1.102}$$

The time of flight given by Eq. (1.102) depends on the path taken $y(x)$. The extremum condition is (the proof that it is a minimum is left to the exercises),

$$\delta T = 0 \tag{1.103}$$

where the variation is the difference between two paths $y(x)$ and $y(x) + \delta y(x)$ that both start at $(x_i, y_i)$ and end at $(x_f, y_f)$. Therefore $\delta y(0) = \delta y(L) = 0$. But we may also write,

$$\delta T = \int_0^L dx \frac{\delta T}{\delta y(x)} \delta y(x) + \int_0^L dx \frac{\delta T}{\delta y'(x)} \delta y'(x). \tag{1.104}$$

Now we integrate by parts to get,

$$\delta T = \int_0^L dx \frac{\delta T}{\delta y(x)} \delta y(x) + \int_0^L dx\, \frac{d}{dx}\left(\frac{\delta T}{\delta y'(x)} \delta y(x)\right) - \int_0^L dx\, \delta y(x) \frac{d}{dx} \frac{\delta T}{\delta y'(x)}. \tag{1.105}$$

The middle term is the integral of a derivative so that it in the term is the bracket evaluated at the end point which is zero, since $\delta y(x)$ vanishes at both end points. Thus the term that remains is,

$$\delta T = \int_0^L dx\, \delta y(x) \left(\frac{\delta T}{\delta y(x)} - \frac{d}{dx} \frac{\delta T}{\delta y'(x)}\right). \tag{1.106}$$

Since $T$ is stationary, we must have $\delta T = 0$ for each path $\delta y(x)$. The path with the shortest time of flight obeys the Euler-Lagrange equation,

$$\frac{\delta T}{\delta y(x)} - \frac{d}{dx} \frac{\delta T}{\delta y'(x)} = 0. \tag{1.107}$$

Alternatively, if we write $T = \int_0^L dx\, f(y(x), y'(x), x)$, then it is also true that,

$$\frac{\partial f(y(x), y'(x), x)}{\partial y(x)} - \frac{d}{dx} \frac{\partial f(y(x), y'(x), x)}{\partial y'(x)} = 0. \tag{1.108}$$

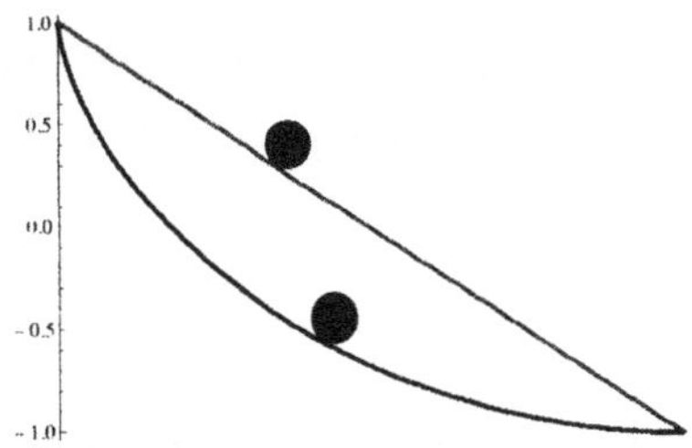


Figure 1.9: The illustration shows how a mass sliding down a straight path lags behind one that is sliding down a cycloidal path when both are released from rest and from the same height.

where now $\frac{\partial}{\partial y(x)}$ is an ordinary partial derivative.

Direct substitution gives,

$$\frac{\delta T}{\delta y'(x)} = \frac{2y'(x)}{2\sqrt{1+y'^2(x)}\sqrt{2g(H-y(x))}} \tag{1.109}$$

$$\frac{\delta T}{\delta y(x)} = 2g\sqrt{1+y'^2(x)}\frac{1}{2\sqrt{2g(H-y(x))}2g(H-y(x))}. \tag{1.110}$$

After simplification, the Euler-Lagrange equation becomes,

$$2+3y'^2(x)+y'^4(x)+2(-H+y(x))y''(x)=0. \tag{1.111}$$

One can see that this equation is a second-order nonlinear differential equation, something we wish to avoid. In fact, it is possible to write the relevant equation much more simply by employing what is known as the Beltrami identity. According to this identity, if the Lagrangian (or the quantity $f$ in this write-up) is explicitly independent of $x$ (which means it depends on $x$ only through $y(x)$) we may write,

$$f(y(x),y'(x)) - y'(x)\frac{\partial f(y(x),y'(x))}{\partial y'(x)} = C, \tag{1.112}$$

where $C$ is a constant.

**Proof of Beltrami identity**: Consider

$$\frac{d}{dx}\left(f(y(x),y'(x),x) - y'(x)\frac{\partial f(y(x),y'(x),x)}{\partial y'(x)}\right)$$

$$= \frac{df(y(x),y'(x),x)}{dx} - y''(x)\frac{\partial f(y(x),y'(x),x)}{\partial y'(x)}$$

$$-y'(x)\frac{d}{dx}\frac{\partial f(y(x),y'(x),x)}{\partial y'(x)}. \quad (1.113)$$

We now use the Lagrange equation Eq. (1.108) to write,

$$\frac{d}{dx}\left(f - y'(x)\frac{\partial f}{\partial y'(x)}\right) = \frac{df}{dx} - y''(x)\frac{\partial f}{\partial y'(x)} - y'(x)\frac{\partial f}{\partial y(x)}. \quad (1.114)$$

Next we note that

$$\frac{df}{dx} = \frac{\partial f}{\partial x} + y'(x)\frac{\partial f}{\partial y(x)} + y''(x)\frac{\partial f}{\partial y'(x)}. \quad (1.115)$$

This means,

$$\frac{d}{dx}\left(f - y'(x)\frac{\partial f}{\partial y'(x)}\right) = \frac{\partial f}{\partial x} \quad (1.116)$$

as required.

This follows from the stronger version of the identity which states that the equation obeyed by $f$ is,

$$\frac{d}{dx}\left(f(y(x),y'(x),x) - y'(x)\frac{\partial f(y(x),y'(x),x)}{\partial y'(x)}\right) - \frac{\partial f(y(x),y'(x),x)}{\partial x} = 0. \quad (1.117)$$

The proof is given in the box. We shall focus only on its application. Therefore,

$$f - y'(x)\frac{\delta f}{\delta y'(x)} = \frac{1}{\sqrt{2g(H - y(x))}\sqrt{1 + y'^2(x)}} = const. \quad (1.118)$$

This means for some constant $A$,

$$(H - y(x))(1 + y'^2(x)) = 2A. \quad (1.119)$$

The solution to this is transparent in the following parametric form:

$$H - y(t) = A(1 - cos(t));\ x(t) = A(t - sin(t)). \quad (1.120)$$

To see this we evaluate,

$$\frac{dy}{dx} = \frac{\frac{dy}{dt}}{\frac{dx}{dt}} = -\frac{sin(t)}{(1-cos(t))} \tag{1.121}$$

or,

$$1+y'^2(x) = 1+\frac{sin^2(t)}{(1-cos(t))^2} = cosec[t/2]^2. \tag{1.122}$$

Thus,

$$(H-y(x))(1+y'^2(x)) = A(1-cos(t))cosec[t/2]^2 = 2A. \tag{1.123}$$

Figure 1.10: A cycloid is the locus of points drawn by the tip of a spoke of a wheel as the wheel rolls along the ground.

We have to now relate $A$ to the given data, namely $H, L$. The initial point may be chosen as $t = 0$ so that $y(0) = H$ and $x(0) = 0$ as required. Now the final point is,

$$H - y(t_f) = A(1-cos(t_f)) = H;\ x(t_f) = A(t_f - sin(t_f)) = L. \tag{1.124}$$

The quantity $A$ has to be determined from the above transcendental equation by eliminating $t_f$. This can only be done numerically.

### 1.5.2 Fermat's Principle in Optics

As a second example, we consider Fermat's principle in optics. This principle is due to the French mathematician Pierre de Fermat. It states that the path taken by light in moving from a starting point to a final point in a refracting medium is the quickest one among all possibilities. If we denote by

$$T = \int_A^B dt = \int_A^B \frac{ds}{v}. \tag{1.125}$$

The velocity $v$ of light in a medium depends on the refractive index. Typically we have either two or more optically homogeneous media separated by boundaries that give rise to spatial dependence of the refractive index, or for example in an optical fiber, it could be that the refractive index is a smooth function of the position. In either case we write $v = \frac{c}{n}$ where $c$ is the speed of light in a vacuum and $n$ is the position-dependent refractive index. So the quantity to be minimized is,

$$T = \frac{1}{c}\int_A^B n\,ds. \tag{1.126}$$

Assume that the ray of light moves from $(x_A, y_A, z_A)$ to $(x_B, y_B, z_B)$. We choose to parameterize the path as $(x(z), y(z), z)$, i.e., in terms of the z-coordinate. In this case, $ds = \sqrt{dx^2 + dy^2 + dz^2} = |dz|\sqrt{1 + \left(\frac{dx}{dz}\right)^2 + \left(\frac{dy}{dz}\right)^2}$

$$T = \frac{1}{c}\int_{z_A}^{z_B} n(x(z),y(z),z)\sqrt{1+\left(\frac{dx}{dz}\right)^2+\left(\frac{dy}{dz}\right)^2}\,|dz|$$

$$\equiv \int_{z_A}^{z_B} dz\,u(z)\,f(x,y;x',y') \tag{1.127}$$

where $dz\,u(z) = |dz|$. The function $u(z)$ has the property that it is $u(z) = +1$ if $z$ is increasing in that segment of the path, and $u(z) = -1$ if $z$ is decreasing in that segment of the path. We have shown many times earlier that the path that minimizes the time taken ('action') is the one that is the solution to the Euler-Lagrange equations.

$$\frac{d}{dz}u(z)\frac{\partial f}{\partial x'(z)} = u(z)\frac{\partial f}{\partial x(z)};\ \frac{d}{dz}u(z)\frac{\partial f}{\partial y'(z)} = u(z)\frac{\partial f}{\partial y(z)} \tag{1.128}$$

We may write,

$$f(x,y;x',y') = \frac{1}{c}\,n(x(z),y(z),z)\sqrt{1+\left(\frac{dx}{dz}\right)^2+\left(\frac{dy}{dz}\right)^2}. \tag{1.129}$$

Thus, the path taken by light obeys the following equations:

$$\frac{d}{dz}\,u(z)\,n(x(z),y(z),z)\,\frac{x'(z)}{\sqrt{1+x'^2(z)+y'^2(z)}}$$

$$= u(z)\frac{\partial n(x(z),y(z),z)}{\partial x(z)}\sqrt{1+x'^2(z)+y'^2(z)} \tag{1.130}$$

$$\frac{d}{dz}\,u(z)\,n(x(z),y(z),z)\,\frac{y'(z)}{\sqrt{1+x'^2(z)+y'^2(z)}}$$

Figure 1.11: A French lawyer, Pierre de Fermat (17 August 1601? to 12 January 1665) was an amateur mathematician who heavily influenced number theory, analytic geometry, and optics. His famous conjecture remained unsolved for more than 300 years and led to the development of algebraic number theory. He introduced the principle of least time in optics, a version of which is found in Lagrangian and Hamiltonian mechanics and is connected to Huygens principle of optics.

$$= u(z)\frac{\partial n(x(z),y(z),z)}{\partial y(z)}\sqrt{1+x'^2(z)+y'^2(z)}. \tag{1.131}$$

We illustrate this method using two concrete examples. The first concerns the derivation of the law of reflection. To do this, we imagine two homogeneous regions. The first is $z<0$, which a is vacuum, and $z>0$, which is a perfect conductor. The interface $z=0$ is a reflecting surface. This means that light only exists in the region $z<0$. Furthermore, we assert that the reflection takes place in the $y=0$ plane. This means $\frac{\partial n}{\partial x}=\frac{\partial n}{\partial y}=0$, so that we may deduce,

$$u(z)\frac{x'(z)}{\sqrt{1+x'^2(z)}}=C \tag{1.132}$$

where $C$ is some constant. This means,

$$x'(z)=m\,u(z) \tag{1.133}$$

where $m$ is some other constant. Consider now two situations, one where the beam is approaching the interface and one where it is receding from the surface. In the former case, $z$ is increasing so that $u(z) = 1$. In the latter case $z$ is decreasing so that $u(z) = -1$. Thus while approaching the interface the path of the light beam is,

$$x(z) = mz + x(0), \tag{1.134}$$

and while receding it is

$$x(z) = -mz + x(0). \tag{1.135}$$

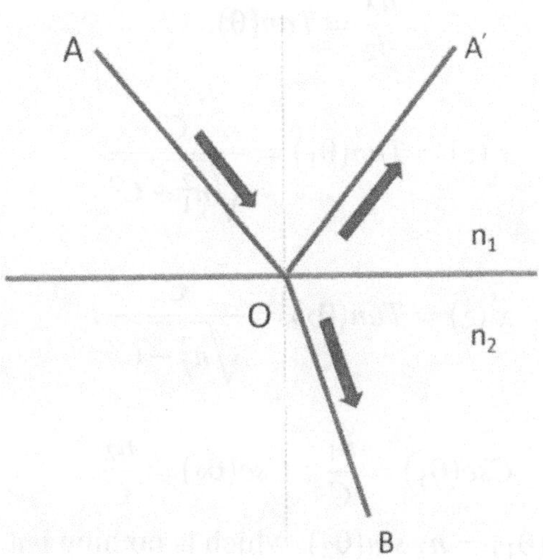


Figure 1.12: Fermat's principle says that the time taken for light to traverse $AOA'$ is a minimum when $O$ is a point on the interface. Similarly time taken to traverse $AOB$ is a minimum.

This shows that the angle of incidence is equal to the angle of reflection, which is nothing but the law of reflection. Next we derive Snell's law of refraction. Here too we consider two regions separated by an interface $z = 0$. The region $z < 0$ has refractive index $n_1$ and the region $z > 0$ has refractive index $n_2$. As before, we assume that the beam always propagates in the $y = 0$ plane. The one difference in this situation is that $u(z) = 1$ always since the beam is always propagating in the direction of increasing $z$. As before, $\frac{\partial n}{\partial x} = \frac{\partial n}{\partial y} = 0$. Putting all this together we get,

$$n(z)\frac{x'(z)}{\sqrt{1 + x'^2(z)}} = C. \tag{1.136}$$

In region $z < 0$ we may write,

$$x'(z) = \frac{C}{\sqrt{n_1^2 - C^2}}. \tag{1.137}$$

In region $z > 0$ we may write,

$$x'(z) = \frac{C}{\sqrt{n_2^2 - C^2}}. \tag{1.138}$$

But we can see that

$$\frac{dx}{dz} = Tan(\theta), \tag{1.139}$$

so that for $z < 0$,

$$x'(z) = Tan(\theta_1) = \frac{C}{\sqrt{n_1^2 - C^2}}. \tag{1.140}$$

and for $z > 0$,

$$x'(z) = Tan(\theta_2) = \frac{C}{\sqrt{n_2^2 - C^2}}. \tag{1.141}$$

Thus,

$$Csc(\theta_1) = \frac{n_1}{C} \; ; \; Csc(\theta_2) = \frac{n_2}{C}. \tag{1.142}$$

From this we get $n_1 \, sin(\theta_1) = n_2 \, sin(\theta_2)$, which is nothing but Snell's law.

### 1.5.3 Least Square Fit

We may use the variational method to approximate a set of data points with a function with prescribed properties. Consider a set of points $\mathfrak{L} = \{(x_i, y_i); i = 1, 2, ..., N\}$ that we may consider are obtained from a set of $N$ measurements of the pair $x, y$. We may wish to relate $y$ to $x$ using some simple function, so that we may be able to infer reasonable values for $y$ at values of $x$ which are not part of the list $\mathfrak{L}$ of $N$ measurements. Or maybe we wish to find an approximate solution of a differential equation by minimizing some measure of error. Typical of the latter type of problems involves solving eigenvalue equations such as $A\psi = \lambda\psi$ (such as Schrödinger's equation) to find an eigenvalue. Let us consider the fitting problem involving the data points first. Let us assume that the function we seek may be expressed as a linear combination of a finite number of 'basis' functions. That is, we shall assume that $f(x) = \sum_{i=1}^{m} c_i \varphi_i(x)$ with known $\varphi_i(x)$ and $m$, is a 'good' approximation of the curve that passes close to each point in the list $\mathfrak{L}$. We have to define what we mean by a 'good' approximation. Intuitively, this is something that would minimize the difference between $y_i$ and $f(x_i)$. This cannot be done at each $x_i$, for that would be practically useless since it would involve using as many basis functions as there are data points. Thus one takes the point of view that the combined error, perhaps defined as $\sum_{i=1}^{N} |y_i - f(x_i)|$ or, in a manner more amenable to algebraic manipulations, $\sum_{i=1}^{N}(y_i - f(x_i))^2$, is to be minimized. It is important to appreciate that the error should be thought of (and defined) as a positive quantity so that minimizing it would mean the difference between $y_i$ and $f(x_i)$ is as close to zero as possible rather than large and negative. Thus we wish to minimize the function $\Delta(c_1, c_2, ..., c_m) = \sum_{i=1}^{N}(y_i - \sum_{j=1}^{m} c_j \varphi_j(x_i))^2$ with respect to the variables $c_j$.

$$\frac{\partial}{\partial c_j}\Delta(c_1, c_2, ..., c_m) = \frac{\partial}{\partial c_j}\sum_{i=1}^{N}(y_i - \sum_{j=1}^{m} c_j \varphi_j(x_i))^2 = 0 \tag{1.143}$$

Thus we see that we have to solve a linear system of $m$ equations (typically much smaller than the number of data points $N$) and obtain the coefficients $c_k$.

$$\sum_{i=1}^{N} y_i \varphi_j(x_i) - \sum_{k=1}^{m} c_k \sum_{i=1}^{N} \varphi_k(x_i)\varphi_j(x_i) = 0 \tag{1.144}$$

This uniquely fixes the function $f(x)$, which passes 'close' to all the points in the list $\mathfrak{L}$. Indeed it is frequently the case that this function does not pass through any of the points in that list, but merely passes close to all of them.

Next we consider the question of solving the eigenvalue problem to find an eigenvalue. Consider a space of functions of a single variable (say). We also assume a suitable inner product has been defined such as $(\psi, \phi) \equiv \int_a^b w(x)\psi(x)\phi(x)dx$ where $w(x)$ is a weight function. Typically, for applications to quantum mechanics we set $w(x) \equiv 1$ and the interval $[a,b]$ could either be finite (if periodic boundary conditions are assumed) or be all of the real line. For applications to Sturm-Liouville problems, $w(x)$ is prescribed in the interval $[a,b]$ (think of Legendre or Hermite polynomials). Here we rewrite the equation $A\psi = \lambda\psi$ in a weaker form as $\lambda = \frac{(\psi, A\psi)}{(\psi,\psi)} = (f, Af)$, where $f(x) = \frac{\psi(x)}{\sqrt{(\psi,\psi)}}$. Here $A$ could be some differential operator such as $\frac{d^2}{dx^2} + u(x)$. As before, choose a set of basis functions so that $f(x) = \sum_{k=1}^{m} c_k \varphi_k(x)$. Note that $(f, f) = 1$ so that we have to impose the constraint $\sum_{k,k'=1}^{m} c_k c_{k'} \int_a^b dx\, w(x)\varphi_k(x)\varphi_{k'}(x) = 1$ while minimizing. This means we have to minimize,

$$\lambda = (f, Af) = \sum_{k,k'=1}^{m} c_k c_{k'} \int_a^b dx\, w(x)\, \varphi_k(x) A\varphi_{k'}(x) \tag{1.145}$$

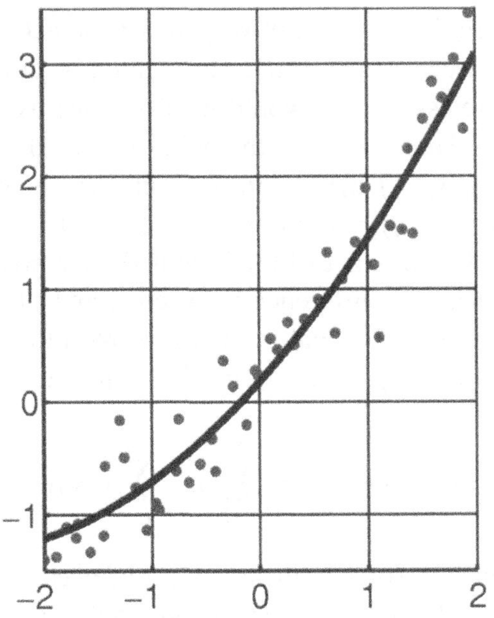


Figure 1.13: A collection of data points being fit by a basis containing polynomials of at most the second degree (source: Wikipedia).

with respect to variables $c_k$ subject to the said constraint. The well-known procedure in calculus is to use the method of Lagrange multipliers. We write,

$$F[c] \equiv \sum_{k,k'=1}^{m} c_k c_{k'} A_{k,k'} - \mu \sum_{k,k'=1}^{m} c_k c_{k'} t_{k,k'}. \tag{1.146}$$

Minimizing means solving for c's where $\frac{\partial}{\partial c_k} F[c] = 0$. This means,

$$\sum_{k'=1}^{m} c_{k'} A_{k,k'} - \mu \sum_{k'=1}^{m} c_{k'} t_{k,k'} = 0, \tag{1.147}$$

or, $Det(A - \mu t) = 0$. This has to be supplemented with the constraint condition namely,

$$\sum_{k,k'=1}^{m} c_k c_{k'} t_{k,k'} = 1 \tag{1.148}$$

From Eq. (1.145) we get

$$\lambda = \sum_{k,k'=1}^{m} c_k c_{k'} A_{k,k'} \tag{1.149}$$

From the eigenvalue equation Eq.(1.147) we get

$$\lambda = \sum_{k,k'=1}^{m} c_k c_{k'} A_{k,k'} = \mu \sum_{k,k'=1}^{m} c_k c_{k'} t_{k,k'} = \mu. \tag{1.150}$$

The last step follows upon enforcing the normalization constraint. Thus finding $\lambda$ is the same as finding an eigenvalue $\mu$ obeying $Det(A - \mu t) = 0$. The matrix $A_{k,k'}$ being finite dimensional ($m$ has to be small in order to make this procedure more useful) is relatively easier to deal with than solving the full problem, which involves an infinite dimensional matrix A (operator). A price must however be paid for this simplification, namely the obtained eigenvalue may be only close to one of the eigenvalues. Not only is the answer not an exact eigenvalue, it is not even guaranteed to be the one we are interested in unless the choice of basis functions was appropriate.

## 1.6 Exercises

**Q.1** Verify all the steps leading up to Eq. (1.13).

**Q.2** Generalize the concept of Legendre transformation to many variables.

**Q.3** In thermodynamics, the entropy is a function of internal energy and volume and number of particles $S(E,N,V)$. Perform a Legendre transformation with respect to energy and obtain the transformed function keeping in mind that temperature $T$ is defined as $\frac{1}{T} = \frac{\partial S}{\partial E}$. Try out all other possibilities. Successively transform $N$ (you should get a chemical potential somewhere) and $V$ (you should get pressure). Try also transforming two variables together and finally all three of them. Can you recognize any of the transformed functions as the standard ones from thermodynamics textbooks?

**Q.4** Verify that the Lagrange equation of Eq. (1.94) is nothing but the familiar time-dependent Schrodinger equation. Show that in this case, $N(t) = \int dx\, \psi^*(x,t)\psi(x,t)$ is independent of time. If we regard $\psi(x,t)$ as a field obeying the classical field equations of this Lagrangian, then the statement that $N(t)$ is time independent is true only in an average sense following Ehrenfest's principle.

**Q.5** The Brachistochrone problem showed that the path which ensures the time of flight (Eq. (1.102)) an extremum is a cycloid. Prove that this extremum is a minimum (the second derivative should be positive).

**Q.6** Consider a rubber band whose ends are tied to two stubs separated by a distance equal to the relaxed length $L$ of the band. When the band is plucked it is going to vibrate. The problem is to find the tension $T(\lambda,t)$ and the net strain energy contained in the band given that at $t = 0$ the displacement of each point $\lambda$ is $x(\lambda,0) = \frac{L}{10} sin(\frac{\lambda\pi}{L})$ (for simplicity, in this question, assume that the band only oscillates along the length of the band). Model the rubber band as a collection of closely spaced masses, where the $(\rho\lambda)$-th mass is located at $(x(\lambda,t), y(\lambda,t))$ connected to each other by springs with constant $k$ such that in the continuum limit the mass per unit length $\rho$ is fixed, and the spring constant goes to infinity such that the ratio of the spring constant and the number of masses is fixed.

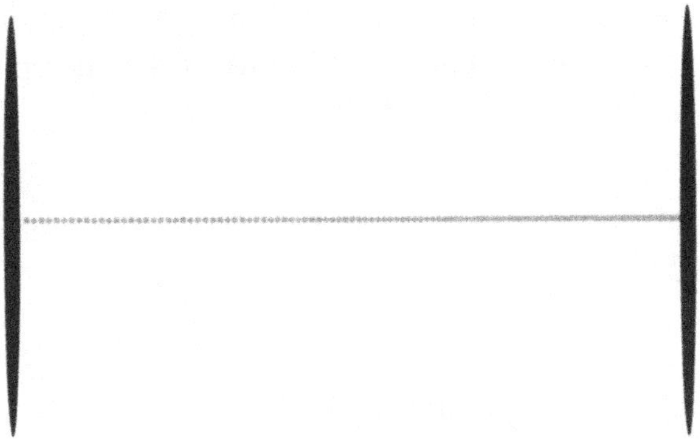

Figure 1.14: The initial state of the rubber band in Q.5. The left half is stretched, the right half is compressed and the ends are fixed.

**Q.7** Imagine a child who holds the center of the band in Q.2 and pulls it in a direction perpendicular to the band by a distance $d \ll L$ and releases the band from rest at $t = 0$. Describe the subsequent motion of the band. What are the physically reasonable boundary conditions?

**Q.8** How would you generalize Q.2 and Q.3 if the system in question were an elastic membrane instead of a rubber band?

**Q.9** Derive the equation for the shape of a slender rope hanging under its own weight supported at two ends at the same level. This shape is called a catenary.

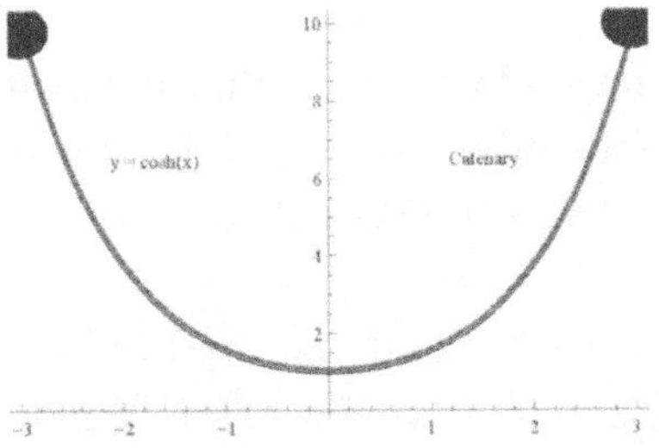


Figure 1.15: A rope with mass uniformly distributed along is length, hangs freely supported at the ends only. The shape made is known as a catenary.

**Q.10** Show that for a fixed value of the perimeter of a closed curve, the shape of the curve that maximizes the enclosed area is a circle. Conversely, show that for a fixed area, the shape that minimizes the perimeter is a circle.

**Q.11** Show that the function $\psi(x)$ that minimizes the energy functional $E = \int dx \psi^*(x)(\frac{p^2}{2m} + V(x))\psi(x)$ and subject to the normalization constraint on the wavefunction obeys Schrodinger's wave equation.

**Q.12** Derive an equation for $\psi(x)$ that minimizes the energy functional $E = \int dx \psi^*(x)\frac{p^2}{2m}\psi(x)$ subject to the normalization constraint and the Pauli blocking constraint that in momentum space, only momenta greater than the Fermi momentum are allowed. This would represent a free electron in a metal for example.

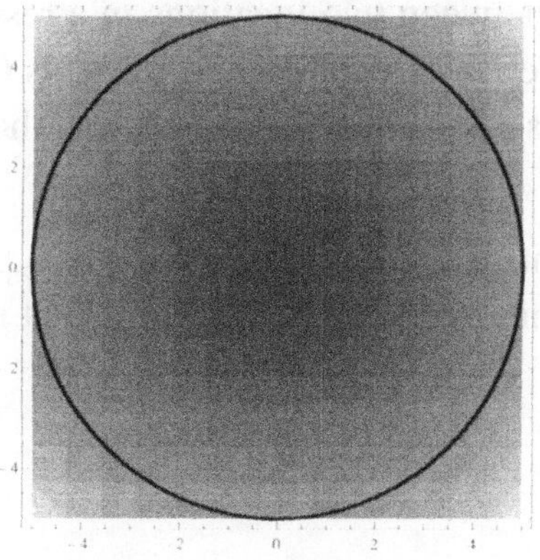

Figure 1.16: A cross section of the cylinder with graded refractive index that approaches unity toward the edge of the cylinder.

**Q.13** Imagine a cylindrical medium which has a graded refractive index, that is, the speed of light changes in a smooth fashion in the medium from point to point according to the formula $v(r) = \frac{r}{a}c + \frac{(a-r)}{a}\frac{c}{n}$, where $a$ is the radius of the cylinder and $n$ is a constant (refractive index at $r = 0$). Using Fermat's principle, find the path of a beam that starts from the origin making an angle $\theta$ with the z-axis.

**Q.14** Long long ago, in a galaxy far far away, there was a world where all beings, including light itself, were confined to the transparent surface of a sphere of radius $R$. This upper hemisphere had a refractive index $n_1$ and the lower had a refractive index $n_2$. Using Fermat's principle, derive Snell's Law in this world. Have fun imagining the multiple refractions that are inevitable in this example.

**Q.15** The Euclidean distance between two points $A$ and $B$ in a plane is $d_E = \sqrt{(x_A - x_B)^2 + (y_A - y_B)^2}$. Show that the shape of the path connecting these two points that minimizes $d_E$ is a straight line. The 'Manhattan distance' or 'taxicab

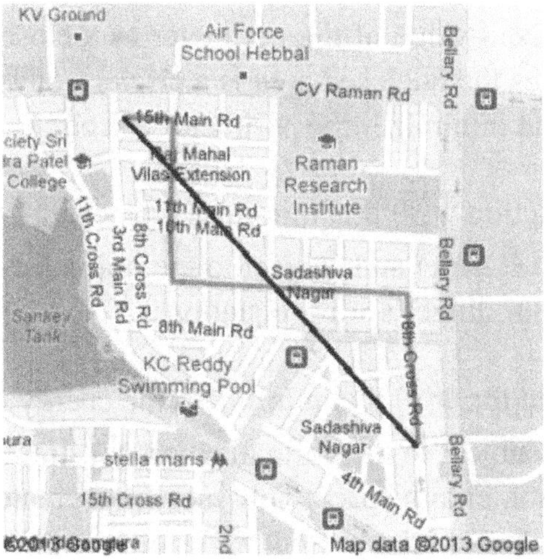


Figure 1.17: The distance between two locations in a residential city block is constrained by the nature of the main roads and cross roads (Q.14). The 'taxicab path' is not unique, whereas the straight line that minimizes the Euelidean distance is.

distance' between two points $A$ and $B$ in a plane is $d_M = |x_A - x_B| + |y_A - y_B|$. What is the shape of the path connecting these two points that minimizes $d_M$?

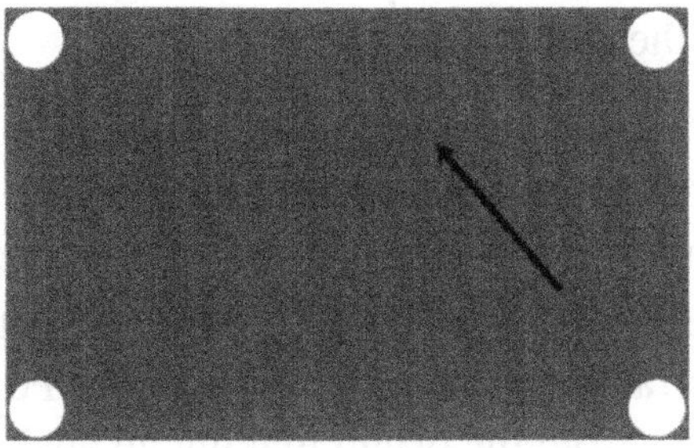

Figure 1.18: The billiard table question.

**Q.16** Imagine a rectangular billiard table of size $ml_0 \times nl_0$ (and $m$ and $n$ are relatively prime integers) with only corner pockets. A billiard ball bounces off the sides of the table elastically and obeys the law of reflection. Show that a ball sent at $45^0$ from a corner will be pocketed in another corner after $m+n-2$ bounces.

**Q.17** Use Lagrange multipliers to solve the problem of a particle constrained to move along a sphere in the presence of gravity. How should the multiplier be used in a Lagrangian approach? Use an elementary approach (Newtonian dynamics) to derive equations of motion for the same problem.

**Q.18** Explain how you would find a fit for a sinusoidal function using polynomials such as $1, x, x^2, \ldots$ Specifically, find an explicit fit for $f(x) = sin(x)$ using polynomials from the basis $\{1, x, x^2\}$ in the interval $x \in [0, \pi]$.

# Chapter 2

# Symmetries and Noether's Theorem

Symmetries are ubiquitous in nature. We admire the symmetrical and intricate patterns on the wing of a butterfly, the simple and symmetrical ripples formed on the surface of still water when a pebble is thrown and so on. Symmetries are not restricted to visual objects. A piece of music where a phrase repeats periodically is also symmetric and pleasing to the ear. We may define symmetry as a transformative property that makes an object possessing such a symmetry look (or feel, sound, etc.) the same after the transformation. Objects that do not look the same after the transformation are not symmetric under that transformation. For example, a palindrome is symmetric when the last letter is replaced by the first, the second to the last by the second, and so on. But it is not symmetric under other transformations such as the interchange of oddly and evenly located alphabets.

We could imagine more abstract kinds of symmetries. Typical among these are mathematical transformations that leave certain physical quantities unchanged. In the present context, what we have in mind are continuous transformations of the generalized coordinates that leave the Lagrangian of the system unchanged. If $q(t)$ denotes the collection of all generalized coordinates, then we may imagine transforming them by the application of some continuous transformation that changes them to $q_s(t)$. Here $s$ is a continuous parameter that facilitates the transformation. We may assert that at $s = 0$ the generalized coordinates are as before namely, $q_{s=0}(t) \equiv q(t)$. Upon changing the value of $s$, the coordinates become something else. To give a concrete example, let us take the generalized coordinates to be the simple two-dimensional Cartesian coordinates $(x(t), y(t))$. In this case we may imagine a continuous transformation that makes $x(t) \to x_\theta(t)$ and $y(t) \to y_\theta(t)$ where $x_\theta(t) = cos(\theta)\, x(t) + sin(\theta)\, y(t)$ and $y_\theta(t) = -sin(\theta)\, x(t) + cos(\theta)\, y(t)$. Here $\theta$ is a continuously changing parameter. In a general context we may imagine this transformation leaving the Lagrangian or the Hamiltonian of the system unchanged. Mathematically, this symmetry just means $L(q_s(t), \dot{q}_s(t)) \equiv L(q(t), \dot{q}(t))$

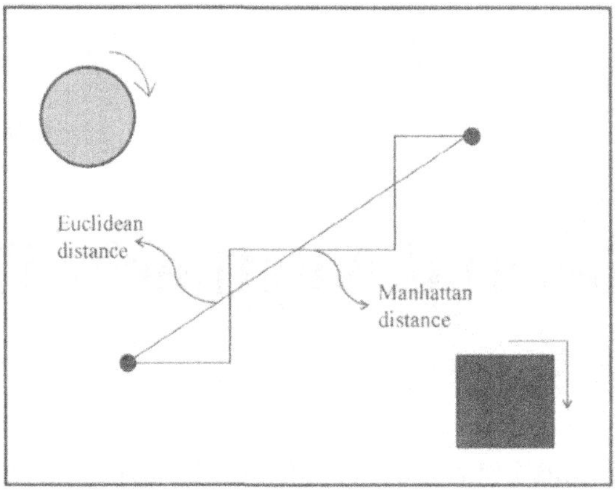


Figure 2.1: Diagram showing two different definitions of distance between two points. The familiar Euclidean distance and the 'Manhattan distance', which is the sum of the lengths of the horizontal and vertical segments that make up the zigzag path. The former is invariant under continuous rotations, but the latter only under discrete rotations by a right angle. This illustrates the notions of continuous and discrete symmetry.

where we may choose $q_{s=0}(t) \equiv q(t)$. In the Hamiltonian formulation we would have $H(q_s, p_s) = H(q, p)$. Therefore, even though the coordinates change under the transformation, the Lagrangian (Hamiltonian) does not. In this case this transformation is called a symmetry of the Lagrangian (Hamiltonian). We shall describe some examples shortly. But this information is sufficient for us to derive a conservation law. First we discuss the Lagrangian framework.

## 2.1 Noether's Theorem in a Lagrangian Setting

As we mentioned earlier, a symmetry of the Lagrangian means that $L(q_s(t), \dot{q}_s(t))$ is independent of $s$ even though $q_s(t)$ depends on $s$. This means we may write,

$$\frac{d}{ds}L(q_s(t), \dot{q}_s(t)) = 0. \tag{2.1}$$

We may use the chain rule of multi-variable calculus to rewrite this as,

$$\frac{d}{ds}L(q_s(t), \dot{q}_s(t)) = \frac{d}{ds}L(q_s(t), \dot{q}_s(t))$$

Figure 2.2: Daughter of the German mathematician Max Noether, Emmy Noether (23 March 1882 to 14 April 1935) is arguably considered to be the greatest female mathematician of all time. The theorem in mathematical physics that bears her name connects two of the most important guiding principles in modern physics: symmetry and conservation laws. She made fundamental contributions to number fields, representation theory of groups, commutative rings, etc.

$$= \left( \frac{dq_s(t)}{ds} \frac{\partial L(q_s(t), \dot{q}_s(t))}{\partial q_s(t)} + \frac{d\dot{q}_s(t)}{ds} \frac{\partial L(q_s(t), \dot{q}_s(t))}{\partial \dot{q}_s(t)} \right) = 0 \tag{2.2}$$

(Note that an expression such as $\frac{dq_s(t)}{ds} \frac{\partial L(q_s(t), \dot{q}_s(t))}{\partial q_s(t)}$ is shorthand for $\sum_{j=1}^{N} \frac{dq_s^j(t)}{ds} \frac{\partial L(q_s(t), \dot{q}_s(t))}{\partial q_s^j(t)}$, assuming there are $N$ generalized coordinates). Now, Lagrange equations tell us that,

$$\frac{\partial L(q_s(t), \dot{q}_s(t))}{\partial q_s(t)} = \frac{d}{dt} \frac{\partial L(q_s(t), \dot{q}_s(t))}{\partial \dot{q}_s(t)}. \tag{2.3}$$

Substituting this into Eq. (2.2) we may rewrite the left-hand side as,

$$0 = \frac{dq_s(t)}{ds} \frac{d}{dt} \frac{\partial L(q_s(t), \dot{q}_s(t))}{\partial \dot{q}_s(t)} + \frac{d\dot{q}_s(t)}{ds} \frac{\partial L(q_s(t), \dot{q}_s(t))}{\partial \dot{q}_s(t)}$$

$$= \frac{d}{dt}\left(\frac{dq_s(t)}{ds}\frac{\partial L(q_s(t),\dot{q}_s(t))}{\partial \dot{q}_s(t)}\right). \tag{2.4}$$

The last result follows since we may interchange the derivatives with respect to $t$ and $s$ so that $\frac{d}{dt}\frac{dq_s(t)}{ds} = \frac{d\dot{q}_s(t)}{ds}$. Therefore, associated with this symmetry, there is a conservation law—the existence of a quantity that is time independent.

$$\frac{d}{dt}Q = 0 \tag{2.5}$$

Here $Q$ is known as the Noether's constant associated with this symmetry. The explicit formula for this may be read out from the earlier equation, namely (without loss of generality we may set $s = 0$),

$$Q = \left(\sum_{j=1}^{N} \frac{dq_s^j(t)}{ds}\frac{\partial L(q_s(t),\dot{q}_s(t))}{\partial \dot{q}_s^j(t)}\right)_{s=0}. \tag{2.6}$$

It is easy to see why only continuous symmetries lead to conservation laws in this framework. We may imagine discrete symmetries—an example would be if the Lagrangian is an even function of the generalized coordinates and their derivatives $L(-q(t), -\dot{q}(t)) = L(q(t), \dot{q}(t))$. There is no continuous parameter associated with this symmetry that we may differentiate with respect to. Hence, the tools of calculus may not be exploited to arrive at a conservation law. One could also do the reverse. Given a conserved quantity, we may wish to examine what symmetry leads to this conservation law. This is ascertained by inverting Eq. (2.6) assuming the Noether constant $Q$ and Lagrangian $L$ are known, leading to an evaluation of $\frac{dq_s^j(t)}{ds}$.

Some examples may illustrate these points. Imagine a particle acted on by a central force in three dimensions. In this case the Lagrangian of the particle is,

$$L(\mathbf{r},\dot{\mathbf{r}}) = \frac{1}{2}m\dot{\mathbf{r}}^{\,2} - V(r). \tag{2.7}$$

Imagine a transformation $\mathbf{r}(t) \to \mathbf{r}_s(t) \equiv M_s\,\mathbf{r}(t)$ where $M_s$ is a $3\times 3$ orthogonal matrix (and $\mathbf{r}(t)$ is a column vector with three rows). Since there is a (continuous) infinity of such orthogonal matrices, we have chosen to parameterize using the index $s$. Then it is clear that since $M_s$ is time independent, $\dot{\mathbf{r}}_s(t) \equiv M_s\,\dot{\mathbf{r}}(t)$, and also since $M_s$ is orthogonal, $\dot{\mathbf{r}}_s^{\,2}(t) \equiv \dot{\mathbf{r}}^{\,2}(t)$ and also $r \equiv |\mathbf{r}(t)| \equiv |\mathbf{r}_s(t)| = r_s$. Therefore we may write,

$$L(\mathbf{r}_s,\dot{\mathbf{r}}_s) = L(\mathbf{r},\dot{\mathbf{r}}). \tag{2.8}$$

Therefore the orthogonal transformation (simple rotation) is a symmetry of the Lagrangian of a free particle in the presence of a central force. We wish to see what kind of conserved quantities emerge. We may write down a constant of the motion since Noether's theorem tells us (from Eq.(2.6)) that $Q$ is a constant where,

$$Q = \left(\frac{d\mathbf{r}_s(t)}{ds}\cdot\frac{\partial L(\mathbf{r}_s(t),\dot{\mathbf{r}}_s(t))}{\partial \dot{\mathbf{r}}_s(t)}\right)_{s=0}. \tag{2.9}$$

Here $\frac{\partial L(\mathbf{r}_s(t),\dot{\mathbf{r}}_s(t))}{\partial \dot{\mathbf{r}}_s(t)}$ is a vector obtained by formally differentiating $L$ with respect to $\dot{\mathbf{r}}_s(t)$. Further progress is not possible without a concrete realization of the matrix $M_s$ in terms of the continuous parameter $s$. We choose to think of $s$ as the angle of rotation about some chosen axis denoted by $\hat{n}$. Here $s$ is the single real parameter that changes whereas $\hat{n}$ is fixed. Therefore for each axis $\hat{n}$ we expect to find a Noether constant. Thus we are going to obtain a family of constants of the motion parameterized by the unit vector $\hat{n}$. Let us write $\mathbf{r} = \mathbf{r}_{\parallel} + \mathbf{r}_{\perp}$, where $\mathbf{r}_{\parallel} = (\mathbf{r}\cdot\hat{n})\hat{n}$ and $\mathbf{r}_{\perp} = \mathbf{r} - (\mathbf{r}\cdot\hat{n})\hat{n}$. Upon rotation by an angle $s$ about the axis $\hat{n}$, it is clear that $\mathbf{r}_{\parallel}$ remains unchanged, but $\hat{r}_{\perp}$, which has two linearly independent components, rotates by an angle $s$ in a plane perpendicular to axis $\hat{n}$.

$$\mathbf{r}_s(t) = \mathbf{r}_{\parallel}(t) + \mathbf{r}_{s,\perp}(t) \tag{2.10}$$

Thus

$$Q = \left( \frac{d\mathbf{r}_{s,\perp}(t)}{ds} \cdot m\,\dot{\mathbf{r}}_s(t) \right)_{s=0}. \tag{2.11}$$

We now postulate the right-handed mutually orthogonal unit vectors that we shall use as a basis : $(\hat{e}_1, \hat{e}_2, \hat{n})$. Thus we may write, $\mathbf{r}_{\perp}(t) = r_{\perp,1}(t)\,\hat{e}_1 + r_{\perp,2}(t)\,\hat{e}_2$. Upon rotation these components become $\mathbf{r}_{\perp,s}(t) = r_{\perp,1,s}(t)\,\hat{e}_1 + r_{\perp,2,s}(t)\,\hat{e}_2$.

$$r_{\perp,1,s}(t) = cos(s) r_{\perp,1}(t) + sin(s) r_{\perp,2}(t) \tag{2.12}$$

$$r_{\perp,2,s}(t) = -sin(s) r_{\perp,1}(t) + cos(s) r_{\perp,2}(t) \tag{2.13}$$

Since $\mathbf{r}_{\perp}$ is perpendicular to $\mathbf{r}_{\parallel}$ we should be able to write,

$$Q = \left( \frac{d\mathbf{r}_{s,\perp}(t)}{ds} \cdot m\,\dot{\mathbf{r}}_{s,\perp}(t) \right)_{s=0}. \tag{2.14}$$

From the equations that describe rotation of the coordinate system (Eq. (2.13)) we may conclude,

$$\frac{d}{ds} r_{\perp,1,s}(t) = r_{\perp,2,s}(t)\ ;\ \frac{d}{ds} r_{\perp,2,s}(t) = -r_{\perp,1,s}(t). \tag{2.15}$$

Hence,

$$\frac{d}{ds}\mathbf{r}_s(t) = r_{\perp,2,s}(t)\,\hat{e}_1 - r_{\perp,1,s}(t)\,\hat{e}_2. \tag{2.16}$$

Therefore,

$$Q = (r_{\perp,2,s}(t)\,\hat{e}_1 - r_{\perp,1,s}(t)\,\hat{e}_2) \cdot m\,(\dot{r}_{\perp,1,s}(t)\,\hat{e}_1 + \dot{r}_{\perp,2,s}(t)\,\hat{e}_2) \tag{2.17}$$

$$= m(r_{\perp,2,s}(t)\,\dot{r}_{\perp,1,s}(t) - r_{\perp,1,s}(t)\,\dot{r}_{\perp,2,s}(t)) = -(\mathbf{r}_s(t)\times m\dot{\mathbf{r}}_s(t))\cdot\hat{n} = (\mathbf{r}\times\mathbf{p})_{-\hat{n}}. \tag{2.18}$$

Thus, Noether's constant is the component of the angular momentum along $-\hat{n}$. Since $\hat{n}$ was any unit vector, this means that the angular momentum vector itself is a conserved quantity. Thus Noether's theorem allows us to deduce conserved quantities starting from known symmetries of the Lagrangian. The idea here is that symmetries being intuitive in nature, even the abstract kind involving mathematical expressions, they are easier to guess than the conserved quantities themselves.

## 2.2 Noether's Theorem in a Hamiltonian Setting

As in the Lagrangian approach, in the Hamiltonian approach too, symmetry means the function that determines the time evolution of the system (in this case the Hamiltonian) is unchanged even though the phase space variables change with respect to some continuous variable $s$. Thus, even though $(p,q) \to (p_s,q_s)$, $H(q_s,p_s) = H(q,p)$ so that,

$$\frac{d}{ds}H(p_s,q_s) = 0. \tag{2.19}$$

We wish to simplify the left-hand side so that it is rewritten as the time derivative of some other quantity, thereby allowing us to derive a conserved quantity. To this end we use the chain rule to write,

$$\frac{d}{ds}H = \frac{dq}{ds}\frac{\partial H}{\partial q} + \frac{dp}{ds}\frac{\partial H}{\partial p}. \tag{2.20}$$

But we know from an earlier chapter that,

$$\frac{dq}{ds} = \frac{\partial G}{\partial p}; \frac{dp}{ds} = -\frac{\partial G}{\partial q}. \tag{2.21}$$

where $G$ is the generator of the transformation $(q,p) \to (q_s,p_s)$. Thus we have,

$$\frac{d}{ds}H = \frac{\partial G}{\partial p}\frac{\partial H}{\partial q} - \frac{\partial G}{\partial q}\frac{\partial H}{\partial p} = \{H,G\}. \tag{2.22}$$

Thus the symmetry implies,

$$\{H,G\} = 0. \tag{2.23}$$

But we also know that

$$\frac{d}{dt}G = \{G,H\}. \tag{2.24}$$

Putting these together we conclude that $G = const$. Thus, a symmetry such as Eq. (2.19) leads to a conserved quantity such as $G$. The reader is encouraged to re-derive the results of the earlier section using the Hamiltonian approach.

## 2.3 Dynamical Symmetries

We have shown earlier how the angular momentum is a generator of rotations. We just showed, using Noether's theorem in the Lagrangian framework, that angular momentum is a conserved quantity (that is, time independent) provided the Lagrangian is invariant under rotations. In the context of the inverse square force, it is well known that there is another vector conserved quantity known as the Laplace-Runge-Lenz vector (LRL vector). Just as simple rotations is the symmetry behind the conservation of angular momentum, we wish to ascertain what transformation is the symmetry that leads to the conservation of the LRL vector. To answer this effectively, we consider Noether's theorem in the Hamiltonian setting. Before we do this, it is proper to write down an expression for the conserved LRL vector. This refers to a situation where a body of mass $m$ orbits a much more massive object so that the force acting on this body is $-\frac{k}{r^2}\hat{r}$. The LRL vector is defined as,

$$\mathbf{A} = \mathbf{p}\times\mathbf{L} - mk\hat{r}, \tag{2.25}$$

where $\mathbf{p}$ is the linear momentum and $\mathbf{L}$ is the angular momentum. We now show by direct computation that this quantity is conserved. Upon performing the time derivative, keeping in mind that the angular momentum vector is conserved we obtain,

$$\begin{aligned}\frac{d}{dt}\mathbf{A} &= -k\frac{\hat{r}}{r^2}\times(\mathbf{r}\times\mathbf{p}) - \frac{mk}{r}\frac{d}{dt}\mathbf{r} + \frac{mk}{r^3}(\mathbf{r}\cdot\frac{d\mathbf{r}}{dt})\mathbf{r} \\ &= -k\frac{1}{r^2}\left[\mathbf{r}(\hat{r}\cdot\mathbf{p}) - \mathbf{p}(\hat{r}\cdot\mathbf{r})\right] - \frac{mk}{r}\frac{d}{dt}\mathbf{r} + \frac{mk}{r^3}(\mathbf{r}\cdot\frac{d\mathbf{r}}{dt})\mathbf{r} = 0.\end{aligned} \tag{2.26}$$

The last result follows from the observation $\mathbf{p} = m\frac{d\mathbf{r}}{dt}$.

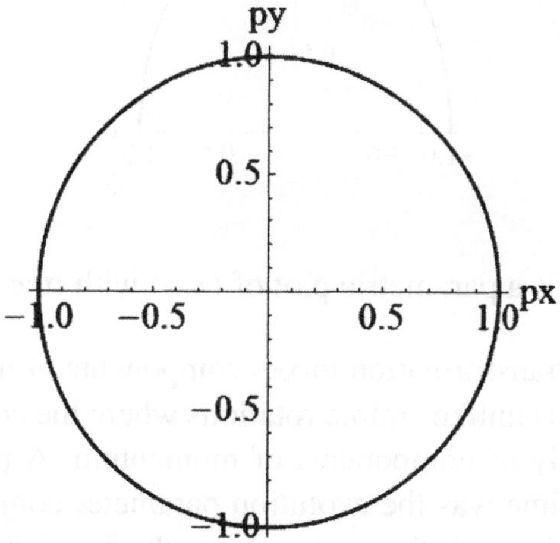


Figure 2.3: This is a parametric plot of $(p_x, p_y)$ with $\alpha$ being the parameter over the same range as in the earlier plot. This reproduces the well-known result that the momentum vector traces out a circle in case of the inverse square force.

Since we have shown that the LRL vector is constant (time independent), we may set the generator of the symmetry responsible for conserving this vector, $G \equiv \alpha = f(A_x, A_y)$, where $f$ is some suitable function of the components. It is useful to choose $\alpha = tan^{-1}\left(\frac{A_y}{A_x}\right)$. There is a good physical reason for this choice. The magnitude of the LRL vector is related to other conserved quantities such as angular momentum and total energy. It is the direction of this vector that is a new conserved quantity distinct from the others just mentioned. The above choice ensures that $\alpha$ represents the angle made by the LRL vector with some chosen x-axis. Now that we know the generator of the symmetry transformation, all that remains is to ascertain the details of the symmetry, namely the s-dependence of the mapping $(q,p) \to (q_s, p_s)$. We write as usual,

$$\frac{dx}{ds} = \frac{\partial \alpha}{\partial p_x}\,;\ \frac{dp_x}{ds} = -\frac{\partial \alpha}{\partial x}\,;\ \frac{dy}{ds} = \frac{\partial \alpha}{\partial p_y}\,;\ \frac{dp_y}{ds} = -\frac{\partial \alpha}{\partial y}. \tag{2.27}$$

These have to be solved using $\alpha = Tan^{-1}(A_y/A_x)$ where,

$$A_x = p_y(xp_y - yp_x) - mk\frac{x}{\sqrt{x^2+y^2}}\,;\ A_y = -p_x(xp_y - yp_x) - mk\frac{y}{\sqrt{x^2+y^2}}. \tag{2.28}$$

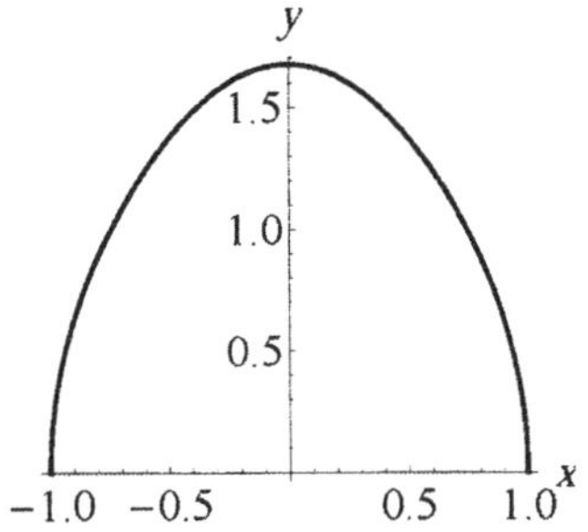


Figure 2.4: This is a parametric plot of $(x,y)$ with $\alpha$ as the parameter.

One may see that this transformation mixes components of momenta with components of position. This is unlike simple rotations where the components of position transform independently of components of momentum. A parametric plot shows the evolution. Just as time was the evolution parameter conjugate to the Hamiltonian in that case, here the evolution parameter is the angular momentum conjugate to the conserved $\alpha$, which is the angle made by the **A** vector on the x-axis.

## 2.4 Symmetries in Field Theories

In case of fields, a myriad of different possibilities emerge as a choice of the symmetry transformation. This is due to the index that counts the number of degrees of freedom that in the case of fields, becomes a continuous variable. A continuous variable can be scaled by any factor, even differentiated upon, thereby allowing many more possibilities. To see some examples consider,

$$L[\theta,\dot{\theta}] = \frac{1}{2}\int_{-\infty}^{\infty} dx \left( \frac{1}{c^2}\left(\frac{\partial\theta(x,t)}{\partial t}\right)^2 - \left(\frac{\partial\theta(x,t)}{\partial x}\right)^2 \right). \tag{2.29}$$

Set $\theta_s(x,t) = \theta(x+s,t)$. It is easy to see that,

$$L[\theta_s,\dot{\theta}_s] = L[\theta,\dot{\theta}]. \tag{2.30}$$

From Noether's theorem it follows that the conserved quantity is,

$$Q = \left( \int_{-\infty}^{\infty} dx \frac{d\theta_s(x,t)}{ds} \frac{\delta L}{\delta\dot{\theta}_s(x,t)} \right)_{s=0}. \tag{2.31}$$

Notice that the symbol $x$ counts the number of degrees of freedom. Since it is a continuous variable, it follows that there are a continuous infinity of degrees of freedom consistent with a field theory. Noether's theorem demands that we sum over all those degrees of freedom in order to obtain a conserved quantity. Thus the conserved quantity is,

$$Q = \frac{1}{c^2}\int_{-\infty}^{\infty} dx \frac{\partial\theta(x,t)}{\partial x}\frac{\partial\theta(x,t)}{\partial t}. \tag{2.32}$$

It is easy to verify directly that this quantity is time independent. A direct differentiation gives,

$$\frac{d}{dt}Q = \frac{1}{c^2}\int_{-\infty}^{\infty} dx \frac{\partial^2\theta(x,t)}{\partial x\partial t}\frac{\partial\theta(x,t)}{\partial t} + \frac{1}{c^2}\int_{-\infty}^{\infty} dx \frac{\partial\theta(x,t)}{\partial x}\frac{\partial^2\theta(x,t)}{\partial t^2}. \tag{2.33}$$

We now use the equation of motion, in this case the wave equation to rewrite the above relation as,

$$\frac{d}{dt}Q = \frac{1}{c^2}\int_{-\infty}^{\infty} dx \frac{\partial^2\theta(x,t)}{\partial x\partial t}\frac{\partial\theta(x,t)}{\partial t} + \int_{-\infty}^{\infty} dx \frac{\partial\theta(x,t)}{\partial x}\frac{\partial^2\theta(x,t)}{\partial x^2} \tag{2.34}$$

$$= \frac{1}{2c^2}\int_{-\infty}^{\infty} dx \frac{\partial}{\partial x}\left(\frac{\partial\theta(x,t)}{\partial t}\right)^2 + \frac{1}{2}\int_{-\infty}^{\infty} dx \frac{\partial}{\partial x}\left(\frac{\partial\theta(x,t)}{\partial x}\right)^2 = 0 \tag{2.35}$$

since we are going to assume that the fields and all their derivatives vanish at infinity. It must be stressed that the conservation law applies only in conjunction with the equation of motion, whereas the symmetry of the Lagrangian is independent of whether the variables obey the Lagrange equations. The former situation where the Lagrange equations are assumed is known as an 'on shell' condition. The latter, more general situation is known as 'off shell'. Thus we say that the symmetry of the Lagrangian is valid off shell whereas Noether's theorem is valid on shell.

As a next example, we consider a scalar field in three dimensions.

$$L[\theta,\dot{\theta}] = \frac{1}{2}\int d^3x \left(\frac{1}{c^2}\left(\frac{\partial\theta(\mathbf{x},t)}{\partial t}\right)^2 - (\nabla_{\mathbf{x}}\theta(\mathbf{x},t))^2\right). \tag{2.36}$$

Now consider the transformation $\theta_s(\mathbf{x},t) = \theta(M_s\mathbf{x},t)$ where $M_s$ is an orthogonal $3\times 3$ matrix. Now we wish to examine,

$$L[\theta_s,\dot{\theta}_s] = \frac{1}{2}\int d^3x \left(\frac{1}{c^2}\left(\frac{\partial\theta(M_s\mathbf{x},t)}{\partial t}\right)^2 - (\nabla_{\mathbf{x}}\theta(M_s\mathbf{x},t))^2\right). \tag{2.37}$$

Set $M_s\mathbf{x} \equiv \mathbf{y}$. Since the Jacobian of the transformation is unity we get $d^3x \equiv d^3y$. Further, $\nabla_{\mathbf{x}} = M_s\nabla_{\mathbf{y}}$. Therefore,

$$L[\theta_s,\dot{\theta}_s] = \frac{1}{2}\int d^3y \left(\frac{1}{c^2}\left(\frac{\partial\theta(\mathbf{y},t)}{\partial t}\right)^2 - (M_s\nabla_{\mathbf{y}}\theta(\mathbf{y},t))^2\right). \tag{2.38}$$

The square of $M_s$ times any vector should be equal to the square of that vector since $M_s$ is an orthogonal matrix. Thus we see that $L[\theta_s,\dot{\theta}_s] = L[\theta,\dot{\theta}]$. From this we may write down Noether's constant as,

$$Q = \int d^3x \left(\frac{d\theta_s(\mathbf{x},t)}{ds}\frac{\delta L}{\delta\dot{\theta}(\mathbf{x},t)}\right)_{s=0}. \tag{2.39}$$

The generalized momentum is easy to write down

$$\frac{\delta L}{\delta\dot{\theta}(\mathbf{x},t)} = \frac{1}{c^2}\left(\frac{\partial\theta(\mathbf{x},t)}{\partial t}\right), \tag{2.40}$$

whereas the derivative with respect to the flow variable is

$$\frac{d}{ds}\theta_s(\mathbf{x},t) = \frac{d}{ds}\theta(M_s\mathbf{x},t) = \frac{dM_s\mathbf{x}}{ds}\frac{d}{dM_s\mathbf{x}}\theta(M_s\mathbf{x},t) = \frac{dM_s\mathbf{x}}{ds}\cdot(\nabla_{\mathbf{y}}\theta(\mathbf{y},t))_{\mathbf{y}=M_s\mathbf{x}}. \tag{2.41}$$

From this we may write Noether's constant as,

$$Q = \int d^3x \left(\frac{dM_s}{ds}\mathbf{x}\right)_{s=0}\cdot(\nabla_{\mathbf{x}}\theta(\mathbf{x},t))\frac{1}{c^2}\frac{\partial\theta(\mathbf{x},t)}{\partial t}. \tag{2.42}$$

Specifically consider rotations about the z-axis.

$$M = \begin{pmatrix} cos(s) & sin(s) & 0 \\ -sin(s) & cos(s) & 0 \\ 0 & 0 & 1 \end{pmatrix} \tag{2.43}$$

Therefore,

$$\left(\frac{dM}{ds}\mathbf{x}\cdot\nabla\right)_{s=0} = \left(\begin{pmatrix} 0 & 1 & 0 \\ -1 & 0 & 0 \\ 0 & 0 & 0 \end{pmatrix}\begin{pmatrix} x \\ y \\ z \end{pmatrix}\right)^T \begin{pmatrix} \frac{\partial}{\partial x} \\ \frac{\partial}{\partial y} \\ \frac{\partial}{\partial z} \end{pmatrix} \tag{2.44}$$

$$= \left(\begin{pmatrix} y \\ -x \\ 0 \end{pmatrix}\right)^T \begin{pmatrix} \frac{\partial}{\partial x} \\ \frac{\partial}{\partial y} \\ \frac{\partial}{\partial z} \end{pmatrix} = y\frac{\partial}{\partial x} - x\frac{\partial}{\partial y}. \tag{2.45}$$

Therefore,

$$Q = \int d^3x \left((y\frac{\partial}{\partial x} - x\frac{\partial}{\partial y})\theta(\mathbf{x},t)\right)\frac{1}{c^2}\frac{\partial\theta(\mathbf{x},t)}{\partial t}. \tag{2.46}$$

Evaluating the rate of change of $Q$ and using the wave equation $\frac{1}{c^2}\frac{\partial^2}{\partial t^2}\theta = \nabla^2\theta$, we get,

$$\frac{dQ}{dt} = \int d^3x \left((y\frac{\partial}{\partial x} - x\frac{\partial}{\partial y})\frac{\partial\theta(\mathbf{x},t)}{\partial t}\right)\frac{1}{c^2}\frac{\partial\theta(\mathbf{x},t)}{\partial t}$$

$$+\int d^3x \left((y\frac{\partial}{\partial x} - x\frac{\partial}{\partial y})\theta(\mathbf{x},t)\right)\nabla^2\theta(\mathbf{x},t). \tag{2.47}$$

This may be rewritten as,

$$\frac{dQ}{dt} = \frac{1}{2c^2}\int d^3x\,(y\frac{\partial}{\partial x} - x\frac{\partial}{\partial y})\left(\frac{\partial\theta(\mathbf{x},t)}{\partial t}\right)^2$$

$$+\int d^3x\,(y\frac{\partial\theta(\mathbf{x},t)}{\partial x} - x\frac{\partial\theta(\mathbf{x},t)}{\partial y})\nabla^2\theta(\mathbf{x},t). \tag{2.48}$$

The first of these terms is clearly zero since $\int dydz\int_{-\infty}^{\infty}dx\frac{\partial}{\partial x}(...)$ $= \int dxdz\int_{-\infty}^{\infty}dy\frac{\partial}{\partial y}(...) = 0$. Thus, only the second term has to be evaluated. Consider,

$$\int d^3x\,y\frac{\partial\theta(\mathbf{x},t)}{\partial x}\nabla^2\theta(\mathbf{x},t) = \tag{2.49}$$

$$\int d^3x\,y\frac{\partial\theta(\mathbf{x},t)}{\partial x}\frac{\partial^2}{\partial x^2}\theta(\mathbf{x},t) + \int d^3x\,y\frac{\partial\theta(\mathbf{x},t)}{\partial x}\frac{\partial^2}{\partial y^2}\theta(\mathbf{x},t)$$

$$+\int d^3x\,y\frac{\partial\theta(\mathbf{x},t)}{\partial x}\frac{\partial^2}{\partial z^2}\theta(\mathbf{x},t) = \frac{1}{2}\int d^3x\,y\frac{\partial}{\partial x}\left(\frac{\partial\theta(\mathbf{x},t)}{\partial x}\right)^2$$

$$+\int d^3x\,y\frac{\partial\theta(\mathbf{x},t)}{\partial x}\frac{\partial^2}{\partial y^2}\theta(\mathbf{x},t) - \frac{1}{2}\int d^3x\,y\frac{\partial}{\partial x}\left(\frac{\partial\theta(\mathbf{x},t)}{\partial z}\right)^2. \tag{2.50}$$

Of these, the first and the last terms on the right-hand side above are zero for reasons similar to the one already alluded to. To simplify the middle term we make use of integration by parts to write,

$$\int d^3x\, y \frac{\partial \theta(\mathbf{x},t)}{\partial x} \frac{\partial^2}{\partial y^2} \theta(\mathbf{x},t) = -\int d^3x\, y\theta(\mathbf{x},t) \frac{\partial}{\partial x} \frac{\partial^2}{\partial y^2} \theta(\mathbf{x},t)$$

$$= -\int d^3x\, [\frac{\partial^2}{\partial y^2}(y\theta(\mathbf{x},t))] \frac{\partial}{\partial x} \theta(\mathbf{x},t)$$

$$= -2\int d^3x\, \frac{\partial}{\partial y}\theta(\mathbf{x},t) \frac{\partial}{\partial x}\theta(\mathbf{x},t) - \int d^3x\, y \frac{\partial^2 \theta(\mathbf{x},t)}{\partial y^2} \frac{\partial}{\partial x}\theta(\mathbf{x},t). \tag{2.51}$$

Notice that the term on the extreme left and the term on the extreme right of the above chain of identities are equal apart from the sign. Thus we may write for this term,

$$\int d^3x\, y \frac{\partial \theta(\mathbf{x},t)}{\partial x} \frac{\partial^2}{\partial y^2} \theta(\mathbf{x},t) = -\int d^3x\, \frac{\partial \theta(\mathbf{x},t)}{\partial y} \frac{\partial \theta(\mathbf{x},t)}{\partial x}. \tag{2.52}$$

Similarly,

$$\int d^3x\, x \frac{\partial \theta(\mathbf{x},t)}{\partial y} \nabla^2 \theta(\mathbf{x},t) = \int d^3x\, x \frac{\partial \theta(\mathbf{x},t)}{\partial y} \frac{\partial^2}{\partial x^2} \theta(\mathbf{x},t)$$

$$= -\int d^3x\, \frac{\partial \theta(\mathbf{x},t)}{\partial y} \frac{\partial \theta(\mathbf{x},t)}{\partial x}. \tag{2.53}$$

Thus,

$$\frac{dQ}{dt} = \int d^3x\, (y \frac{\partial \theta(\mathbf{x},t)}{\partial x} - x \frac{\partial \theta(\mathbf{x},t)}{\partial y}) \nabla^2 \theta(\mathbf{x},t) = -\int d^3x\, \frac{\partial \theta(\mathbf{x},t)}{\partial y} \frac{\partial \theta(\mathbf{x},t)}{\partial x}$$

$$+ \int d^3x\, \frac{\partial \theta(\mathbf{x},t)}{\partial y} \frac{\partial \theta(\mathbf{x},t)}{\partial x} = 0. \tag{2.54}$$

As a last example we choose the Lagrangian,

$$L[\mathbf{A}, \dot{\mathbf{A}}] = \frac{1}{2} \int d^3x \sum_{i=1}^{3} \left( \frac{1}{c^2} \left( \frac{\partial A_i(\mathbf{x},t)}{\partial t} \right)^2 - (\nabla_{\mathbf{x}} A_i(\mathbf{x},t))^2 \right). \tag{2.55}$$

Here $\mathbf{A} \equiv (A_1, A_2, A_3)$ is a three-dimensional vector in three-dimensional space. We now consider the transformation $\mathbf{A}(\mathbf{x},t) \to \mathbf{A}_s(\mathbf{x},t) = M_s \mathbf{A}(\mathbf{x},t)$ where $M_s$ is an orthogonal $3 \times 3$ matrix. It is clear that $L[\mathbf{A}_s, \dot{\mathbf{A}}_s] = L[\mathbf{A}, \dot{\mathbf{A}}]$. From this we may deduce the conserved quantity as,

$$Q = \int d^3x \left( \frac{dM_s}{ds} \mathbf{A}(\mathbf{x},t) \right)_{s=0} \cdot \frac{1}{c^2} \frac{\partial \mathbf{A}(\mathbf{x},t)}{\partial t}. \tag{2.56}$$

As before, we specialize to rotation about the z-axis and this tells us that this quantity is nothing but $\int d^3x(\mathbf{A}\times\partial_t\mathbf{A})_z$. This is true for each direction, therefore the vector conserved quantity is,

$$\mathbf{Q}=\int d^3x\,(\mathbf{A}(\mathbf{x},t)\times\partial_t\mathbf{A}(\mathbf{x},t)). \tag{2.57}$$

■ The idea of symmetries leading to conservation laws as described above is confined to continuous symmetries. However, there are examples where discrete symmetries also lead to conservation laws. There does not appear to be a general formalism to explore this, but we give an example nonetheless. Consider a Hamiltonian with a periodic potential, say, in one dimension $H(x,p)=\frac{p^2}{2m}+V(x)$ such that $V(x+a)=V(x)$. Since $a$ is a fixed quantity, this is a discrete symmetry rather than a continuous one. It is therefore not possible to directly use the formalism developed so far. However, we may revert to Newtonian mechanics to answer this question quite simply. The total energy is conserved since the Hamiltonian is time independent. Therefore,

$$E=\frac{1}{2}mv^2(x)+V(x). \tag{2.58}$$

The implication is that the trajectory $x(t)$ is assumed to be invertible (even if the potential $V(x)$ is periodic, the motion need not be) or at most not invertible at a finite number of points, which may be ignored since they contribute a vanishingly small interval while integrating with respect to these variables. Thus we may also write $t(x)$ without any ambiguity. Therefore $\dot{x}(t)\equiv\dot{x}(t(x))\equiv v(x)$. Now we invoke the condition on $V(x)$ that it is periodic. This implies, $v^2(x+a)=v^2(x)$. We now introduce the quantity which has dimensions of momentum,

$$\pi(x)\equiv\frac{1}{a}\int_x^{x+a}dy\,|m\,v(y)|. \tag{2.59}$$

Here, $x$ is the time-dependent location of the particle, hence $\pi(x)$ is potentially time dependent. We now show that it is not. In other words, it is a constant of the motion. To see this differentiate with time,

$$\frac{d}{dt}\pi(x)=\frac{1}{a}\dot{x}(t)\,m\,(|v(x+a)|-|v(x)|)=0. \tag{2.60}$$

Thus this quantity is conserved and is different from the total energy, which is another conserved quantity. While the total energy is conserved for any time-independent Hamiltonian, the present quantity (analogous to 'crystal momentum' in solid-state physics) is conserved solely as a result of the spatial periodicity of the potential energy. This quantity is also familiar as the classical action between end points $x$ and $x+a$, $\pi(x)\equiv S=\int_x^{x+a}|p(y)|dy\equiv\int_x^{x+a}\sqrt{2m(E-V(y))}dy$. We conclude with the observation that in the limit when there is no periodic potential, viz.

$a \to 0$, the conserved quantity in Eq. (2.59) becomes identical to the (magnitude of) usual momentum.

## 2.5 Exercises

**Q.1** We know that if the Lagrangian changes by a total time derivative under some transformation, the action is unchanged. Even in this situation there is a conserved quantity. What is it? (Assume a form of the transformation, $L(q_s, \dot{q}_s) - L(q, \dot{q}) = \frac{F(q_s, \dot{q}_s)}{dt}$).

**Q.2** Give an example wherein a rotational symmetry of a Lagrangian becomes a translational symmetry upon making a canonical transformation of the coordinates.

**Q.3** A Lagrangian $L(q, \dot{q})$ is invariant under a scale transformation $q_s(t) = sq(t)$. Find such a Lagrangian. What is the Noether constant associated with this symmetry?

**Q.4** Generalize the above question when there are $n$ generalized coordinates.

**Q.5** Imagine a potential energy function $V(x, y)$ that is unchanged when the rotation about the z-axis is by $\frac{2\pi}{n}$ and multiples thereof ($n$ is an integer). In this case, find the conserved quantity associated with this symmetry.

**Q.6** Imagine a Lagrangian $L(x^\mu(s), \dot{x}^\mu(s)) \equiv \frac{1}{2}m\dot{x}^\mu(s)\dot{x}_\mu(s) - V(x^\mu x_\mu)$. This is invariant under Lorentz transformations. Find the Noether constant associated with a Lorentz boost in the x-direction. What does this constant mean?

**Q.7** Consider a particle moving on a d-dimensional hypersphere. Write down the Lagrangian for this. What symmetries does this Lagrangian possess? What are the corresponding Noether constants?

# Chapter 3

# The Electromagnetic Field and Stress Energy Tensor

The most familiar example of a relativistic field theory is the electromagnetic field. We wish to think of the Maxwell equations as the Lagrange equations of a suitable Lagrangian. The aim is to find this suitable Lagrangian. Next by examining the symmetries of the Lagrangian we may write down conserved quantities via Noether's theorem. But first we begin with a description of the relativistic nature of the electromagnetic field. We wish to convince the reader that hidden in the equations that describe everyday phenomena are the seeds of special relativity. Historically, the observation that these equations are not compatible with Galilean relativity is what led to the development of special relativity.

## 3.1 Relativistic Nature of the Electromagnetic Field

We show in this section that the equations of electromagnetism (Maxwell's equations) are compatible with Lorentz transformations but not Galilean transformations. We purposely avoid using four-vector notation from the start as it hides some details about how the transformations actually work. Many students find these quite mysterious and it is hoped that an explicit demonstration component-wise, though clumsy is nevertheless illuminating. The electromagnetic field is described by two vectors **E** and **B**. These are determined by the solutions of Maxwell's equation that involve sources of these field, namely, current and change densities **j** and $\rho$, respectively.

First we show that$j^{\mu} = (\rho, j_x, j_y, j_z)$ transforms as a four-vector in the sense of

Figure 3.1: James Clerk Maxwell (13 June 1831 to 5 November 1879) was a Scottish mathematician and the leading figure of one of the greatest revolutions in physics—the theory of electromagnetism. He unified the equations governing electricity and magnetism into one framework and studied the properties of electromagnetic waves. He contributed greatly to the kinetic theory of gases and is associated with the Maxwell-Boltzmann distribution. He laid the foundation of color photography.

special relativity. This means (for boosts in the x-direction),

$$\rho^{'}(\mathbf{r}^{'},t^{'}) = \gamma(\rho(\mathbf{r},t) - \frac{v}{c^2} j_x(\mathbf{r},t)); j^{'}_x(\mathbf{r}^{'},t^{'}) = \gamma(j_x(\mathbf{r},t) - v\rho(\mathbf{r},t)) \tag{3.1}$$

and the y and z components are unchanged. Remember that $x^{\mu} = (x^0, x^1, x^2, x^3) = (t, x, y, z)$ whereas, $x_{\mu} = (x_0, x_1, x_2, x_3) = (t, -x, -y, -z)$. To derive the transformation law of the four-current let us start with the equation of continuity (using Einstein's summation convention),

$$\frac{\partial}{\partial x^{\mu}} j^{\mu} = \frac{\partial}{\partial t}\rho + \nabla \cdot \mathbf{j} = 0. \tag{3.2}$$

Since we expect the Lorentz transformation to be linear we write,

$$\rho^{'}(\mathbf{r}^{'},t^{'}) = c_{00}\, \rho(\mathbf{r},t) + c_{01}\, j_x(\mathbf{r},t);\quad j^{'}_x(\mathbf{r}^{'},t^{'}) = c_{11}\, j_x(\mathbf{r},t) + c_{10}\, \rho(\mathbf{r},t). \tag{3.3}$$

We know that,

$$x^{'0} = \gamma(x^0 - \frac{vx^1}{c^2});\ x^{'1} = \gamma(x^1 - vx^0);\ x^{'2} = x^2;\ x^{'3} = x^3. \tag{3.4}$$

Inverted, this reads,

$$x^0 = \gamma(x^{'0} + \frac{vx^{'1}}{c^2}); x^1 = \gamma(x^{'1} + vx^{'0}). \tag{3.5}$$

(In passing we note, $x_0' = \gamma(x_0 + \frac{vx_1}{c^2}); -x_1' = \gamma(-x_1 - vx_0)$). This means,

$$\frac{\partial}{\partial x'^0} = \frac{\partial x^0}{\partial x'^0}\frac{\partial}{\partial x^0} + \frac{\partial x^1}{\partial x'^0}\frac{\partial}{\partial x^1} = \gamma(\frac{\partial}{\partial x^0} + v\frac{\partial}{\partial x^1}); -\frac{\partial}{\partial x'^1} = \gamma(-\frac{\partial}{\partial x^1} - \frac{v}{c^2}\frac{\partial}{\partial x^0}) \quad (3.6)$$

$$\frac{\partial}{\partial x'^2} = \frac{\partial}{\partial x^2}; \frac{\partial}{\partial x'^3} = \frac{\partial}{\partial x^3}. \quad (3.7)$$

We want the equation of continuity to read the same in all frames. Hence we must have,

$$\frac{\partial}{\partial x^{0'}}\rho' + \frac{\partial}{\partial x^{1'}}j_x' = \frac{\partial}{\partial x^0}\rho + \frac{\partial}{\partial x^1}j_x. \quad (3.8)$$

Inserting Eq. (3.3) and Eq. (3.6) into Eq. (3.8) we get,

$$\gamma(\frac{\partial}{\partial x^0} + v\frac{\partial}{\partial x^1})[c_{00}\,\rho(\mathbf{r},t) + c_{01}\,j_x(\mathbf{r},t)] \quad (3.9)$$

$$-\gamma(-\frac{\partial}{\partial x^1} - \frac{v}{c^2}\frac{\partial}{\partial x^0})[c_{11}\,j_x(\mathbf{r},t) + c_{10}\,\rho(\mathbf{r},t)] = \frac{\partial}{\partial x^0}\rho + \frac{\partial}{\partial x^1}j_x. \quad (3.10)$$

This leads to the following equations:

$$\gamma c_{00} + \gamma\frac{v}{c^2}c_{10} = 1\ ;\ vc_{00} + c_{10} = 0 \quad (3.11)$$

$$c_{01} + \frac{v}{c^2}c_{11} = 0\ ;\ \gamma v c_{01} + \gamma c_{11} = 1. \quad (3.12)$$

Thus $c_{00} = c_{11} = \gamma$, $c_{10} = -v\gamma$, $c_{01} = -\frac{v}{c^2}\gamma$. Now we wish to see how the electric and magnetic fields should transform under Lorentz transformations. Consider the two equations (Gauss's Law and Ampere's Law),

$$\nabla.\mathbf{E} = 4\pi\rho; \nabla\times\mathbf{B} = \frac{4\pi}{c}\mathbf{j} + \frac{1}{c}\frac{\partial\mathbf{E}}{\partial t} \quad (3.13)$$

In the Lorentz transformed frame it is,

$$\nabla'.\mathbf{E}' = 4\pi\rho'; \nabla'\times\mathbf{B}' = \frac{4\pi}{c}\mathbf{j}' + \frac{1}{c}\frac{\partial\mathbf{E}'}{\partial t'}. \quad (3.14)$$

We may now substitute the transformed operators into Eq. (3.14)

$$\frac{\partial}{\partial x'^1}E_x' + \frac{\partial}{\partial x'^2}E_y' + \frac{\partial}{\partial x'^3}E_z' = 4\pi\rho, \quad (3.15)$$

and,

$$\frac{\partial}{\partial x'^2}B_z' - \frac{\partial}{\partial x'^3}B_y' = (\nabla'\times\mathbf{B}')_x = \frac{4\pi}{c}j_x' + \frac{1}{c}\frac{\partial E_x'}{\partial x'^0} \quad (3.16)$$

$$\frac{\partial}{\partial x'^3}B_x' - \frac{\partial}{\partial x'^1}B_z' = (\nabla'\times\mathbf{B}')_y = \frac{4\pi}{c}j_y' + \frac{1}{c}\frac{\partial E_y'}{\partial x'^0} \quad (3.17)$$

$$\frac{\partial}{\partial x'^1}B'_y - \frac{\partial}{\partial x'^2}B'_x = (\nabla' \times \mathbf{B}')_z = \frac{4\pi}{c}j'_z + \frac{1}{c}\frac{\partial E'_z}{\partial x'^0}. \tag{3.18}$$

Now we substitute Eq. (3.1) into the right hand side of Eq. (3.14) and Eq. (3.15) and reexpress $\rho, j$ in terms of the unprimed electric and magnetic fields.

$$\gamma(\frac{\partial}{\partial x^1} + \frac{v}{c^2}\frac{\partial}{\partial x^0})E'_x + \frac{\partial}{\partial x^2}E'_y + \frac{\partial}{\partial x^3}E'_z = 4\pi\gamma(\rho - (\frac{v}{c^2})j_x) \tag{3.19}$$

$$\frac{\partial}{\partial x^2}B'_z - \frac{\partial}{\partial x^3}B'_y = \frac{4\pi}{c}\gamma(j_x - v\rho) + \frac{1}{c}\gamma(\frac{\partial}{\partial x^0} + v\frac{\partial}{\partial x^1})E'_x \tag{3.20}$$

$$\frac{\partial}{\partial x^3}B'_x - \gamma(\frac{\partial}{\partial x^1} + \frac{v}{c^2}\frac{\partial}{\partial x^0})B'_z = \frac{4\pi}{c}j_y + \frac{1}{c}\gamma(\frac{\partial}{\partial x^0} + v\frac{\partial}{\partial x^1})E'_y \tag{3.21}$$

$$\gamma(\frac{\partial}{\partial x^1} + \frac{v}{c^2}\frac{\partial}{\partial x^0})B'_y - \frac{\partial}{\partial x^2}B'_x = \frac{4\pi}{c}j_z + \frac{1}{c}\gamma(\frac{\partial}{\partial x^0} + v\frac{\partial}{\partial x^1})E'_z \tag{3.22}$$

But we also know that,

$$\frac{\partial}{\partial x^1}E_x + \frac{\partial}{\partial x^2}E_y + \frac{\partial}{\partial x^3}E_z = 4\pi\rho \tag{3.23}$$

and

$$\frac{\partial}{\partial x^2}B_z - \frac{\partial}{\partial x^3}B_y = \frac{4\pi}{c}j_x + \frac{1}{c}\frac{\partial E_x}{\partial x^0} \tag{3.24}$$

$$\frac{\partial}{\partial x^3}B_x - \frac{\partial}{\partial x^1}B_z = \frac{4\pi}{c}j_y + \frac{1}{c}\frac{\partial E_y}{\partial x^0} \tag{3.25}$$

$$\frac{\partial}{\partial x^1}B_y - \frac{\partial}{\partial x^2}B_x = \frac{4\pi}{c}j_z + \frac{1}{c}\frac{\partial E_z}{\partial x^0} \tag{3.26}$$

We may eliminate $\rho, j$ from Eq. (3.20) using Eq. (3.24). For example if we replace $\rho$ and $j_x$ from Eq. (3.20) ,

$$\frac{\partial}{\partial x^2}B'_z - \frac{\partial}{\partial x^3}B'_y =$$

$$\gamma(\frac{\partial}{\partial x^2}B_z - \frac{\partial}{\partial x^3}B_y - \frac{1}{c}\frac{\partial E_x}{\partial x^0}) - \frac{v}{c}\gamma(\frac{\partial}{\partial x^1}E_x + \frac{\partial}{\partial x^2}E_y + \frac{\partial}{\partial x^3}E_z) + \frac{1}{c}\gamma((\frac{\partial}{\partial x^0} + v\frac{\partial}{\partial x^1})E'_x) \tag{3.27}$$

Since each derivative is independent we must have,

$$E_x = E'_x \tag{3.28}$$

Similarly,

$$B_x = B'_x \tag{3.29}$$

and

$$B'_z = \gamma(B_z - \frac{v}{c}E_y) \tag{3.30}$$

$$B_y' = \gamma(B_y + \frac{v}{c}E_z) \tag{3.31}$$

. If we look at Eq. (3.18),

$$\gamma(\frac{\partial}{\partial x^1} + \frac{v}{c^2}\frac{\partial}{\partial x^0})E_x' + \frac{\partial}{\partial x^2}E_y' + \frac{\partial}{\partial x^3}E_z'$$

$$= \gamma(\frac{\partial}{\partial x^1}E_x + \frac{\partial}{\partial x^2}E_y + \frac{\partial}{\partial x^3}E_z) - \gamma\frac{v}{c}(\frac{\partial}{\partial x^2}B_z - \frac{\partial}{\partial x^3}B_y - \frac{1}{c}\frac{\partial E_x}{\partial x^0}), \tag{3.32}$$

we find after matching term by term,

$$E_y' = \gamma(E_y - \frac{v}{c}B_z) \tag{3.33}$$

$$E_z' = \gamma(E_z + \frac{v}{c}B_y). \tag{3.34}$$

Thus we know how all the components of the electric and magnetic fields transform under Lorentz transformation. Now consider the quadratic expressions,

$$E_x'^2 + E_y'^2 + E_z'^2 = E_x^2 + \gamma^2(E_y - \frac{v}{c}B_z)^2 + \gamma^2(E_z + \frac{v}{c}B_y)^2 \tag{3.35}$$

$$B_x'^2 + B_y'^2 + B_z'^2 = B_x^2 + \gamma^2(B_y - \frac{v}{c}E_z)^2 + \gamma^2(B_z - \frac{v}{c}E_y)^2. \tag{3.36}$$

Taking the difference, we find $\mathbf{E}'^2 - \mathbf{B}'^2 = \mathbf{E}^2 - \mathbf{B}^2$. Thus the difference $\mathbf{E}^2 - \mathbf{B}^2$ is a Lorentz invariant. Similarly, we may show that $\mathbf{E}.\mathbf{B}$ is also a Lorentz invariant. We know that the electric and magnetic fields can be expressed in terms of potentials. Now we wish to ascertain how the potentials transform under Lorentz transformations.

$$\mathbf{E} = -\nabla\phi - \frac{1}{c}\frac{\partial}{\partial t}\mathbf{A} \tag{3.37}$$

$$\mathbf{B} = \nabla \times \mathbf{A} \tag{3.38}$$

Focus on the electric field,

$$E_x' = -\frac{\partial}{\partial x'^1}\phi' - \frac{1}{c}\frac{\partial}{\partial x'^0}A_x' = \gamma(\frac{\partial}{\partial x^1} - \frac{v}{c^2}\frac{\partial}{\partial x^0}\phi') - \frac{1}{c}\gamma(\frac{\partial}{\partial x^0} + v\frac{\partial}{\partial x^1})A_x' \tag{3.39}$$

$$E_y' = -\frac{\partial}{\partial x'^2}\phi' - \frac{1}{c}\frac{\partial}{\partial x'^0}A_y' = -\frac{\partial}{\partial x^2}\phi' - \frac{1}{c}\gamma(\frac{\partial}{\partial x^0} + v\frac{\partial}{\partial x^1})A_y' \tag{3.40}$$

$$E_z' = -\frac{\partial}{\partial x'^3}\phi' - \frac{1}{c}\frac{\partial}{\partial x'^0}A_z' = -\frac{\partial}{\partial x^3}\phi' - \frac{1}{c}\gamma(\frac{\partial}{\partial x^0} + v\frac{\partial}{\partial x^1})A_z'. \tag{3.41}$$

But we know that,

$$E_x' = E_x = -\frac{\partial}{\partial x^1}\phi - \frac{1}{c}\frac{\partial}{\partial x^0}A_x \tag{3.42}$$

Figure 3.2: A self-taught English mathematician and physicist, Oliver Heaviside (18 May 1850 to 3 February 1925) was heavily influenced by Maxwell's treatise on electromagnetism and gave Maxwell's equations the vector form that we commonly see today. He studied electric circuits and discovered a technique for solving differential equations.

$$E_y^{'} = \gamma(E_y - \frac{v}{c}B_z) = \gamma(-\frac{\partial}{\partial x^2}\phi - \frac{1}{c}\frac{\partial}{\partial x^0}A_y - \frac{v}{c}\frac{\partial}{\partial x^1}A_y + \frac{v}{c}\frac{\partial}{\partial x^2}A_x) \tag{3.43}$$

$$E_z^{'} = \gamma(E_z + \frac{v}{c}B_y) = \gamma(-\frac{\partial}{\partial x^3}\phi - \frac{1}{c}\frac{\partial}{\partial x^0}A_z + \frac{v}{c}\frac{\partial}{\partial x^3}A_x - \frac{v}{c}\frac{\partial}{\partial x^1}A_z). \tag{3.44}$$

After equating the two sets of equations we find,

$$-\frac{\partial}{\partial x^1}\phi - \frac{1}{c}\frac{\partial}{\partial x^0}A_x = (-\gamma\frac{\partial}{\partial x^1}\phi^{'} - \gamma\frac{v}{c^2}\frac{\partial}{\partial x^0}\phi^{'}) - \frac{1}{c}\gamma(\frac{\partial}{\partial x^0}A_x^{'} + v\frac{\partial}{\partial x^1}A_x^{'}). \tag{3.45}$$

or

$$A_x = \gamma(A_x^{'} + \frac{v}{c}\phi^{'}); \phi = \gamma(\phi^{'} + \frac{v}{c}A_x^{'}); A_y = A_y^{'}; A_z = A_z^{'}. \tag{3.46}$$

The inverse is,

$$A_x^{'} = \gamma(A_x - \frac{v}{c}\phi); \phi^{'} = \gamma(\phi - \frac{v}{c}A_x); A_y^{'} = A_y; A_z^{'} = A_z. \tag{3.47}$$

Thus $(\frac{\phi}{c}, A_x, A_y, Az) = A^\mu$ is a contra-variant four-vector. Since we have obtained a four-vector, we may construct a rank two tensor by taking derivatives with respect to the coordinates. Define $F^{\mu\nu} = \partial^\mu A^\nu - \partial^\nu A^\mu$ where $\partial^\mu \equiv \frac{\partial}{\partial x_\mu}$. These quantities may be thought of as components of a $4 \times 4$ matrix whose components are as follows:

$$F^{\mu\nu} = \begin{pmatrix} 0 & -E_x & -E_y & -E_z \\ E_x & 0 & -B_z & B_y \\ E_y & B_z & 0 & -B_x \\ E_z & -B_y & B_x & 0 \end{pmatrix}. \tag{3.48}$$

We see that $F^{i,0} = E_i$ where $i = 1,2,3$ and $E_1 \equiv E_x$, etc. This is will be used in the subsequent example.

■ A Lorentz transformation from $(x,t)$ to $(x',t')$ preserves the indefinite metric (no fixed sign), namely $x^2 - c^2t^2 = x'^2 - c^2t'^2$. We wish to make Lorentz four-vectors resemble Euclidean vectors so that a Lorentz transformation becomes an orthogonal transformation and we may exploit symmetries under orthogonal transformations. This means that the time components of four-vectors all get a multiplicative factor of $i$. In the preceding discussion we saw that $F^{k,0} = E_k$. In Euclidean space, $F^{k,0} \to iF^{k,0}$, so that $E_k \to iE_k$. The Euclidean field tensor then becomes,

$$F^{\mu\nu} = \begin{pmatrix} 0 & -iE_x & -iE_y & -iE_z \\ iE_x & 0 & -B_z & B_y \\ iE_y & B_z & 0 & -B_x \\ iE_z & -B_y & B_x & 0 \end{pmatrix}. \tag{3.49}$$

This matrix is such that the function $P(\lambda) = Det[F - \lambda\mathbf{1}]$ is unchanged under orthogonal transformations (similarity transformation with orthogonal matrices) of the matrix $F$. In this case, $P(\lambda)$ is the following polynomial:

$$P(\lambda) = \lambda^4 - \lambda^2(\mathbf{E}^2 - \mathbf{B}^2) - (\mathbf{E}\cdot\mathbf{B})^2. \tag{3.50}$$

Since the above should be unchanged under orthogonal transformations for each $\lambda$, it follows that $\mathbf{E}^2 - \mathbf{B}^2$ and $\mathbf{E}\cdot\mathbf{B}$ are unchanged under the orthogonal transformation in the Euclidean space (with imaginary time). This transformation is nothing but the usual Lorentz transformation in actual time. Hence $\mathbf{E}\cdot\mathbf{B}$ and $\mathbf{E}^2 - \mathbf{B}^2$ are Lorentz invariants.

■ Show that the Lorentz force equation (in special relativity) can be written in a covariant form,

$$\frac{dp^\alpha}{d\tau} = \frac{q}{c}\, u_\beta\, F^{\alpha\beta}, \tag{3.51}$$

where $p^\alpha = mu^\alpha = m\frac{dx^\alpha}{d\tau}$ is the four-momentum and $x^\mu = (ct,x,y,z)$ and $x_\mu = (ct,-x,-y,-z)$. Also $c^2\, d\tau^2 = dx^\mu\, dx_\mu$ is the proper time. The easiest way to show this is to multiply both sides of Eq. (3.51) by $d\tau$ (and later divide by $dt$) and specialize to the case when $\alpha = i = 1,2,3$ in which case, $F^{i0} = -\partial_i\phi - \frac{1}{c}\frac{\partial}{\partial t}A^i =$

$E_{i=x,y,z}$ and $F^{12} = -B^3 = -B_z$ and all cyclic permutations, with $F^{ij} = -F^{ji}$. What is the physical meaning of Eq. (3.51) when $\alpha = 0$ ?

Here is the procedure. After multiplying by $d\tau$, dividing by $dt$, and setting $\alpha = k = 1, 2, 3$, we get ($u_\beta \, d\tau = dx_\beta$),

$$\frac{dp^k}{dt} = \frac{q}{c}\frac{dx_\beta}{dt} F^{k\beta} = \frac{q}{c}\frac{dx_0}{dt} F^{k0} + \frac{q}{c}\sum_j \frac{dx_j}{dt} F^{kj}. \tag{3.52}$$

But $\frac{dx_j}{dt} = -\frac{dx^j}{dt} = -v_j$. Hence,

$$\frac{dp^1}{dt} = q\, E_x + \frac{q}{c}\, v_y\, B_z - \frac{q}{c}\, v_z\, B_y = qE_x + \frac{q}{c}(\mathbf{v} \times \mathbf{B})_x. \tag{3.53}$$

All other components are similar; hence, the Lorentz force is $q(\mathbf{E} + \frac{\mathbf{v}}{c} \times \mathbf{B})$. When $\alpha = 0$,

$$dp^0 = \frac{q}{c}\, dx_\beta\, F^{0\beta} = \sum_{j=1,2,3} \frac{q}{c}\, dx^j\, F^{j0} = \frac{q}{c} \mathbf{dx} \cdot \mathbf{E}. \tag{3.54}$$

But $p^0 = \mathfrak{E}/c$ where $\mathfrak{E}$ is the energy of the particle. Hence,

$$d\mathfrak{E} = \mathbf{dx} \cdot (q\mathbf{E}). \tag{3.55}$$

This is saying that the change in energy of the particle is because of the work done by the electric field in displacing the particle by an amount $d\mathbf{x}$. The displacement is perpendicular to the magnetic force, hence the magnetic force does not do any work.

■ Show that the equation Eq. (3.51) in the earlier question is nothing but the Lagrange equation of the Lagrangian,

$$L(x, \dot{x}) = -mc\sqrt{\dot{x}^\mu(\tau)\dot{x}_\mu(\tau)} + \frac{q}{c}\, \dot{x}^\mu(\tau)\, A_\mu(x). \tag{3.56}$$

We start with,

$$\begin{aligned} \frac{\partial L}{\partial \dot{x}^\alpha} &= -mc\, \frac{\partial}{\partial \dot{x}^\alpha}\sqrt{\eta_{\mu\nu}\, \dot{x}^\mu(\tau)\dot{x}^\nu(\tau)} + \frac{q}{c}\frac{\partial \dot{x}^\mu}{\partial \dot{x}^\alpha} A_\mu(x) \\ &= -mc\, \dot{x}_\alpha \left(\eta_{\mu\nu}\, \dot{x}^\mu(\tau)\dot{x}^\nu(\tau)\right)^{-\frac{1}{2}} + \frac{q}{c} A_\alpha(x) \end{aligned} \tag{3.57}$$

$$\frac{\partial L}{\partial x^\alpha} = \frac{q}{c}\, \dot{x}^\mu(\tau)\, \frac{\partial A_\mu(x)}{\partial x^\alpha}. \tag{3.58}$$

The definition of proper time is such that $\dot{x}^\mu \dot{x}_\mu = c^2$. Hence,

$$\frac{d}{d\tau}\frac{\partial L}{\partial \dot{x}^\alpha} = -m\,\ddot{x}_\alpha + \frac{q}{c}\frac{d}{d\tau}A_\alpha(x) = -mc\,\ddot{x}_\alpha + \frac{q}{c}\,\dot{x}^\mu(\tau)\,\frac{\partial A_\alpha(x)}{\partial x^\mu}. \tag{3.59}$$

Equating the two we get,

$$m\,\ddot{x}_\alpha = \frac{q}{c}\,\dot{x}^\mu(\tau)\,\left(\frac{\partial A_\alpha(x)}{\partial x^\mu} - \frac{\partial A_\mu(x)}{\partial x^\alpha}\right), \tag{3.60}$$

or

$$\frac{dp_\alpha}{d\tau} = \frac{q}{c}\,u^\mu(\tau)\,F_{\alpha\mu}, \tag{3.61}$$

where $p_\alpha = mu_\alpha$ and $u_\alpha = \dot{x}_\alpha$.

## 3.2 Lagrangian of the EM Field

Consider the four Maxwell equations in CGS units.

$$\nabla\cdot\mathbf{E} = 4\pi\rho\ ;\ \nabla\cdot\mathbf{B} = 0 \tag{3.62}$$

$$\nabla\times\mathbf{E} = -\frac{1}{c}\frac{\partial\mathbf{B}}{\partial t}\ ;\ \nabla\times\mathbf{B} = \frac{4\pi}{c}\mathbf{J} + \frac{1}{c}\frac{\partial\mathbf{E}}{\partial t} \tag{3.63}$$

We wish to think of these as the Lagrange equations of a suitable Lagrangian. For this we have to identify suitable generalized coordinates. It is well known that these equations may be simplified and reduced considerably by working with potentials—scalar and vector potentials. They are defined as $\mathbf{E} = -\nabla\phi - \frac{1}{c}\frac{\partial\mathbf{A}}{\partial t}$ and $\mathbf{B} = \nabla\times\mathbf{A}$. The four Maxwell equations reduce to two.

$$-\nabla^2\phi - \frac{1}{c}\partial_t\nabla\cdot\mathbf{A} = 4\pi\rho \tag{3.64}$$

$$\nabla(\nabla\cdot\mathbf{A}) - \nabla^2\mathbf{A} + \frac{1}{c}\frac{\partial}{\partial t}\left(\nabla\phi + \frac{1}{c}\frac{\partial\mathbf{A}}{\partial t}\right) = \frac{4\pi}{c}\mathbf{J} \tag{3.65}$$

We identify the generalized coordinates as $q_i \to (\phi(\mathbf{r}), \mathbf{A}(\mathbf{r}))$ where the vector $\mathbf{r}$ plays the role of the index $i$. Just as we would have written $L(Q,\dot{Q}) = \sum_i L_i(Q,\dot{Q})$ if we had many degrees of freedom, we may suspect that the Lagrangian would be of the form,

$$L = \int d^3r\,\mathfrak{L}(Q,\dot{Q}) + \int d^3r\,\rho(\mathbf{r},t)\phi(\mathbf{r},t) + \frac{1}{c}\int d^3r\,\mathbf{J}(\mathbf{r},t)\cdot\mathbf{A}(\mathbf{r},t) \tag{3.66}$$

where $i \to \int d^3r$ and $Q \to (\phi(\mathbf{r},t), \mathbf{A}(\mathbf{r},t))$ and $\dot{Q} \to \partial_t\phi(\mathbf{r},t), \partial_t\mathbf{A}(\mathbf{r},t)$. We have explicitly separated the source terms since they appear naturally. To see this, we write the Lagrange equations,

$$\partial_t \frac{\delta L}{\delta \partial_t \phi(\mathbf{r},t)} = \frac{\delta L}{\delta \phi(\mathbf{r},t)}; \ \partial_t \frac{\delta L}{\delta \partial_t \mathbf{A}(\mathbf{r},t)} = \frac{\delta L}{\delta \mathbf{A}(\mathbf{r},t)}. \tag{3.67}$$

An examination of the first of the equations suggests that we have to choose $\mathfrak{L}$ to be independent of $\partial_t\phi$. A choice of Lagrangian such as,

$$L = -\frac{1}{4\pi}\int d^3r' \ \phi(\mathbf{r}',t)(4\pi\rho(\mathbf{r}',t) + \frac{1}{2}\nabla'^2\phi(\mathbf{r}',t) + \frac{1}{c}\partial_t\nabla'\cdot\mathbf{A}(\mathbf{r}',t))$$
$$+L'(\mathbf{A}, \partial_t\mathbf{A}) \tag{3.68}$$

reproduces Gauss's Law. To see this, we differentiate with respect to $\phi(\mathbf{r},t)$ and set equal to zero,

$$\frac{\delta L}{\delta\phi(\mathbf{r},t)} =$$
$$-\frac{1}{4\pi}\int d^3r' \ \frac{\delta\phi(\mathbf{r}',t)}{\delta\phi(\mathbf{r},t)}(4\pi\rho(\mathbf{r}',t) + \frac{1}{2}\nabla'^2\phi(\mathbf{r}',t) + \frac{1}{c}\partial_t\nabla'\cdot\mathbf{A}(\mathbf{r}',t))$$
$$-\frac{1}{4\pi}\int d^3r' \ \phi(\mathbf{r}',t)\frac{\delta}{\delta\phi(\mathbf{r},t)}(4\pi\rho(\mathbf{r}',t) + \frac{1}{2}\nabla'^2\phi(\mathbf{r}',t) + \frac{1}{c}\partial_t\nabla'\cdot\mathbf{A}(\mathbf{r}',t)). \tag{3.69}$$

If we were dealing with systems with a finite number of degrees of freedom we would write $\frac{\delta q_i(t)}{\delta q_j(t)} = \delta_{i,j}$, in the present case we should instead write, $\frac{\delta\phi(\mathbf{r},t)}{\delta\phi(\mathbf{r}',t)} = \delta(\mathbf{r}-\mathbf{r}')$, namely the Dirac delta function. The second term reads as follows,

$$\int d^3r' \ \phi(\mathbf{r}',t)\frac{\delta}{\delta\phi(\mathbf{r},t)}(\frac{1}{2}\nabla'^2\phi(\mathbf{r}',t)) = \int d^3r' \ \phi(\mathbf{r}',t)\frac{1}{2}\nabla'^2\delta(\mathbf{r}-\mathbf{r}')$$
$$= \int d^3r' \ (\nabla'^2\phi(\mathbf{r}',t))\frac{1}{2}\delta(\mathbf{r}-\mathbf{r}') = \frac{1}{2}(\nabla^2\phi(\mathbf{r},t)). \tag{3.70}$$

The last result follows from integration by parts : $\int f\nabla^2 g = -\int \nabla f\cdot\nabla g = \int g\nabla^2 f$. Thus this term added to the first term reproduces Gauss's Law. The part of the Lagrangian involving the vector potential may be deduced as follows. First, it is easy to suspect that,

$$\partial_t \frac{\delta L}{\delta\partial_t\mathbf{A}(\mathbf{r},t)} = \partial_t\frac{1}{c^2}\frac{\partial}{\partial t}\mathbf{A}. \tag{3.71}$$

From Eq. (3.65) we have,

$$\partial_t\frac{1}{c^2}\frac{\partial}{\partial t}\mathbf{A} = \frac{4\pi}{c}\mathbf{J} - \nabla(\nabla\cdot\mathbf{A}) + \nabla^2\mathbf{A} - \nabla\frac{1}{c}\frac{\partial}{\partial t}\phi. \tag{3.72}$$

Combining with the Lagrange equations we get,

$$\frac{\delta L}{\delta \mathbf{A}(\mathbf{r},t)} = \frac{4\pi}{c}\mathbf{J} - \nabla(\nabla\cdot\mathbf{A}) + \nabla^2\mathbf{A} - \nabla\frac{1}{c}\frac{\partial}{\partial t}\phi \tag{3.73}$$

$$\frac{\delta L}{\delta \partial_t \mathbf{A}(\mathbf{r},t)} = \frac{1}{c^2}\partial_t \mathbf{A}(\mathbf{r},t). \tag{3.74}$$

This may be integrated to give,

$$\begin{aligned}
4\pi L = & \frac{4\pi}{c}\int d^3r'\, \mathbf{J}(\mathbf{r}',t)\cdot\mathbf{A}(\mathbf{r}',t) - \frac{1}{2}\int d^3r'\, \mathbf{A}(\mathbf{r}',t)\cdot\nabla'(\nabla'\cdot\mathbf{A}(\mathbf{r}',t)) \\
& + \frac{1}{2c^2}\int d^3r'\, (\partial_t \mathbf{A}(\mathbf{r}',t))^2 + \frac{1}{2}\int d^3r'\, \mathbf{A}(\mathbf{r}',t)\cdot\nabla'^2\mathbf{A}(\mathbf{r}',t) \\
& - \int d^3r'\, \mathbf{A}(\mathbf{r}',t)\cdot\nabla'\frac{1}{c}\frac{\partial}{\partial t}\phi(\mathbf{r}',t) + L'(\phi).
\end{aligned} \tag{3.75}$$

A comparison of Eq. (3.68) and Eq. (3.75) shows that the overall Lagrangian may be written as,

$$\begin{aligned}
4\pi L = & -\int d^3r'\, \phi(\mathbf{r}',t)(4\pi\rho(\mathbf{r}',t) + \frac{1}{2}\nabla'^2\phi(\mathbf{r}',t) + \frac{1}{c}\partial_t\nabla'\cdot\mathbf{A}(\mathbf{r}',t)) + \\
& + \frac{4\pi}{c}\int d^3r'\, \mathbf{J}(\mathbf{r}',t)\cdot\mathbf{A}(\mathbf{r}',t) - \frac{1}{2}\int d^3r'\, \mathbf{A}(\mathbf{r}',t)\cdot\nabla'(\nabla'\cdot\mathbf{A}(\mathbf{r}',t)) \\
& + \frac{1}{2c^2}\int d^3r'\, (\partial_t \mathbf{A}(\mathbf{r}',t))^2 + \frac{1}{2}\int d^3r'\, \mathbf{A}(\mathbf{r}',t)\cdot\nabla'^2\mathbf{A}(\mathbf{r}',t).
\end{aligned} \tag{3.76}$$

It is left to the reader to verify that this may be written more compactly as,

$$\begin{aligned}
L = & -\int d^3r'\, \phi(\mathbf{r}',t)\rho(\mathbf{r}',t) + \frac{1}{c}\int d^3r'\, \mathbf{J}(\mathbf{r}',t)\cdot\mathbf{A}(\mathbf{r}',t) \\
& + \frac{1}{8\pi}\int d^3r' (\mathbf{E}^2(\mathbf{r}',t) - \mathbf{B}^2(\mathbf{r}',t)).
\end{aligned} \tag{3.77}$$

One may alternatively describe the dynamics using the Hamiltonian. As is well known, the two are related via a Legendre transformation. We have to first identify the canonical momentum. This is defined as,

$$\mathbf{P}_A(\mathbf{r},t) = \frac{\delta L}{\delta \partial_t \mathbf{A}(\mathbf{r},t)}. \tag{3.78}$$

Since the Lagrangian does not depend on the time derivative of the scalar potential, there is no need to introduce the canonical momentum in that case. Using the Lagrangian we just derived for the electromagnetic field, we may conclude that,

$$\mathbf{P}_A(\mathbf{r},t) = -\frac{1}{4\pi c}\mathbf{E}(\mathbf{r},t). \tag{3.79}$$

Legendre's transformation tells us that,

$$H(\phi,\mathbf{A};\mathbf{P}_A) = \int d^3r\ \mathbf{P}_A(\mathbf{r},t)\cdot\partial_t\mathbf{A}(\mathbf{r},t) - L. \tag{3.80}$$

Therefore,

$$H(\phi,\mathbf{A};\mathbf{P}_A) = \frac{1}{4\pi}\int d^3r'\ \phi(\mathbf{r}',t)(4\pi\rho(\mathbf{r}',t) - \nabla'\cdot\mathbf{E}(\mathbf{r}',t))$$

$$-\frac{1}{c}\int d^3r'\ \mathbf{J}(\mathbf{r}',t)\cdot\mathbf{A}(\mathbf{r}',t) + \frac{1}{8\pi}\int d^3r'(\mathbf{E}^2(\mathbf{r}',t) + \mathbf{B}^2(\mathbf{r}',t)). \tag{3.81}$$

Here $\mathbf{E}$ is the same as $\mathbf{P}_A$ apart from a factor of $-1/c$ and of course, $\mathbf{B} = \nabla\times\mathbf{A}$. This expression is 'off-shell'. This means it is valid independent of whether the fields obey Maxwell equations. Indeed, the (Hamilton) equations of motion of this Hamiltonian would be the Maxwell equations. They are,

$$0 = \frac{\delta H(\phi,\mathbf{A};\mathbf{P}_A)}{\delta\phi(\mathbf{r}',t)} \tag{3.82}$$

$$\frac{\partial}{\partial t}\mathbf{A} = \frac{\delta H(\phi,\mathbf{A};\mathbf{P}_A)}{\delta\mathbf{P}_A(\mathbf{r}',t)} \tag{3.83}$$

and,

$$\frac{\partial}{\partial t}\mathbf{P}_A = -\frac{\delta H(\phi,\mathbf{A};\mathbf{P}_A)}{\delta\mathbf{A}(\mathbf{r}',t)}. \tag{3.84}$$

It is well known that the Hamiltonian of a time-independent system is nothing but the total energy. First we note that on-shell, the first term in Eq. (3.81) is zero due to Gauss' Law. Next we may see that there are two pieces, the first involving the external current. This is nothing but the Joule heating energy. The second one is the energy contained in the electromagnetic field.

### 3.2.1 Gauge Symmetry and Conservation Laws

We now wish to examine the symmetries of the Lagrangian and see what kind of conserved quantities emerge. The first symmetry we wish to study is the gauge symmetry. This is a somewhat different kind of symmetry in that it is a functional symmetry. Unlike the others we have encountered so far that involve a few parameters, this involves a whole function. Thus instead of obtaining a few conserved quantities, we obtain a local conservation law.

Gauge symmetry means we impose the condition that the action or the time integral of the Lagrangian in Eq. (3.77) is invariant under the transformation $\phi(\mathbf{r},t)\rightarrow$

$\phi(\mathbf{r},t) - \frac{1}{c}\partial_t\xi(\mathbf{r},t)$ and $\mathbf{A}(\mathbf{r},t) \to \mathbf{A}(\mathbf{r},t) + \nabla\xi(\mathbf{r},t)$. Now we wish to see what kind of conservation law this implies. Thus using Eq. (3.77) we may deduce,

$$L(\phi - \frac{1}{c}\partial_t\xi, \mathbf{A} + \nabla\xi; \partial_t\mathbf{A} + \nabla\partial_t\xi) - L(\phi, \mathbf{A}; \partial_t\mathbf{A}) = \tag{3.85}$$

$$\int d^3r^{'}\, \frac{1}{c}\partial_t\xi(\mathbf{r}^{'},t)4\pi\rho(\mathbf{r}^{'},t) + \frac{4\pi}{c}\int d^3r^{'}\, \mathbf{J}(\mathbf{r}^{'},t)\cdot\nabla\xi(\mathbf{r}^{'},t). \tag{3.86}$$

We impose the condition that the action $S = \int_{t_i}^{t_f} dt\, L$ be invariant under these gauge transformations. This transformation keeps the end points fixed so that $\xi(\mathbf{r},t_i) = \xi(\mathbf{r},t_f) = 0$. It means,

$$0 = S(after - gauge - trnsf) - S(before - gauge - trnsf)$$

$$= \int_{t_i}^{t_f} dt\ (L(\phi - \frac{1}{c}\partial_t\xi, \mathbf{A} + \nabla\xi; \partial_t\mathbf{A} + \nabla\partial_t\xi) - L(\phi, \mathbf{A}; \partial_t\mathbf{A})) = 0. \tag{3.87}$$

After integrating by parts in both space and time, these observations imply,

$$0 = -\int d^3r^{'} \int_{t_i}^{t_f} dt\ (\frac{4\pi}{c}\partial_t\rho(\mathbf{r}^{'},t) + \frac{4\pi}{c}\ \nabla\cdot\mathbf{J}(\mathbf{r}^{'},t))\xi(\mathbf{r}^{'},t). \tag{3.88}$$

Since $\xi(\mathbf{r}^{'},t)$ is arbitrary, we must have,

$$\partial_t\rho(\mathbf{r},t) + \nabla\cdot\mathbf{J}(\mathbf{r},t) = 0. \tag{3.89}$$

Notice that the resulting statement about conservation is a local one. This is nothing but the statement of charge conservation which follows upon integrating over a volume bounded by a surface where the component of the current density normal to the surface, vanishes. Therefore, imposition of gauge symmetry on an action with sources automatically implies charge conservation. We have seen that for the action to be gauge invariant, it is sufficient for the sources to obey a conservation law and the source-free part of the action depend on the electric and magnetic fields rather than the potentials directly. A term such as $a(x)\mathbf{E}(x)\cdot\mathbf{B}(x)$ is both quadratic in the gauge-invariant fields and is also Lorentz invariant provided $a(x)$ is a Lorentz scalar. Thus it may be added to the source-free Lagrangian density to describe what are known as axions. If $a(x)$ is a constant, then this extra term does not contribute to the dynamical equations, but it does so when it is a variable. Recent developments in topological insulators have made this an important subject in condensed matter physics.

## 3.3 Stress Energy Tensor of the EM Field

We just showed that a local conservation law for sources may be derived by using functional symmetries of the full Lagrangian. Now we show that a generalized version of the continuity equation for the EM field alone may be derived by taking into account the lack of an explicit dependence of the Lagrangian density of a source-free electromagnetic field on space and time (the dependence is only implicit through the fields). To do this we start with the Lagrangian of the electromagnetic field without sources,

$$L = \frac{1}{8\pi}\int d^3r\,(\mathbf{E}^2 - \mathbf{B}^2). \tag{3.90}$$

Thus the Lagrangian density is,

$$\mathfrak{L} = \frac{1}{8\pi}(\mathbf{E}^2 - \mathbf{B}^2). \tag{3.91}$$

This Lagrangian density is explicitly independent of the position and time coordinates (it depends on $(\mathbf{x},t)$ only through $\mathbf{E}$ and $\mathbf{B}$). One may regard the Lagrangian density as a function of the derivatives of the four-vector potential $\partial_\nu A_\rho$. In fact, the Lagrange equations may be rewritten as,

$$\partial_\nu \frac{\delta\mathfrak{L}}{\delta\partial_\nu A_\rho} = \frac{\delta\mathfrak{L}}{\delta A_\rho}, \tag{3.92}$$

where summation over repeated indices is implied. Here $\partial_\nu \equiv \frac{\partial}{\partial x^\nu}$. First we rewrite the Lagrangian density in a four-vector notation.

$$\mathfrak{L} = -\frac{1}{16\pi}F^{\mu\nu}F_{\mu\nu}, \tag{3.93}$$

where $F_{\mu\nu} = \partial_\mu A_\nu - \partial_\nu A_\mu$. The reader may easily verify that this is correct by writing out all the components. Since $\mathfrak{L}$ is explicitly independent of $A_\rho$, the resulting Lagrange equation is nothing but,

$$\partial^\nu\partial_\nu A_\rho = 0. \tag{3.94}$$

The stress energy tensor, also known as the energy momentum tensor is a quantity that may be thought of as a $4\times 4$ matrix that has the property that its four-divergence vanishes whenever the Lagrangian density does not explicitly depend on the position and time coordinates. At this stage we prefer to derive an expression for this in a more general manner. For instance, the Lagrangian could also depend on the fields themselves as this happens when photons are regarded as being massive. This is not entirely a hypothetical situation as it is realized when light interacts with matter. For massive photons, a Lagrangian known as Proca's Lagrangian may be introduced:

$$\mathfrak{L}[\partial A, A] = -\frac{1}{16\pi}F^{\mu\nu}F_{\mu\nu} + \frac{(mc)^2}{8\pi\hbar^2}A^\mu A_\mu. \tag{3.95}$$

The equation for motion would then become Proca's equation,

$$\partial_\mu \partial^\mu A^\nu - \partial^\nu(\partial_\mu A^\mu) = -\frac{(mc)^2}{\hbar^2} A^\nu. \tag{3.96}$$

The reason why we consider this now is because the Lagrangian depends both on the derivative of the fields and the fields themselves, which makes it slightly more general than the usual electromagnetic field. Now consider the following fact. The Lagrangian density $\mathfrak{L}[A, \partial A]$ is explicitly independent of the position and time coordinates. This means the Lagrangian depends on the position and time on through the fields. Therefore,

$$\partial_\mu \mathfrak{L}[A, \partial A] \equiv (\partial_\mu A_\rho(x)) \frac{\delta \mathfrak{L}[A, \partial A]}{\delta A_\rho(x)} + (\partial_\mu \partial_\nu A_\rho(x)) \frac{\delta \mathfrak{L}[A, \partial A]}{\delta \partial_\nu A_\rho(x)}. \tag{3.97}$$

Using the Lagrange equation for the first term on the right-hand side we get,

$$\begin{aligned} \partial_\mu \mathfrak{L}[A, \partial A] &\equiv (\partial_\mu A_\rho(x)) \partial_\nu \frac{\delta \mathfrak{L}[A, \partial A]}{\delta \partial_\nu A_\rho(x)} + (\partial_\mu \partial_\nu A_\rho(x)) \frac{\delta \mathfrak{L}[A, \partial A]}{\delta \partial_\nu A_\rho(x)} \\ &= \partial_\nu [(\partial_\mu A_\rho(x)) \frac{\delta \mathfrak{L}[A, \partial A]}{\delta \partial_\nu A_\rho(x)}]. \end{aligned} \tag{3.98}$$

We may now rewrite this as,

$$\partial_\nu \left( [(\partial_\mu A_\rho(x)) \frac{\delta \mathfrak{L}[A, \partial A]}{\delta \partial_\nu A_\rho(x)}] - \mathfrak{L}[A, \partial A] \delta^\nu_\mu \right) = 0. \tag{3.99}$$

Therefore, there exists a tensor called the energy momentum tensor or stress energy tensor of the electromagnetic field that may be written as,

$$\tilde{T}^\nu_\mu = (\partial_\mu A_\rho) \frac{\partial \mathfrak{L}}{\partial(\partial_\nu A_\rho)} - \delta^\nu_\mu \mathfrak{L}, \tag{3.100}$$

such that it four-divergence vanishes.

$$\partial_\nu \tilde{T}^\nu_\mu = 0 \tag{3.101}$$

The only problem with this definition in Eq. (3.100), it is not symmetrical in the indices $\mu, \nu$. To make it symmetrical, we have to add another appropriate rank two tensor that also obeys the same conservation law. To find out which one, we have to first rewrite the Lagrangian in a four-vector notation. Henceforth, we focus on massless photons. We choose to define $A^\mu = (\phi, A_x, A_y, A_z)$. This means we have to multiply the time component by $c$ in the position four-vector also : $x^\mu = (ct, x, y, z)$.

$$\mathbf{E} = -\nabla\phi - \frac{1}{c}\dot{\mathbf{A}}\ ; \quad \mathbf{B} = \nabla \times \mathbf{A} \tag{3.102}$$

Or,

$$E_i = -(\partial_i A^0) - (\partial_0 A^i). \tag{3.103}$$

If $n(1,2) = 3, n(2,3) = 1, n(3,1) = 2$, then,

$$B_{n(i,j)} = \partial_i A^j - \partial_j A^i \tag{3.104}$$

$$\mathbf{B}^2 = \frac{1}{2}\sum_{i,j}(\partial_i A^j - \partial_j A^i)^2 \tag{3.105}$$

$$\mathfrak{L} = \frac{1}{8\pi}\Big(\sum_i(-(\partial_i A^0) - (\partial_0 A^i))^2 - \frac{1}{2}\sum_{i,j}(\partial_i A^j - \partial_j A^i)^2\Big)$$

$$= -\frac{1}{16\pi}(\partial_\mu A_\nu - \partial_\nu A_\mu)(\partial^\mu A^\nu - \partial^\nu A^\mu) = -\frac{1}{16\pi}F_{\mu\nu}F^{\mu\nu} \tag{3.106}$$

where $F_{\mu\nu} = (\partial_\mu A_\nu - \partial_\nu A_\mu)$. Now it is quite easy to calculate $\tilde{T}^\nu_\mu$.

$$\frac{\partial \mathfrak{L}}{\partial(\partial_\nu A_\rho)} = -\frac{1}{4\pi}\frac{\partial F_{\mu\sigma}}{\partial(\partial_\nu A_\rho)}F^{\mu\sigma} = -\frac{1}{4\pi}F^{\nu\rho} \tag{3.107}$$

Thus,

$$\tilde{T}^\nu_\mu = -\frac{1}{4\pi}(\partial_\mu A_\rho)F^{\nu\rho} + \delta^\nu_\mu\,\frac{1}{16\pi}F_{\sigma\rho}F^{\sigma\rho}. \tag{3.108}$$

As we can see, this is not symmetric in the indices $\mu, \nu$. To make it symmetric, we first write this with both the indices on top on the left.

$$\tilde{T}^{\mu\nu} = -\frac{1}{4\pi}(\partial^\nu A^\rho)F^\mu_\rho + \eta^{\mu\nu}\,\frac{1}{16\pi}F_{\sigma\rho}F^{\sigma\rho} \tag{3.109}$$

Then we add the following new tensor,

$$\tilde{S}^{\mu\nu} = \frac{1}{4\pi}(\partial^\rho A^\nu)F^\mu_\rho. \tag{3.110}$$

Then this becomes,

$$T^{\mu\nu} = \tilde{T}^{\mu\nu} + \tilde{S}^{\mu\nu} = -\frac{1}{4\pi}F^{\nu\rho}F^\mu_\rho + \frac{1}{16\pi}\eta^{\mu\nu}\,F_{\sigma\rho}F^{\sigma\rho}, \tag{3.111}$$

which is clearly symmetric as the product of F's is symmetric. The only thing that remains is to show that this procedure does not violate the conservation law. For this we first start with the Lagrange equation:

$$\partial_\nu\frac{\partial \mathfrak{L}}{\partial(\partial_\nu A_\mu)} - \frac{\partial \mathfrak{L}}{\partial A_\mu} = 0. \tag{3.112}$$

The RHS is zero because $\mathfrak{L}$ depends only on the derivatives of $A$ and not directly on $A$. This is true only for Lagrangian without sources. This means,

$$\partial_\nu F^{\nu\mu} = 0. \tag{3.113}$$

First we rewrite,

$$\tilde{T}^\mu_\nu = -\frac{1}{4\pi}(\partial_\nu A_\rho)F^{\mu\rho} + \delta^\mu_\nu \frac{1}{16\pi} F_{\sigma\rho}F^{\sigma\rho} \tag{3.114}$$

$$\tilde{S}^\mu{}_\nu = \frac{1}{4\pi}(\partial_\rho A_\nu)F^{\mu\rho}. \tag{3.115}$$

It is easy to see that $\partial_\mu \tilde{S}^\mu{}_\nu = 0$.

$$\partial_\mu \tilde{S}^\mu{}_\nu = \frac{1}{4\pi}(\partial_\rho \partial_\mu A_\nu)F^{\mu\rho} + \frac{1}{4\pi}(\partial_\rho A_\nu)\partial_\mu F^{\mu\rho} \tag{3.116}$$

But $\partial_\rho \partial_\mu$ is symmetric under the exchange of $\mu, \rho$, and $F^{\mu\rho}$ is antisymmetric. Since both these indices are being summed over, the answer is zero. The term $\partial_\mu F^{\mu\rho} = 0$ from the Lagrange equation. Hence $\partial_\mu \tilde{S}^\mu{}_\nu = 0$. Now we have to verify that $\partial_\mu \tilde{T}^\mu_\nu = 0$.

$$\partial_\mu \tilde{T}^\mu_\nu = -\frac{1}{4\pi}(\partial_\nu \partial_\mu A_\rho)F^{\mu\rho} - \frac{1}{4\pi}(\partial_\nu A_\rho)\partial_\mu F^{\mu\rho} + \frac{1}{8\pi}(\partial_\nu F_{\sigma\rho})F^{\sigma\rho} \tag{3.117}$$

We know that $\partial_\mu F^{\mu\rho} = 0$. If we write

$$\partial_\mu A_\rho = \frac{1}{2}(\partial_\mu A_\rho - \partial_\rho A_\mu) + \frac{1}{2}(\partial_\mu A_\rho + \partial_\rho A_\mu), \tag{3.118}$$

we should retain only the antisymmetric part as this is multiplying $F^{\mu\rho}$ in the first term, which is antisymmetric. Hence,

$$\partial_\mu \tilde{T}^\mu_\nu = -\frac{1}{8\pi}(\partial_\nu F_{\mu\rho})F^{\mu\rho} + \frac{1}{8\pi}(\partial_\nu F_{\sigma\rho})F^{\sigma\rho} = 0. \tag{3.119}$$

Now we show that $T^{0\nu}$ has an easily identifiable physical meaning.

$$T^{0\nu} = -\frac{1}{4\pi}F^{\nu\rho}F^0_\rho + \frac{1}{16\pi}\eta^{0\nu} F_{\sigma\rho}F^{\sigma\rho} \tag{3.120}$$

Here,

$$T^{00} = -\frac{1}{4\pi}F^{0i}F^0_i + \frac{1}{16\pi} F_{\sigma\rho}F^{\sigma\rho} \tag{3.121}$$

and, say,

$$T^{03} = -\frac{1}{4\pi}F^{3\rho}F^0_\rho = -\frac{1}{4\pi}F^{31}F^0_1 - \frac{1}{4\pi}F^{32}F^0_2. \tag{3.122}$$

We know, $F^{0i} = (\partial^0 A^i - \partial^i A^0) = (\frac{1}{c}\frac{\partial}{\partial t}A^i + \nabla_i \phi) = -E_i$ and $F^0_i = (-\frac{1}{c}\frac{\partial}{\partial t}A^i - \nabla_i \phi) = E_i$. Also $-\frac{1}{8\pi}(E^2 - B^2) = \frac{1}{16\pi} F_{\sigma\rho}F^{\sigma\rho}$. Hence,

$$T^{00} = \frac{1}{4\pi}\mathbf{E}^2 - \frac{1}{8\pi}(\mathbf{E}^2 - \mathbf{B}^2) = \frac{1}{8\pi}(\mathbf{E}^2 + \mathbf{B}^2) = u, \tag{3.123}$$

which is nothing but the energy density of the electromagnetic field in CGS units. Now if we examine,

$$F^{31} = \partial^3 A^1 - \partial^1 A^3 = -\frac{\partial A_x}{\partial z} + \frac{\partial A_z}{\partial x} = -B_y \tag{3.124}$$

$$F^{32} = \partial^3 A^2 - \partial^2 A^3 = -\frac{\partial A_y}{\partial z} + \frac{\partial A_z}{\partial y} = B_x. \tag{3.125}$$

Hence,

$$T^{03} = \frac{1}{4\pi} B_y E_x - \frac{1}{4\pi} B_x E_y = \frac{1}{4\pi} (\mathbf{E} \times \mathbf{B})_z. \tag{3.126}$$

Thus $T^{0i}$ is proportional to the $i-th$ component of the Poynting vector $\mathbf{S} = \frac{c}{4\pi}(\mathbf{E} \times \mathbf{B})$. The conservation law:

$$0 = \partial_\nu T^{0\nu} = \frac{1}{c}\frac{\partial}{\partial t} T^{00} + \sum_{i=1(x),2(y),3(z)} \nabla_i T^{0i} = \frac{1}{c}\frac{\partial u}{\partial t} + \frac{1}{c}\nabla \cdot \mathbf{S} \tag{3.127}$$

or

$$\frac{\partial u}{\partial t} + \nabla \cdot \mathbf{S} = 0. \tag{3.128}$$

This equation may be derived directly from the original form of the Maxwell equations without using four vectors. Start with these two (in free space $\mathbf{J} = 0$):

$$\nabla \times \mathbf{E} = -\frac{1}{c}\frac{\partial \mathbf{B}}{\partial t} \; ; \; \nabla \times \mathbf{B} = \frac{1}{c}\frac{\partial \mathbf{E}}{\partial t}. \tag{3.129}$$

Take the dot product of the first one with **B** and the second with **E** and subtract. Then we get,

$$\mathbf{B} \cdot \nabla \times \mathbf{E} = -\frac{1}{c}\mathbf{B} \cdot \frac{\partial \mathbf{B}}{\partial t} \; ; \; \mathbf{E} \cdot \nabla \times \mathbf{B} = \frac{1}{c}\mathbf{E} \cdot \frac{\partial \mathbf{E}}{\partial t}. \tag{3.130}$$

Subtract the first from the second,

$$\mathbf{E} \cdot \nabla \times \mathbf{B} - \mathbf{B} \cdot \nabla \times \mathbf{E} = \frac{1}{c}\mathbf{E} \cdot \frac{\partial \mathbf{E}}{\partial t} + \frac{1}{c}\mathbf{B} \cdot \frac{\partial \mathbf{B}}{\partial t} = \frac{1}{2c}\frac{\partial}{\partial t}(\mathbf{E}^2 + \mathbf{B}^2) = \frac{4\pi}{c}\frac{\partial u}{\partial t}. \tag{3.131}$$

But,

$$\mathbf{E} \cdot \nabla \times \mathbf{B} - \mathbf{B} \cdot \nabla \times \mathbf{E} = -\nabla \cdot (\mathbf{E} \times \mathbf{B}). \tag{3.132}$$

Hence,

$$\frac{\partial u}{\partial t} + \nabla \cdot \mathbf{S} = 0. \tag{3.133}$$

Thus the Poynting vector is nothing but the energy flux, or the momentum of radiation flowing into or out of a volume. This leads to an increase or decrease in energy of radiation. But for the electromagnetic field, not only is the energy conserved but momentum is itself conserved. The rate of change of total momentum in a volume is because of momentum flux flowing in and out of the system. Just as $T^{0i}$ is the energy flux, $T^{1,i}, T^{2,i}, T^{3,i}$ are the components of the momentum flux.

$$\partial_\nu T^{\nu i} = \frac{1}{c}\frac{\partial}{\partial t}T^{0i} + \nabla_j T^{ji} = 0 \tag{3.134}$$

But $T^{0i} = \frac{1}{4\pi}(\mathbf{E}\times\mathbf{B})_i = \frac{1}{c}S_i$. Define $(T^{1a}, T^{2a}, T^{3a}) = \mathbf{T}^a$.

$$\frac{1}{c^2}\frac{\partial S_a}{\partial t} + \nabla\cdot\mathbf{T}^a = 0 \tag{3.135}$$

Here $u$ is the energy (density), $\mathbf{S}$ is the energy flux, and $\mathbf{T}^a$ is the flux of the energy flux. This is because not only is the total energy in a volume conserved, the total momentum of the EM radiation ($\int d^3r\mathbf{S}$) is also conserved.

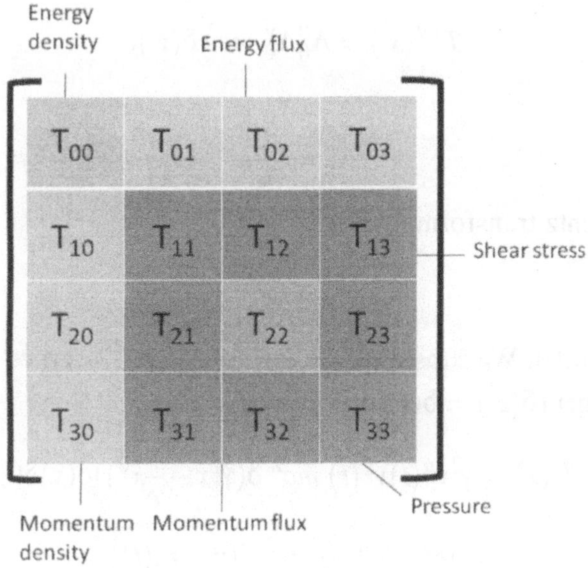


Figure 3.3: The meaning of the various components of the stress energy tensor.

■ There are other possible symmetries that one may consider. Some of them lead to trivial conservation laws. Consider a rotation in three dimensions so that $\mathbf{r}\to\mathbf{r}'\equiv M\,\mathbf{r}$, $\mathbf{A}(\mathbf{r},t)\to\mathbf{A}'(\mathbf{r}',t)\equiv M\,\mathbf{A}(\mathbf{r},t)$, and $\phi(\mathbf{r},t)\to\phi'(\mathbf{r}',t)=\phi(\mathbf{r},t)$, where $M$ is an orthogonal matrix independent of position and time. Choosing an appropriate gauge such as $\phi\equiv 0$ and $\nabla\cdot\mathbf{A}=0$ (radiation gauge), it is easy to convince oneself that the conserved quantity has the expression $\mathbf{P}=\int d^3r\,\partial_t\mathbf{A}\times\mathbf{A}$. It is also easy to convince oneself that this quantity vanishes identically. Hence, this does not yield a constant of the motion.

■ Find the energy momentum tensor of a particle of rest mass $m$ moving with velocity $\mathbf{v}$. The simplest way to do this is to use the tensor nature of this quantity. This tensor, being of rank two, transforms as the product of two coordinates under Lorentz transformation,

$$T^{\mu\nu}(x) = \Lambda^{\mu}_{\rho}\Lambda^{\nu}_{\sigma}\, T'^{\rho\sigma}(x'), \tag{3.136}$$

where $\Lambda^{\mu}_{\rho} \equiv \frac{\partial x'^{\mu}}{\partial x^{\rho}}$ and summation over repeated indices is implied. Let us imagine that the reference frame of the label $x'$ is one where the particle is at rest. In this case, the tensor has only one component viz. the time–time component equal to the energy density.

$$T'^{\rho\sigma}(x) = \delta_{\rho,0}\delta_{\sigma,0}\, mc^2\, \delta(\mathbf{r}') \tag{3.137}$$

Now imagine that we view this particle moving with some velocity $v$ in the positive x-direction, then the energy momentum tensor in this frame would be,

$$T^{\mu\nu}(x') = \Lambda^{\mu}_{0}\Lambda^{\nu}_{0}\, mc^2\, \delta(\mathbf{r}'). \tag{3.138}$$

Now,

$$\Lambda^{\mu}_{0} = \frac{\partial x^{\mu}}{\partial x'^{0}} \tag{3.139}$$

We write the Lorentz transformation as

$$\Lambda^{\mu}_{0} = \gamma v^{\mu}, \tag{3.140}$$

where $v^{\mu} = (1, \frac{v}{c}, 0, 0)$. We substitute the formulas for $\mathbf{r}' = (\gamma(x - \frac{v}{c}x^0), y, z)$ and $\Lambda^{\mu}_{0}$ in Eq. (3.138) to get ($\delta(\mathbf{r}') = \delta(x')\delta(y')\delta(z')$),

$$T^{\mu\nu}(x) = \gamma^2 v^{\mu}(t)v^{\nu}(t)\, mc^2\, \delta(\gamma(x - \frac{v}{c}x^0))\delta(y)\delta(z)$$

$$= \gamma v^{\mu}(t)v^{\nu}(t)\, mc^2\, \delta(\mathbf{r} - \mathbf{r}_0(t)). \tag{3.141}$$

The last result follows from the observation $\mathbf{r}_0 \equiv (\frac{v}{c}x^0, 0, 0) = (vt, 0, 0)$ and $\delta(\gamma X) = \frac{\delta(X)}{\gamma}$. Thus in general we may write for a particle of rest mass $m$ moving with velocity $\mathbf{v}(t) = \frac{d}{dt}\mathbf{r}_0(t)$,

$$T^{\mu\nu}(x) = v^{\mu}(t)v^{\nu}(t)\, \frac{mc^2}{\sqrt{1 - \frac{\mathbf{v}(t)^2}{c^2}}}\, \delta(\mathbf{r} - \mathbf{r}_0(t)) \tag{3.142}$$

where $v^{\mu}(t) \equiv (1, \frac{\mathbf{v}(t)}{c})$.

■ In this example, we consider the stress energy tensor of a (perfect) fluid in thermodynamic equilibrium. In the rest frame of the fluid, the stress energy tensor is purely diagonal, with the time component being the rest energy density and the spatial components related to the pressure.

$$\tilde{T}(x') = \begin{pmatrix} \rho c^2 & 0 & 0 & 0 \\ 0 & p & 0 & 0 \\ 0 & 0 & p & 0 \\ 0 & 0 & 0 & p \end{pmatrix} \quad (3.143)$$

Being a rank two tensor, its components transform as the product of two position four-vectors would. Therefore,

$$T^{\mu\nu}(x) = \Lambda^{\mu}_{\rho}\Lambda^{\nu}_{\sigma}\,\tilde{T}^{\rho\sigma}(x')$$

$$= \Lambda^{\mu}_{0}\Lambda^{\nu}_{0}\,\rho\, c^2 + \sum_{i=1}^{3}\Lambda^{\mu}_{i}\Lambda^{\nu}_{i}\, p \quad (3.144)$$

where $\Lambda^{\mu}_{\nu} = \frac{\partial x^{\mu}}{\partial x'^{\nu}}$. But,

$$u^{\mu} = c\Lambda^{\mu}_{0} \quad (3.145)$$

is the four-velocity. Also,

$$\eta^{\mu\nu} = \Lambda^{\mu}_{\rho}\eta^{\rho\sigma}\Lambda^{\nu}_{\sigma} \quad (3.146)$$

with summation convention over repeated indices is implied. In the right-hand side we set $\eta^{\rho\sigma} = diag(-1,1,1,1)$ to obtain,

$$\eta^{\mu\nu} = -\Lambda^{\mu}_{0}\Lambda^{\nu}_{0} + \sum_{i=1,2,3}\Lambda^{\mu}_{i}\Lambda^{\nu}_{i}. \quad (3.147)$$

Thus,

$$T^{\mu\nu}(x) = u^{\mu}u^{\nu}\,\rho + (\eta^{\mu\nu} + \frac{u^{\mu}u^{\nu}}{c^2})\, p = (\rho + \frac{p}{c^2})u^{\mu}u^{\nu} + p\,\eta^{\mu\nu}. \quad (3.148)$$

The above expression is the energy momentum tensor of a perfect fluid (no heat conduction and no viscosity, so in the comoving frame $T$ is diagonal). In astrophysical applications, the special case $p = 0$ is known as 'dust'. Now we move to a slightly different topic, namely solving for the equations of motion using Green function methods.

## 3.4 Solution of Maxwell's Equations Using Green's Functions

At this stage it is appropriate to study some specific solutions to Maxwell's equations using the Green function concept that is so ubiquitous in field theory. Green functions are used to solve inhomogeneous partial differential equations of the type,

$$T(\partial_{\nu}, \partial_{\nu}\partial_{\mu})u(x) = f(x), \quad (3.149)$$

subject to appropriate boundary conditions. Here $x$ is a d-dimensional vector and $T$ is some operator that is at most second order. The idea is to first obtain the 'Green function', which is nothing but the solution to,

$$T(\partial_\nu, \partial_\nu \partial_\mu) G(x, x') = \delta(x - x') \tag{3.150}$$

subject to the same boundary conditions, then one may simply write,

$$u(x) = \int d^d x' \; G(x, x') f(x'). \tag{3.151}$$

We now present an explanation of the choice of gauge. This notion is easy to follow in electrostatics. Consider the problem of finding the electric potential for a system of static charges and nothing else (no conductors and so on). Every student understands that the electric field $\mathbf{E} = -\nabla\phi$ only determines the scalar potential $\phi$ up to an additive constant, which is nothing but the integration constant obtained while inverting the relation. Choosing $\phi$ to vanish at infinity or at some other point is an example of a gauge choice. The reason is $\phi(\mathbf{r}) = \int d^3r' \frac{(\nabla' \cdot \mathbf{E}(\mathbf{r}'))}{4\pi|\mathbf{r}-\mathbf{r}'|} + C$ is the most general solution for the potential in this case (two solutions $\phi$ and $\phi'$ that obey $\nabla^2\phi = \nabla^2\phi' = -\nabla \cdot \mathbf{E}$ differ at most by a constant). In magnetostatics, there is much more freedom. Inverting the relation $\mathbf{B}(\mathbf{r}) = \nabla \times \mathbf{A}(\mathbf{r})$ we get, $\mathbf{A}(\mathbf{r}) = \int d^3r' \frac{\nabla' \times \mathbf{B}(\mathbf{r}')}{4\pi|\mathbf{r}-\mathbf{r}'|} + \nabla\xi$. This is because unlike in electrostatics, here we have two possible solutions $\mathbf{A}$ and $\mathbf{A}'$ such that $\nabla(\nabla \cdot \mathbf{A}) - \nabla^2\mathbf{A} = \nabla(\nabla \cdot \mathbf{A}') - \nabla^2\mathbf{A}' = \nabla \times \mathbf{B}$. This is obeyed even when $\mathbf{A} - \mathbf{A}' = \nabla\xi$ for some function $\xi$, whereas in electrostatics $\phi - \phi'$ had to be a constant. Some standard gauge choices are as follows.

**Coulomb gauge or Transverse gauge** $\nabla \cdot \mathbf{A} = 0$
**Lorentz gauge** $\nabla \cdot \mathbf{A} + \frac{1}{c}\frac{\partial\phi}{\partial t} = 0$
**Weyl gauge** $\phi = 0$

To be more specific we list several concrete examples that are quite familiar to the reader. We start with the Poisson equation for the electric potential

$$\nabla^2\phi = -4\pi\rho, \tag{3.152}$$

whose solution in terms of the Green function is

$$\phi(\mathbf{x}) = -4\pi \int d^3x' \; G(\mathbf{x}, \mathbf{x}')\rho(\mathbf{x}') \tag{3.153}$$

where

$$\nabla^2 G(\mathbf{x},\mathbf{x}') = \delta(\mathbf{x}-\mathbf{x}'). \tag{3.154}$$

In magnetostatics, it is the vector potential that obeys a Poisson equation (in CGS units),

$$\nabla^2 \mathbf{A} = -\frac{4\pi}{c}\mathbf{J} \tag{3.155}$$

$$\mathbf{A}(\mathbf{x}) = -\frac{4\pi}{c}\int d^3x'\; G(\mathbf{x},\mathbf{x}')\mathbf{J}(\mathbf{x}'). \tag{3.156}$$

In the general case of a four-dimensional Poisson equation (inhomogeneous wave equation)

$$\nabla^2 A^\mu - \frac{1}{c^2}\frac{\partial^2}{\partial t^2}A^\mu = -\frac{4\pi}{c}J^\mu, \tag{3.157}$$

then

$$A^\mu(x) = -\frac{4\pi}{c}\int d^4x'\; G(x,x')J^\mu(x'), \tag{3.158}$$

where

$$\nabla^2 G(x,x') - \frac{1}{c^2}\frac{\partial^2}{\partial t^2}G(x,x') = \delta(x-x'). \tag{3.159}$$

### 3.4.1 Gauss's Law in Electrostatics

Imagine free space where some finite region of space contains charges defined by density $\rho(\mathbf{x})$. In this case we may assert that at infinity the potential tends to zero (or some constant). In this case, the Green function that is consistent with this condition is,

$$G(\mathbf{x}-\mathbf{x}') = -\frac{1}{4\pi\,|\mathbf{x}-\mathbf{x}'|}. \tag{3.160}$$

We have to show that this obeys the defining equation:

$$\nabla^2 G(\mathbf{x}-\mathbf{x}') = \delta(\mathbf{x}-\mathbf{x}'). \tag{3.161}$$

It is clear that this Green function vanishes at infinity. We just have to show that it obeys the equation for a Green function. For this we first consider a region $R : \mathbf{x} \in R$ that excludes $\mathbf{x}'$. It is easy to see (for example in Cartesian coordinates) that,

$$\nabla^2 G(\mathbf{x}-\mathbf{x}') = 0. \tag{3.162}$$

Now imagine a region that includes $\mathbf{x}'$. We may focus on a small sphere $\Omega_\varepsilon$ of radius $\varepsilon$ with center at $\mathbf{x}'$ since outside the region, the equality is satisfied as we have already seen. Since the defining property of the Dirac delta function is

$$\int_{\Omega_\varepsilon} d^3x\, \delta(\mathbf{x}-\mathbf{x}')f(\mathbf{x}) = f(\mathbf{x}'), \tag{3.163}$$

for smooth (many times differentiable) functions $f$ we must also have (upon multiplying Eq. (3.161) by $f(\mathbf{x}')$ and integrating over $\mathbf{x}'$),

$$\int_{\Omega_\varepsilon} d^3x\, f(\mathbf{x})\, \nabla^2 G(\mathbf{x}-\mathbf{x}') = f(\mathbf{x}'). \tag{3.164}$$

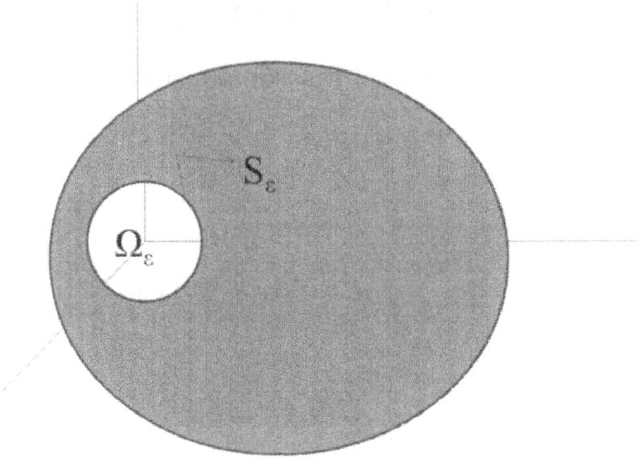


Figure 3.4: The origin coincides with $\mathbf{x}'$. The Laplacian acting on the Green function $\Omega_\varepsilon$ is trivially zero. Within this region some care has to be exercised.

Since $f(\mathbf{x})$ is a smooth function in the neighborhood of $\mathbf{x}'$ we may approximate $f(\mathbf{x})$ as $f(\mathbf{x}')$ and take it outside the integral. Consider the left-hand side (set $\mathbf{x}-\mathbf{x}' = \mathbf{R}$),

$$f(\mathbf{x}') \int_{\Omega_\varepsilon} d^3x\, \nabla^2 G(\mathbf{x}-\mathbf{x}') = f(\mathbf{x}') \int_{S_\varepsilon} dS\, \hat{R}\cdot\nabla G(\mathbf{R}). \tag{3.165}$$

The last result follows from Gauss's theorem where $S_\varepsilon$ is the surface of $\Omega_\varepsilon$. But,

$$\hat{R}\cdot\nabla G(\mathbf{R}) = \frac{\partial}{\partial R} G(R) = \frac{1}{4\pi R^2}, \tag{3.166}$$

and $da = R^2\, d\Omega$. This means,

$$f(\mathbf{x}') \int_{\Omega_\varepsilon} d^3x\, \nabla^2 G(\mathbf{x}-\mathbf{x}') = f(\mathbf{x}') \int_{S_\varepsilon} d\Omega\, R^2 \frac{1}{4\pi R^2} = f(\mathbf{x}') \tag{3.167}$$

as required. An alternative proof uses a Fourier transform.

$$G(\mathbf{R}) = \int \frac{d^3q}{(2\pi)^3} e^{i\mathbf{q}\cdot\mathbf{R}}\, \tilde{G}(\mathbf{q}) \tag{3.168}$$

Its inverse is,

$$\tilde{G}(\mathbf{q}) = \int d^3R\, e^{-i\mathbf{q}.\mathbf{R}}\, G(\mathbf{R}). \tag{3.169}$$

Substituting the expression in Eq. (3.160) for $G(\mathbf{R})$ we get,

$$\tilde{G}(\mathbf{q}) = -\int d^3R\, e^{-i\mathbf{q}.\mathbf{R}}\, \frac{1}{4\pi R}. \tag{3.170}$$

In spherical coordinates, $d^3R = R^2 d\Omega$ and,

$$\int d\Omega\, e^{-i\mathbf{q}\cdot\mathbf{R}} = 4\pi\, \frac{sin(qR)}{qR} \tag{3.171}$$

$$\tilde{G}(\mathbf{q}) = -\int_0^\infty R^2 dR\, 4\pi\, \frac{sin(qR)}{qR}\, \frac{1}{4\pi R} = -\frac{1}{q^2}. \tag{3.172}$$

Therefore,

$$G(\mathbf{x}-\mathbf{x}') = -\int \frac{d^3q}{(2\pi)^3} \frac{e^{i\mathbf{q}\cdot(\mathbf{x}-\mathbf{x}')}}{q^2}. \tag{3.173}$$

We are now going to show that for any smooth function $f(\mathbf{x})$, the following identity holds.

$$\int d^3x'\, \nabla^2 G(\mathbf{x}-\mathbf{x}') f(\mathbf{x}') = f(\mathbf{x}) \tag{3.174}$$

Performing the integral over $\mathbf{x}'$ we get,

$$\int d^3x'\, \nabla^2 G(\mathbf{x}-\mathbf{x}') f(\mathbf{x}')$$

$$= -\nabla^2 \int \frac{d^3q}{(2\pi)^3} \frac{e^{i\mathbf{q}\cdot\mathbf{x}}}{q^2}\, \tilde{f}(\mathbf{q}) = \int \frac{d^3q}{(2\pi)^3} e^{i\mathbf{q}\cdot\mathbf{x}}\, \tilde{f}(\mathbf{q}) = f(\mathbf{x}). \tag{3.175}$$

Thus the electric potential of a system of charges that occupy a finite region of space with no other constraints present is given by

$$\phi(\mathbf{x}) = \int d^3x'\, \frac{\rho(\mathbf{x}')}{|\mathbf{x}-\mathbf{x}'|}. \tag{3.176}$$

More interesting situations arise when there are conductors present. This means the regions occupied by the conductors have a constant potential. The problem is to determine the electric potential outside these conductors. In this case the image method is employed where typically one asserts that the potential outside the conductors are determined by an equivalent problem of having fictitious 'image charges' $\rho_{im}(\mathbf{x}')$ that are chosen so that the surface of the conductor (which is common to both the interior of the conductor, which we are not interested in, and the exterior, which we are interested in) is held at a fixed potential. Thus we may write,

$$\phi(\mathbf{x}) = \int_{exterior} d^3x' \, \frac{\rho(\mathbf{x}')}{|\mathbf{x}-\mathbf{x}'|} + \int_{interior} d^3y' \, \frac{\rho_{im}(\mathbf{y}')}{|\mathbf{x}-\mathbf{y}'|}. \tag{3.177}$$

Since the points $\mathbf{y}'$ lie in the interior of the conductor and $\mathbf{x}$ lies outside the conductor, the image term does not contribute to the Laplacian of $\phi(\mathbf{x})$. The image charge distribution is then determined by applying the constraint,

$$\phi(\mathbf{x})\Big|_{\mathbf{x}\in S_c} = \phi_0 = \int_{exterior} d^3x' \, \frac{\rho(\mathbf{x}')}{|\mathbf{x}(u,v)-\mathbf{x}'|} + \int_{interior} d^3y' \, \frac{\rho_{im}(\mathbf{y}')}{|\mathbf{x}(u,v)-\mathbf{y}'|} \tag{3.178}$$

assuming that $\mathbf{x}(u,v)$ parameterizes the surface of the conductor. The above equation has to be inverted to obtain $\rho_{im}(\mathbf{y}')$ and then used to obtain the electric potential outside the conductor. Clearly the usefulness of this technique depends on the simplicity of the surface.

■ Consider a spherical grounded conductor of radius $a$ with center at the origin. Imagine that a charge $q$ is placed at a distance $l > a$ from the origin on the z-axis. The problem is to find the potential at all points outside the conductor (since inside the conductor it is zero). This is a well-known problem in elementary physics that may be solved by the general method outlined earlier,

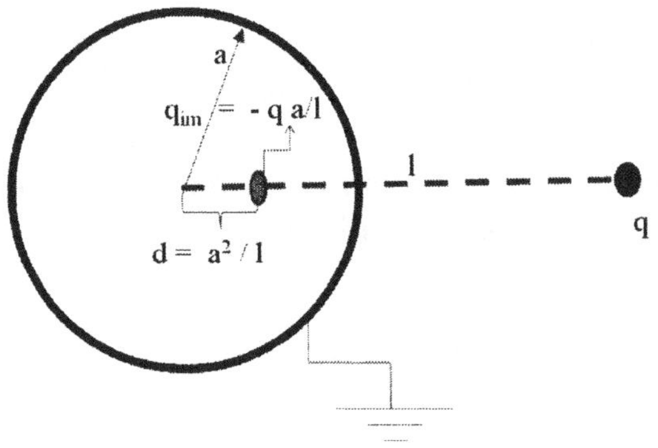


Figure 3.5: Diagram illustrating the image method.

$$0 = \frac{q}{|\mathbf{x}(u,v)-l\hat{k}|} + \frac{q_{im}}{|\mathbf{x}(u,v)-d\hat{k}|}. \tag{3.179}$$

$q_{im}$ and $d < a$ may be determined by solving the above equations by choosing $\mathbf{x}(u,v) = a\hat{k}$ and $\mathbf{x}(u,v) = -a\hat{k}$

$$0 = \frac{q}{|a-l|} + \frac{q_{im}}{|a-d|} \; ; \; 0 = \frac{q}{|-a-l|} + \frac{q_{im}}{|-a-d|}, \tag{3.180}$$

or

$$\frac{l-a}{a-d} = \frac{a+l}{a+d}, \tag{3.181}$$

or $d = \frac{a^2}{l}$.

$$q_{im} = -q\frac{|a+d|}{|a+l|} = -\frac{qa}{l}. \tag{3.182}$$

Thus the full potential outside the conductor may written down as follows.

$$\phi(\mathbf{x}) = \frac{q}{|\mathbf{x} - \hat{k}\, l|} + \frac{q_{im}}{|\mathbf{x} - d\, \hat{k}|} \tag{3.183}$$

We now present an alternative perspective to the problem of finding the potential in some region. This uses Green's theorem of calculus. Imagine a region of space with some localized charge density $\rho(\mathbf{r})$. Imagine further that in the vicinity of these charges, there are conductors, with each at some specified potential. The problem is to find the electric potential at any point outside the conductors. Imagine a region $\Omega$ that excludes both the point $\mathbf{r}$ and the interior of all the conductors. The boundary $S$ of this region may be represented as a small spherical surface of radius $\varepsilon$ centered at $\mathbf{r}$ so that $\Omega$ lies outside this small sphere, and also by the boundaries of all the conductors. Set $V(\mathbf{r}') \equiv -\frac{1}{4\pi|\mathbf{r}'-\mathbf{r}|}$. In this case we may write

$$\int_{\Omega} d^3r' \; (\phi(\mathbf{r}')\nabla'^2 V(\mathbf{r}') - V(\mathbf{r}')\nabla'^2\phi(\mathbf{r}'))$$

$$= \int_S da' \; (\phi(\mathbf{r}')\frac{\partial V(\mathbf{r}')}{\partial n'} - V(\mathbf{r}')\frac{\phi(\mathbf{r}')}{\partial n'}), \tag{3.184}$$

where $\frac{\partial}{\partial n'}$ is the derivative outward normal to each of the surfaces that bound $\Omega$. Since $\mathbf{r}$ is excluded from $\Omega$, $\nabla'^2 V \equiv 0$ and $\nabla'^2\phi(\mathbf{r}') = -4\pi\rho(\mathbf{r}')$. Therefore,

$$\int_{\Omega} d^3r' \; (4\pi\rho(\mathbf{r}')V(\mathbf{r}'))$$

$$= \int_{S_{conductors}} da' \; (\phi(\mathbf{r}')\frac{\partial V(\mathbf{r}')}{\partial n'} - V(\mathbf{r}')\frac{\partial\phi(\mathbf{r}')}{\partial n'})$$

$$+ \int_{S_\varepsilon} da' \; (\phi(\mathbf{r}')\frac{\partial V(\mathbf{r}')}{\partial n'} - V(\mathbf{r}')\frac{\partial\phi(\mathbf{r}')}{\partial n'}). \tag{3.185}$$

The integral over the small spherical surface of radius $\varepsilon$ may be performed by first observing that the potential and its derivatives are continuous at the point $\mathbf{r}$ so that these terms may be taken outside the integration. Since we are talking about the inward normal to the small spherical surface,

$$\int_{S_\varepsilon} da' \,(\phi(\mathbf{r}')\frac{\partial V(\mathbf{r}')}{\partial n'} - V(\mathbf{r}')\frac{\partial \phi(\mathbf{r}')}{\partial n'}) =$$

$$-\phi(\mathbf{r})\, 4\pi\varepsilon^2 \,\frac{1}{4\pi\varepsilon^2} - \frac{\phi(\mathbf{r})}{\partial n} 4\pi\varepsilon^2 \,\frac{-1}{4\pi\varepsilon} = -\phi(\mathbf{r}). \tag{3.186}$$

The last result follows from the observation that,

$$\frac{\partial V}{\partial n'} = -\frac{1}{4\pi\varepsilon^2}. \tag{3.187}$$

Thus we finally have,

$$\phi(\mathbf{r}) = \int_\Omega d^3r' \,\frac{\rho(\mathbf{r}')}{|\mathbf{r}' - \mathbf{r}|} + \int_{S_{conductors}} da' \,(\phi(\mathbf{r}')\frac{\partial V(\mathbf{r}')}{\partial n'} - V(\mathbf{r}')\frac{\partial \phi(\mathbf{r}')}{\partial n'}). \tag{3.188}$$

We see here that not only does one have to know the potential on the surface of each conductor, one also has to know the normal component of the electric field. However, the problem has a unique solution with only the potentials specified, only this method does not allow us to find it. The Green function method can allow us to find the complete answer.

### 3.4.2 Lienard-Wiechert Potentials

Next, we wish to determine the potentials of a moving charge. The charge density of a moving charge $q$, which is at position $\mathbf{r}_0(t)$ at time $t$, is given by $\rho(\mathbf{r},t) = q\,\delta(\mathbf{r}-\mathbf{r}_0(t))$ and the current density is given by, $\mathbf{j}(\mathbf{r},t) = q\,\mathbf{v}_0(t)\,\delta(\mathbf{r}-\mathbf{r}_0(t))$ where $\mathbf{v}_0(t) \equiv \dot{\mathbf{r}}_0(t)$ is the velocity and $\delta(\mathbf{r})$ is the three-dimensional delta function. We have to solve (we have used the Lorentz gauge),

$$\nabla^2\phi - \frac{1}{c^2}\frac{\partial^2}{\partial t^2}\phi = -4\pi\, q\, \delta(\mathbf{r} - \mathbf{r}_0(t)) \tag{3.189}$$

$$\nabla^2\mathbf{A} - \frac{1}{c^2}\frac{\partial^2}{\partial t^2}\mathbf{A} = -\frac{4\pi}{c}\, q\, \mathbf{v}_0(t)\, \delta(\mathbf{r} - \mathbf{r}_0(t)). \tag{3.190}$$

To solve, we write

$$X(\mathbf{x},t) = \int \frac{d^3k}{(2\pi)^3}\, e^{i\mathbf{k}\cdot\mathbf{r}}\, \tilde{X}(\mathbf{k},t), \tag{3.191}$$

where $X$ could stand for either $\phi$ or $\mathbf{A}$.

$$-k^2\tilde{\phi} - \frac{1}{c^2}\frac{\partial^2}{\partial t^2}\tilde{\phi} = -4\pi\, q\, e^{-i\mathbf{k}\cdot\mathbf{r}_0(t)} \tag{3.192}$$

$$-k^2\tilde{\mathbf{A}} - \frac{1}{c^2}\frac{\partial^2}{\partial t^2}\tilde{\mathbf{A}} = -\frac{4\pi}{c}\, q\, \mathbf{v}_0(t)\, e^{-i\mathbf{k}\cdot\mathbf{r}_0(t)} \tag{3.193}$$

The solutions are,

$$\tilde{\phi}(\mathbf{k},t) = \int_{-\infty}^{t} dt_1\, \frac{4\pi c\, q\, e^{-i\mathbf{k}\cdot\mathbf{r}_0(t_1)} sin(ck(t-t_1))}{k}. \tag{3.194}$$

A similar equation for $\mathbf{A}$ may also be written down by the reader. The equation was second order in time and we have implicitly assumed some boundary/initial conditions that will be justified later. In real space we may write,

$$\phi(\mathbf{r},t) = \int_{-\infty}^{t} dt_1 \int \frac{d^3k}{(2\pi)^3}\, e^{i\mathbf{k}\cdot(\mathbf{r}-\mathbf{r}_0(t_1))}\, \frac{4\pi c\, q\, sin(ck(t-t_1))}{k}. \tag{3.195}$$

The equation for $\mathbf{A}(\mathbf{r},t)$ is obtained by replacing the speed of light in the numerator of the above expression by $\mathbf{v}_0(t_1)$. Using the result $d^3k \equiv k^2 dk\, d\Omega$ and $\int d\Omega\, e^{i\mathbf{k}\cdot\mathbf{R}} = 4\pi\, \frac{sin(kR)}{kR}$, we get,

$$\phi(\mathbf{r},t) = \int_{-\infty}^{t} dt_1 \int_0^{\infty} \frac{dk}{(2\pi)^3}\, \frac{sin(k|\mathbf{r}-\mathbf{r}_0(t_1)|)\, sin(ck(t-t_1))}{|\mathbf{r}-\mathbf{r}_0(t_1)|}\, (4\pi)^2 c\, q. \tag{3.196}$$

The integrand is an even function of $k$, hence we may extend to $-\infty$ and divide by two. Integrating,

$$\int_0^{\infty} dk\, sin(ka) sin(kb) = -\frac{\pi}{2}\,(\delta(a+b) - \delta(a-b)). \tag{3.197}$$

Since $t > t_1$, only one of the Delta functions survives.

$$\phi(\mathbf{r},t) = \int_{-\infty}^{t} d\, ct_1\, \frac{\delta(|\mathbf{r}-\mathbf{r}_0(t_1)| - c(t-t_1))}{|\mathbf{r}-\mathbf{r}_0(t_1)|}\, q \tag{3.198}$$

Let $t_*$ be the solution to the equation $|\mathbf{r}-\mathbf{r}_0(t_*)| - c(t-t_*) = 0$. This integral has to be evaluated with some care. Consider the Heaviside step function $\theta(|\mathbf{r}-\mathbf{r}_0(t_1)| - c(t-t_1))$. Its derivative may be written as

$$\frac{d}{d\, ct_1}\theta(|\mathbf{r}-\mathbf{r}_0(t_1)| - c(t-t_1)) = \delta(|\mathbf{r}-\mathbf{r}_0(t_1)| - c(t-t_1))(1 + \frac{d}{d\, ct_1}|\mathbf{r}-\mathbf{r}_0(t_1)|). \tag{3.199}$$

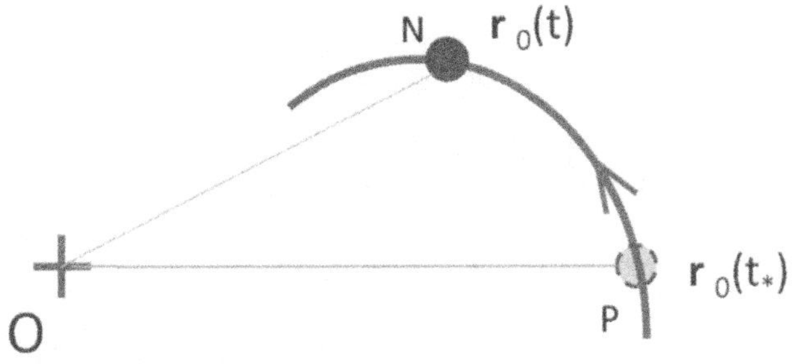


Figure 3.6: Diagram illustrating the retardation effect, the charged particle moving in the trajectory shown influences a test object at $O$ not from the present location $N$ but from an earlier location $P$.

Define $\mathbf{R}(t_1) = \mathbf{r} - \mathbf{r}_0(t_1)$. Then,

$$\frac{d}{d\,ct_1}|\mathbf{r} - \mathbf{r}_0(t_1)| = \frac{d}{d\,ct_1}(\mathbf{R}(t_1)\cdot\mathbf{R}(t_1))^{\frac{1}{2}} = -\frac{1}{R(t_1)}\mathbf{R}(t_1)\cdot\frac{\mathbf{v}_0(t_1)}{c} \tag{3.200}$$

since $\frac{d}{d\,ct_1}\mathbf{R}(t_1) = -\frac{\mathbf{v}_0(t_1)}{c}$. Thus,

$$\frac{d}{d\,ct_1}\theta(|\mathbf{r} - \mathbf{r}_0(t_1)| - c(t - t_1)) = \delta(|\mathbf{r} - \mathbf{r}_0(t_1)| - c(t - t_1))(1 - \frac{\mathbf{R}(t_*)}{R(t_*)}\cdot\frac{\mathbf{v}_0(t_*)}{c}). \tag{3.201}$$

Substituting the above expression for the delta function into Eq. (3.198) we get,

$$\phi(\mathbf{r},t) = \int_{-\infty}^{t} d\,ct_1\;\frac{\frac{d}{d\,ct_1}\theta(|\mathbf{r} - \mathbf{r}_0(t_1)| - c(t - t_1))}{(1 - \frac{1}{R(t_*)}\mathbf{R}(t_*)\cdot\frac{\mathbf{v}_0(t_*)}{c})R(t_*)}\,q = \frac{q}{(R(t_*) - \mathbf{R}(t_*)\cdot\frac{\mathbf{v}_0(t_*)}{c})}. \tag{3.202}$$

Thus the Lienard-Wiechert potentials are as follows.

$$\phi(\mathbf{r},t) = \frac{q}{(R(t_*) - \mathbf{R}(t_*)\cdot\frac{\mathbf{v}_0(t_*)}{c})} \tag{3.203}$$

$$\mathbf{A}(\mathbf{r},t) = \frac{q\frac{\mathbf{v}_0(t_*)}{c}}{(R(t_*) - \mathbf{R}(t_*)\cdot\frac{\mathbf{v}_0(t_*)}{c})} \tag{3.204}$$

Now we try to understand this more intuitively. When $\mathbf{r}_0$ is independent of time, we reproduce Coulomb's law of electrostatics where there is an electric potential but no magnetic (vector) potential. Moreover, this potential is just $\phi = q/R$ where $R = |\mathbf{r} - \mathbf{r}_0|$. Hence the choice of initial conditions in solving for these potentials were indeed correct. The general case tells us that the potential at time $t$ depends not on the position of the particle at time $t$ (which is $\mathbf{r}_0(t)$), but on the position at an earlier time $t_* = t - \frac{R(t_*)}{c}$. This time is the present time $t$ minus the time light would have taken to travel the distance from the particle at $\mathbf{r}_0(t_*)$ to the point of interest, namely $\mathbf{r}$. This is known as the retardation effect and the Lienard-Wiechert potentials are retarded potentials. The information of the location of the particle takes a finite time to arrive at the point of interest. Hence the delay or retardation. From these potentials one may go ahead and evaluate the electric and magnetic fields. For this it is preferable to start with the expression in Eq. (3.198) and its vector counterpart,

$$\mathbf{A}(\mathbf{r},t) = \int_{-\infty}^{t} d\, ct_1 \, \frac{\delta(|\mathbf{r}-\mathbf{r}_0(t_1)| - c(t-t_1))}{|\mathbf{r}-\mathbf{r}_0(t_1)|} \, q \, \frac{\mathbf{v}_0(t_1)}{c}. \tag{3.205}$$

The magnetic field is evaluated as,

$$\begin{aligned}\mathbf{B}(\mathbf{r},t) &= \nabla \times \mathbf{A}(\mathbf{r},t) \\ &= -\int_{-\infty}^{t} d\, ct_1 \, \frac{\delta(|\mathbf{r}-\mathbf{r}_0(t_1)| - c(t-t_1))}{|\mathbf{r}-\mathbf{r}_0(t_1)|^3} \, q \, \frac{(\mathbf{r}-\mathbf{r}_0(t_1)) \times \mathbf{v}_0(t_1)}{c} \\ &+ \int_{-\infty}^{t} d\, ct_1 \, \frac{\delta'(|\mathbf{r}-\mathbf{r}_0(t_1)| - c(t-t_1))}{|\mathbf{r}-\mathbf{r}_0(t_1)|^2} \, q \, \frac{(\mathbf{r}-\mathbf{r}_0(t_1)) \times \mathbf{v}_0(t_1)}{c}.\end{aligned} \tag{3.206}$$

Set $c\tau = |\mathbf{r}-\mathbf{r}_0(t_1)| - c(t-t_1)$. Then,

$$\frac{dc\tau}{dct_1} = (1 - \frac{\mathbf{v}_0(t_1)\cdot\mathbf{R}(t_1)}{cR(t_1)}). \tag{3.207}$$

Therefore,

$$\begin{aligned}\mathbf{B}(\mathbf{r},t) &= \nabla \times \mathbf{A}(\mathbf{r},t) \\ &= -\frac{1}{|\mathbf{r}-\mathbf{r}_0(t_*)|^3 (1 - \frac{\mathbf{v}_0(t_*)\cdot\mathbf{R}(t_*)}{cR(t_*)})} \, q \, \frac{(\mathbf{r}-\mathbf{r}_0(t_*)) \times \mathbf{v}_0(t_*)}{c} \\ &+ \int_{-\infty}^{t} d\, ct_1 \, \frac{d}{dct_1} \delta(|\mathbf{r}-\mathbf{r}_0(t_1)| - c(t-t_1)) \, \frac{q\, (\mathbf{r}-\mathbf{r}_0(t_1)) \times \mathbf{v}_0(t_1)}{(R(t_1) - \frac{\mathbf{v}_0(t_1)\cdot\mathbf{R}(t_1)}{c}) |\mathbf{r}-\mathbf{r}_0(t_1)| c}.\end{aligned} \tag{3.208}$$

Using integration by parts we get,

$$\begin{aligned}\mathbf{B}(\mathbf{r},t) = \nabla \times \mathbf{A}(\mathbf{r},t) &= -q \frac{1 - \frac{v_0^2(t_*)}{c^2}}{(R(t_*) - \frac{\mathbf{R}(t_*)\cdot\mathbf{v}_0(t_*)}{c})^3} \frac{\mathbf{R}(t_*) \times \mathbf{v}_0(t_*)}{c} \\ &- \frac{q}{c^2 (R(t_*) - \frac{\mathbf{R}(t_*)\cdot\mathbf{v}_0(t_*)}{c})^3} (\mathbf{R}(t_*)\cdot\mathbf{a}_0(t_*)) \frac{\mathbf{R}(t_*) \times \mathbf{v}_0(t_*)}{c}\end{aligned}$$

$$-\frac{q}{c^2(R(t_*)-\frac{\mathbf{R}(t_*)\cdot\mathbf{v}_0(t_*)}{c})^2}(\mathbf{R}(t_*)\times\mathbf{a}_0(t_*)). \tag{3.209}$$

The electric field is given by,

$$\mathbf{E}(\mathbf{r},t)=q\frac{1-\frac{v_0^2(t_*)}{c^2}}{(R(t_*)-\frac{\mathbf{R}(t_*)\cdot\mathbf{v}_0(t_*)}{c})^3}(\mathbf{R}(t_*)-\frac{\mathbf{v}_0(t_*)}{c}R(t_*))$$
$$+\frac{q}{c^2(R(t_*)-\frac{\mathbf{R}(t_*)\cdot\mathbf{v}_0(t_*)}{c})^3}(\mathbf{R}(t_*)\cdot\mathbf{a}_0(t_*))(\mathbf{R}(t_*)-\frac{\mathbf{v}_0(t_*)}{c}R(t_*))$$
$$-\frac{q}{c^2(R(t_*)-\frac{\mathbf{R}(t_*)\cdot\mathbf{v}_0(t_*)}{c})^2}R(t_*)\,\mathbf{a}_0(t_*). \tag{3.210}$$

From the above two equations it is clear that,

$$\mathbf{B}(\mathbf{r},t)=\frac{\mathbf{R}(t_*)\times\mathbf{E}(\mathbf{r},t)}{R(t_*)}. \tag{3.211}$$

Therefore, the magnetic field is perpendicular to the electric field at all times.

## 3.5 Diffraction Theory

One may enquire as to the nature of the electromagnetic field emanating from localized sources. We have found the answer to such a question already in the time-independent case. Now we wish to study the question of propagation of electromagnetic radiation. This naturally leads to the phenomenon of diffraction. Let us start with the Maxwell equations and derive an expression for the potentials. We use the decomposition $\mathbf{E}=-\nabla\phi-\frac{1}{c}\frac{\partial\mathbf{A}}{\partial t}$ and $\mathbf{B}=\nabla\times\mathbf{A}$. Inserting these into Gauss's Law we get,

$$-\nabla\cdot(\nabla\phi+\frac{1}{c}\frac{\partial\mathbf{A}}{\partial t})=4\pi\rho, \tag{3.212}$$

and from Ampere's Law we get,

$$\frac{1}{c^2}\frac{\partial^2\mathbf{A}}{\partial t^2}-\nabla^2\mathbf{A}=\frac{4\pi}{c}\mathbf{J}-\nabla(\frac{1}{c}\frac{\partial}{\partial t}\phi+(\nabla\cdot\mathbf{A})). \tag{3.213}$$

It is convenient to use the Lorentz gauge condition where we set,

$$(\frac{1}{c}\frac{\partial}{\partial t}\phi+(\nabla\cdot\mathbf{A}))=0. \tag{3.214}$$

Then we get,

$$\frac{1}{c^2}\frac{\partial^2\mathbf{A}}{\partial t^2}-\nabla^2\mathbf{A}=\frac{4\pi}{c}\mathbf{J} \tag{3.215}$$

$$(\nabla^2\phi - \frac{1}{c^2}\frac{\partial^2\phi}{\partial t^2}) = -4\pi\rho. \tag{3.216}$$

Gauss's Law is not independent of this as it can be linked to Ampere's Law through equation of continuity. Now we consider the situation where the source $\mathbf{J}(\mathbf{r},t)$ is localized in space but is oscillating with a fixed frequency $\omega$. First we want to find the Green function in this case. For this we choose the current to be $\mathbf{J}(\mathbf{r},t) = \mathbf{J}(\mathbf{r},0)cos(\omega t)$. Then the solution will also be of the form $\mathbf{A}(\mathbf{r},t) = \mathbf{A}(\mathbf{r},0)\ cos(\omega t)$.

$$-\frac{\omega^2}{c^2}\mathbf{A}(\mathbf{r},0) - \nabla^2\mathbf{A}(\mathbf{r},0) = \frac{4\pi}{c}\mathbf{J}(\mathbf{r},0) \tag{3.217}$$

One typically uses Green functions to solve this equation. First consider the related problem (where $k = \frac{\omega}{c}$)

$$(k^2 + \nabla^2)G(\mathbf{r}-\mathbf{r}') = \delta(\mathbf{r}-\mathbf{r}'), \tag{3.218}$$

subject to the boundary condition that $G(\pm\infty) = 0$. The operator $\nabla^2 + k^2$ is known as the Helmholtz operator and the above equations are Helmholtz equations. The solution is easiest with Fourier transform,

$$G(\mathbf{r}-\mathbf{r}') = \int \frac{d^3K}{(2\pi)^3}\ e^{i\mathbf{K}\cdot(\mathbf{r}-\mathbf{r}')}\ \mathfrak{G}(\mathbf{K}) \tag{3.219}$$

$$\delta(\mathbf{r}-\mathbf{r}') = \int \frac{d^3K}{(2\pi)^3}\ e^{i\mathbf{K}\cdot(\mathbf{r}-\mathbf{r}')}. \tag{3.220}$$

Substituting into Eq. (3.218) we get,

$$(k^2 - K^2)\mathfrak{G}(\mathbf{K}) = 1. \tag{3.221}$$

The Green function may be written as,

$$G(\mathbf{r}-\mathbf{r}') = \int \frac{d^3K}{(2\pi)^3}\ e^{i\mathbf{K}\cdot(\mathbf{r}-\mathbf{r}')}\ \frac{1}{(k^2 - K^2)}. \tag{3.222}$$

This formal expression has to be suitably interpreted near the singularity $|K| = k$ in such a way that the resulting Green function respects the boundary conditions at infinity. Set $\mathbf{R} = \mathbf{r} - \mathbf{r}'$. This means,

$$G(\mathbf{R}) = \int \frac{d^3K}{(2\pi)^3}\ e^{i\mathbf{K}\cdot\mathbf{R}}\ \frac{1}{(k^2 - K^2)}$$

$$= \int_0^\infty \frac{K^2\ dK}{(2\pi)^3}\ (4\pi)\frac{sin(KR)}{KR}\ \frac{1}{(k^2 - K^2)} \tag{3.223}$$

after integrating over the solid angle. Since the integrand is an even function of $K$, we may extend the integration to $-\infty$ after dividing by two.

$$G(\mathbf{R}) = \int_{-\infty}^{\infty} \frac{K^2\, dK}{(2\pi)^3}\,(2\pi)\frac{sin(KR)}{KR}\,\frac{1}{(k^2-K^2)} \tag{3.224}$$

To perform this integration we use the residue method from complex analysis. The integration is to be interpreted as the principal part. This means,

$$G(\mathbf{R}) = \int_{-\infty}^{\infty} \frac{dK}{(2\pi)^2}\,\frac{e^{iKR}}{iR}\,\frac{K}{(k^2-K^2)}$$

$$= \frac{1}{2iR}\int_{-\infty}^{\infty} \frac{dK}{(2\pi)^2}\,\frac{e^{iKR}}{(k-K)} - \frac{1}{2iR}\int_{-\infty}^{\infty} \frac{dK}{(2\pi)^2}\,\frac{e^{iKR}}{k+K}$$

$$G(\mathbf{R}) = -\frac{cos(kR)}{4\pi R}. \tag{3.225}$$

Notice that the Helmholtz operator is even in the parameter $k$ and hence we expect the solution to also have this property. Hence we avoid using complex expressions such as $e^{ikR}$ for this reason and also because potentials are real. The full solution may then be written as,

$$\mathbf{A}(\mathbf{r},t) = \int d^3r'\;\frac{cos(k|\mathbf{r}-\mathbf{r}'|)cos(\omega t)}{4\pi|\mathbf{r}-\mathbf{r}'|}\frac{4\pi}{c}\mathbf{J}(\mathbf{r}',0) \tag{3.226}$$

$$\phi(\mathbf{r},t) = \int d^3r'\;\frac{cos(k|\mathbf{r}-\mathbf{r}'|)cos(\omega t)}{4\pi|\mathbf{r}-\mathbf{r}'|}4\pi\,\rho(\mathbf{r}',0). \tag{3.227}$$

Thus if one knows the current distribution at the source, we are able extract the fields at some point away from the source. However, in practice this is not convenient as the current distributions may be unknown. It is therefore more desirable to relate the field at a distant point to the field distribution near some aperture rather than to the source current. As we did in the case of electrostatics, here too we invoke Green's theorem. Let $\Omega$ represent a region bounded by a surface containing an aperture where the amplitude of light is nonzero. The amplitude on the rest of the surface is assumed to be zero as this is an opaque screen. This region $\Omega$ also excludes the point of interest $\mathbf{r}$ by a small sphere of radius $\varepsilon$ around it. This situation is depicted in figure 3.7.

$$\int_{\Omega} d^3r'\;(\phi(\mathbf{r}')\nabla'^2 G(\mathbf{r}'-\mathbf{r}) - G(\mathbf{r}'-\mathbf{r})\nabla'^2\phi(\mathbf{r}'))$$

$$= \int_S da\;(\phi(\mathbf{r}')\frac{\partial G(\mathbf{r}'-\mathbf{r})}{\partial n'} - G(\mathbf{r}'-\mathbf{r})\frac{\partial\phi(\mathbf{r}')}{\partial n'}) \tag{3.228}$$

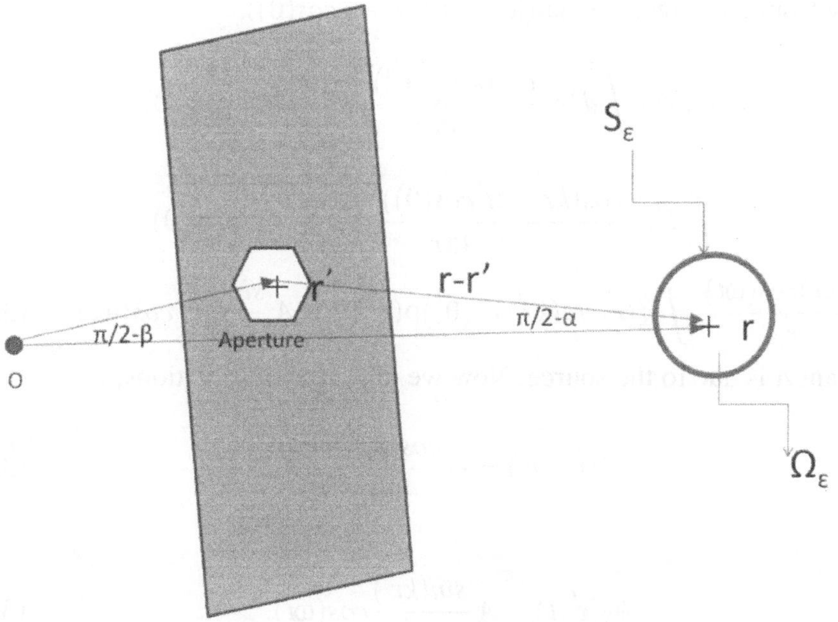


Figure 3.7: Diagram illustrating the regions involved in deriving the Kirchoff Fresnel Integral. Here $O$ is both the source and the origin of the coordinate system.

The surface $S = S_\varepsilon + S_{aperture-sheet}$. We now make use of the identity $(\nabla^2 + k^2)\phi(\mathbf{r}) = 0$ and $(\nabla^2 + k^2)G(\mathbf{r} - \mathbf{r}') = 0$. This means the left-hand side of Eq. (3.228) is zero. The right-hand side may be split into two parts,

$$0 = \int_{S_\varepsilon} da\ (\phi(\mathbf{r}')\frac{\partial G(\mathbf{r}' - \mathbf{r})}{\partial n'} - G(\mathbf{r}' - \mathbf{r})\frac{\partial \phi(\mathbf{r}')}{\partial n'})$$

$$+ \int_{S_{aperture}} da\ (\phi(\mathbf{r}')\frac{\partial G(\mathbf{r}' - \mathbf{r})}{\partial n'} - G(\mathbf{r}' - \mathbf{r})\frac{\partial \phi(\mathbf{r}')}{\partial n'}). \tag{3.229}$$

As usual the integral over the small sphere is,

$$\int_{S_\varepsilon} da\ (\phi(\mathbf{r}')\frac{\partial G(\mathbf{r}' - \mathbf{r})}{\partial n'} - G(\mathbf{r}' - \mathbf{r})\frac{\partial \phi(\mathbf{r}')}{\partial n'}) = -\phi(\mathbf{r}). \tag{3.230}$$

Thus,

$$\phi(\mathbf{r}) = \int_{S_{aperture}} da\ (\phi(\mathbf{r}')\frac{\partial G(\mathbf{r}' - \mathbf{r})}{\partial n'} - G(\mathbf{r}' - \mathbf{r})\frac{\partial \phi(\mathbf{r}')}{\partial n'}) \tag{3.231}$$

The value of the potential at the aperture may be approximated by the value without the screen. To obtain the value of the potential at the location of the aperture, we use the earlier result (Eq. (3.227)). In that equation we assume that the source $\rho(\mathbf{r}',0)$ is far away from the screen so that($|\mathbf{r}-\mathbf{r}'| \approx r - r'cos(\theta)$),

$$\phi_S(\mathbf{r},t) = \int d^3r' \frac{cos(k|\mathbf{r}-\mathbf{r}'|)cos(\omega t)}{4\pi|\mathbf{r}-\mathbf{r}'|} 4\pi\, \rho(\mathbf{r}',0)$$

$$\approx \int d^3r' \frac{cos(kr - kr'cos(\theta))cos(\omega t)}{4\pi r} 4\pi\, \rho(\mathbf{r}',0)$$

$$= \frac{sin(kr)cos(\omega t)}{r} \int d^3r'\, sin(kr'cos(\theta))\rho(\mathbf{r}',0) = A\, \frac{sin(kr)}{kr} cos(\omega t). \quad (3.232)$$

The constant $A$ is due to the source. Now we make the observations,

$$G(\mathbf{r}-\mathbf{r}') = -\frac{cos(k|\mathbf{r}-\mathbf{r}'|)}{4\pi|\mathbf{r}-\mathbf{r}'|} \quad (3.233)$$

and

$$\phi_S(\mathbf{r}',t) = A\, \frac{sin(kr')}{kr'} cos(\omega t). \quad (3.234)$$

These two relations have to be inserted into Eq. (3.231). We assume that the aperture is located in the $z = 0$ plane. We then neglect $1/r'$ in comparison with $k$. This means ($z' = \mathbf{r}' \cdot \hat{k}$, $\hat{n} = -\hat{k}$)

$$\frac{\partial G(\mathbf{r}-\mathbf{r}')}{\partial n'} \approx -\frac{k\, sin(k|\mathbf{r}-\mathbf{r}'|)}{4\pi|\mathbf{r}-\mathbf{r}'|} \frac{\partial|\mathbf{r}-\mathbf{r}'|}{\partial z'}$$

$$\frac{\partial \phi_S(\mathbf{r}',t)}{\partial n'} = -A\, \frac{cos(kr')}{r'} \frac{\partial r'}{\partial z'} \quad (3.235)$$

$$\phi(\mathbf{r}) =$$

$$\int_{S_{aperture}} da\, A\, \frac{(-sin(kr')sin(k|\mathbf{r}-\mathbf{r}'|)\frac{\partial|\mathbf{r}-\mathbf{r}'|}{\partial z'} - cos(k|\mathbf{r}-\mathbf{r}'|)cos(kr')\frac{\partial r'}{\partial z'})}{4\pi r'|\mathbf{r}-\mathbf{r}'|}. \quad (3.236)$$

The above is a version of the Fresnel-Kirchoff integral for the diffraction produced by an aperture illuminated by a point source. The novel features are the angles—$cos(\alpha) = \frac{\partial|\mathbf{r}-\mathbf{r}'|}{\partial z'} = \frac{z'-z}{|\mathbf{r}-\mathbf{r}'|}$ is the cosine of the angle made by a ray from a point on the aperture to the point of interest (screen) and $cos(\beta) = \frac{\partial r'}{\partial z'} = \frac{z'}{r'}$ is the cosine of the angle made by a ray from the source to a point on the aperture. A limiting case of the above formula is known as Fraunhofer diffraction when both the distance from the source to the aperture and the screen to the aperture are large compared to the dimensions of the aperture. Assuming that the aperture lies in the x-y plane, we have,

$$r' = (x'^2 + y'^2 + z'^2)^{\frac{1}{2}} = z'\ (1 + \frac{x'^2 + y'^2}{z'^2})^{\frac{1}{2}} \approx (z' + \frac{x'^2 + y'^2}{2z'})$$

$$|\mathbf{r} - \mathbf{r}'| = ((x - x')^2 + (y - y')^2 + (z - z')^2)^{\frac{1}{2}} \approx (z - z') + \frac{(x - x')^2 + (y - y')^2}{2(z - z')}. \tag{3.237}$$

This means that in case of Fraunhofer diffraction, we may also assume the angles involved are small ($cos(\alpha) \approx -1$ and $cos(\beta) \approx 1$),

$$\phi(\mathbf{r}) = -\int_{S_{aperture}} da\, A\, \frac{cos(kz + k\frac{x'^2+y'^2}{2z'} + k\frac{(x-x')^2+(y-y')^2}{2(z-z')})}{4\pi z'|z - z'|}. \tag{3.238}$$

The aperture size is typically small enough so that we may assume that $|x'| \ll |x|$ and $|y'| \ll |y|$. This means,

$$\phi(\mathbf{r}) \approx -Re\frac{A}{4\pi z'|z - z'|}\ e^{ikz} e^{ik\frac{x^2+y^2}{2(z-z')}} \iint_{S_{aperture}} dx'dy'\, e^{-ik\frac{xx'+yy'}{(z-z')}}. \tag{3.239}$$

The vector potential has a similar expression. The vector potential only provides information about the state of polarization of light and has no direct role to play here. Hence, most treatments just treat the fields as scalar quantities. The above expression Eq. (3.239) is the well-known Fraunhofer diffraction pattern. It is nothing but the Fourier transform of the aperture function.

■ Calculate the Fraunhofer diffraction pattern from a square aperture of width $W$. In this case the integration extends from $-\frac{W}{2}, \frac{W}{2}$.

$$\phi(\mathbf{r}) \approx -Re\frac{A}{4\pi z'|z - z'|}\ e^{ikz} e^{ik\frac{x^2+y^2}{2(z-z')}} \int_{-\frac{W}{2}}^{\frac{W}{2}} dx' \int_{-\frac{W}{2}}^{\frac{W}{2}} dy'\, e^{-ik\frac{xx'+yy'}{(z-z')}}$$

$$= -Re\frac{A}{4\pi z'|z - z'|}\ e^{ikz} e^{ik\frac{x^2+y^2}{2(z-z')}} \int_{-\frac{W}{2}}^{\frac{W}{2}} dx'\, e^{-ik\frac{x}{(z-z')}x'} \int_{-\frac{W}{2}}^{\frac{W}{2}} dy'\, e^{-ik\frac{y}{(z-z')}y'} \tag{3.240}$$

This means,

$$\phi(\mathbf{r}) \approx -|z - z'|\frac{A}{\pi z'}\ cos(kz)\ \frac{sin(k\frac{xW}{2(z-z')})sin(k\frac{yW}{2(z-z')})}{k^2xy}. \tag{3.241}$$

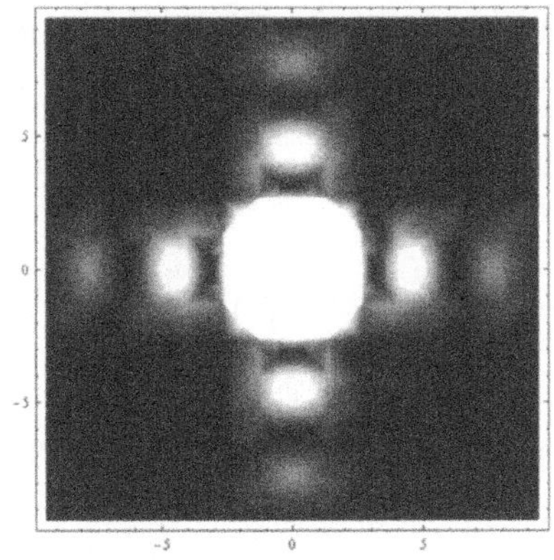


Figure 3.8: Density plot shows the Fraunhofer intensity distribution from a square aperture; the brighter colors have higher intensity.

The above equation gives the diffracted field at distant points from the aperture, assuming the source is also far from the aperture and the screen. We may also study the reverse situation where these distances may not be considered large to enable the approximations just made. This is called Fresnel diffraction. One of its main predictions is that the center of the shadow of a circular opaque disk is not dark but bright. This is called Arago's spot and confirms the wave nature of light. A description of this is left to the exercises.

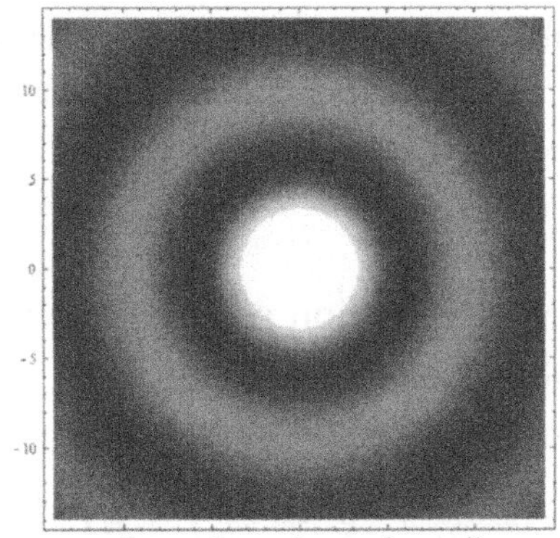


Figure 3.9: The shadow of an opaque circular planar object illuminated by coherent light is not dark but has a bright spot (or disk, here) at the center, known as Arago's spot. This confirms the wave nature of light.

■ X-ray diffraction in crystals: Imagine a plane wave (X-ray) with vector **k**

$$\psi_{inc}(\mathbf{r},t) = A\, e^{i(\mathbf{k}\cdot\mathbf{r}-\omega t)} \tag{3.242}$$

incident on an atom located at $\mathbf{R_n}$. The scattered amplitude $\psi_{scatt}(\mathbf{r},t)$ emerging from the atom is proportional to the incident amplitude at the location of the atom viz. $\psi_{inc}(\mathbf{R_n},t)$, and has the form of a spherical wave.

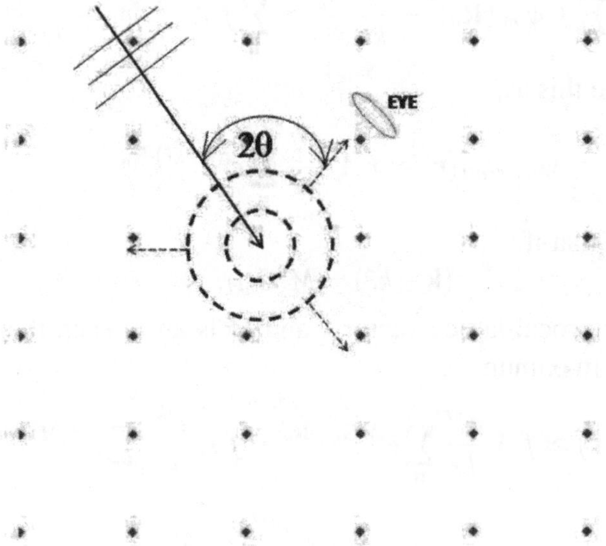


Figure 3.10: Bragg's Law: Diffraction of plane wave X-rays from atoms in a crystal.

$$\psi_{scatt}(\mathbf{r},t) = f\psi_{inc}(\mathbf{R_n},t)\frac{e^{ik|\mathbf{r}-\mathbf{R_n}|}}{k|\mathbf{r}-\mathbf{R_n}|} \tag{3.243}$$

Here $f$ is called the atomic form factor, which determines the strength of the scattering. The term 'form' factor is also indicative of this being potentially a function of the angles which reflects an angular distribution (or lack thereof) of charges in the atom. In this example, we consider only the isotropic case. If there are many atoms scattering, all the scattered waves add up to give,

$$\psi_{tot,scatt}(\mathbf{r},t) = \sum_{\mathbf{n}} f\,\psi_{inc}(\mathbf{R_n},t)\frac{e^{ik|\mathbf{r}-\mathbf{R_n}|}}{k|\mathbf{r}-\mathbf{R_n}|}. \tag{3.244}$$

It may be verified that each of the equations Eq. (3.242), Eq. (3.243), and Eq. (3.244) obeys the wave equation (Homework: Show this),

$$\nabla^2\psi = \frac{1}{c^2}\frac{\partial^2\psi}{\partial t^2} \tag{3.245}$$

since $\omega = ck$. Now imagine that the point $\mathbf{r}$ on the screen is far away from the crystal. This means $|\mathbf{r}| \gg |\mathbf{R_n}|$. Therefore,

$$|\mathbf{r} - \mathbf{R_n}| \approx (r^2 - 2\mathbf{r}\cdot\mathbf{R_n})^{\frac{1}{2}} \approx r - \hat{r}\cdot\mathbf{R_n}. \tag{3.246}$$

In the denominator of Eq. (3.244) we may simply write $|\mathbf{r} - \mathbf{R_n}| \approx r$. But in the phase part we have to be more careful. Since the time part is a simple exponential we simply drop it (the modulus squared of the amplitude, which is the intensity, will not involve it)

$$\psi_{tot,scatt}(\mathbf{r}) \approx \sum_{\mathbf{n}} f\ \psi_{inc}(\mathbf{R}_n)\frac{e^{ik(r-\hat{r}\cdot\mathbf{R_n})}}{kr} \approx \sum_{\mathbf{n}} f\ A\ e^{i\mathbf{k}\cdot\mathbf{R_n}}\frac{e^{ik(r-\hat{r}\cdot\mathbf{R_n})}}{kr}. \tag{3.247}$$

We may now rewrite this as,

$$\psi_{tot,scatt}(\mathbf{r}) \approx f\ A\ \frac{e^{ikr}}{kr}\sum_{\mathbf{n}} e^{i(\mathbf{k}-k\hat{r})\cdot\mathbf{R_n}}. \tag{3.248}$$

From this it is clear that if

$$(\mathbf{k} - k\hat{r}) = M\ \mathbf{G}_{hkl}, \tag{3.249}$$

where $\mathbf{G}_{hkl}$ is a reciprocal lattice vector [1] and $M$ is an integer. then the scattered amplitude will be a maximum.

$$\psi_{tot,scatt}(\mathbf{r}) \approx f\ A\ \frac{e^{ikr}}{kr}\sum_{\mathbf{n}} e^{iM\mathbf{G}_{hkl}\cdot\mathbf{R_n}} = f\ A\ \frac{e^{ikr}}{kr}\sum_{\mathbf{n}} e^{i2\pi M\times integer}$$

$$= N_{tot} f\ A\ \frac{e^{ikr}}{kr} \tag{3.250}$$

where $N_{tot}$ is the total number of atoms in the crystal. Eq. (3.249) is known as Laue's criterion. If the angle between the observation direction $\hat{r}$ and the incident direction $\hat{k}$ is $2\theta$, we may square Eq. (3.249) to get,

$$(k^2 + k^2 - 2k^2 cos(2\theta)) = 4k^2\ sin^2(\theta) = M^2\ |\mathbf{G}_{hkl}|^2. \tag{3.251}$$

This means

$$2k\ sin(\theta) = M|\mathbf{G}_{hkl}|. \tag{3.252}$$

Multiply both sides by $\lambda = \frac{2\pi}{k}$ and divide by $|\mathbf{G}_{hkl}|$ to get

$$2d_{hkl}\ sin(\theta) = M\lambda, \tag{3.253}$$

where $d_{hkl} = \frac{2\pi}{|\mathbf{G}_{hkl}|}$ is the distance between adjacent lattice planes. The above relation is called Bragg's Law of X-ray diffraction. In the exercises we see some applications of this law.

[1]Points on a lattice may be written as $\mathbf{R_n} = n\mathbf{a} + m\mathbf{b} + p\mathbf{c}$, where $n, m, p$ are integers and $\mathbf{a}, \mathbf{b}$, and $\mathbf{c}$ are non-coplanar basis vectors. Also $\mathbf{G}_{hkl} = h\mathbf{A} + k\mathbf{B} + l\mathbf{C}$, with $h, k, l$ integers, is reciprocal to $\mathbf{R}_n$ in the sense that $\mathbf{G}_{hkl}\cdot\mathbf{R}_n = 2\pi \times integer$. To ensure this, we must have $\mathbf{A} = 2\pi\frac{\mathbf{b}\times\mathbf{c}}{\mathbf{a}\cdot\mathbf{b}\times\mathbf{c}}$ and cyclic permutations thereof.

## 3.6 Exercises

**Q.1** Show that **E.B** is a Lorentz invariant.

**Q.2** By expanding the vector potential in a plane wave basis

$$\mathbf{A}(\mathbf{r},t) = Re[\sum_{\mathbf{k}} \mathbf{a}(\mathbf{k})\ e^{i(\mathbf{k}.\mathbf{r}-\omega_k t)}], \tag{3.254}$$

show that the vector $\mathbf{P} = \int d^3r \partial_t \mathbf{A}(\mathbf{r},t) \times \mathbf{A}(\mathbf{r},t)$ vanishes identically.

**Q.3** Show that the hybrid spherical plane wave

$$\psi(\mathbf{r},t) = A\ e^{i(kr-\omega t)} \tag{3.255}$$

and the plane wave with inverse square law

$$\psi(\mathbf{r},t) = A\ \frac{e^{i(\mathbf{k}\cdot\mathbf{r}-\omega t)}}{r} \tag{3.256}$$

do not obey the wave equation in three dimensions and hence are not acceptable as wave descriptions.

**Q.4** Verify that Eq. (3.77) follows from the earlier steps.

**Q.5** Using the image method, find the electric potential in a region enclosed by two conducting grounded wedges if a charge $q$ resides at a distance $d$ from the apex as shown.

**Q.6** Consider the possibility of generalizing the notion of an equipotential surface to magnetostatics. This would correspond to a surface on which the vector potential is a constant vector. Imagine a sphere of radius $R$ on which the vector potential is $\mathbf{A}_0$— a constant vector on the upper hemisphere and is zero on the lower hemisphere. Find the vector potential elsewhere outside the sphere.

**Q.7** Imagine a charge $q$ that is moving in a circle of fixed radius $R$ with frequency $\omega$, find the electric and magnetic fields outside the charge (taking into account retardation effects).

**Q.8** Imagine an (ideal) electric dipole $\vec{p}$ situated at the origin that is reversing orientation after each half-time period $T/2$. Find an expression for the electric and magnetic fields at a distance from the dipole taking into account retardation effects. Hint: First prove that the charge density of an ideal dipole situated at the origin is $\rho(\mathbf{r},t) = (\mathbf{p}(t)\cdot\nabla)\delta^3(\mathbf{r})$ with $\mathbf{p}(t) = p(cos(\frac{2\pi t}{T})\hat{k} + sin(\frac{2\pi t}{T})\hat{j})$.

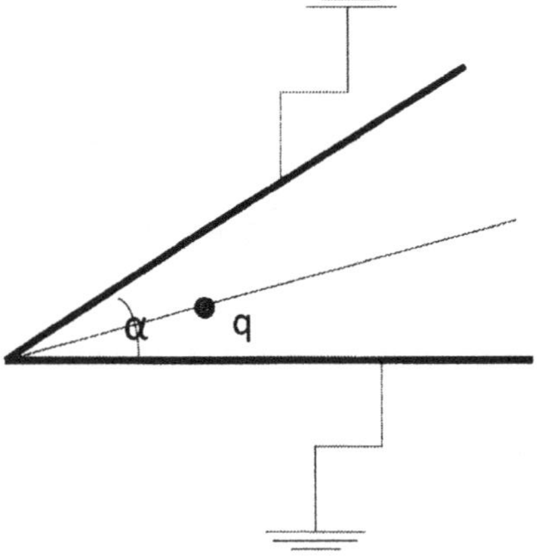


Figure 3.11: Find the potential in the region enclosed by the grounded conducting wedge.

**Q.9** Imagine a rod of rest length $l_0$ and charge density $\lambda$, accelerating in a straight line in such a way that the direction of motion and the end points are collinear. Find the electric and magnetic field everywhere taking into account retardation effects.

**Q.10** Find the intensity distribution from two mutually perpendicular infinitely long thin slits illuminated by a monochromatic coherent source far from the screen.

**Q.11** The Arago spot, Fresnel bright spot, or Poisson spot is an illuminated point that appears at the center of a circular object's shadow due to Fresnel diffraction. Use the most general form of the Kirchoff Fresnel formula (Fresnel theory) to prove this. Hint: The answer is a Bessel function.

**Q.12** Graphene is a two-dimensional sheet of carbon atoms arranged in a honeycomb structure. This lattice can be thought of as two interpenetrating triangular lattices. Lattice $A$ contains carbon atoms at locations

$$\vec{r}^{\,A}(n_1, n_2) = n_1\vec{a}_1 + n_2\vec{a}_2, \tag{3.257}$$

$\vec{a}_1 = \sqrt{3}a_{cc}\,(\frac{\sqrt{3}}{2}\hat{i} + \frac{1}{2}\hat{j})$, and $\vec{a}_2 = \sqrt{3}a_{cc}\,\hat{j}$, and lattice $B$ also contains carbon atoms at locations

$$\vec{r}^{\,B}(n_1, n_2) = a_{cc}\hat{i} + n_1\vec{a}_1 + n_2\vec{a}_2 \tag{3.258}$$

where $a_{cc} = 1.42$ Å is the nearest carbon–carbon distance and $n_1, n_2 = 0, \pm 1, \pm 2, ...$ are integers. Draw a neat labeled diagram of this lattice showing lattice A and lattice

B atoms, the basis vectors, and the angles and so on. According to Bragg's Law, a bright spot will be seen at an angle $2\theta_{hk}$ between the incident direction and the direction of observation, whenever

$$2d_{hk}\,sin(\theta_{hk}) = \lambda. \tag{3.259}$$

Find $d_{hk}$, the distance between consecutive Bragg lines (not planes) in this two-dimensional Graphene lattice. If the X-ray wavelength is $\lambda = a_{cc}$, find the values of $\theta_{hk}$ for $(h,k) = (1,0)$ and $(h,k) = (1,-1)$. Show that the bright spots have an intensity equal proportional to

$$I_{hk} \sim cos^2\left(\frac{1}{2}\mathbf{G}_{hk} \cdot \mathbf{t}\right), \tag{3.260}$$

where $\mathbf{t}$ is the distance vector between sublattice $A$ and sublattice $B$. Find $\mathbf{G}_{hk}$ and $\mathbf{d}$ and the ratio of the intensities between the bright spot at $(h,k) = (1,0)$ and the bright spot at $(h,k) = (1,-1)$.

# Chapter 4

# Elasticity Theory and Fluid Mechanics

This chapter deals with the description of elastic bodies. Elasticity is a phenomenon wherein a body responds to external forces not (only) by accelerating (as a rigid body would have done) but by deforming—changing its shape or size or both. Elasticity, unlike plasticity, is reversible. Elastic objects that change their shape or size in response to external forces (or 'stresses') revert back to their original state upon removal of those forces. Plastic objects on the other hand do not. A fluid is an elastic material that flows. This means that the material will suffer acceleration whenever anisotropic stresses are applied. We first study elasticity theory followed by fluid mechanics. These descriptions are meant to be simple introductions only. The reader is referred to the bibliography for more specialized treatises.

## 4.1 Stress and Strain Tensor of Deformable Bodies

In this section, we discuss stress and strain in (elastically) deformable bodies. Stress is nothing but the force per unit area acting at some point on a chosen surface within a deformable body. Strain is response to stress, namely deformation of the body. Quantitatively it is a ratio between the deformation length and the original length (either in the same direction or in a perpendicular direction). We now go on to provide more details.

### 4.1.1 Stress

Imagine a point inside a deformable body. Think of an imaginary surface that passes through this point. On this imaginary surface we may imagine are forces (per unit area) acting both perpendicular to it and parallel to it (parallel has two independent components). The perpendicular force per unit area is called normal stress and the two components parallel to the surface constitute shear stresses. For a surface whose unit normal is along the z-axis, we may denote $\sigma_{zz}$ as the normal stress and $\sigma_{zx}$ and $\sigma_{zy}$ as the two components of shear stress. Thus, in general, stress is a tensor ($3 \times 3$ matrix) that changes from point to point : $\sigma_{ij}(\mathbf{x})$. The force acting on a surface whose unit normal is $\hat{n}$ and area is $dA$ is given by $d\mathbf{F} = dA\ \hat{n} \cdot \sigma(\mathbf{x})$ (matrix times a vector is a vector). Now imagine there is a closed surface. The net force acting on that surface is

$$\mathbf{F}_{stress} = \oint_S d\mathbf{F} = \oint_S dA\ \hat{n} \cdot \sigma(\mathbf{x}). \tag{4.1}$$

The net force due to stress should be compensated by a bulk force in order to maintain equilibrium. We postulate that a force per unit volume $\mathbf{f}_b$ also acts on the volume so that

$$\int \mathbf{f}_b\ dV = -\mathbf{F}_{stress}. \tag{4.2}$$

We now apply Gauss's theorem on the right-hand side of Eq. (4.1) to get,

$$\mathbf{F}_{stress} = \oint_S d\mathbf{F} = \oint_S dA\ \hat{n} \cdot \sigma(\mathbf{x}) = \int dV\ \nabla \cdot \sigma(\mathbf{x}). \tag{4.3}$$

Combining this with Eq. (4.2) we get, $\int dV\ \ (\nabla \cdot \sigma(\mathbf{x}) + \mathbf{f}_b) = 0$, but since the region can be anything, we must have,

$$\nabla \cdot \sigma(\mathbf{x}) + \mathbf{f}_b(\mathbf{x}) = 0. \tag{4.4}$$

(Note that since $\sigma$ is a matrix, $\nabla \cdot \sigma$ is a vector).

Therefore, in order to determine stress with given body forces, we have to solve for $\sigma$ using Eq. (4.4) together with the boundary condition

$$\hat{n} \cdot \sigma(\mathbf{x})\,|_{\mathbf{x} \in S} = \mathbf{p}_S \tag{4.5}$$

where $\mathbf{p}_S = \frac{d\mathbf{F}}{dA}$ is the force per unit area acting on the surface $S$ of the body which has unit normal $\hat{n}$. Thus, given $\mathbf{f}_b(\mathbf{x})$ the body force per unit volume, $\mathbf{p}_S$ the surface force per unit area, and the unit normal $\hat{n}$ to the surface $S$ bounding the body, the unknown, namely, the stress tensor $\sigma(\mathbf{x})$ may be determined. The stress equation (Eq. (4.4)) and the boundary conditions Eq. (4.5) have to be solved together with given body forces and surface forces to determine the stress. This is typically done in conjunction with the strain function where the stress–strain relation and the nature of the strain function imposes further limitations on the possible forms of the stress function. We shall discuss these issues subsequently.

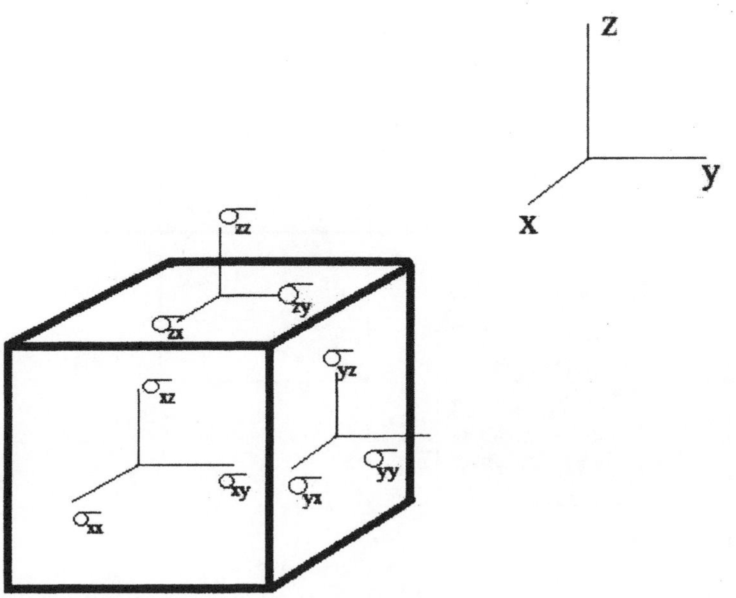


Figure 4.1: Diagram showing the meaning of normal and shear stress.

### 4.1.2 Strain

Imagine that each point **x** in a body undergoes a displacement given by a displacement vector $\mathbf{D}(\mathbf{x}) \equiv (u_x, u_y, u_z)$. The problem of elasticity theory is to find this vector using some reasonable assumptions about the material in question. To this end we define strain as

$$\varepsilon_{i,j} = \frac{1}{2}\left(\frac{\partial u_i}{\partial x_j} + \frac{\partial u_j}{\partial x_i}\right), \tag{4.6}$$

where $i, j = x, y, z$. Here, just like stress, the strain $\varepsilon_{i,j}$ is a $3 \times 3$ matrix. It is desirable to get an intuitive feeling for what this quantity means. For example,

$$\varepsilon_{xx} = \frac{\partial u_x}{\partial x} \tag{4.7}$$

represents the the ratio of the change $du_x$ (in the x-direction) in the distance between two points that were originally separated by a distance $dx$. This is called normal strain. This is made clear in the diagram (Fig. 4.2). The off-diagonal part is nothing but the sum of the angles $\alpha$ and $\beta$. Thus the off-diagonal $\varepsilon_{xy}$ represents change in the shape of the parallelogram, whereas $\varepsilon_{xx}$ represents the change in size of the parallelogram. Thus the strain tensor encodes both the change in size and change in shape of the object.

For isotropic linear materials, the strain tensor should be relatable to the applied stress tensor (tensor means matrix) $\sigma_{i,j}$. In this case we write that strain is propor-

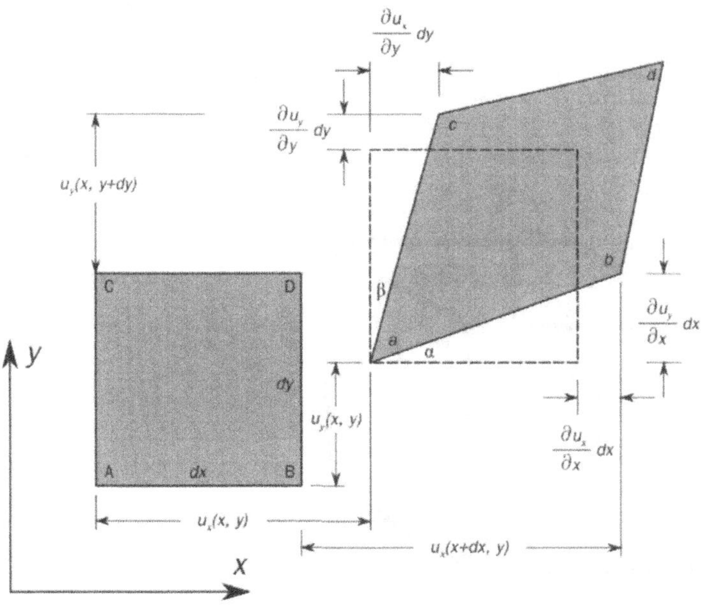


Figure 4.2: Depicts the definitions of normal and shear strain (source: Wikipedia).

tional to stress.

$$\varepsilon_{i,j} = a\,\sigma_{i,j} + b\,\sigma\,\delta_{i,j} \tag{4.8}$$

Here $\sigma = \sum_{i=1}^{3} \sigma_{i,i}$ is a scalar and $a, b$ are material-dependent but stress-independent constants. The above expression Eq. (4.8) ensures that if I double each component of the stress (there are nine components), then each component of the strain also doubles. We have to determine $a$ and $b$. To do this, imagine that there is only one component of stress $\sigma_{xx}$ and all others are zero. The material will expand in the x-direction but compress in the $y$ and $z$ directions as a result of the stress ($\sigma = \sigma_{xx}$ in this case).

$$\varepsilon_{x,x} = a\,\sigma_{x,x} + b\,\sigma_{xx} \tag{4.9}$$

$$\varepsilon_{y,y} = b\,\sigma_{xx}\ ;\ \varepsilon_{z,z} = b\,\sigma_{xx} \tag{4.10}$$

Young's modulus $E$ is defined as (stress upon strain),

$$E = \frac{\sigma_{xx}}{\varepsilon_{xx}}. \tag{4.11}$$

Poisson's ratio is defined as the ratio of the strains in two orthogonal directions.

$$\nu = -\frac{\varepsilon_{y,y}}{\varepsilon_{x,x}} \tag{4.12}$$

The minus sign indicates stretching in one direction also means compression in a lateral direction (not by the same amount but a fraction $\nu$). Thus,

$$\varepsilon_{x,x} = (a+b)\,\sigma_{x,x} = \frac{1}{E}\,\sigma_{x,x} \tag{4.13}$$

$$-\nu\, \varepsilon_{x,x} = \varepsilon_{y,y} = b\, \sigma_{x,x} = b\, E\, \varepsilon_{x,x} \tag{4.14}$$

$$\varepsilon_{z,z} = b\, \sigma_{xx}. \tag{4.15}$$

Thus $b = -\frac{\nu}{E}$ and $a + b = \frac{1}{E}$. This means $a = \frac{(1+\nu)}{E}$. Therefore the stress-strain relation becomes

$$\varepsilon_{i,j} = \frac{(1+\nu)}{E}\sigma_{i,j} - \frac{\nu}{E}\sigma\, \delta_{i,j}. \tag{4.16}$$

Thus the fundamental equations of linear elasticity theory are the relation between stress and body force (Eq. (4.4)), the surface force equation Eq. (4.5), the relation between strain and displacements Eq. (4.6), and the linear stress–strain relation (which is an approximation valid only for linear isotropic materials) Eq. (4.16).

### 4.1.3 The Stress Function Method

Here we discuss various methods used to solve for the strain with given body and surface forces. In two dimensions, a systematic method, namely the stress function method may be used. Of course, what we have in mind is not really a two-dimensional object but a problem where there is symmetry along one of the directions so that variations of strain along that direction may be ignored. This situation is known as plane strain (plane stress is another different possibility, which is relegated to the exercises). This is common, for example, in situations such as a wall with lateral pressure applied on it, a tunnel or a cylindrical tube with internal pressure, and so on. First we assume that the body forces are derivable from a potential $\mathbf{f}_b = -\nabla V$. The stress–body force equation in two dimensions becomes,

$$\frac{\partial \sigma_{xx}}{\partial x} + \frac{\partial \sigma_{yx}}{\partial y} = \frac{\partial V}{\partial x} \tag{4.17}$$

$$\frac{\partial \sigma_{xy}}{\partial x} + \frac{\partial \sigma_{yy}}{\partial y} = \frac{\partial V}{\partial y}. \tag{4.18}$$

An important special case is when $V(x,y) = \rho g y$, namely the gravitational potential. The following substitutions automatically obey these equations.

$$\sigma_{xx}(x,y) = V(x,y) + \frac{\partial^2 \phi(x,y)}{\partial y^2} \tag{4.19}$$

$$\sigma_{yy}(x,y) = V(x,y) + \frac{\partial^2 \phi(x,y)}{\partial x^2} \tag{4.20}$$

$$\sigma_{xy}(x,y) = -\frac{\partial^2 \phi(x,y)}{\partial x \partial y}. \tag{4.21}$$

In order to derive an equation for $\phi$, we must use the constraint provided by a requirement that the strain be related to derivatives of a displacement function. Therefore,

$$\varepsilon_{xx} = \frac{\partial u_x}{\partial x} \; ; \; \varepsilon_{yy} = \frac{\partial u_y}{\partial y} \; ; \; \varepsilon_{xy} = \frac{1}{2}\left(\frac{\partial u_x}{\partial y} + \frac{\partial u_y}{\partial x}\right). \tag{4.22}$$

From this it is easy to verify that,

$$\frac{\partial^2 \varepsilon_{xx}}{\partial y^2} + \frac{\partial^2 \varepsilon_{yy}}{\partial x^2} = 2\frac{\partial^2}{\partial x \partial y}\varepsilon_{xy}. \tag{4.23}$$

The above equation is known as a compatibility condition. It is compatible with the requirement that the components of the strain tensor be related to derivatives of displacements. Now we derive the relation between stress and strain in the special situation of plane strain. In this situation, the normal strain in the z-direction is zero. From Eq. (4.16) we get,

$$0 = \varepsilon_{zz} = \frac{(1+\nu)}{E}\sigma_{zz} - \frac{\nu}{E}(\sigma_{xx} + \sigma_{yy} + \sigma_{zz}). \tag{4.24}$$

This means the normal stress in the z-direction is proportional to the sum of the normal stresses in the other two directions.

$$\sigma_{zz} = \nu(\sigma_{xx} + \sigma_{yy}) \tag{4.25}$$

Substituting these into Eq. (4.16) we obtain the relation between stress and strain for plane strain.

$$\varepsilon_{xx} = \frac{1}{E}((1-\nu^2)\sigma_{xx} - \nu(1+\nu)\sigma_{yy}) \; ; \; \varepsilon_{yy} = \frac{1}{E}((1-\nu^2)\sigma_{yy} - \nu(1+\nu)\sigma_{xx}) \tag{4.26}$$

$$\varepsilon_{xy} = \frac{(1+\nu)}{E}\sigma_{xy} \tag{4.27}$$

Substituting these into the compatibility condition gives us

$$\begin{aligned} &\frac{\partial^2}{\partial y^2}\frac{1}{E}((1-\nu^2)\sigma_{xx} - \nu(1+\nu)\sigma_{yy}) \\ &+\frac{1}{E}\frac{\partial^2}{\partial x^2}((1-\nu^2)\sigma_{yy} - \nu(1+\nu)\sigma_{xx}) \\ &= 2\frac{\partial^2}{\partial x \partial y}\frac{(1+\nu)}{E}\sigma_{xy}. \end{aligned} \tag{4.28}$$

The idea now is to eliminate the shear stress and write the above equation only in terms of the normal stresses and body forces. To this end, we differentiate Eqs.

(4.18), the first with respect to $x$ and the second with respect to $y$ and adding we get,

$$\frac{\partial^2\sigma_{xx}}{\partial x^2}+\frac{\partial^2\sigma_{yy}}{\partial y^2}+2\frac{\partial^2\sigma_{yx}}{\partial y\partial x}=\frac{\partial^2 V}{\partial x^2}+\frac{\partial^2 V}{\partial y^2}. \tag{4.29}$$

Substituting the above formula for the mixed partial derivatives of the shear strain into Eq. (4.28) we get,

$$(1-\nu)(\frac{\partial^2}{\partial x^2}+\frac{\partial^2}{\partial y^2})(\sigma_{xx}+\sigma_{yy})=\frac{\partial^2 V}{\partial x^2}+\frac{\partial^2 V}{\partial y^2}. \tag{4.30}$$

Thus the normal stresses have been related to the body forces. We may now substitute the expression for these stresses in terms of the stress function $\phi$ to get the plane strain equation for the stress function,

$$(1-\nu)(\frac{\partial^2}{\partial x^2}+\frac{\partial^2}{\partial y^2})^2\phi(x,y)=-(1-2\nu)(\frac{\partial^2 V}{\partial x^2}+\frac{\partial^2 V}{\partial y^2}). \tag{4.31}$$

■ When body forces are absent or constant, the stress equation simply becomes,

$$(\frac{\partial^2}{\partial x^2}+\frac{\partial^2}{\partial y^2})^2\phi(x,y)=0. \tag{4.32}$$

For objects with a rectangular cross section, the solution to this is in terms of suitable polynomials in $x$ and $y$. For example, let us investigate what situation the choice $\phi(x,y)=-\frac{F}{d^3}xy^2(3d-2y)$ corresponds to. It is given that the object occupies a region $0\leq y\leq d$ and $x\geq 0$. This function clearly obeys the stress equation. It is further given that body forces are absent. This means,

$$\sigma_{xx}(x,y)=\frac{\partial^2\phi(x,y)}{\partial y^2}=-\frac{6F}{d^2}x+\frac{12F}{d^3}xy \tag{4.33}$$

$$\sigma_{yy}(x,y)=\frac{\partial^2\phi(x,y)}{\partial x^2}=0 \tag{4.34}$$

$$\sigma_{xy}(x,y)=-\frac{\partial^2\phi(x,y)}{\partial x\partial y}=\frac{6F}{d^3}y(y-d). \tag{4.35}$$

Normal stress in the x-direction vanishes at $x=0$ and at $y=\frac{d}{2}$, the shear stress vanishes at the ends $y=0$ and $y=d$. If one wishes to find the forces acting on the surface $y=0$, the unit outward normal would be $-\hat{j}$. The force per unit area acting on this surface would be,

$$\mathbf{F}(y=0)=-\sigma_{yx}(x,0)\hat{\imath}-\sigma_{yy}(x,0)\hat{\jmath}=0. \tag{4.36}$$

Similarly, on the surface $y = d$ the force is also zero. The situation is such that the part of the cross section $0 < y < d/2$ is being pushed in the negative x-direction, whereas the part $d/2 < y < d$ is being pulled in the positive x-direction (plot $\sigma_{xx}$ vs. $y$).

■ Imagine a cuboid of sides $L_x, L_y, L_z$ made of an isotropic linear elastic material. We wish to determine its shape as a result of the deformation it suffers on account of its weight. The body force is the weight per unit volume $\mathbf{f}_b = -\frac{Mg}{V}\hat{k}$, where $V = L_xL_yL_z$. Therefore,

$$\partial_x\sigma_{xz} + \partial_y\sigma_{yz} + \partial_z\sigma_{zz} - \frac{Mg}{V} = 0 \tag{4.37}$$

$$\partial_x\sigma_{xx} + \partial_y\sigma_{yx} + \partial_z\sigma_{zx} = \partial_x\sigma_{xy} + \partial_y\sigma_{yy} + \partial_z\sigma_{zy} = 0 \tag{4.38}$$

at each point inside the material. At this stage it is clear that of the nine components, all the shear components of stress and strain vanish in this problem as this is a question involving only normal stress and strain. Also the component of the stress $\hat{n}\cdot\sigma$ on the bottom surface ($\hat{n} = -\hat{k}$) is nothing but $-\sigma_{zz}\hat{k}$ should be the force acting on the bottom surface per unit area $\mathbf{p}_S = \hat{k}\frac{Mg}{L_xL_y}$ assuming it is uniform.

$$\sigma_{zz}(z = 0) = -\frac{Mg}{L_xL_y} \tag{4.39}$$

Let us assume that all other components of the stresses vanish on the surfaces. This means the unique solution to the stress question is

$$\sigma_{zz}(z) = \frac{Mg}{L_xL_y}(\frac{z}{L_z} - 1), \tag{4.40}$$

and all other components of the stress vanish identically. Using the stress-strain relation we get,

$$\varepsilon_{xx} = \varepsilon_{yy} = -\frac{\nu}{E}\,\sigma \tag{4.41}$$

$$\varepsilon_{zz} = \frac{(1+\nu)}{E}\,\sigma_{zz} - \frac{\nu}{E}\,\sigma. \tag{4.42}$$

But $\sigma = 0 + 0 + \sigma_{zz}$.

$$\frac{\partial u_x}{\partial x} = \frac{\partial u_y}{\partial y} = -\frac{\nu}{E}\frac{Mg}{L_xL_y}(\frac{z}{L_z} - 1) \tag{4.43}$$

$$\frac{\partial u_z}{\partial z} = \frac{1}{E}\frac{Mg}{L_xL_y}(\frac{z}{L_z} - 1) \tag{4.44}$$

These may be integrated to yield,

$$u_z = \frac{1}{E}\frac{Mg}{L_xL_y}\left(\frac{z^2}{2L_z} - z\right) + c_1(x,y) \tag{4.45}$$

$$u_x = -\frac{\nu}{E}\frac{Mg}{L_xL_y}\left(\frac{xz}{L_z} - x\right) + c_2(y,z) \tag{4.46}$$

$$u_y = -\frac{\nu}{E}\frac{Mg}{L_xL_y}\left(\frac{zy}{L_z} - y\right) + c_3(x,z). \tag{4.47}$$

Since shear strains are zero, we must have

$$\varepsilon_{xy} = \frac{1}{2}\left(\frac{\partial u_x}{\partial y} + \frac{\partial u_y}{\partial x}\right) = 0; \tag{4.48}$$

$$\varepsilon_{yz} = \frac{1}{2}\left(\frac{\partial u_y}{\partial z} + \frac{\partial u_z}{\partial y}\right) = 0; \tag{4.49}$$

$$\varepsilon_{zx} = \frac{1}{2}\left(\frac{\partial u_z}{\partial x} + \frac{\partial u_x}{\partial z}\right) = 0, \tag{4.50}$$

or

$$c_{2,y}(y,z) + c_{3,x}(x,z) = 0; \quad -\frac{\nu}{E}\frac{Mgy}{V} + c_{3,z}(x,z) + c_{1,y}(x,y) = 0; \tag{4.51}$$

$$c_{1,x}(x,y) - \frac{\nu}{E}\frac{Mgx}{V} + c_{2,z}(y,z) = 0. \tag{4.52}$$

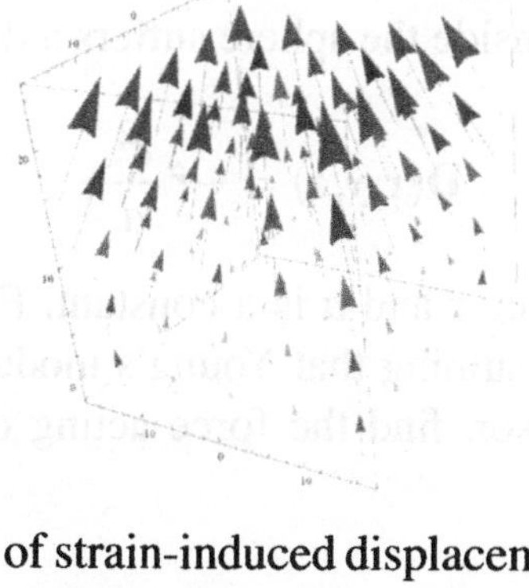

Figure 4.3: The distribution of strain-induced displacement in the material is shown. All arrows with the same color have the same magnitude of displacement.

This means that $c_3(x,z) = xf(z) + g(z)$ and $c_2(y,z) = -yf(z) + h(z)$. Further $f(z) = f'(0)z + f(0)$, $g(z) = g'(0)z + g(0)$ and $h(z) = h'(0)z + h(0)$. Also we impose the condition that the line $x = y = 0$ displaces vertically but not sideways. This means $u_x(x = y = 0) = u_y(x = y = 0) = 0$. So that $g(z) \equiv h(z) \equiv 0$.

$$-\frac{\nu}{E}\frac{Mgy}{V} + (xf^{'}(0)) + c_{1,y}(x,y) = 0; \tag{4.53}$$

$$c_{1,x}(x,y) - \frac{\nu}{E}\frac{Mgx}{V} + (-yf^{'}(0)) = 0 \tag{4.54}$$

Since $c_{1,x,y} = c_{1,y,x}$, we must also have $f^{'}(0) = 0$.

$$-\frac{\nu}{E}\frac{Mgy}{V} + c_{1,y}(x,y) = 0; \; c_{1,x}(x,y) - \frac{\nu}{E}\frac{Mgx}{V} = 0 \tag{4.55}$$

Thus,

$$c_1(x,y) = \frac{\nu}{E}\frac{Mg(x^2+y^2)}{2V} \tag{4.56}$$

since we expect $c_1(0,0) = 0$. The solution that corresponds to a displacement vector that is radially outward in the x-y plane is $c_2 \equiv c_3 \equiv 0$. Thus, the full solution is,

$$u_z = \frac{1}{E}\frac{Mg}{L_xL_y}(\frac{z^2}{2L_z} - z) + \frac{\nu}{E}\frac{Mg(x^2+y^2)}{2V} \tag{4.57}$$

$$u_x = -\frac{\nu}{E}\frac{Mg}{L_xL_y}(\frac{xz}{L_z} - x) \tag{4.58}$$

$$u_y = -\frac{\nu}{E}\frac{Mg}{L_xL_y}(\frac{zy}{L_z} - y). \tag{4.59}$$

■ Imagine a sphere of mass $M$ and radius $R$ that is strained in the following fashion. Each point $(x,y,z)$ inside the sphere suffers a displacement by an amount

$$\mathbf{D}(x,y,z) = -\hat{r}\,\frac{r^2}{a}, \tag{4.60}$$

where $\hat{r}$ is the radial unit vector and $a$ is a constant. Find the strain tensor. From this, find the stress tensor assuming that Young's modulus and Poisson's ratio are known. From the stress tensor, find the force acting on the outer surface of the sphere at radius $R$.

$$\mathbf{D}(x,y,z) = -(x\hat{i} + y\hat{j} + z\hat{k})\,\frac{(x^2+y^2+z^2)^{\frac{1}{2}}}{a} \tag{4.61}$$

$$\frac{\partial u_x}{\partial x} = -\frac{(2x^2+y^2+z^2)}{ar};\ \frac{\partial u_y}{\partial y} = -\frac{(x^2+2y^2+z^2)}{ar};\ \frac{\partial u_z}{\partial z} = -\frac{(x^2+y^2+2z^2)}{ar} \tag{4.62}$$

$$\frac{\partial u_x}{\partial y} = \frac{\partial u_y}{\partial x} = -\frac{xy}{ar};\ \frac{\partial u_z}{\partial x} = \frac{\partial u_x}{\partial z} = -\frac{zx}{ar};\ \frac{\partial u_y}{\partial z} = \frac{\partial u_z}{\partial y} = -\frac{yz}{ar} \tag{4.63}$$

The strain tensor is therefore,

$$\varepsilon = \begin{pmatrix} -\frac{(2x^2+y^2+z^2)}{ar} & -\frac{xy}{ar} & -\frac{zx}{ar} \\ -\frac{xy}{ar} & -\frac{(x^2+2y^2+z^2)}{ar} & -\frac{yz}{ar} \\ -\frac{zx}{ar} & -\frac{zy}{ar} & -\frac{(x^2+y^2+2z^2)}{ar} \end{pmatrix}. \tag{4.64}$$

From the stress–strain relation we get,

$$\sum_{i=1,2,3} \varepsilon_{i,i} = \frac{(1-2\nu)}{E}\sigma, \tag{4.65}$$

or

$$\sigma = -\frac{E}{(1-2\nu)}\frac{4r}{a} \tag{4.66}$$

$$\sigma_{i,j} = \frac{E}{(1+\nu)}\left(\varepsilon_{i,j} - \frac{\nu}{(1-2\nu)}\frac{4r}{a}\delta_{i,j}\right). \tag{4.67}$$

The $j$-th component of the force per unit area acting on a surface with unit normal $\hat{n}$ is $f_j = (\hat{n}\cdot\sigma)_j$; in Cartesian coordinates it is (since $\hat{n} = (x/r, y/r, z/r)$)

$$f_j(x,y,z) = \frac{x}{r}\sigma_{x,j} + \frac{y}{r}\sigma_{y,j} + \frac{z}{r}\sigma_{z,j}. \tag{4.68}$$

Thus,

$$\begin{aligned} f_x(x,y,z) &= \frac{x}{r}\sigma_{x,x} + \frac{y}{r}\sigma_{y,x} + \frac{z}{r}\sigma_{z,x} \\ &= \frac{x}{r}\sigma_{x,x} + \frac{y}{r}\sigma_{y,x} + \frac{z}{r}\sigma_{z,x}. \end{aligned} \tag{4.69}$$

But,

$$\sigma_{x,x} = \frac{E}{(1+\nu)}\left(-\frac{(2x^2+y^2+z^2)}{ar} - \frac{\nu}{(1-2\nu)}\frac{4r}{a}\right); \tag{4.70}$$

$$\sigma_{y,x} = -\frac{E}{(1+\nu)}\frac{xy}{ar};\ \sigma_{z,x} = -\frac{E}{(1+\nu)}\frac{xz}{ar} \tag{4.71}$$

therefore,

$$f_x(x,y,z) = \frac{x}{r}\frac{E}{(1+\nu)}\left(-\frac{(2x^2+y^2+z^2)}{ar} - \frac{\nu}{(1-2\nu)}\frac{4r}{a}\right)$$

$$-\frac{y}{r}\frac{E}{(1+\nu)}\frac{xy}{ar}-\frac{z}{r}\frac{E}{(1+\nu)}\frac{xz}{ar}=-\frac{E}{(1+\nu)}\frac{2x}{a}\frac{1}{(1-2\nu)}. \tag{4.72}$$

Thus the force acting per unit area is,

$$\mathbf{f}=-\frac{E}{(1+\nu)}\frac{2x}{a}\frac{1}{(1-2\nu)}\mathbf{r}. \tag{4.73}$$

It points radially inward in accordance with expectations.

■ Imagine a disk of radius $R$ is strained so that the displacement of each point $(r,\theta)$ on the disk is

$$\mathbf{D}(r,\theta)=\lambda\,\hat{\theta}\,r, \tag{4.74}$$

where $\lambda$ is some constant. Find the shear strains and shear stresses.

Write $\hat{\theta}=-sin(\theta)\hat{\imath}+cos(\theta)\hat{\jmath}$. Then it is clear that in Cartesian coordinates,

$$\mathbf{D}(x,y)=\lambda\,(-y\hat{\imath}+x\hat{\jmath}). \tag{4.75}$$

From this it is easy to see that the strain tensor vanishes identically as does the stress tensor.

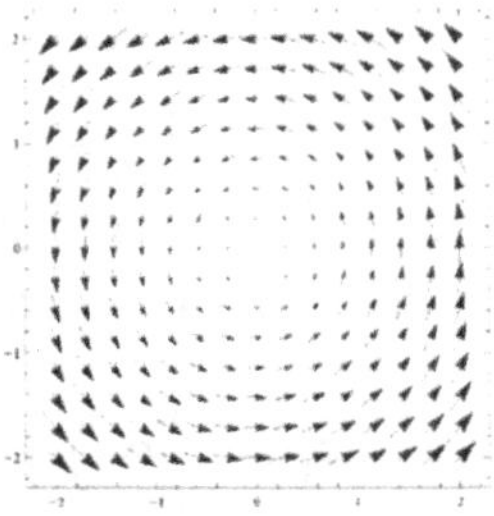

Figure 4.4: If at every radius, each point is shifted by an equal amount tangentially, there is no strain suffered by the material.

This may seem puzzling, but may be clarified using the following argument. Each point $\mathbf{r}$ is displaced by $\mathbf{D}$ making the new vector $\mathbf{r}^{'}=\mathbf{r}+\mathbf{D}$. Thus, $\mathbf{r}=x\hat{\imath}+y\hat{\jmath}$; then

$$\begin{aligned} x^{'}\hat{\imath}+y^{'}\hat{\jmath} &= x\hat{\imath}+y\hat{\jmath}+\lambda\,(-sin(\theta)\hat{\imath}+cos(\theta)\hat{\jmath})\,r \\ &= x\hat{\imath}+y\hat{\jmath}+\lambda\,(-y\,\hat{\imath}+x\,\hat{\jmath}), \end{aligned} \tag{4.76}$$

or

$$x^{'}=x-\lambda y\,;\; y^{'}=y+\lambda x$$

$$\begin{pmatrix} x^{'} \\ y^{'} \end{pmatrix}=\begin{pmatrix} 1 & -\lambda \\ \lambda & 1 \end{pmatrix}\begin{pmatrix} x \\ y \end{pmatrix} \tag{4.77}$$

$$M = \begin{pmatrix} 1 & -\lambda \\ \lambda & 1 \end{pmatrix} ; M^T = \begin{pmatrix} 1 & \lambda \\ -\lambda & 1 \end{pmatrix}. \tag{4.78}$$

One may see that $MM^T = M^T M = 1 + \lambda^2$. This means $M$ is proportional to an orthogonal matrix. Thus the transformation is nothing but a simple overall rotation of all points by the same angle, followed by an overall scaling of the distance from the center by the same factor. In order for strain to be present, different points have to move by different amounts (either radially or tangentially). Here they don't, hence the strain vanishes.

■ This example is about simple torsion. Imagine a cylinder subject to forces (per unit area) on the circular top and bottom cross sections form $\mathbf{f}_{top}(r,\phi,z) = G\,\alpha\, r\hat{\phi}$ and $\mathbf{f}_{bottom}(r,\phi,z) = -G\,\alpha\, r\hat{\phi}$, where $(r,\phi,z)$ are cylindrical coordinates and $G,\alpha$ are constants. There are no forces acting on the cylindrical surface. It is further given that all bulk forces are absent. Also, $G$ is known as the modulus of rigidity and $\alpha$ is the angle of twist per unit length of the shaft. The problem as usual, is to find the stress, strain, and the nature of displacements in the cylinder.

This means that since bulk forces are absent,

$$\nabla\cdot\sigma = 0. \tag{4.79}$$

Further,

$$\hat{r}\cdot\sigma = 0, \tag{4.80}$$

but for the top and bottom cross sections,

$$\hat{k}\cdot\sigma = \mathbf{f}_{top}(r,\phi,z) \tag{4.81}$$

$$(-\hat{k})\cdot\sigma = \mathbf{f}_{bottom}(r,\phi,z). \tag{4.82}$$

One could go about finding the tensor $\sigma$ consistent with these constraints in a systematic manner, but we prefer instead to guess an answer and verify that it works.

We wish to test the following answer,

$$\sigma(x,y,z) = -G\,\alpha\, y\,(\hat{\imath}\otimes\hat{k}+\hat{k}\otimes\hat{\imath}) + G\,\alpha\, x\,(\hat{\jmath}\otimes\hat{k}+\hat{k}\otimes\hat{\jmath}). \tag{4.83}$$

The above notation simply means $\sigma(x,y,z)$ has a component $\sigma_{xz}$ which is nothing but the coefficient of $\hat{\imath}\otimes\hat{k}$, which is identical to the coefficient of $\hat{k}\otimes\hat{\imath}$ and so on. This formidable guess is made less so by realizing at the outset that this tensor is uniform in the z-direction viz. it respects cylindrical symmetry. Further, if one takes the dot product with $\hat{k}$ we get

$$\hat{k}\cdot\sigma(x,y,z) = -G\,\alpha\, y\,\hat{\imath} + G\,\alpha\, x\,\hat{\jmath} = G\,\alpha\, r\hat{\phi}, \tag{4.84}$$

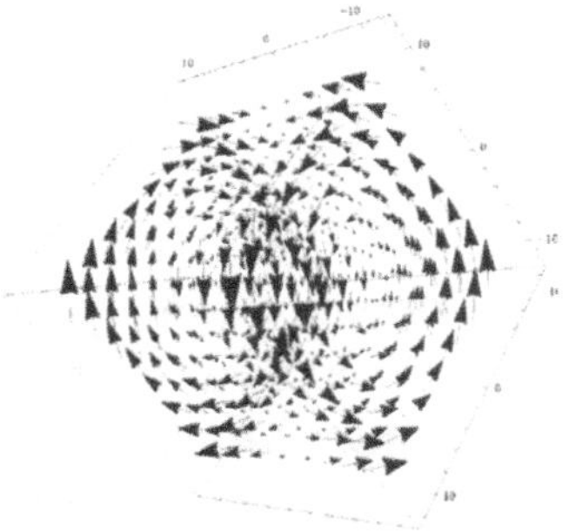

Figure 4.5: A cylinder is subject to twisting forces in opposite directions at opposite ends. The displacement vectors suffered by the cylinder are shown. Note that the sense at one end is opposite to the sense at the other end, and there is no displacement at the center.

which is as it should be (note that $\hat{k}\cdot\hat{\imath}\otimes\hat{k} \equiv (\hat{k}\cdot\hat{\imath})\hat{k} = 0$, whereas $\hat{k}\cdot\hat{k}\otimes\hat{\imath} \equiv (\hat{k}\cdot\hat{k})\hat{\imath} = \hat{\imath}$). On the surface of the cylinder $\hat{n} = \frac{x}{r}\hat{\imath} + \frac{y}{r}\hat{\jmath}$,

$$\hat{n}\cdot\sigma(x,y,z) = -G\,\alpha\,y\,(\frac{x}{r}\hat{k}) + G\,\alpha\,x\,(\frac{y}{r}\hat{k}) = 0. \tag{4.85}$$

Also it is obvious by inspection that Eq. (4.83) obeys Eq. (4.79). From the stress-strain relation Eq. (4.16) we get the following formula for strain (since $\sigma = Tr(\sigma) \equiv 0$):

$$\varepsilon = \frac{(1+\nu)}{E}\,(-G\,\alpha\,y\,(\hat{\imath}\otimes\hat{k} + \hat{k}\otimes\hat{\imath}) + G\,\alpha\,x\,(\hat{\jmath}\otimes\hat{k} + \hat{k}\otimes\hat{\jmath})). \tag{4.86}$$

This means

$$\varepsilon_{xx} = \varepsilon_{yy} = \varepsilon_{zz} = \varepsilon_{xy} = \varepsilon_{yx} = 0, \tag{4.87}$$

and

$$\varepsilon_{xz} = \varepsilon_{zx} = \frac{(1+\nu)}{E}\,(-G\,\alpha\,y) \tag{4.88}$$

$$\varepsilon_{yz} = \varepsilon_{zy} = \frac{(1+\nu)}{E}\,G\,\alpha\,x. \tag{4.89}$$

These may be integrated to give the displacement,

$$D_z = 0; D_x = \frac{(1+\nu)}{E}\,(-G\,\alpha\,zy);\; D_y = \frac{(1+\nu)}{E}\,G\,\alpha\,xz. \tag{4.90}$$

## 4.2 Strain Energy

We wish to derive the equilibrium condition,

$$\nabla\cdot\sigma(\mathbf{x}) + \mathbf{f}_b(\mathbf{x}) = 0 \tag{4.91}$$

as the consequence of minimizing some function of the displacements which we then identify with the elastic strain energy. First we make the following observations. From Hooke's Law, the stress tensor is a linear combination of the components of the strain tensor,

$$\sigma_{ij}(\mathbf{x}) = M^{ijkl}\varepsilon_{kl}(\mathbf{x}), \tag{4.92}$$

where summation over $k, l$ is implied. We further assume that the body forces are derivable from a potential, so that $\mathbf{f}_b(\mathbf{x}) = -\nabla V(\mathbf{x})$. Thus the equilibrium condition Eq. (4.91) may be written as,

$$\nabla_i M^{ijkl}\varepsilon_{kl}(\mathbf{x}) - \nabla_j V(\mathbf{x}) = 0. \tag{4.93}$$

This equation may be thought of as a consequence of a variational principle,

$$\int_\Omega d^3x\,(\delta u_j(\mathbf{x}))\,(\nabla_i M^{ijkl}\varepsilon_{kl}(\mathbf{x}) - \nabla_j V(\mathbf{x})) = 0, \tag{4.94}$$

for each variation $\delta u_j(\mathbf{x})$ in the displacement vector of the elastic object. In order to simplify the proceedings, we assume that this variation vanishes on the boundary of $\Omega$. Since this variation is otherwise arbitrary, the term in the parenthesis should vanish at each point. Now we rewrite the variational integral,

$$\int_\Omega d^3x\,(\delta u_j(\mathbf{x}))\,\nabla_i M^{ijkl}\varepsilon_{kl}(\mathbf{x}) - \int_\Omega d^3x\,(\delta u_j(\mathbf{x}))\,\nabla_j V(\mathbf{x}) =$$

$$= -\int_\Omega d^3x\,(\delta\nabla_i u_j(\mathbf{x}))\,M^{ijkl}\varepsilon_{kl}(\mathbf{x}) + \int_\Omega d^3x\,\delta(\nabla_j u_j(\mathbf{x}))\,V(\mathbf{x}). \tag{4.95}$$

The last relation follows from integration by parts and setting the boundary term to zero, since the variation in the displacement vanishes on the boundary:

$$\begin{aligned}\int_\Omega d^3x\,(\nabla_i(\ldots))(M^{i\ldots}\varepsilon_{kl}(\mathbf{x})) &= -\int_\Omega d^3x\,((\ldots))(M^{i\ldots}\nabla_i\varepsilon_{kl}(\mathbf{x})) \\ &\quad + \int_S dS_i\,((\ldots))(M^{i\ldots}\varepsilon_{kl}(\mathbf{x})).\end{aligned} \tag{4.96}$$

Here the surface term is set to zero. Since $M^{ijkl} = M^{jikl}$, $(\nabla_j u_j(\mathbf{x})) = Tr(\varepsilon(\mathbf{x}))$, and the strain tensor may be expressed in terms of the derivatives of the displacement

$$\varepsilon_{kl}(\mathbf{x}) = \frac{1}{2}\left(\frac{\partial u_k(\mathbf{x})}{\partial x_l} + \frac{\partial u_l(\mathbf{x})}{\partial x_k}\right), \tag{4.97}$$

the above equation becomes

$$-\int_\Omega d^3x\,(\delta\varepsilon_{ij}(\mathbf{x}))\,M^{ijkl}\varepsilon_{kl}(\mathbf{x}) + \int_\Omega d^3x\,\delta Tr(\varepsilon(\mathbf{x}))\,V(\mathbf{x}) = 0. \tag{4.98}$$

This may be rewritten as

$$-\delta\int_\Omega d^3x\,\mathfrak{U}(\mathbf{x}) = 0, \tag{4.99}$$

where $\mathfrak{U}(\mathbf{x})$ is the elastic energy density,

$$\mathfrak{U}(\mathbf{x}) = \frac{1}{2}\varepsilon_{ij}(\mathbf{x})\, M^{ijkl}\varepsilon_{kl}(\mathbf{x}) - Tr(\varepsilon(\mathbf{x}))\ V(\mathbf{x}). \tag{4.100}$$

Alternatively,

$$\mathfrak{U}(\mathbf{x}) = \frac{1}{2}\varepsilon_{ij}(\mathbf{x})\sigma^{ij}(\mathbf{x}) - Tr(\varepsilon(\mathbf{x}))\ V(\mathbf{x}). \tag{4.101}$$

If body forces are absent then the elastic strain energy density stored in the body is given by

$$\mathfrak{U}_{strain}(\mathbf{x}) = \frac{1}{2}\varepsilon_{ij}(\mathbf{x})\sigma^{ij}(\mathbf{x}). \tag{4.102}$$

# 4.3 Euler and Navier Stokes Equations

In this section, we introduce the equations of fluid dynamics. Fluids are elastic media that do not support shear stresses. The response of an elastic solid to shear stress is to deform, whereas the response of a fluid to shear stress is to move (accelerate). We start with the particle description where we think of a fluid as composed of individual atoms and then take the continuum limit to derive the relevant equations.

## 4.3.1 Equation of Continuity and Current Algebra

Imagine a collection of $N$ particles each of mass $m$. The position and velocity of the i-th particle will be denoted by $(\mathbf{r}_i, \mathbf{v}_i)$. By definition, $\mathbf{v}_i(t) = \frac{d}{dt}\mathbf{r}_i(t)$. The number density of particles may be defined as,

$$\rho(\mathbf{r},t) = \sum_{i=1}^{N} \delta(\mathbf{r} - \mathbf{r}_i(t)). \tag{4.103}$$

Dimensionally, this is then the number of particles per unit volume at location $\mathbf{r}$. The (number) current density, may similarly be defined as,

$$\mathbf{J}(\mathbf{r},t) = \sum_{i=1}^{N} \mathbf{v}_i(t)\ \delta(\mathbf{r} - \mathbf{r}_i(t)). \tag{4.104}$$

This is the the number of particles crossing per unit area per unit time in the direction of $\mathbf{J}$. These two are not independent and may be related through kinematics. Differentiating the particle density we get (together with chain rule of calculus),

$$\frac{\partial}{\partial t}\rho(\mathbf{r},t) = -\sum_{i=1}^{N} \frac{d\mathbf{r}_i(t)}{dt} \cdot \nabla_{\mathbf{r}}\delta(\mathbf{r} - \mathbf{r}_i(t)) \tag{4.105}$$

Figure 4.6: Considered to be one of the greatest mathematicians, Leonhard Euler's (15 April 1707 to 18 September 1783) work had an everlasting influence on calculus, number theory, analysis, fluid dynamics, optics, astronomy, and geometry. He laid the foundation of graph theory and his work in complex numbers yielded an equation often termed as the most beautiful mathematical identity viz. $e^{i\pi} + 1 = 0$.

Taking the divergence of the current we get,

$$\nabla \cdot \mathbf{J}(\mathbf{r},t) = \sum_{i=1}^{N} \mathbf{v}_i(t) \cdot \nabla_{\mathbf{r}} \delta(\mathbf{r} - \mathbf{r}_i(t)). \tag{4.106}$$

Adding these two we get,

$$\frac{\partial}{\partial t}\rho(\mathbf{r},t) + \nabla \cdot \mathbf{J}(\mathbf{r},t) = 0. \tag{4.107}$$

This is nothing but the equation of continuity. It may be shown that the components of current density and the particle density obey a closed algebra involving Poisson brackets. In order to accomplish this we write $\mathbf{v}_i = \frac{\mathbf{p}_i}{m}$ so that we may use the canonical Hamiltonian formulation. The general Poisson bracket between two variables $A(Q,P)$ and $B(Q,P)$ where $Q = \{\mathbf{r}_i : i = 1,2,...,N\}$ and $Q = \{\mathbf{p}_i : i = 1,2,...,N\}$ is

$$\{A,B\} = \sum_{i=1}^{N} \left( \frac{\partial A}{\partial \mathbf{r}_i} \cdot \frac{\partial B}{\partial \mathbf{p}_i} - \frac{\partial A}{\partial \mathbf{p}_i} \cdot \frac{\partial B}{\partial \mathbf{r}_i} \right). \tag{4.108}$$

The Poisson bracket between two $\rho$'s vanishes identically as may be expected since $\rho$ depends only on $Q$ but not on $P$

$$\{\rho(\mathbf{r},t), \rho(\mathbf{r}',t)\} = 0. \tag{4.109}$$

Now we examine the bracket between $\rho$ and $\mathbf{J}$, specifically $J_l$ the l-th component (here $r_{i,l}$ refers to the l-th component of the position vector of the i-th particle, $r_l$ refers to the l-th component of the fixed vector $\mathbf{r}$ whereas $\mathbf{r}_k$ refers to the dynamical variable corresponding to the position of the k-th particle). In the next few identities we make use of the chain rule,

$$\frac{\partial\delta(\mathbf{r}-\mathbf{r}_k(t))}{\partial r_{i,l}} = [\nabla_{r_l}\delta(\mathbf{r}-\mathbf{r}_k(t))]\frac{\partial(r_l - r_{k,l})}{\partial r_{i,l}} = -\nabla_{r_l}\delta(\mathbf{r}-\mathbf{r}_k(t))\ \delta_{k,i}. \tag{4.110}$$

Therefore,

$$\{\rho(\mathbf{r},t), J_l(\mathbf{r}',t)\} = \frac{1}{m}\sum_{i=1}^{N}\left(\frac{\partial\sum_{k=1}^{N}\delta(\mathbf{r}-\mathbf{r}_k(t))}{\partial r_{i,l}}\cdot\frac{\partial p_{i,l}\,\delta(\mathbf{r}'-\mathbf{r}_i(t))}{\partial p_{i,l}}\right)$$

$$= -\frac{1}{m}\sum_{i=1}^{N}(\nabla_{r_l}\delta(\mathbf{r}-\mathbf{r}_i(t)))\ \delta(\mathbf{r}'-\mathbf{r}_i(t))$$

$$= -\frac{1}{m}(\nabla_{r_l}\delta(\mathbf{r}-\mathbf{r}'))\ \sum_{i=1}^{N}\delta(\mathbf{r}'-\mathbf{r}_i(t)) = -\frac{1}{m}(\nabla_{r_l}\delta(\mathbf{r}-\mathbf{r}'))\ \rho(\mathbf{r}',t). \tag{4.111}$$

Thus, the second identity of current algebra reads as follows:

$$\{\rho(\mathbf{r},t), J_l(\mathbf{r}',t)\} = -\frac{1}{m}(\nabla_{r_l}\delta(\mathbf{r}-\mathbf{r}'))\ \rho(\mathbf{r}',t). \tag{4.112}$$

The final identity concerns the bracket between two different components of the current.

$$\{J_a(\mathbf{r},t), J_b(\mathbf{r}',t)\} = \sum_{i=1}^{N}\sum_{l=1,2,3}\left(\frac{\partial J_a(\mathbf{r},t)}{\partial r_{i,l}}\frac{\partial J_b(\mathbf{r}',t)}{\partial p_{i,l}} - \frac{\partial J_a(\mathbf{r},t)}{\partial p_{i,l}}\frac{\partial J_b(\mathbf{r}',t)}{\partial r_{i,l}}\right)$$

$$= \sum_{i=1}^{N}\sum_{l=1,2,3}\left(\frac{p_{i,a}(t)}{m}\ \frac{\partial\delta(\mathbf{r}-\mathbf{r}_i(t))}{\partial r_{i,l}}\ \frac{1}{m}\ \frac{\partial p_{i,b}(t)}{\partial p_{i,l}}\delta(\mathbf{r}'-\mathbf{r}_i(t))\right)$$

$$-\sum_{i=1}^{N}\sum_{l=1,2,3}\left(\frac{1}{m}\frac{\partial p_{i,a}(t)}{\partial p_{i,l}}\ \delta(\mathbf{r}-\mathbf{r}_i(t))\frac{p_{i,b}(t)}{m}\ \frac{\partial\delta(\mathbf{r}'-\mathbf{r}_i(t))}{\partial r_{i,l}}\right)$$

$$= -\frac{1}{m}J_a(\mathbf{r}',t)\ (\nabla_{r_b}\delta(\mathbf{r}-\mathbf{r}')) + \frac{1}{m}J_b(\mathbf{r},t)\ (\nabla_{r'_a}\delta(\mathbf{r}'-\mathbf{r})) \tag{4.113}$$

Thus the final identity of current algebra reads as follows.

$$\{J_a(\mathbf{r},t), J_b(\mathbf{r}',t)\} = -\frac{1}{m}J_a(\mathbf{r}',t)\ (\nabla_{r_b}\delta(\mathbf{r}-\mathbf{r}')) + \frac{1}{m}J_b(\mathbf{r},t)\ (\nabla_{r'_a}\delta(\mathbf{r}'-\mathbf{r})) \tag{4.114}$$

The identities of current algebra, namely Eq. (4.109), Eq. (4.112), and Eq. (4.114) tell us that Poisson brackets (later on in quantum mechanics, commutators) between local currents and densities obey an algebra reminiscent of a Lie algebra where the brackets are linear combinations of those variables themselves. But they are not simple. It would be desirable to make them simpler in order to make them more useful. To this end we propose the following definition for the local velocity of the system of particles (rather than one individual particle). It is defined to be that vector $\mathbf{v}(\mathbf{r},t)$ that obeys

$$\mathbf{J}(\mathbf{r},t) = \rho(\mathbf{r},t)\ \mathbf{v}(\mathbf{r},t). \tag{4.115}$$

Thus the velocity of the system of particles is nothing but the ratio of the local current to the local density. Of course this definition is meaningful only at points where the density is non-vanishing. At locations where the density vanishes, the velocity is ill defined. Given this definition we wish to determine the algebra between $\rho, \mathbf{v}$. We substitute this formula for $\mathbf{J}$ in the current algebra identities to get (we use the identities such as $\{AB,CD\} = A\{B,CD\} + \{A,CD\}B$ and $\{A,BC\} = \{A,B\}C + B\{A,C\}$),

$$-\frac{1}{m}(\nabla_{\mathbf{r}}\delta(\mathbf{r}-\mathbf{r}^{'}))\ \rho(\mathbf{r}^{'},t) = \{\rho(\mathbf{r},t),\mathbf{J}(\mathbf{r}^{'},t)\} = \{\rho(\mathbf{r},t),\rho(\mathbf{r}^{'},t)\ \mathbf{v}(\mathbf{r}^{'},t)\} =$$

$$\{\rho(\mathbf{r},t),\rho(\mathbf{r}^{'},t)\}\ \mathbf{v}(\mathbf{r}^{'},t) + \rho(\mathbf{r}^{'},t)\ \{\rho(\mathbf{r},t),\mathbf{v}(\mathbf{r}^{'},t)\}. \tag{4.116}$$

From this we may conclude that,

$$\{\rho(\mathbf{r},t),\mathbf{v}(\mathbf{r}^{'},t)\} = -\frac{1}{m}(\nabla_{\mathbf{r}}\delta(\mathbf{r}-\mathbf{r}^{'})). \tag{4.117}$$

We make the following observation for future reference. We may set $\mathbf{v}(\mathbf{r}^{'},t) = -\frac{1}{m}\nabla_{\mathbf{r}^{'}}\Pi(\mathbf{r}^{'},t)$ to obtain

$$\{\Pi(\mathbf{r}^{'},t),\rho(\mathbf{r},t)\} = \delta(\mathbf{r}-\mathbf{r}^{'}). \tag{4.118}$$

Now we substitute the formula for current in terms of velocity into Eq. (4.114),

$$-\frac{1}{m}J_a(\mathbf{r}^{'},t)\ (\nabla_{r_b}\delta(\mathbf{r}-\mathbf{r}^{'})) + \frac{1}{m}J_b(\mathbf{r},t)\ (\nabla_{r^{'}_a}\delta(\mathbf{r}^{'}-\mathbf{r}))$$

$$= \{\rho(\mathbf{r},t)\ v_a(\mathbf{r},t),\rho(\mathbf{r}^{'},t)\ v_b(\mathbf{r}^{'},t)\}$$

$$= \rho(\mathbf{r},t)\ \{v_a(\mathbf{r},t),\rho(\mathbf{r}^{'},t)\}\ v_b(\mathbf{r}^{'},t) + \rho(\mathbf{r}^{'},t)\ \{\rho(\mathbf{r},t),v_b(\mathbf{r}^{'},t)\}\ v_a(\mathbf{r},t)$$

$$+\rho(\mathbf{r}^{'},t)\ \rho(\mathbf{r},t)\ \{v_a(\mathbf{r},t),v_b(\mathbf{r}^{'},t)\}. \tag{4.119}$$

Here we have used the identity in Eq. (4.109). Now we use the identity Eq. (4.117) above to get,

$$Q = -\frac{1}{m}J_a(\mathbf{r}^{'},t)\ (\nabla_{r_b}\delta(\mathbf{r}-\mathbf{r}^{'})) + \frac{1}{m}J_b(\mathbf{r},t)\ (\nabla_{r^{'}_a}\delta(\mathbf{r}^{'}-\mathbf{r}))$$

$$= \rho(\mathbf{r},t)\ \{v_a(\mathbf{r},t),\rho(\mathbf{r}',t)\}\ v_b(\mathbf{r}',t) + \rho(\mathbf{r}',t)\ \{\rho(\mathbf{r},t),v_b(\mathbf{r}',t)\}\ v_a(\mathbf{r},t)$$
$$+\rho(\mathbf{r}',t)\ \rho(\mathbf{r},t)\ \{v_a(\mathbf{r},t),v_b(\mathbf{r}',t)\}$$
$$= \rho(\mathbf{r},t)\ \frac{1}{m}(\nabla_{r'_a}\delta(\mathbf{r}'-\mathbf{r}))\ v_b(\mathbf{r}',t) - \rho(\mathbf{r}',t)\ \frac{1}{m}(\nabla_{r_b}\delta(\mathbf{r}-\mathbf{r}'))\ v_a(\mathbf{r},t)$$
$$+\rho(\mathbf{r}',t)\ \rho(\mathbf{r},t)\ \{v_a(\mathbf{r},t),v_b(\mathbf{r}',t)\}. \tag{4.120}$$

Now we use identities such as

$$(\nabla_{r'_a}\delta(\mathbf{r}'-\mathbf{r}))\ v_b(\mathbf{r}',t) = (\nabla_{r'_a}\delta(\mathbf{r}'-\mathbf{r})\ v_b(\mathbf{r}',t)) - \delta(\mathbf{r}'-\mathbf{r})\ (\nabla_{r'_a}v_b(\mathbf{r}',t))$$
$$= (\nabla_{r'_a}\delta(\mathbf{r}'-\mathbf{r}))\ v_b(\mathbf{r},t) - \delta(\mathbf{r}'-\mathbf{r})\ (\nabla_{r'_a}v_b(\mathbf{r}',t)). \tag{4.121}$$

Inserting this into the preceding identities and comparing with the definition of $Q$ we obtain,

$$0 = -\rho(\mathbf{r},t)\ \frac{1}{m}\delta(\mathbf{r}'-\mathbf{r})\ (\nabla_{r_a}v_b(\mathbf{r},t)) + \rho(\mathbf{r},t)\ \frac{1}{m}\delta(\mathbf{r}-\mathbf{r}')\ (\nabla_{r_b}v_a(\mathbf{r},t))$$
$$+\rho(\mathbf{r}',t)\ \rho(\mathbf{r},t)\ \{v_a(\mathbf{r},t),v_b(\mathbf{r}',t)\}. \tag{4.122}$$

Now comes the crucial step. If we make the ansatz, $v_a(\mathbf{r},t) = -\frac{1}{m}\nabla_a\Pi(\mathbf{r},t)$ we get,

$$0 = \rho(\mathbf{r},t)\ \frac{1}{m}\delta(\mathbf{r}'-\mathbf{r})\ (\nabla_a\frac{1}{m}\nabla_b\Pi(\mathbf{r},t))$$
$$-\rho(\mathbf{r},t)\ \frac{1}{m}\delta(\mathbf{r}-\mathbf{r}')\ (\nabla_b\frac{1}{m}\nabla_a\Pi(\mathbf{r},t))$$
$$+\rho(\mathbf{r}',t)\ \rho(\mathbf{r},t)\ \{-\frac{1}{m}\nabla_a\Pi(\mathbf{r},t),-\frac{1}{m}\nabla'_b\Pi(\mathbf{r}',t)\}. \tag{4.123}$$

Since $\nabla_a\nabla_b \equiv \nabla_b\nabla_a$, so long as the regions we are dealing with have $\rho(\mathbf{r},t) \neq 0$, we may conclude,

$$\{\Pi(\mathbf{r},t),\Pi(\mathbf{r}',t)\} = 0. \tag{4.124}$$

The condition $\rho(\mathbf{r},t) \neq 0$ is important since we divided by this quantity to arrive at the above result. Thus Eq. (4.118) and Eq. (4.124) indicate that the velocity is derivable from a (scalar) potential and the potential is conjugate to the density. Now we move on to discuss the fundamental equations of fluids.

## 4.3.2 The Euler Equation

Imagine that the collection of particles is subject to an external force $\mathbf{f}(\mathbf{r},t)$. Newton's second law states that for particles $i = 1,2,...,N$ each of mass $m$,

$$m\ddot{\mathbf{r}}_i(t) = \mathbf{f}(\mathbf{r}_i(t),t). \tag{4.125}$$

To utilize this, consider differentiating Eq. (4.104).

$$\begin{aligned}\frac{\partial}{\partial t}\mathbf{J}(\mathbf{r},t) &= \sum_{i=1}^{N} \ddot{\mathbf{r}}_i(t)\ \delta(\mathbf{r}-\mathbf{r}_i(t)) - \sum_{i=1}^{N} \dot{\mathbf{r}}_i(t)\ (\dot{\mathbf{r}}_i(t)\cdot\nabla)\ \delta(\mathbf{r}-\mathbf{r}_i(t))\\ &= \frac{1}{m}\sum_{i=1}^{N} \mathbf{f}(\mathbf{r}_i(t),t)\ \delta(\mathbf{r}-\mathbf{r}_i(t)) - \sum_{i=1}^{N} \dot{\mathbf{r}}_i(t)\ (\dot{\mathbf{r}}_i(t)\cdot\nabla)\ \delta(\mathbf{r}-\mathbf{r}_i(t))\end{aligned} \tag{4.126}$$

Now we make a few formal observations. Assume that the Dirac delta functions in Eq. (4.103) and Eq. (4.104) have been rendered un-singular (using, say, the idea $\delta(x) \approx \frac{\varepsilon/\pi}{x^2+\varepsilon^2}$ and $\delta^3(\mathbf{r}) = \delta(x)\delta(y)\delta(z)$). In this case it makes sense to speak of $\delta(0)$. We now insert $\mathbf{r}_i(t)$ in place of $\mathbf{r}$ in Eq. (4.103) and Eq. (4.104);

$$\rho(\mathbf{r}_i(t),t) = \delta(0) \tag{4.127}$$

and

$$\mathbf{J}(\mathbf{r}_i(t),t) = \dot{\mathbf{r}}_i(t)\ \delta(0). \tag{4.128}$$

Thus we may write,

$$\dot{\mathbf{r}}_i(t) = \frac{\mathbf{J}(\mathbf{r}_i(t),t)}{\rho(\mathbf{r}_i(t),t)}. \tag{4.129}$$

We substitute this into the right-hand side of Eq. (4.126) to get,

$$\begin{aligned}\frac{\partial}{\partial t}\mathbf{J}(\mathbf{r},t) &= \frac{1}{m}\sum_{i=1}^{N} \mathbf{f}(\mathbf{r}_i(t),t)\ \delta(\mathbf{r}-\mathbf{r}_i(t))\\ &- \sum_{k=1,2,3} \nabla_k \sum_{i=1}^{N} \frac{\mathbf{J}(\mathbf{r}_i(t),t)}{\rho(\mathbf{r}_i(t),t)} \frac{J_k(\mathbf{r}_i(t),t)}{\rho(\mathbf{r}_i(t),t)}\ \delta(\mathbf{r}-\mathbf{r}_i(t)).\end{aligned} \tag{4.130}$$

The gradient operator is placed outside since it acts on $\mathbf{r}$ and there is only one function that depends on $\mathbf{r}$. Now we replace $\mathbf{r}_i(t)$ with $\mathbf{r}$ in all places except in the delta function and use Eq. (4.103).

$$\frac{\partial}{\partial t}\mathbf{J}(\mathbf{r},t) = \frac{1}{m}\mathbf{f}(\mathbf{r},t)\ \rho(\mathbf{r},t) - \sum_{k=1,2,3} \nabla_k \left(\frac{\mathbf{J}(\mathbf{r},t)}{\rho(\mathbf{r},t)}\ J_k(\mathbf{r},t)\right). \tag{4.131}$$

Now we make the substitution $\mathbf{J} = \rho\mathbf{v}$ to get,

$$\frac{\partial}{\partial t}\rho(\mathbf{r},t)\mathbf{v}(\mathbf{r},t) = \frac{1}{m}\mathbf{f}(\mathbf{r},t)\,\rho(\mathbf{r},t) - \sum_{k=1,2,3} \nabla_k(\mathbf{v}(\mathbf{r},t)\,\rho(\mathbf{r},t)v_k(\mathbf{r},t)) \tag{4.132}$$

We perform the time derivative and the gradient operator in the above equation combining with the continuity equation Eq. (4.107) to get,

$$-\mathbf{v}(\mathbf{r},t)(\mathbf{v}(\mathbf{r},t)\cdot\nabla\rho(\mathbf{r},t)) - \mathbf{v}(\mathbf{r},t)\rho(\mathbf{r},t)\nabla\cdot\mathbf{v}(\mathbf{r},t) + \rho(\mathbf{r},t)\frac{\partial}{\partial t}\mathbf{v}(\mathbf{r},t)$$

$$= \frac{1}{m}\mathbf{f}(\mathbf{r},t)\,\rho(\mathbf{r},t) - \rho(\mathbf{r},t)(\mathbf{v}(\mathbf{r},t)\cdot\nabla)\mathbf{v}(\mathbf{r},t) \tag{4.133}$$

$$-\mathbf{v}(\mathbf{r},t)\,(\mathbf{v}(\mathbf{r},t)\cdot\nabla)\rho(\mathbf{r},t) - \mathbf{v}(\mathbf{r},t)\,\rho(\mathbf{r},t)(\nabla\cdot\mathbf{v}(\mathbf{r},t)). \tag{4.134}$$

This may be simplified so that we obtain the celebrated Euler equation.

$$\frac{\partial}{\partial t}\mathbf{v}(\mathbf{r},t) = \frac{1}{m}\mathbf{f}(\mathbf{r},t) - (\mathbf{v}(\mathbf{r},t)\cdot\nabla)\mathbf{v}(\mathbf{r},t) \tag{4.135}$$

The external force $\mathbf{f}$ acting on this collection of particles (called a 'fluid') may be thought of as arising due to, say, a pressure, which means the force per unit volume is the negative gradient of this scalar quantity. The force per unit volume is $\mathbf{f}(\mathbf{r})\rho(\mathbf{r})$. This has to be equal to $-\nabla p(\mathbf{r})$. There could also be another source of external force, namely the weight of the particles—in this case we add a contribution $m\mathbf{g}$. Therefore, the Euler equation for a fluid under pressure $p$ and in a uniform gravitational field is,

$$\frac{\partial}{\partial t}\mathbf{v}(\mathbf{r},t) = -\frac{\nabla p}{m\rho} + \mathbf{g} - (\mathbf{v}(\mathbf{r},t)\cdot\nabla)\mathbf{v}(\mathbf{r},t). \tag{4.136}$$

This has to be supplemented with the equation of continuity,

$$\frac{\partial}{\partial t}\rho(\mathbf{r},t) + \nabla\cdot(\rho(\mathbf{r},t)\mathbf{v}(\mathbf{r},t)) = 0. \tag{4.137}$$

These have to be solved subject to appropriate boundary conditions and other other simplifying assumptions to obtain the nature of the flow. In many cases we may simplify the solution by choosing $\mathbf{v}(\mathbf{r},t) = -\nabla\Pi(\mathbf{r},t)$. As we pointed out earlier, in regions where the density vanishes, the velocity need not be irrotational (of course when the particles are electrically charged and magnetic fields are present, the velocity need not be irrotational anywhere. We have not considered this case here). Thus the above system of equations admits both irrotational solutions for the velocity field as well as solutions that correspond to vortices.

At this stage we introduce terminology that is commonly used to describe various simplifying assumptions.

**Incompressible**: $\rho(\mathbf{r},t) = const.$ in both space and time (if it is constant in space, it also has to be constant in time since the total mass of the system is conserved).

**Irrotational**: The velocity is expressible as the gradient of a scalar $\mathbf{v}(\mathbf{r},t) = -\nabla\Pi(\mathbf{r},t)$. We have seen earlier that this is the norm in regions where the density does not vanish (assuming there are no magnetic fields and such).

**Steady State**: The velocity and density are independent of time but depend only on position.

We now consider these situations in the subsequent examples.

■ An incompressible fluid is one whose density is constant in both space and time. In this case, the continuity equation implies $\nabla \cdot \mathbf{v} = 0$. Since this density is not zero, the discussion of earlier sections implies that the velocity is derivable from a potential $\mathbf{v}(\mathbf{r},t) = -\frac{1}{m}\nabla_{\mathbf{r}}\Pi(\mathbf{r},t)$, thus $\Pi$ obeys the Laplace equation $\nabla^2\Pi = 0$. Specifically, let us consider the problem of a fluid in two spatial dimensions impinging on an obstacle in the shape of a wedge with angle $\alpha$ in the first quadrant. The boundary condition is that the velocity normal to the obstacle surfaces vanishes at the surfaces. Laplace's equation in two dimensions reads as follows.

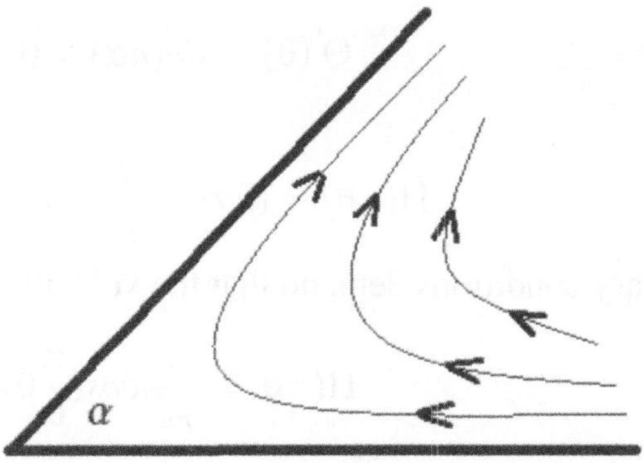


Figure 4.7: Flow past an angular obstruction.

$$(\frac{1}{r}\frac{\partial}{\partial r}r\frac{\partial}{\partial r} + \frac{1}{r^2}\frac{\partial^2}{\partial \theta^2})\Pi(\mathbf{r}) = 0 \tag{4.138}$$

We now use the method of separation of variables to write down some characteristic solution. $\Pi(\mathbf{r}) \equiv R(r)\Theta(\theta)$.

$$(\frac{r}{R(r)}\frac{\partial}{\partial r}r\frac{\partial}{\partial r}R(r) + \frac{\Theta''(\theta)}{\Theta(\theta)}) = 0 \tag{4.139}$$

This means for some constant $n$,

$$\frac{\Theta''(\theta)}{\Theta(\theta)} = -n^2 \tag{4.140}$$

and,

$$\frac{r}{R(r)}\frac{\partial}{\partial r}r\frac{\partial}{\partial r}R(r) = n^2 \; ; \; r^2R''(r) + rR'(r) - n^2\, R(r) = 0. \tag{4.141}$$

Set $R(r) = r^m$; this means,

$$m(m-1) + m - n^2 = 0, \tag{4.142}$$

$m = \pm n$. The condition that the velocity vanishes on the boundary of the obstacle is,

$$\left(\frac{\partial\Theta(\theta)}{\partial\theta}\right)_{\theta=0} = \left(\frac{\partial\Theta(\theta)}{\partial\theta}\right)_{\theta=\alpha} = 0 \tag{4.143}$$

$$\Theta(\theta) = cos(n\theta) \tag{4.144}$$

$$\Theta'(\theta) = sin(n\alpha) = 0. \tag{4.145}$$

$n = \frac{\pi}{\alpha}$.

$$\Pi(r,\theta) = (a\, r^{\frac{\pi}{\alpha}} + \frac{b}{r^{\frac{\pi}{\alpha}}})\; cos(\frac{\pi}{\alpha}\theta) \tag{4.146}$$

If the boundary conditions demand that the velocity at infinity vanishes, then

$$\Pi(r,\theta) = \frac{b}{r^{\frac{\pi}{\alpha}}}\; cos(\frac{\pi}{\alpha}\theta) \tag{4.147}$$

$$\mathbf{v}(r,\theta) = \frac{\pi}{\alpha}\frac{b}{r^{\frac{\pi}{\alpha}+1}}(\hat{r}\; cos(\frac{\pi}{\alpha}\theta) + \hat{\theta}\; sin(\frac{\pi}{\alpha}\theta)). \tag{4.148}$$

Now we turn to the Euler equation. We assume that the flow is in steady state. This means the velocity is explicitly time independent. Further, there is no gravity. This means from the Euler equation Eq. (4.136) we may write,

$$-\frac{\nabla p}{m\rho} - (\mathbf{v}(\mathbf{r},t)\cdot\nabla)\mathbf{v}(\mathbf{r},t) = 0. \tag{4.149}$$

Using the polar expression for the velocity in this equation we obtain,

$$-\frac{\nabla p}{m\rho} =$$

$$(\frac{\pi}{\alpha}+1)\frac{\pi}{\alpha}\frac{b}{r^{\frac{\pi}{\alpha}+1}}(\,cos(\frac{\pi}{\alpha}\theta))\frac{\pi}{\alpha}\frac{b}{r^{(\frac{\pi}{\alpha}+2)}}(\hat{r}\,cos(\frac{\pi}{\alpha}\theta)+\hat{\theta}\,sin(\frac{\pi}{\alpha}\theta))$$

$$-\frac{\pi}{\alpha}\frac{b}{r^{\frac{\pi}{\alpha}+1}}(\frac{sin(\frac{\pi}{\alpha}\theta)}{r})\frac{\pi}{\alpha}\frac{b}{r^{\frac{\pi}{\alpha}+1}}(\hat{\theta}\,cos(\frac{\pi}{\alpha}\theta)-\hat{r}\,sin(\frac{\pi}{\alpha}\theta))$$

$$-\frac{\pi}{\alpha}\frac{b}{r^{\frac{\pi}{\alpha}+1}}(\frac{sin(\frac{\pi}{\alpha}\theta)}{r})\frac{\pi}{\alpha}\frac{b}{r^{\frac{\pi}{\alpha}+1}}(-\hat{r}\,\frac{\pi}{\alpha}\,sin(\frac{\pi}{\alpha}\theta)+\hat{\theta}\,\frac{\pi}{\alpha}\,cos(\frac{\pi}{\alpha}\theta))=0. \tag{4.150}$$

While deriving this, we should not forget that $\frac{\partial}{\partial\theta}\hat{r}=\hat{\theta}$ and $\frac{\partial}{\partial\theta}\hat{\theta}=-\hat{r}$. This means $\partial_\theta p\equiv 0$ and therefore the equations may be integrated to yield,

$$p(r)=-\frac{m\rho}{2}\frac{\pi^2}{\alpha^2}\frac{b^2}{r^{(\frac{2\pi}{\alpha}+2)}}. \tag{4.151}$$

Thus we have uniquely determined both the velocity field and the pressure as a function of position.

# 4.4 Bernoulli's Equation for an Incompressible Fluid

Just as one may derive the law of conservation of energy in point particle mechanics starting from the dynamical equations for the coordinates and momenta when conservative forces are involved, in fluid mechanics this is also possible. The simplest situation where it is possible is when the fluid is incompressible and the flow is irrotational. This means that the density is constant in time and space, hence we have the two equations (equation of continuity and Euler equation, henceforth we shall not make a distinction between number density and mass density, setting $mass = 1$)

$$\nabla\cdot\mathbf{v}=0 \tag{4.152}$$

$$\frac{\partial}{\partial t}\mathbf{v}(\mathbf{r},t)=-\frac{\nabla p}{\rho}+\mathbf{g}-(\mathbf{v}(\mathbf{r},t)\cdot\nabla)\mathbf{v}(\mathbf{r},t). \tag{4.153}$$

For irrotational flows, the velocity may be written as $\mathbf{v}=-\nabla\Pi$; therefore, the Euler equation reads as follows:

$$-\frac{\partial}{\partial t}\nabla\Pi(\mathbf{r},t)=-\frac{\nabla p}{\rho}+\mathbf{g}-(\nabla\Pi(\mathbf{r},t)\cdot\nabla)\nabla\Pi(\mathbf{r},t). \tag{4.154}$$

This may be rewritten as,

$$0 = \nabla(\frac{\partial}{\partial t}\Pi(\mathbf{r},t) - \frac{p}{\rho} + \mathbf{g}\cdot\mathbf{r} - \frac{1}{2}(\nabla\Pi(\mathbf{r},t))^2). \tag{4.155}$$

This means

$$(\frac{\partial}{\partial t}\Pi(\mathbf{r},t) - \frac{p}{\rho} + \mathbf{g}\cdot\mathbf{r} - \frac{1}{2}(\nabla\Pi(\mathbf{r},t))^2) = f(t), \tag{4.156}$$

where $f(t)$ depends at most, only on time but not on position. Further, in case of steady flows there is no explicit time dependence, so that we may write (setting $\mathbf{g} = -\hat{k}g$),

$$\frac{p}{\rho} + gz + \frac{1}{2}v^2 = const. \tag{4.157}$$

This is the well-known Bernoulli's principle (or equation).

Figure 4.8: Daniel Bernoulli (8 February 1700 to 17 March 1782) was one of the great mathematicians from the Bernoulli family. His work in fluid dynamics is of critical importance in present-day aerodynamics. He contributed to mechanics, probability theory, economic theory and kinetic theory, of gases.

■ Show that the flow in Eq. (4.148) is consistent with Bernoulli's principle.

This may be approached as follows. Since there is no gravity we must have,

$$\frac{1}{2}v^2 + \frac{p}{\rho} = const. \tag{4.158}$$

or,

$$\frac{\pi^2}{2\alpha^2}\frac{b^2}{r^{\frac{2\pi}{\alpha}+2}} + \frac{p}{\rho} = const. = 0. \tag{4.159}$$

We choose the constant to be zero since we require pressure and velocity to vanish at $r = \infty$. The pressure obtained is the same as Eq. (4.151).

# 4.5 Navier Stokes Equation

Until now we have been content at discussing conservative forces, namely those derivable from a scalar potential. As in classical mechanics of point particles, a notable exception occurs in friction forces. Most fluids also experience a force reminiscent of friction—both due to surfaces of obstacles that they have to flow around and also internally, a phenomenon known as viscosity. We now consider the latter contribution, since the former contribution requires knowledge of the specific boundary conditions and so on. Imagine a layer of moving particles each of mass $m$, labeled as the k-th layer moving with a net drift velocity $v_k$. There is a layer on top of this also containing such particles moving with velocity $v_{k+1}$ and a layer below labeled $k-1$ containing particles moving with velocity $v_{k-1}$. Particles from a small thickness $dl$ from the $k+1$-th layer enter the $k$-th layer. The momentum entering the $k$-th layer from the $k+1$-th layer is $(\rho A dl)v_{k+1}$, where $\rho$ is the (assumed uniform) density and $A$ is the cross-sectional area. Similarly, a momentum $(\rho A dl)v_{k-1}$ enters the $k$-th layer from the $k-1$-th layer. Each layer $k$ also supplies a momentum $(\rho A dl)v_k$ to each of its adjacent layers. This means that the net momentum gained by the $k$-th layer is,

$$dP_k = (\rho A dl)v_{k+1} + (\rho A dl)v_{k-1} - 2(\rho A dl)v_k. \tag{4.160}$$

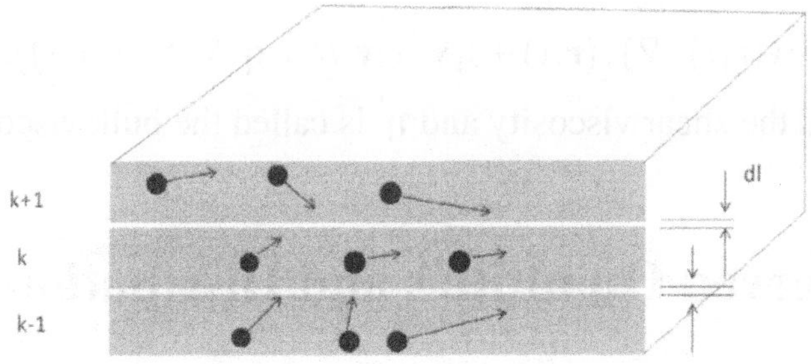


Figure 4.9: Viscosity is due to net momentum flow from adjacent layers.

Therefore,

$$m\frac{dv_k}{dt} = \alpha(v_{k+1} - v_k) + \alpha(v_{k-1} - v_k). \tag{4.161}$$

In the continuum limit we make the following associations $v_k \to v(x)$, $v_{k\pm 1} \to v(x \pm \Delta x)$, where $\Delta x$ is some kind of lattice spacing. We write the mass of each particle as $m = \Delta x\, \rho$ where $\rho$ is nonsingular. We also set, $\alpha \equiv \frac{\eta}{\Delta x}$ where $\eta$ is a nonsingular constant. We may now write the force equation as,

$$\frac{\partial v(x,t)}{\partial t} = \frac{\eta}{\rho}\frac{\partial^2 v(x,t)}{\partial x^2}. \tag{4.162}$$

Here $\eta$ is called the coefficient of viscosity. We may recognize the above equation as the diffusion equation. Thus, viscosity leads to diffusive flow. In three dimensions, we have to include such a term in each direction for each component of the velocity, so that the viscosity contribution then becomes, $\frac{\eta}{\rho}(\frac{\partial^2}{\partial x^2}+\frac{\partial^2}{\partial y^2}+\frac{\partial^2}{\partial z^2})\mathbf{v} \equiv \frac{\eta}{\rho}\nabla^2\mathbf{v}$ on the right-hand side of the Euler equation, making it the Navier-Stokes equation. Thus the Navier-Stokes equation (NS equation) is,

$$\frac{\partial}{\partial t}\mathbf{v}(\mathbf{r},t) = -\frac{\nabla p}{\rho} + \mathbf{g} - (\mathbf{v}(\mathbf{r},t)\cdot\nabla)\mathbf{v}(\mathbf{r},t) + \frac{\eta}{\rho}\nabla^2\mathbf{v}(\mathbf{r},t). \tag{4.163}$$

This equation, together with the continuity equation, represent some of the most important equations of fluid mechanics. Strictly speaking, the above equation is valid only for an incompressible fluid (since mass $m$ was assumed time independent). In general, for compressible flows we have to add an additional term proportional to the gradient of the divergence $\nabla(\nabla\cdot\mathbf{v})$ since the two possible ways of making a vector by taking two derivatives of $\mathbf{v}$ are $\nabla^2\mathbf{v}$ and $\nabla(\nabla\cdot\mathbf{v})$ and this new term would vanish for incompressible flows. For a general compressible flow, we should be writing

$$\begin{aligned}\rho(\mathbf{r},t)\frac{\partial}{\partial t}\mathbf{v}(\mathbf{r},t) &= -\nabla p + \rho(\mathbf{r},t)\mathbf{g}\\ &-\rho(\mathbf{r},t)(\mathbf{v}(\mathbf{r},t)\cdot\nabla)\mathbf{v}(\mathbf{r},t) + \eta\nabla^2\mathbf{v}(\mathbf{r},t) + \eta'\,\nabla(\nabla\cdot\mathbf{v}(\mathbf{r},t)),\end{aligned} \tag{4.164}$$

where $\eta$ is called the shear viscosity and $\eta'$ is called the bulk viscosity.

## 4.6 Conserved Quantities and Dissipation Rates

The continuity equation provides one kind of conservation law. Integrating this equation over a region $\Omega$ where the fluid is present gives us,

$$\frac{\partial}{\partial}\int_\Omega d^3r\,\rho(\mathbf{r},t) + \int_\Omega d^3r\,\nabla\cdot(\rho(\mathbf{r},t)\mathbf{v}(\mathbf{r},t)) = 0. \tag{4.165}$$

Applying Gauss's theorem on the second term gives us,

$$\frac{\partial}{\partial t}N(t) + \int_S dA\,(\rho(\mathbf{r},t)\mathbf{v}(\mathbf{r},t))\cdot\hat{n} = 0. \tag{4.166}$$

Here $N(t) = \int_\Omega d^3r\,\rho(\mathbf{r},t)$ and $(\rho(\mathbf{r},t)\mathbf{v}(\mathbf{r},t))\cdot\hat{n}$ is the number of particles per unit area per unit time escaping from the surface $S$ (if this quantity is negative, particles are flowing in). The symbol $S$ represents the surface bounding $\Omega$ and $\hat{n}$ is the unit normal to the surface. Note that $S$ can be the union of several disjoint pieces if solid (impenetrable) boundaries are present in the fluid. This means the time rate of change of $N(t)$—the number of particles in the volume $\Omega$ is the negative of this quantity, which is the content of Eq. (4.166). To obtain a conservation law we imagine a situation where the net flux of particles through the surface $S$ is zero, which means $N(t)$ is independent of time—a conserved quantity.

Now we wish to study momentum and energy transport in a fluid. For this we write down a general form for the NS equations.

$$\rho(\mathbf{r},t)\left(\frac{\partial}{\partial t}+(\mathbf{v}(\mathbf{r},t)\cdot\nabla)\right)\mathbf{v}(\mathbf{r},t)=\mathbf{f}_{ext}(\mathbf{r},t)+\eta\;\nabla^2\mathbf{v}(\mathbf{r},t)+\eta'\;\nabla(\nabla\cdot\mathbf{v}(\mathbf{r},t)) \tag{4.167}$$

Here $f_{ext}(\mathbf{r},t)$ is the sum of all forces per unit volume acting on the fluid, external to the fluid. This includes forces originating from pressure, body forces, and so on. We may rewrite the above equation for each component as follows,

$$f_{j,ext}(\mathbf{r},t)=\frac{\partial}{\partial t}(\rho(\mathbf{r},t)v_j(\mathbf{r},t))+\nabla\cdot\mathbf{T}_j, \tag{4.168}$$

where

$$\mathbf{T}_j(\mathbf{r},t)=(v_j(\mathbf{r},t)\rho(\mathbf{r},t)\mathbf{v}(\mathbf{r},t)-\eta\;\nabla v_j(\mathbf{r},t)-\eta'\;\nabla_j\mathbf{v}(\mathbf{r},t)). \tag{4.169}$$

We have used the equation of continuity to derive this formula. We now integrate Eq. (4.168) over a volume where the fluid is present. This leads to an expression for the net force on the fluid due to all sources external to the fluid (using Gauss's theorem for the divergence of $\mathbf{T}_j$),

$$F_{j,ext}(t)=\frac{\partial}{\partial t}P_j(t)+\int_\Omega d^3r\,\nabla\cdot\mathbf{T}_j=\frac{\partial}{\partial t}P_j(t)+\int_S dA\;\mathbf{T}_j\cdot\hat{n} \tag{4.170}$$

where $\hat{n}$ is the normal to the surface $S$ that bounds the volume $\Omega$ where the fluid is present and,

$$P_j(t)=\int_\Omega d^3r\;\rho(\mathbf{r},t)v_j(\mathbf{r},t) \tag{4.171}$$

is the $j$-th component of net momentum of the fluid. The interpretation of Eq. (4.170) is as follows. The net external force acting on the fluid in region $\Omega$ is due to an explicit time rate of change of the net momentum of the fluid and also because momentum per unit time is flowing in and out of the fluid (the term with $\mathbf{T}_j$). This distinction can be made clearer by examining a situation where the densities and velocities do not depend on time $t$ but only depend on position $\mathbf{r}$. In this case too, force can act on the fluid since even for steady flows, momentum flows in and out of the region occupied by the fluid as given by the term involving $\mathbf{T}_j$. Put differently, even in order to maintain a steady flow, forces may have to act on the fluid depending on other conditions such as presence of rigid boundaries and so on.

Lastly we consider energy transport. The kinetic energy density is given by,

$$\mathfrak{K} = \frac{1}{2}\rho(\mathbf{r},t)\mathbf{v}^2(\mathbf{r},t). \tag{4.172}$$

We now derive a rate equation for this quantity.

$$\partial_t \mathfrak{K} = -\frac{1}{2}\nabla\cdot(\mathbf{v}^2(\mathbf{r},t)\rho(\mathbf{r},t)\mathbf{v}(\mathbf{r},t)) + \eta^{'}\,(\mathbf{v}(\mathbf{r},t)\cdot\nabla)(\nabla\cdot\mathbf{v}(\mathbf{r},t))$$
$$+\eta\;\mathbf{v}(\mathbf{r},t)\cdot\nabla^2\mathbf{v}(\mathbf{r},t) + \mathbf{v}(\mathbf{r},t)\cdot\mathbf{f}_{ext}(\mathbf{r},t) \tag{4.173}$$

This very general result is not particularly useful unless we make further simplifying assumptions. To this end we assume that the fluid is incompressible and the fluid is at rest at all boundaries of the fluid. These two assumptions lead us to write down the rate of change of kinetic energy by integrating over the volume $\Omega$ (since now divergence of the velocity is zero).

$$\partial_t K = \eta\int_\Omega d^3r\;\mathbf{v}(\mathbf{r},t)\cdot\nabla^2\mathbf{v}(\mathbf{r},t) + \int_\Omega d^3r\;\mathbf{v}(\mathbf{r},t)\cdot\mathbf{f}_{ext}(\mathbf{r},t) \tag{4.174}$$

The second term on the right-hand side of Eq. (4.174) is the rate of work done on the fluid by forces external to it. The first term is energy dissipated per unit time due to viscosity. To see that this term is negative, we may rewrite it as follows.

$$\eta\int_\Omega d^3r\;\mathbf{v}(\mathbf{r},t)\cdot\nabla^2\mathbf{v}(\mathbf{r},t) = \sum_{j=1}^{3}\eta\int_\Omega d^3r\;v_j(\mathbf{r},t)\nabla^2 v_j(\mathbf{r},t)$$
$$= \sum_{j=1}^{3}\eta\int_\Omega d^3r\;\nabla\cdot(v_j(\mathbf{r},t)\nabla v_j(\mathbf{r},t))$$
$$-\sum_{j=1}^{3}\eta\int_\Omega d^3r\;(\nabla v_j(\mathbf{r},t))^2 = -\sum_{j=1}^{3}\eta\int_\Omega d^3r\;(\nabla v_j(\mathbf{r},t))^2 \tag{4.175}$$

The last result follows from the application of Gauss's theorem and the observation that the velocity field vanishes on the boundary of $\Omega$.

## 4.7 Turbulence

In all the examples we have studied so far, we have studied special solutions of the Euler and continuity equations. But we have not addressed the question of stability of those solutions. A steady (time-independent) solution for the velocity field that obeys the prescribed boundary conditions will not necessarily be realized in nature if it is unstable to perturbations. This instability manifests itself as turbulence. The transition to turbulence is determined by a phenomenological dimensionless number called the Reynolds number. For concreteness, we consider the situation of an incompressible fluid of density $\rho$ and viscosity $\eta$, flowing with velocity $u$ past a solid of a fixed shape but variable size characterized by a length $l$. The Reynolds number of the flow pattern (denoted by $Re$) around this body is defined as

$$Re = \frac{\rho u l}{\eta} \tag{4.176}$$

This dimensionless quantity determines roughly whether the flow is laminar (unidirectional) or turbulent (with vortices, for example). Depending upon the shape of the object, there are critical Reynolds numbers that separate the laminar flows (low $Re$ relative to the critical) from the turbulent flows (high $Re$). We may rewrite the NS equation purely in terms of dimensionless quantities. Define $\mathbf{r}' = \frac{\mathbf{r}}{l}$ so that $\nabla' = l\nabla$ and $t' = \frac{tu}{l}$ and $\frac{\partial}{\partial t'} = \frac{l}{u}\frac{\partial}{\partial t}$, $\mathbf{v}'(\mathbf{r}',t') = \frac{\mathbf{v}(\mathbf{r},t)}{u}$ where $\mathbf{v}$ is the velocity field that appears in the NS equation, and acceleration due to gravity or some other force analogous to it would scale as $g' = g\frac{l}{u^2}$. Similarly we write $\rho' = \rho l^3$, $p' = p\frac{l^3}{mu^2}$. Substituting the inverted version of these relations into the NS equations we may rewrite, purely in terms of dimensionless quantities,

$$\frac{\partial}{\partial t'}\mathbf{v}'(\mathbf{r}',t') = -\frac{\nabla' p'}{\rho'} + \mathbf{g}' - (\mathbf{v}'(\mathbf{r}',t')\cdot\nabla')\mathbf{v}'(\mathbf{r}',t') + \frac{1}{Re}\nabla'^2\mathbf{v}'(\mathbf{r}',t'). \tag{4.177}$$

Thus we see that the Reynolds number $Re = \frac{u\rho'}{l^2\eta} = \frac{\rho u l}{\eta}$ enters naturally into the formalism.

**Examples:**

■ We consider the problem of flow around spherical and cylindrical obstacles. The idea is to calculate the drag force on the obstacle. A different perspective allows us to consider the same problem as finding the drag force on a moving object of a spherical or cylindrical shape in a stationary fluid.

In the first perspective, the boundary condition we use is that far from the obstacle, the flow is unidirectional and uniform with velocity $\mathbf{u}$. On the surface of the sphere or cylinder, the velocity field vanishes. We use Eq. (4.163) (with $\mathbf{g} = 0$ and $\dot{\mathbf{v}} = 0$)

Figure 4.10: Turbulence is one of the great unsolved problems of Newtonian mechanics, due to non-linearities in Navier-Stokes equation. In Richard Feynman's words, "It always bothers me ... why should it take infinite amount of logic to figure out what one tiny piece of space-time is going to do?"

to get,

$$\nabla^2\mathbf{v}(\mathbf{r}) = \frac{\nabla p}{\eta} + \frac{\rho}{\eta}(\mathbf{v}(\mathbf{r})\cdot\nabla)\mathbf{v}(\mathbf{r}). \tag{4.178}$$

This may also be rewritten using dimensionless quantities

$$-\frac{\nabla' p'}{\rho'} - (\mathbf{v}'(\mathbf{r}',t')\cdot\nabla')\mathbf{v}'(\mathbf{r}',t') + \frac{1}{Re}\nabla'^2\mathbf{v}'(\mathbf{r}',t') = 0. \tag{4.179}$$

For small Reynolds numbers, we may expect all quantities to have an expansion of the form

$$\mathbf{v}'(\mathbf{r}',t') = \hat{u} + Re\ \mathbf{v}_1'(\mathbf{r}',t') + Re^2\ \mathbf{v}_2'(\mathbf{r}',t') + \ldots\ldots \tag{4.180}$$

Similarly,

$$\frac{p'}{\rho'} \equiv p'' = p_0'' + Re\ p_1'' + Re^2\ p_2'' + \ldots\ldots \tag{4.181}$$

Hence,

$$-\nabla' p_0'' + \nabla'^2\mathbf{v}_1'(\mathbf{r}',t') = 0 \tag{4.182}$$

$$-\nabla'\ p_1'' - (\hat{u}\cdot\nabla')\ \mathbf{v}_1'(\mathbf{r}',t') + \nabla'^2\ \mathbf{v}_2'(\mathbf{r}',t') = 0. \tag{4.183}$$

These have to be supplemented with the incompressibility condition, namely,

$$\nabla_1'\cdot\mathbf{v}_1'(\mathbf{r}',t') = 0\ ;\ \nabla_2'\cdot\mathbf{v}_1'(\mathbf{r}',t') = 0, \ldots \tag{4.184}$$

since these conditions are valid term by term. This means,

$$\nabla'^2 p_0'' = \nabla'^2 p_1'' = 0. \tag{4.185}$$

Now we go on to apply these ideas to compute the drag force acting on a solid sphere and a solid cylinder assuming the flow is streamline and has small Reynolds number.

**Case of a Sphere**

We imagine a solid sphere with center at the origin and radius $a$ immersed in a fluid that has velocity at infinity equal to $\mathbf{u} = u\,\hat{k}$. To analyze this, it is better to work with spherical polar coordinates and spherical polar unit vectors as basis. The identities associated with these are given in the boxes at the end of this discussion. We purposely use a somewhat inelegant and brute-force approach for two reasons—one is to show that clever tricks that simplify the analysis are invaluable when available. Second, a proper justification of these tricks ultimately rests on a detailed verification. Also these tricks work only for small Reynolds numbers; at larger values they fail and one is forced to use the general method. These are the general formulas valid for all types of functions of the coordinates. Now we make the assumption that the azimuthal coordinate dependence and the component are both absent. This means $v_\phi \equiv 0$ and $\frac{\partial}{\partial\phi} \equiv 0$. This is consistent with the expectation that the velocity field flows around the sphere in such a way that it has no component that winds around the direction of the velocity of the fluid at infinity.

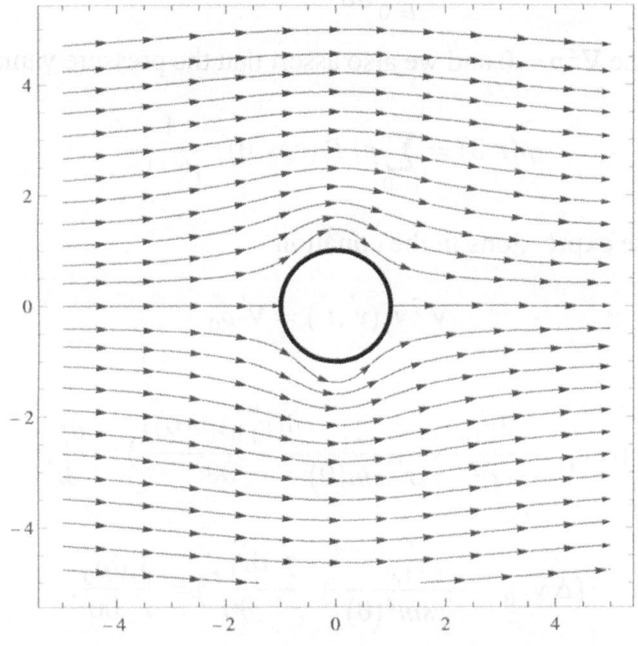


Figure 4.11: Velocity field around a sphere.

Imposing the incompressibility requirement we get,

$$\frac{1}{r^2}\frac{\partial(r^2 v_r)}{\partial r} + \frac{1}{rsin(\theta)}\frac{\partial}{\partial\theta}(v_\theta sin(\theta)) = 0. \tag{4.186}$$

This means that the radial component is always related to the tangential component.

$$v_r(r,\theta) = -\frac{1}{r^2}\int_a^r dr' \; \frac{r'}{sin(\theta)}\frac{\partial}{\partial\theta}(v_\theta(r',\theta)sin(\theta)) \tag{4.187}$$

We adopt the boundary condition that at the surface of the sphere the fluid is at rest since it is assumed that there is no slipping between the surface and the sphere. We are going to set

$$v_\theta(r',\theta) = \sum_{l=0}^{\infty}\frac{\partial}{\partial\theta}P_l(cos(\theta))V_{l,\theta}(r'). \tag{4.188}$$

But,

$$\frac{1}{sin(\theta)}\frac{\partial}{\partial\theta}sin(\theta)\frac{\partial}{\partial\theta}P_l(cos(\theta)) = -l(l+1)P_l(cos(\theta)) \tag{4.189}$$

$$v_r(r,\theta) = \sum_{l=0}^{\infty}\frac{l(l+1)}{r^2}(\int_a^r dr' \; V_{l,\theta}(r')r') \; P_l(cos(\theta)) \tag{4.190}$$

$$v_\theta(r,\theta) = \sum_{l=0}^{\infty}\frac{\partial}{\partial\theta}P_l(cos(\theta))V_{l,\theta}(r). \tag{4.191}$$

Furthermore, since $\nabla^2 p = 0$ and we also assert that the pressure vanishes at infinity,

$$p(r,\theta) = \sum_{l=0}^{\infty} p_l \; P_l(cos(\theta)) \; \frac{1}{r^{(l+1)}}. \tag{4.192}$$

Substituting these expressions in the equation

$$\nabla'^2\mathbf{v}_1'(\mathbf{r}',t') = \nabla' p_0'' \tag{4.193}$$

we get

$$(\Delta' v_{1,r}' - \frac{2v_{1,r}'}{r'^2} - \frac{2}{r'^2 sin(\theta)}\frac{\partial(v_{1,\theta}' sin(\theta))}{\partial\theta}) = \frac{\partial p_0''}{\partial r'}. \tag{4.194}$$

and,

$$(\Delta' v_{1,\theta}' - \frac{v_{1,\theta}'}{r'^2 sin^2(\theta)} + \frac{2}{r'^2}\frac{\partial v_{1,r}'}{\partial\theta}) = \frac{1}{r'}\frac{\partial p_0''}{\partial\theta}. \tag{4.195}$$

Since the partial derivatives are symmetric $\frac{\partial}{\partial r'}\frac{\partial}{\partial\theta}p_0'' = \frac{\partial}{\partial\theta}\frac{\partial}{\partial r'}p_0''$ we get,

$$\frac{\partial}{\partial\theta}(\Delta' v_{1,r}' - \frac{2v_{1,r}'}{r'^2} - \frac{2}{r'^2 sin(\theta)}\frac{\partial(v_{1,\theta}' sin(\theta))}{\partial\theta})$$

$$= \frac{\partial}{\partial r'}r'(\Delta' v_{1,\theta}' - \frac{v_{1,\theta}'}{r'^2 sin^2(\theta)} + \frac{2}{r'^2}\frac{\partial v_{1,r}'}{\partial\theta}). \tag{4.196}$$

Written out in full this constraint expands out to these terms:

$$\sum_{l=0}^{\infty}\left(\frac{1}{r'^2}\frac{d}{dr'}r'^2\frac{d}{dr'}-\frac{l(l+1)}{r'^2}\right)\frac{l(l+1)}{r'^2}\left(\int_a^{r'}dr''\ V_{l,\theta}(r'')r''\right)\frac{d}{d\theta}P_l(cos(\theta))$$

$$-\frac{2}{r'^2}\sum_{l=0}^{\infty}\frac{l(l+1)}{r'^2}\left(\int_a^{r'}dr''\ V_{l,\theta}(r'')r''\right)\frac{d}{d\theta}P_l(cos(\theta))$$

$$+\sum_{l=0}^{\infty}V_{l,\theta}(r')\frac{2}{r'^2}l(l+1)\ \frac{d}{d\theta}P_l(cos(\theta))=\sum_{l=0}^{\infty}\frac{d}{d\theta}P_l(cos(\theta))\frac{d}{dr'}\frac{1}{r'}\frac{d}{dr'}r'^2\frac{d}{dr'}V_{l,\theta}(r')$$

$$+\sum_{l=0}^{\infty}\frac{d^3}{d\theta^3}P_l(cos(\theta))\frac{d}{dr'}\frac{1}{r'}V_{l,\theta}(r')$$

$$+\sum_{l=0}^{\infty}\frac{d}{dr'}\frac{2}{r'}\frac{l(l+1)}{r'^2}\left(\int_a^{r'}dr''\ V_{l,\theta}(r'')r''\right)\frac{d}{d\theta}P_l(cos(\theta))$$

$$+cot(\theta)\sum_{l=0}^{\infty}\frac{d^2}{d\theta^2}P_l(cos(\theta))\frac{d}{dr'}\frac{1}{r'}V_{l,\theta}(r')-\frac{1}{sin^2(\theta)}\sum_{l=0}^{\infty}\frac{d}{d\theta}P_l(cos(\theta))\frac{d}{dr'}\frac{1}{r'}V_{l,\theta}(r').\quad(4.197)$$

The higher derivatives of $P_l(cos(\theta))$ being linearly independent of the lower ones, should drop out. This is going to happen only if we assert that $P_l''(cos(\theta))\equiv 0$, so that $l=1$ is the only term present. This means,

$$\left(\frac{1}{r'^2}\frac{d}{dr'}r'^2\frac{d}{dr'}-\frac{2}{r'^2}\right)\frac{2}{r'^2}\left(\int_a^{r'}dr''\ V_{1,\theta}(r'')r''\right)-\frac{2}{r'^2}\frac{2}{r'^2}\left(\int_a^{r'}dr''\ V_{1,\theta}(r'')r''\right)$$

$$+V_{1,\theta}(r')\frac{4}{r'^2}=\frac{d}{dr'}\frac{1}{r'}\frac{d}{dr'}r'^2\frac{d}{dr'}V_{1,\theta}(r')$$

$$-\frac{d}{dr'}\frac{1}{r'}V_{1,\theta}(r')+\frac{d}{dr'}\frac{2}{r'}\frac{2}{r'^2}\left(\int_a^{r}dr''\ V_{1,\theta}(r'')r''\right)-\frac{d}{dr'}\frac{1}{r'}V_{1,\theta}(r').\quad(4.198)$$

One may solve this systematically by setting $r'=\frac{1}{q'}$ and expressing $V_{1,\theta}$ as a simple polynomial in $q'$. The solution to this would be,

$$V_{1,\theta}(r')=u\left(1-\frac{3a}{4r'}-\frac{a^3}{4r'^3}\right).\quad(4.199)$$

The overall constant is fixed by demanding that at $r=\infty$ the velocity vector is $\mathbf{v}=u\,\hat{k}$.

$$v_r(r,\theta)=u\ cos(\theta)\left(1-\frac{3a}{2r}+\frac{a^3}{2r^3}\right)\quad(4.200)$$

$$v_\theta(r,\theta)=-u\ sin(\theta)\left(1-\frac{3a}{4r}-\frac{a^3}{4r^3}\right)\quad(4.201)$$

In order to facilitate the integration, we write the velocity field in a mixed representation where the components are in polar coordinates but the unit vectors are in Cartesian coordinates.

$$\mathbf{v} = u\,cos(\theta)\,\left(1-\frac{3a}{2r}+\frac{a^3}{2r^3}\right)\hat{r} - u\,sin(\theta)\,\left(1-\frac{3a}{4r}-\frac{a^3}{4r^3}\right)\hat{\theta} \tag{4.202}$$

But,

$$\hat{r} = sin(\theta)cos(\phi)\,\hat{\imath} + sin(\theta)sin(\phi)\,\hat{\jmath} + cos(\theta)\hat{k} \tag{4.203}$$

$$\hat{\theta} = cos(\theta)cos(\phi)\,\hat{\imath} + cos(\theta)sin(\phi)\,\hat{\jmath} - sin(\theta)\hat{k}. \tag{4.204}$$

Therefore, a fully Cartesian form would be,

$$\mathbf{v} = \left(-\frac{3a}{4r^3}+\frac{3a^3}{4r^5}\right)u\,z(x\,\hat{\imath}+y\,\hat{\jmath})$$

$$+(u\,z^2\left(-\frac{3a}{4r^3}+\frac{3a^3}{4r^5}\right) - u\,r^2\left(\frac{3a}{4r^3}+\frac{a^3}{4r^5}\right)+u)\,\hat{k}. \tag{4.205}$$

Now we wish to calculate the force acting on the surface $r = a$ due to the fluid. The forces are due to pressure and viscosity. The force per unit volume, including both these contributions acting on the fluid, is writable as

$$\mathbf{f}_{tot}(\mathbf{r}) = -\nabla p + \eta\nabla^2\mathbf{v}. \tag{4.206}$$

In order to obtain the force acting on a surface, it is better to proceed as follows. Consider some component $j = x, y, z$ of the force,

$$f_{j,tot}(\mathbf{r}) = -\nabla_j p + \eta\nabla^2 v_j. \tag{4.207}$$

We rewrite this as the divergence of some vector.

$$f_{j,tot}(\mathbf{r}) = \nabla\cdot(-\hat{e}_j p + \eta\nabla v_j) \tag{4.208}$$

The j-component of the total force acting on some volume may be written as

$$F_{j,tot} = \int_\Omega d^3r\, f_{j,tot}(\mathbf{r}) = \int_\Omega d^3r\,\nabla\cdot(-\hat{e}_j p + \eta\nabla v_j)$$

$$= \int_S dA(-\hat{e}_j p + \eta\nabla v_j)\cdot\hat{n}, \tag{4.209}$$

where $S$ is the surface(s) bounding $\Omega$. Therefore, we see that the term $\sigma_j = (-\hat{e}_j p + \eta\nabla v_j)\cdot\hat{n}$ has the interpretation of the j-th component of the force per unit area acting on a surface whose outward normal is $\hat{n}$. Now we calculate the net force acting on the surface $r = a$. From Eq. (4.182) we see that (after restoring dimensional quantities),

$$\nabla p = \eta\nabla^2\mathbf{v}. \tag{4.210}$$

Since the velocity field is divergence-free due to incompressibility, we must have $\nabla^2 p = 0$, since from the preceding discussion only $l = 1$ is being considered. The pressure may be written as

$$p(r,\theta) = \frac{p_1}{r^2} cos(\theta), \tag{4.211}$$

and the velocity field as,

$$\mathbf{v} = u\, cos(\theta) \left(1 - \frac{3a}{2r} + \frac{a^3}{2r^3}\right) \hat{r} - u\, sin(\theta) \left(1 - \frac{3a}{4r} - \frac{a^3}{4r^3}\right) \hat{\theta}. \tag{4.212}$$

Thus,

$$\nabla p = -\frac{2p_1}{r^3} cos(\theta)\, \hat{r} - \frac{p_1}{r^3} sin(\theta)\, \hat{\theta} \tag{4.213}$$

and,

$$\Delta \mathbf{v} = \frac{3au\, cos(\theta)}{r^3} \hat{r} + \frac{3sin(\theta) u\, a}{2r^3} \hat{\theta}. \tag{4.214}$$

Therefore, $p_1 = -\frac{3}{2}\eta a u$ and

$$v_z = (u\, z^2 \left(-\frac{3a}{4r^3} + \frac{3a^3}{4r^5}\right) - u\, r^2 \left(\frac{3a}{4r^3} + \frac{a^3}{4r^5}\right) + u). \tag{4.215}$$

The z-component of the force acting on the sphere is, $F_{z,tot} = \int_S dA(-cos(\theta)p + \eta \frac{\partial}{\partial r} v_z)$, which after substitution of pressure and velocity becomes,

$$F_{z,tot} = \int_S dA(\frac{3}{2a}\eta u\, cos^2(\theta) + \eta(u\, cos^2(\theta) \left(-\frac{3}{2a}\right) - u \left(-\frac{3}{2a}\right))) \tag{4.216}$$

or

$$F_{z,tot} = F_{drag} = 6\pi\eta u a. \tag{4.217}$$

This is the famous Stokes formula for the drag of a sphere in a viscous fluid. This derivation appears quite formidable and some simplification is called for. But this comes at the expense of making educated guesses that are not always obvious to the inexperienced. We now explore this simpler approach for the case of a cylinder.

$$\nabla^2 f \equiv \Delta f = \frac{1}{r^2}\frac{\partial}{\partial r}\left(r^2\frac{\partial f}{\partial r}\right) + \frac{1}{r^2 sin(\theta)}\frac{\partial}{\partial \theta}\left(sin(\theta)\frac{\partial f}{\partial \theta}\right)$$
$$+\frac{1}{r^2 sin^2(\theta)}\frac{\partial^2 f}{\partial \phi^2} \tag{4.218}$$

The gradient is

$$\nabla p = \frac{\partial p}{\partial r}\hat{r} + \frac{1}{r}\frac{\partial p}{\partial \theta}\hat{\theta} + \frac{1}{r sin(\theta)}\frac{\partial p}{\partial \phi}\hat{\phi}. \tag{4.219}$$

The vector laplacian is given by

$$\nabla^2 \mathbf{v} \equiv \Delta \mathbf{v} =$$

$$(\Delta v_r - \frac{2v_r}{r^2} - \frac{2}{r^2 sin(\theta)}\frac{\partial(v_\theta sin(\theta))}{\partial \theta} - \frac{2}{r^2 sin(\theta)}\frac{\partial v_\phi}{\partial \phi})\hat{r}$$
$$+(\Delta v_\theta - \frac{v_\theta}{r^2 sin^2(\theta)} + \frac{2}{r^2}\frac{\partial v_r}{\partial \theta} - \frac{2cos(\theta)}{r^2 sin^2(\theta)}\frac{\partial v_\phi}{\partial \phi})\hat{\theta}$$
$$+(\Delta v_\phi - \frac{v_\phi}{r^2 sin^2(\theta)} + \frac{2}{r^2 sin(\theta)}\frac{\partial v_r}{\partial \phi} + \frac{2cos(\theta)}{r^2 sin^2(\theta)}\frac{\partial v_\theta}{\partial \phi})\hat{\phi}. \tag{4.220}$$

The material derivative is given by

$$(\mathbf{v}\cdot\nabla)\mathbf{v} = \left(v_r\frac{\partial v_r}{\partial r} + \frac{v_\theta}{r}\frac{\partial v_r}{\partial \theta} + \frac{v_\phi}{r sin(\theta)}\frac{\partial v_r}{\partial \phi} - \frac{v_\theta^2 + v_\phi^2}{r}\right)\hat{r}$$

$$+\left(v_r\frac{\partial v_\theta}{\partial r} + \frac{v_\theta}{r}\frac{\partial v_\theta}{\partial \theta} + \frac{v_\phi}{r sin(\theta)}\frac{\partial v_\theta}{\partial \phi} + \frac{v_\theta v_r}{r} - \frac{v_\phi^2 cot(\theta)}{r}\right)\hat{\theta}$$

$$+\left(v_r\frac{\partial v_\phi}{\partial r} + \frac{v_\theta}{r}\frac{\partial v_\phi}{\partial \theta} + \frac{v_\phi}{r sin(\theta)}\frac{\partial v_\phi}{\partial \phi} + \frac{v_\phi v_r}{r} - \frac{v_\phi v_\theta cot(\theta)}{r}\right)\hat{\phi} \tag{4.221}$$

and the divergence is given by

$$\nabla\cdot\mathbf{v} = \frac{1}{r^2}\frac{\partial(r^2 v_r)}{\partial r} + \frac{1}{r sin(\theta)}\frac{\partial}{\partial \theta}(v_\theta sin(\theta)) + \frac{1}{r sin(\theta)}\frac{\partial v_\phi}{\partial \phi}. \tag{4.222}$$

**Case of a Cylinder**

To study this case, as before, we have to use the identities for cylindrical coordinates given in the box at the end. As before, the equations to be solved are,

$$\nabla p = \eta \nabla^2 \mathbf{v} \tag{4.223}$$

subject to the incompressibility constraint,

$$\nabla \cdot \mathbf{v} = 0 \tag{4.224}$$

and the boundary conditions $\mathbf{v}(r = \infty) = \mathbf{u}$. Naturally, in this problem, we are going to assume that there is no variation in the z-direction and we may assume without loss of generality that the z-component of the velocity of the fluid is also zero. Instead of following the brute-force method adopted in case of a sphere, we prefer a simpler route, viz., we assert that the pressure is linear in the velocity of the fluid at infinity. The reason is that small Reynolds number flow equations are linear PDEs and the solutions also depend linearly on the parameters on the boundary, such as the velocity at infinity and so on. But pressure is a scalar quantity and velocity is a vector quantity. The way to get a scalar from a vector is to take the dot product with another vector, in this case the position vector is the only option. The coefficient is then some function of the magnitude of the position vector.

$$p(r,\phi) = (\mathbf{u} \cdot \mathbf{r}) f(r) \tag{4.225}$$

Here, $\mathbf{u}$ is the velocity of the fluid at infinity and the function $f(r)$ vanishes at infinity rapidly enough so that $p(\infty, \phi) = 0$. In this case, the steady-state incompressible NS equation Eq. (4.223) may be rewritten as (we choose the x axis to be along $\mathbf{u}$),

$$\frac{\partial(ur\cos(\phi) f(r))}{\partial r} = \eta(\nabla^2 v_r - \frac{v_r}{r^2} - \frac{2}{r^2}\frac{\partial v_\phi}{\partial \phi})$$

$$\frac{1}{r}\frac{\partial(ur\cos(\phi) f(r))}{\partial \phi} = \eta(\nabla^2 v_\phi - \frac{v_\phi}{r^2} + \frac{2}{r^2}\frac{\partial v_r}{\partial \phi}). \tag{4.226}$$

In this case, the incompressibility requirement says $\nabla \cdot \mathbf{v} \equiv 0$, or,

$$\frac{1}{r}\frac{\partial(r v_r)}{\partial r} + \frac{1}{r}\frac{\partial v_\phi}{\partial \phi} = 0. \tag{4.227}$$

This means the radial component is related to the tangential component,

$$v_r = -\frac{1}{r}\int_a^r dr' \frac{\partial v_\phi(r',\phi)}{\partial \phi} \tag{4.228}$$

$$\nabla^2 f \equiv \Delta f = \frac{1}{r}\left(\frac{\partial}{\partial r} r \frac{\partial}{\partial r} f + \frac{\partial}{\partial \phi}\frac{1}{r}\frac{\partial f}{\partial \phi} + r\frac{\partial^2 f}{\partial z^2}\right). \tag{4.229}$$

Therefore,

$$\frac{\partial(urcos(\phi)f(r))}{\partial r}$$

$$= \eta(\frac{1}{r}\left(\frac{\partial}{\partial r}r\frac{\partial}{\partial r}v_r + \frac{\partial}{\partial\phi}\frac{1}{r}\frac{\partial v_r}{\partial\phi}\right) - \frac{v_r}{r^2} - \frac{2}{r^2}\frac{\partial v_\phi}{\partial\phi}), \tag{4.230}$$

and

$$\frac{1}{r}\frac{\partial(urcos(\phi)f(r))}{\partial\phi}$$

$$= \eta(\frac{1}{r}\left(\frac{\partial}{\partial r}r\frac{\partial}{\partial r}v_\phi + \frac{\partial}{\partial\phi}\frac{1}{r}\frac{\partial v_\phi}{\partial\phi}\right) - \frac{v_\phi}{r^2} + \frac{2}{r^2}\frac{\partial v_r}{\partial\phi}). \tag{4.231}$$

We set

$$v_\phi(r,\phi) = w'_{\phi,1}(r)cos(\phi) + w'_{\phi,2}(r)sin(\phi) \tag{4.232}$$

$$v_r(r,\phi) = -\frac{1}{r}(-w_{\phi,1}(r)sin(\phi) + w_{\phi,2}(r)cos(\phi)). \tag{4.233}$$

For large $r$ we must have $\mathbf{v}(\infty,\phi) \equiv \mathbf{u}$ or $v_r(\infty,\phi) = \mathbf{u}\cdot\hat{r} = u\ cos(\phi)$ and $v_\phi(\infty,\phi) = \mathbf{u}\cdot\hat{\phi} = -u\ sin(\phi)$. But,

$$v_\phi(\infty,\phi) = w'_{\phi,1}(\infty)cos(\phi) + w'_{\phi,2}(\infty)sin(\phi) = -u\ sin(\phi) \tag{4.234}$$

$$v_r(\infty,\phi) = -(-w_{\phi,1}(\infty)sin(\phi) + w_{\phi,2}(\infty)cos(\phi)) = u\ cos(\phi). \tag{4.235}$$

Thus $w'_{\phi,1}(\infty,\phi) \equiv 0$ and $w'_{\phi,2}(\infty) = -u$. Since $w'_{\phi,1}(a,\phi) = w'_{\phi,2}(a,\phi) = 0$, this allows us to suspect that perhaps $w_{\phi,1}(r,\phi) \equiv 0$. We shall proceed under this assumption for now. Now we multiply Eq. (4.231) by $r$ and differentiate Eq. (4.230) with respect $\phi$ and Eq. (4.231) with respect $r$ and equate. We also set

$$v_\phi(r,\phi) = w'_{\phi,2}(r)sin(\phi) \tag{4.236}$$

$$v_r(r,\phi) = -\frac{1}{r}w_{\phi,2}(r)cos(\phi) \tag{4.237}$$

so that

$$\frac{1}{r}\left(\frac{\partial}{\partial r}r\frac{\partial}{\partial r}(\frac{1}{r}w_{\phi,2}(r)) - \frac{1}{r}(\frac{1}{r}w_{\phi,2}(r))\right)$$

$$-\frac{(\frac{1}{r}w_{\phi,2}(r))}{r^2} + \frac{2}{r^2}w'_{\phi,2}(r)$$

$$= \left(\frac{\partial^2}{\partial r^2}r\frac{\partial}{\partial r}w'_{\phi,2}(r) - \frac{\partial}{\partial r}\frac{1}{r}w'_{\phi,2}(r)\right)$$

$$-\frac{\partial}{\partial r}\frac{w'_{\phi,2}(r)}{r} + \frac{\partial}{\partial r}\frac{2}{r}(\frac{1}{r}w_{\phi,2}(r)). \tag{4.238}$$

The general solution to this is $w_{\phi,2}(r) = \frac{C_1}{r} + rC_2 + rC_3Log(r)$. Clearly, since $v_\phi \sim w'_{\phi,2}(r)$ and $v_r \sim \frac{w_{\phi,2}(r)}{r}$ and these have to be finite at $r = \infty$, we must set $C_4 \equiv 0$. Thus,

$$v_\phi(r,\phi) = (-\frac{C_1}{r^2} + C_2 + C_3 + C_3Log(r))\ sin(\phi) \tag{4.239}$$

$$v_r(r,\phi) = -(\frac{C_1}{r^2} + C_2 + C_3Log(r))\ cos(\phi). \tag{4.240}$$

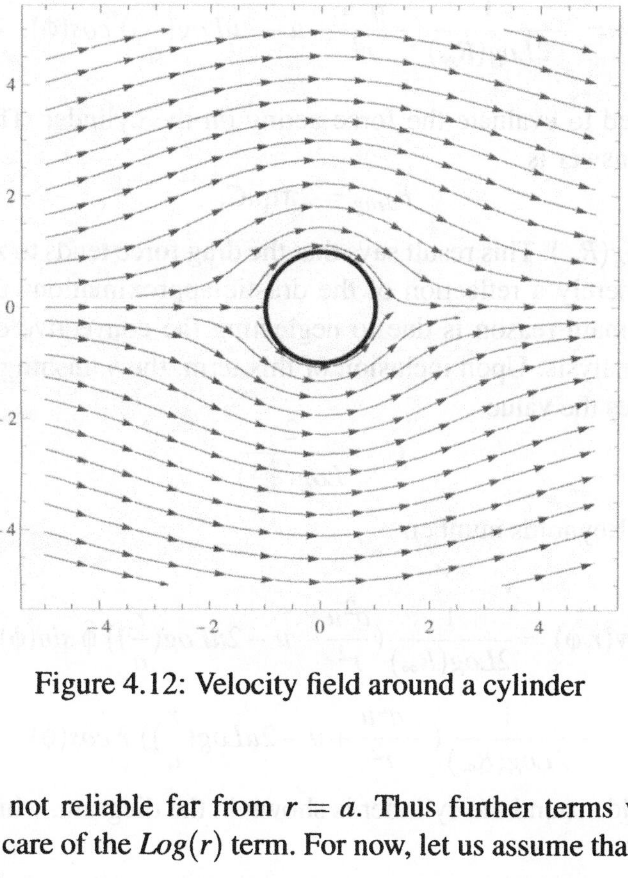


Figure 4.12: Velocity field around a cylinder

This solution is not reliable far from $r = a$. Thus further terms will have to be included to take care of the $Log(r)$ term. For now, let us assume that infinity means some distance $R_\infty$.

$$v_\phi(\infty,\phi) \approx (C_3Log(R_\infty))\ sin(\phi) = -u\ sin(\phi) \tag{4.241}$$

$$v_r(\infty,\phi) \approx -(C_3Log(R_\infty))\ cos(\phi) = u\ cos(\phi) \tag{4.242}$$

Also at $r = a$ the velocity should vanish.

$$0 = (-\frac{C_1}{a^2} + C_2 + C_3 + C_3Log(a))\ sin(\phi) \tag{4.243}$$

$$0 = -(\frac{C_1}{a^2} + C_2 + C_3Log(a))\ cos(\phi) \tag{4.244}$$

This means,

$$C_1 = -\frac{a^2 u}{2Log(R_\infty)}; \; C_2 = \frac{u(1+2Log(a))}{2Log(R_\infty)}; \; C_3 = -\frac{u}{Log(R_\infty)}. \tag{4.245}$$

This in turn means,

$$v_\phi(r,\phi) = \frac{1}{2Log(R_\infty)}\left(\frac{a^2 u}{r^2} - u - 2uLog(\frac{r}{a})\right) sin(\phi) \tag{4.246}$$

$$v_r(r,\phi) = -\frac{1}{2Log(R_\infty)}\left(-\frac{a^2 u}{r^2} + u - 2uLog(\frac{r}{a})\right) cos(\phi). \tag{4.247}$$

One may proceed to evaluate the force acting on the cylinder. This is left to the exercises. The answer is

$$F_{drag} = 2\pi\eta u C, \tag{4.248}$$

where $C = 2/Log(R_\infty)$. This result says that the drag force tends to zero since $R_\infty \to \infty$. But this is merely a reflection of the drastic approximations used. Oseen has shown that the main reason is due to neglecting the convective derivative $\mathbf{v} \cdot \nabla \mathbf{v}$ in the simpler analysis. Upon inclusion of this term, the vanishing coefficient $C$ is 'tamed' and takes the value

$$C = \frac{2}{Log(\frac{7.4}{Re})}, \tag{4.249}$$

where $Re$ is the Reynolds number.

$$\mathbf{v}(r,\phi) = \frac{1}{2Log(R_\infty)}\left(\frac{a^2 u}{r^2} - u - 2uLog(\frac{r}{a})\right) \hat{\phi} \; sin(\phi)$$
$$- \frac{1}{2Log(R_\infty)}\left(-\frac{a^2 u}{r^2} + u - 2uLog(\frac{r}{a})\right) \hat{r} \; cos(\phi) \tag{4.250}$$

The velocity field around the cylinder is shown in the diagram, which is uniform in the z-direction.

The cylindrical Laplacian is

$$\nabla^2 f \equiv \Delta f = \frac{1}{r}\left(\frac{\partial}{\partial r} r \frac{\partial}{\partial r} f + \frac{\partial}{\partial \phi}\frac{1}{r}\frac{\partial f}{\partial \phi} + r\frac{\partial^2 f}{\partial z^2}\right). \quad (4.251)$$

The gradient is

$$\nabla p = \frac{\partial p}{\partial r}\hat{r} + \frac{1}{r}\frac{\partial p}{\partial \phi}\hat{\phi} + \frac{\partial p}{\partial z}\hat{z}. \quad (4.252)$$

The vector Laplacian is given by

$$\Delta \mathbf{v} = (\nabla^2 v_r - \frac{v_r}{r^2} - \frac{2}{r^2}\frac{\partial v_\phi}{\partial \phi})\hat{r} + (\nabla^2 v_\phi - \frac{v_\phi}{r^2} + \frac{2}{r^2}\frac{\partial v_r}{\partial \phi})\hat{\phi} + (\nabla^2 v_z)\hat{z}. \quad (4.253)$$

The material derivative is given by

$$(\mathbf{v}\cdot\nabla)\mathbf{v} = (v_r\frac{\partial v_r}{\partial r} + \frac{v_\phi}{r}\frac{\partial v_r}{\partial \phi} - \frac{v_\phi^2}{r} + v_z\frac{\partial v_r}{\partial z})\hat{r}$$

$$+(v_r\frac{\partial v_\phi}{\partial r} + \frac{v_\phi}{r}\frac{\partial v_\phi}{\partial \phi} + \frac{v_\phi v_r}{r} + v_z\frac{\partial v_\phi}{\partial z})\hat{\phi}$$

$$+(v_r\frac{\partial v_z}{\partial r} + \frac{v_\theta}{r}\frac{\partial v_z}{\partial \phi} + v_z\frac{\partial v_z}{\partial z})\hat{z}, \quad (4.254)$$

and the divergence is given by

$$\nabla\cdot\mathbf{v} = \frac{1}{r}\frac{\partial(rv_r)}{\partial r} + \frac{1}{r}\frac{\partial v_\phi}{\partial \phi} + \frac{\partial v_z}{\partial z}. \quad (4.255)$$

## 4.8 Exercises

**Q.1** Why is the strain tensor defined in the way it is? What is wrong with the choice,

$$\varepsilon_{ij} = \frac{\partial u_i}{\partial x_j}?$$

In other words why is it important to ensure that $\varepsilon_{ij}$ is symmetric? What about the stress tensor? What happens if the stress tensor is not symmetric? (Think of torques acting on a small volume.)

**Q.2** Develop a formalism for the problem where there is symmetry along one of the directions so that variations of *stress* along that direction may be ignored (this is called plane stress similar to plane strain discussed in the chapter).

**Q.3** Find the strain components of a long rod (of length $l$) standing vertically in a gravitational field.

**Q.4** Find the strain components of a hollow sphere (of inner radius $R_1$ and outer radius $R_2$) with a pressure $p_1$ inside and a pressure $p_2$ outside.

**Q.5** Find the strain components of a cylindrical pipe (of external radius $R_2$ and internal radius $R_1$) with pressure $p$ inside and no pressure outside.

**Q.6** Show that the force per unit area acting on a surface of a viscous fluid may be written in component form as

$$t_i = -pn_i + \eta \sum_j n_j \left(\frac{\partial v_i}{\partial x_j} + \frac{\partial v_j}{\partial x_i}\right), \tag{4.256}$$

where $\hat{n}$ is the unit normal to the surface. Show that, in vector notation, this is nothing but

$$\mathbf{t} = -p\hat{n} + \eta(2(\hat{n}\cdot\nabla)\mathbf{v} + \hat{n}\times(\nabla\times\mathbf{v})). \tag{4.257}$$

Show that this can also be written as

$$\mathbf{t} = \mathbb{T}\cdot\hat{n}, \tag{4.258}$$

where $\mathbb{T}$ is the stress tensor of the fluid. Find an expression for this quantity. Show that it is a symmetric tensor. Using this and various vector identities, show that the net force exerted on a finite volume of fluid by the surrounding fluid is

$$\int_S \mathbf{t}dS = \int_V (-\nabla p + \eta\nabla^2\mathbf{v})\, dV, \tag{4.259}$$

where $S$ is the surface bounding the volume $V$.

**Q.7** Show that in case of a unidirectional shear flow $\mathbf{v} = (u(y),0,0)$, the force per unit area acting on a surface with unit normal $\hat{n} = (0,1,0)$ is,

$$\mathbf{t} = (\eta\frac{du}{dy}, -p, 0). \tag{4.260}$$

**Q.8** Find the strain energy associated with the cuboid in the worked-out example in the this chapter. Do the same for the elastic sphere.

**Q.9** In case of purely rotational flow, $\mathbf{v}(r) = u_\theta(r)\hat{e}_\theta$ (where $(r,\theta)$ are plane polar coordinates), find the force acting on a surface with unit normal in (i) $\hat{e}_r$ direction and (ii) in $\hat{e}_\theta$ direction.

**Q.10** Show that the angular momentum density of a fluid is $\vec{S}(\mathbf{r},t) = m\mathbf{r} \times \mathbf{J}(\mathbf{r},t) = m\mathbf{r} \times \rho(\mathbf{r},t)\mathbf{v}(\mathbf{r},t)$. The angular momentum of a volume $V$ of the fluid is $\mathbf{L}(t) \equiv \int_V d^3r\, \vec{S}(\mathbf{r},t)$. Using the equation of continuity and Navier Stokes equation, find the rate of change of this quantity and hence the torque acting on the fluid.

**Q.11** Verify Eq. (4.248).

**Q.12** Generate a sequence of equations in powers of the Reynolds number to describe the flow around a sphere. Try to find the drag force on the sphere up to second order in the Reynolds number.

# Chapter 5

# Toward Quantum Fields: Scalar and Spinor Fields

Thus far we have encountered classical fields such as the electromagnetic field and the current and density fields of fluids and so on. Now we explore the idea that any classical field equations, such as Maxwell's equations, wave equations and so on, are obtainable as the equations of motion of a suitable Lagrangian/Hamiltonian. This would be particularly useful in investigating the nature of the quantum mechanical version of these theories as the machinery for studying quantum mechanics using Hamiltonians, and as we shall see Lagrangians as well, is quite well developed. We start with the wave equation for a scalar field.

$$\nabla^2\phi = \frac{1}{c^2}\frac{\partial^2}{\partial t^2}\phi \tag{5.1}$$

We wish to think of this as the Lagrange equation of a suitable Lagrangian, or equivalently, as the Hamilton equation of a suitable Hamiltonian. The aim now is to find these functions. Typically, a Lagrangian is a quadratic function of the time derivative of the generalized coordinate. This leads to Lagrange equations of the form $\ddot{q} = ....$ We now choose to identify the $\phi(\mathbf{x}) \rightarrow q_i$ - $i^{th}$ generalized coordinate. Thus, just as $L = const\frac{1}{2}\sum_i \dot{q}_i^2 - V(q)$ we may write

$$L = const.\frac{1}{2}\int d^3r\ (\partial_t\phi(\mathbf{x},t))^2 - V(\phi), \tag{5.2}$$

where integration over $\mathbf{x}$ has replaced summation over the index $i$. The Lagrange equation is

$$\partial_t\frac{\delta L}{\delta\dot{\phi}(\mathbf{x},t)} = const.\,(\partial_t^2\phi(\mathbf{x},t)) = const.\,c^2\nabla^2\phi. \tag{5.3}$$

The last line follows from the wave equation. From the above result and Lagrange's equations we may conclude,

$$\frac{\partial L}{\partial \phi(\mathbf{x},t)} = -\frac{\partial V}{\partial \phi(\mathbf{x},t)} = const.\, c^2 \nabla^2 \phi. \tag{5.4}$$

At this stage we choose the constant to be $const. = 1/c^2$ and thus,

$$V(\phi) = \frac{1}{2} \int d^3r \, (\nabla \phi)^2. \tag{5.5}$$

Thus the Lagrangian becomes

$$L = \frac{1}{2c^2} \int d^3r \, (\partial_t \phi(\mathbf{x},t))^2 - \frac{1}{2} \int d^3r \, (\nabla \phi)^2. \tag{5.6}$$

We saw earlier that any equation that involves time derivatives of some field may be recast in the form of Lagrange equations of a Lagrangian. It was shown earlier that the non-relativistic Schrodinger equation may be written as the Lagrange equations of

$$L[\psi, \dot{\psi}; \psi^*, \dot{\psi}^*] = \int dx \, (\psi^*(x,t) \, i\hbar \partial_t \psi(x,t) - \psi(x,t) \, i\hbar \partial_t \psi^*(x,t)$$
$$+ \frac{\hbar^2}{2m} \psi^*(x,t) \partial_x^2 \psi(x,t) - V(x,t) \psi^*(x,t) \psi(x,t)). \tag{5.7}$$

Similarly, the Dirac equation may be thought of as the Lagrange equation of the following Lagrangian.

$$L[\psi, \dot{\psi}; \psi^\dagger, \dot{\psi}^\dagger] = \int d^3r \, (\psi^\dagger(\mathbf{x},t) \, i\hbar \partial_t \psi(\mathbf{x},t) - i\hbar \, (\partial_t \psi^\dagger(\mathbf{x},t)) \psi(\mathbf{x},t)$$
$$- \psi^\dagger(\mathbf{x},t)(-i\hbar c \alpha \cdot \nabla + \beta m c^2) \psi(\mathbf{x},t)), \tag{5.8}$$

where $\beta, \alpha_x, \alpha_y, \alpha_z$ are the four $4 \times 4$ Dirac matrices and $\psi(\mathbf{x},t)$ is a column vector with four rows. The last two assertions require some comment.

On the one hand, the Schrodinger equation and the Dirac equation are supposed to represent the fundamental equations of quantum mechanics. On the other hand, the Lagrange equations are the fundamental equations of classical mechanics. How can they be one and the same? How can one reconcile this apparent discrepancy? The point of view that will be advocated from now on is that we shall always be dealing with a many-particle system. It is well known that there are serious problems with the one-particle description of the Dirac equations due to the ubiquitous existence of anti-particles—a feature absent in non-relativistic quantum mechanics. Hence the point of view shall be that $\psi(\mathbf{x},t)$, instead of being the wave function, is a field not dissimilar to the electromagnetic field that is all pervasive whose excitations we identify as particles (photons in case of EM field and electrons in case of Dirac equation) and antiparticles (in case of Dirac equation). Thus the equation for $\psi(\mathbf{x},t)$, namely the Dirac equation, is nothing but the equation of motion for the operator corresponding to $\psi(\mathbf{x},t)$ as it is in Heisenberg's picture. There is another somewhat unrelated reason for introducing this concept of a Lagrangian of a field. It stems from its use in the statistical mechanics of second-order phase transitions. For example, the Ginzburg Landau Lagrangian of a superconductor in a magnetic field is given by,

$$L_{GL} = \int d^3r \, (\frac{1}{2m}|(-i\hbar\nabla - \frac{2e}{c}\mathbf{A})\psi(\mathbf{x},t)|^2 + \frac{\mathbf{B}^2}{2})$$
$$+ \int d^3r \, (\alpha\, |\psi(\mathbf{x},t)|^2 + \frac{\beta}{2}\, |\psi(\mathbf{x},t)|^4). \tag{5.9}$$

Here the interpretation is that $\psi(\mathbf{x},t)$ is a complex-order parameter whose nonzero value signifies the presence of a superconducting phase in the system. One studies both the classical solutions (corresponding to zero temperature) as well as fluctuations around these solutions (which take into account finite temperature). A similar Lagrangian is possible for a collection of bosons interacting via a hard core potential assuming they are all in the same single particle state. This is the so-called Gross-Pitaevskii equation.

$$L_{GP}[\psi,\dot{\psi};\psi^*,\dot{\psi}^*] = \int d^3x \, (\psi^*(\mathbf{x},t)\; i\hbar\partial_t\psi(\mathbf{x},t) - \psi(\mathbf{x},t)\; i\hbar\partial_t\psi^*(\mathbf{x},t)$$

$$+\frac{\hbar^2}{2m}\psi^*(\mathbf{x},t)\nabla^2\psi(\mathbf{x},t) - V(\mathbf{x},t)\psi^*(\mathbf{x},t)\psi(\mathbf{x},t) - g\,(\psi^*(\mathbf{x},t)\psi(\mathbf{x},t))^2) \tag{5.10}$$

Here, $\psi(\mathbf{x},t)$ has the interpretation of the wavefunction of a single boson so that the wavefunction of a collection of N bosons is $\Psi(\mathbf{x}_1,...,\mathbf{x}_N) = \psi(\mathbf{x}_1)\psi(\mathbf{x}_2)...\psi(\mathbf{x}_N)$. The above has to be combined with the constraint that $\int |\psi(\mathbf{x},t)|^2\, d^3r = 1$. The classical equations determine the evolution of a condensate of bosons in a harmonic trap, for example.

The main focus of this book is nonrelativistic physics. But we should be doing some minimum justice to particle physics by providing some Lagrangians that a reader can find out more about in the references. The Lagrangian of an electromagnetic field is simple; it is purely quadratic in the field variables (four-vector potential). Physically this means a photon does not directly interact with itself (i.e., others of its kind). However, there are other fields, such as Yang Mills fields and Gluon fields, that have a direct interaction between the particles. To introduce this we have to elevate the vector potentials to a matrix form. This means we assert that there are $a = 1,2,..,M$ number of vector fields $A^a_\mu$ and an equal number of square matrices

$t^a$ such that they obey a group property under commutation $[t_a, t_b] = if^{abc}t_c$ where summation over the index $c$ is implied. The matrix four-vector is now

$$A_\mu = \sum_a t_a A^a_\mu. \tag{5.11}$$

The main outcome of this upgrade is that now the components of the four-vector do not have to commute. This means, $[A_\mu, A_\nu] \neq 0$, in general. Examples of such matrices $t_a$ include the three $SU(2)$ generators (the Pauli matrices for $SU(N)$ it is $N^2 - 1$ number of generators)

$$t_1 = \begin{pmatrix} 0 & 1 \\ 1 & 0 \end{pmatrix}; t_2 = \begin{pmatrix} 0 & -i \\ i & 0 \end{pmatrix}; t_3 = \begin{pmatrix} 1 & 0 \\ 0 & -1 \end{pmatrix}, \tag{5.12}$$

or the eight $SU(3)$ generators (Gell-Mann matrices),

$$t_1 = \begin{pmatrix} 0 & 1 & 0 \\ 1 & 0 & 0 \\ 0 & 0 & 0 \end{pmatrix}; t_2 = \begin{pmatrix} 0 & -i & 0 \\ i & 0 & 0 \\ 0 & 0 & 0 \end{pmatrix}$$

$$t_3 = \begin{pmatrix} 1 & 0 & 0 \\ 0 & -1 & 0 \\ 0 & 0 & 0 \end{pmatrix}; t_4 = \begin{pmatrix} 0 & 0 & 1 \\ 0 & 0 & 0 \\ 1 & 0 & 0 \end{pmatrix}$$

$$t_5 = \begin{pmatrix} 0 & 0 & -i \\ 0 & 0 & 0 \\ i & 0 & 0 \end{pmatrix}; t_6 = \begin{pmatrix} 0 & 0 & 0 \\ 0 & 0 & 1 \\ 0 & 1 & 0 \end{pmatrix}$$

and,

$$t_7 = \begin{pmatrix} 0 & 0 & 0 \\ 0 & 0 & -i \\ 0 & i & 0 \end{pmatrix}; t_8 = \begin{pmatrix} \frac{1}{\sqrt{3}} & 0 & 0 \\ 0 & \frac{1}{\sqrt{3}} & 0 \\ 0 & 0 & -\frac{2}{\sqrt{3}} \end{pmatrix}. \tag{5.13}$$

The field tensor is now no longer just what was given in the case of electromagnetism, but now it involves a nonlinear term signifying interaction of the field with itself,

$$F_{\mu\nu} = \partial_\mu A_\nu - \partial_\nu A_\mu + ig[A_\mu, A_\nu], \tag{5.14}$$

where $g$ represents the coupling of the field with itself. These fields are called non-Abelian (non-commutative) fields. The Lagrangian density now becomes

$$\mathcal{L}_G = -\frac{1}{16\pi} Tr(F^{\mu\nu} F_{\mu\nu}), \tag{5.15}$$

where the trace is over the matrices $t_a$. Such gauge fields (bosons) couple to matter fields (fermions) through the minimal coupling procedure. For example, in order to describe the coupling quarks with gluons, we have to write

$$\mathcal{L}_{QCD} = i\bar{\psi}_u(\partial_\mu - ig_s A_\mu)\gamma^\mu \psi_u + i\bar{\psi}_d(\partial_\mu - ig_s A_\mu)\gamma^\mu \psi_d - \frac{1}{16\pi} Tr(F^{\mu\nu} F_{\mu\nu}), \tag{5.16}$$

where $\psi_u$ stands for the field of the up quark and $\psi_d$ stands for the field of the down quark. Since both $\gamma^\mu$ and $A_\mu$ are present together, each of these quark fields are actually column vectors with $4(N^2-1)$ entries. Thus, a host of different Lagrangian densities may be written down in terms of fields that describe various physical systems, both relativistic as well as non-relativistic. Solving for their properties is another matter, however. We shall content ourselves in merely describing the models so that the interested reader may then choose to investigate how to solve one of these models by consulting relevant literature on each of these vast topics. We shall now revert to simple non-relativistic physics.

## 5.1 Some Solutions of the Schrodinger Equation

We pointed out earlier that the Lagrangian of the Schrodinger equation possesses symmetry under a global phase transformation, namely $\psi \to e^{i\theta}\psi$. This leads to the total probability being conserved. Now we consider some solutions to the Schrodinger equation that are of special interest in many body physics. For instance, one may think of the Green function of the Schrodinger equation in the presence of a delta function impulsive potential both in space and time.

Figure 5.1: Erwin Rudolf Josef Alexander Schrodinger (12 August 1887 to 4 January 1961), was an Austrian physicist who developed a number of fundamental results in the field of quantum theory, which formed the basis of wave mechanics: he formulated the wave equation (stationary and time-dependent Schrodinger equation) and revealed the identity of his development of the formalism and matrix mechanics.

$$(i\hbar\frac{\partial}{\partial t}+\frac{\hbar^2}{2m}\frac{\partial^2}{\partial x^2}-V(x,t))\psi(x,t)=0 \tag{5.17}$$

Now imagine that $V(x,t)$ contains two pieces; first is some static potential and an impulse so that $V(x,t)=V_0(x)+\chi\,\delta(x-x_0)\delta(t-t_0)$.

$$(i\hbar\frac{\partial}{\partial t}+\frac{\hbar^2}{2m}\frac{\partial^2}{\partial x^2}-V(x,t))\psi(x,t)=0 \tag{5.18}$$

The idea is to find how the wave function changes in space and time for times $t>t_0$. For $t<t_0$, we assume that the system is an stationary state with energy $E_\nu$.

$$\psi(x,t\leq t_0)=e^{-i\frac{E_\nu}{\hbar}(t-t_0)}\,\varphi_\nu(x), \tag{5.19}$$

where

$$(-\frac{\hbar^2}{2m}\frac{d^2}{dx^2}+V_0(x))\varphi_\nu(x)=E_\nu\,\varphi_\nu(x). \tag{5.20}$$

Since for times $t>t_0$ the Hamiltonian is again the static one, we may use the basis states of the static Hamiltonian to write

$$\psi(x,t>t_0)=\sum_{\nu'}c_{\nu'}\,e^{-i\frac{E_{\nu'}}{\hbar}(t-t_0)}\,\varphi_{\nu'}(x). \tag{5.21}$$

Integrating the time-dependent Schrodinger equation from time $t_0-\varepsilon$ to $t_0+\varepsilon$ we get,

$$i\hbar(\psi(x,t_0+\varepsilon)-\psi(x,t_0-\varepsilon))=\chi\,\delta(x-x_0)\psi(x,t_0). \tag{5.22}$$

Substituting the basis expanded wavefunction in the above constraint we get,

$$(\sum_{\nu'}c_{\nu'}\,\varphi_{\nu'}(x)-\varphi_\nu(x))=\frac{\chi}{i\hbar}\,\delta(x-x_0)\varphi_\nu(x_0). \tag{5.23}$$

Multiplying by the complex conjugate of a basis wavefunction an integrating over $x$ and using orthonormality of the wavefunctions we get,

$$c_{\nu'}=\frac{\chi}{i\hbar}\,\varphi^*_{\nu'}(x_0)\varphi_\nu(x_0)+\delta_{\nu,\nu'}. \tag{5.24}$$

Therefore,

$$\psi(x,t>t_0)=e^{-i\frac{E_\nu}{\hbar}(t-t_0)}\,\varphi_\nu(x)+\sum_{\nu'}\frac{\chi}{i\hbar}\,\varphi^*_{\nu'}(x_0)\varphi_\nu(x_0)\,e^{-i\frac{E_{\nu'}}{\hbar}(t-t_0)}\,\varphi_{\nu'}(x). \tag{5.25}$$

If one starts with a free particle with a plane wave as a basis then $\varphi_\nu(x)=\frac{1}{\sqrt{L}}e^{ikx}$ and $E_\nu=\frac{\hbar^2k^2}{2m}$ so that,

$$\psi(x,t>t_0)=e^{-i\frac{\hbar k^2}{2m}(t-t_0)}\,\frac{1}{\sqrt{L}}e^{ikx}+\frac{\chi}{i\hbar}\,\frac{e^{ikx_0}}{\sqrt{L}}\,\frac{1}{L}\sum_{k'}e^{-i\frac{\hbar k'^2}{2m}(t-t_0)}\,e^{ik'(x-x_0)}. \tag{5.26}$$

Following the description in the box, we write $\sum_{k'}(...) = \frac{L}{2\pi}\int_{-\infty}^{\infty} dk'(...)$ so that,

> To sum over states that form a continuum, we typically employ the following procedure. Consider first a particle in a box of length $L$. Its states are labeled $n = 1,2,3,...$ an integer with corresponding momenta $k_n = \frac{\pi n}{L}$. Thus, summation over these states means evaluating a sum such as $S = \sum_{n=1,2,...} f(k_n) = \frac{L}{\pi}\sum_{n=0}^{\infty} f(\Delta k + n\Delta k)\Delta k \approx \frac{L}{\pi}\int_0^{\infty} dk\, f(\Delta k + k) \approx \frac{L}{2\pi}\int_{-\infty}^{\infty} dk\, f(k)$ where $\Delta x = \frac{\pi}{L}$. We assume that $f(k)$ is an even function of $k$ since for particle in a box $-k$ is the same state as $k$.

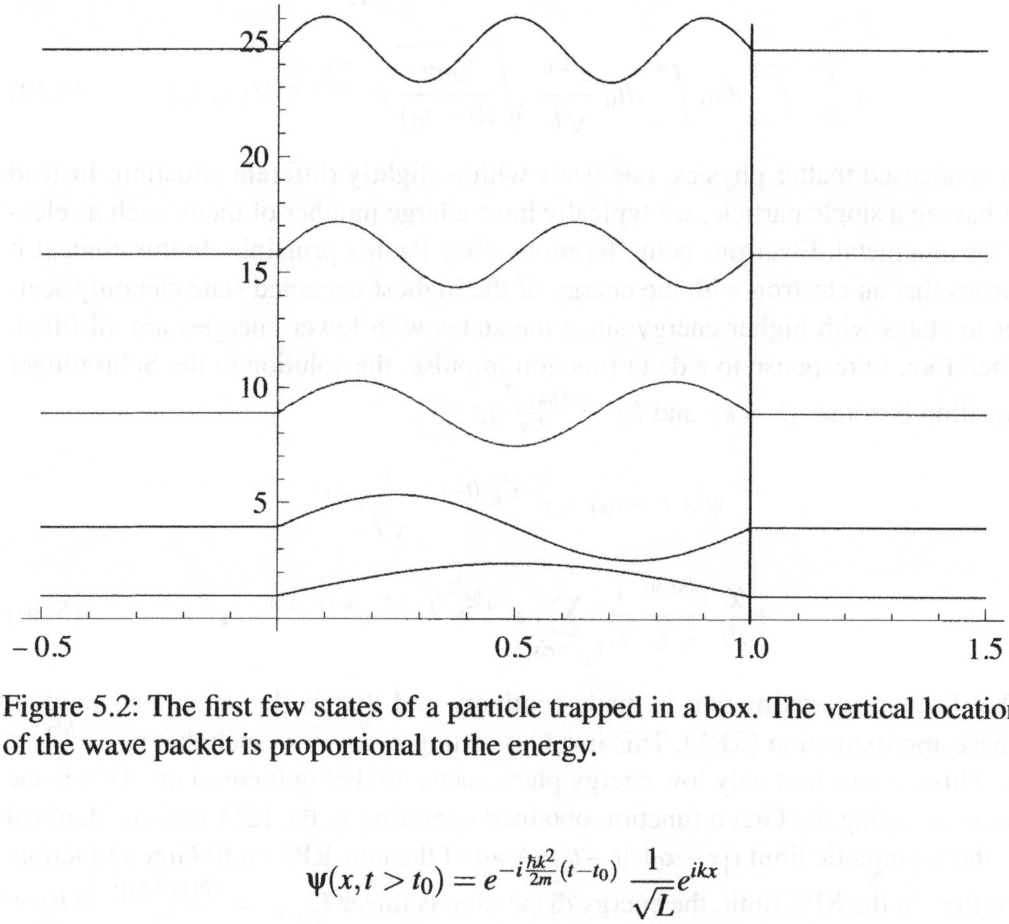


Figure 5.2: The first few states of a particle trapped in a box. The vertical location of the wave packet is proportional to the energy.

$$\psi(x,t>t_0) = e^{-i\frac{\hbar k^2}{2m}(t-t_0)}\frac{1}{\sqrt{L}}e^{ikx}$$
$$+\frac{1}{2\pi}\frac{\chi}{i\hbar}\frac{e^{ikx_0}}{\sqrt{L}}\int_{-\infty}^{\infty} dk'\, e^{-i\frac{\hbar k'^2}{2m}(t-t_0)}\, e^{ik'(x-x_0)}$$
$$= e^{-i\frac{\hbar k^2}{2m}(t-t_0)}\frac{1}{\sqrt{L}}e^{ikx} + \frac{1}{2\pi}\frac{\chi}{i\hbar}\frac{e^{ikx_0}}{\sqrt{L}}\sqrt{\frac{\pi}{\frac{i}{2m}\hbar(t-t_0)}}\, e^{-\frac{m(x-x_0)^2}{2\hbar i(t-t_0)}} \tag{5.27}$$

The term $G(x-x_0,t-t_0)$ where

$$G(x-x_0,t-t_0)=\sqrt{\frac{m}{ih(t-t_0)}}\,e^{-\frac{m(x-x_0)^2}{2\hbar i(t-t_0)}} \tag{5.28}$$

is known as the Green function of a free particle. The reason why it is called that is because the wavefunction of the system with any time-dependent potential $V(x,t)$ may be obtained by linearly combining the above solutions with the external potential as the weights.

$$\psi_V(x,t>t_0)=e^{-i\frac{\hbar k^2}{2m}(t-t_0)}\,\frac{1}{\sqrt{L}}e^{ikx}$$

$$+\frac{1}{ih}\int_{-\infty}^{\infty}dx_0\int_{-\infty}^{\infty}dt_0\,\frac{e^{ikx_0}}{\sqrt{L}}\sqrt{\frac{2m\pi}{i(t-t_0)}}\,e^{-\frac{m(x-x_0)^2}{2\hbar i(t-t_0)}}V(x_0,t_0) \tag{5.29}$$

In condensed matter physics, one deals with a slightly different situation. Instead of having a single particle, we typically have a large number of them, such as electrons in a metal. Electrons being fermions obey Pauli's principle. In this context it means that an electron with the energy of the highest occupied state can only scatter to states with higher energy since the states with lower energies are all filled. Therefore, in response to a delta function impulse, the solution to the Schrodinger equation becomes ($k=k_F$ and $E_F=\frac{(\hbar k_F)^2}{2m}$),

$$\psi(x,t>t_0)=e^{-i\frac{E_F}{\hbar}(t-t_0)}\,\frac{1}{\sqrt{L}}e^{ik_Fx}$$

$$+\frac{\chi}{i\hbar}\frac{e^{ik_Fx_0}}{\sqrt{L}}\frac{1}{L}\sum_{E_{k'}>E_F}e^{-i\frac{\hbar k'^2}{2m}(t-t_0)}\,e^{ik'(x-x_0)}. \tag{5.30}$$

One further approximation is made routinely and that is the celebrated random phase approximation (RPA). This involves setting $k_F,m\to\infty$ such that $v_F=\frac{\hbar k_F}{m}<\infty$. This ensures that only low energy phenomena are being focused on. This is the same as saying the Green function obtained operating in the RPA limit is identical to the asymptotic limit ($|x-x_0|,|t-t_0|\to\infty$) of the non-RPA (full) Green function. Further, in the RPA limit, the energy dispersion is linear $E_{k_F+k}=\frac{(\hbar(k_F+k))^2}{2m}\approx E_F+v_F\hbar k$ for $|k|\ll k_F$ for momenta close to the positive side of the highest occupied state and similarly for the negative side we have, $E_{-k_F+k}=\frac{(\hbar(-k_F+k))^2}{2m}\approx E_F-v_F\hbar k$. Therefore, the asymptotically exact Green function of a free fermion in the presence of a filled Fermi sea would be,

$$\psi(x,t>t_0)=e^{-i\frac{E_F}{\hbar}(t-t_0)}\,\frac{1}{\sqrt{L}}e^{ik_Fx}$$

$$+\frac{\chi}{i\hbar}\frac{e^{ik_Fx_0}}{\sqrt{L}}\,e^{-iE_F(t-t_0)}\,e^{ik_F(x-x_0)}\frac{1}{L}\sum_{k'>0}e^{ik'[(x-x_0)-v_F(t-t_0)]}$$

$$+\frac{\chi}{i\hbar}\,e^{-iE_F(t-t_0)}\,e^{-ik_F(x-x_0)}\frac{e^{ik_Fx_0}}{\sqrt{L}}\frac{1}{L}\sum_{k'<0}e^{iv_F\hbar k'(t-t_0)}\,e^{ik'[(x-x_0)+v_F(t-t_0)]}. \quad (5.31)$$

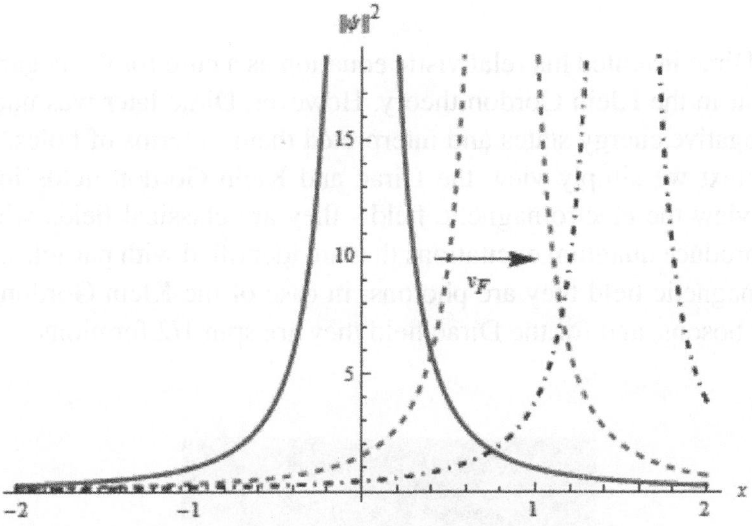


Figure 5.3: The probability density of a right-moving wave packet. It moves with a velocity $v_F$ to the right, and is peaked at the classical distance–time relation of a free particle $x = v_F t$

or,

$$\psi(x,t>t_0) = e^{-i\frac{E_F}{\hbar}(t-t_0)}\,\frac{1}{\sqrt{L}}e^{ik_Fx}$$

$$-\frac{\chi}{i\hbar}\frac{e^{ik_Fx_0}}{\sqrt{L}}\,e^{-i\frac{E_F}{\hbar}(t-t_0)}\,e^{ik_F(x-x_0)}\frac{1}{2\pi i}\,\frac{1}{(x-x_0)-v_F(t-t_0)}$$

$$+\frac{\chi}{i\hbar}\frac{e^{ik_Fx_0}}{\sqrt{L}}\,e^{-i\frac{E_F}{\hbar}(t-t_0)}\,e^{-ik_F(x-x_0)}\frac{1}{2\pi i}\,\frac{1}{(x-x_0)+v_F(t-t_0)}. \quad (5.32)$$

We shall have occasion to use this later on when we encounter Luttinger liquids. For later use, we note the following terminology—the Green function of right movers is

$$G_R(x-x_0,t-t_0) = -e^{-i\frac{E_F}{\hbar}(t-t_0)}\,e^{ik_F(x-x_0)}\frac{1}{2\pi i}\,\frac{1}{(x-x_0)-v_F(t-t_0)}, \quad (5.33)$$

and for left movers it is

$$G_L(x-x_0,t-t_0) = e^{-i\frac{E_F}{\hbar}(t-t_0)}\, e^{-ik_F(x-x_0)} \frac{1}{2\pi i} \frac{1}{(x-x_0)+v_F(t-t_0)}. \tag{5.34}$$

## 5.2 Some Properties of the Dirac Equation and Klein Gordon Equations

Originally, Dirac invented his relativistic equation as a cure for the negative energy states present in the Klein Gordon theory. However, Dirac later was unable to get rid of the negative energy states and interpreted them in terms of holes. But in the modern context we simply view the Dirac and Klein Gordon fields in the same way as we view the electromagnetic field—they are classical fields which, when quantized, produce quantum excitations that are identified with particles. In case of the electromagnetic field they are photons; in case of the Klein Gordon field they are spinless bosons, and for the Dirac field they are spin 1/2 fermions.

Figure 5.4: One of the founding figures of modern quantum mechanics, Paul Dirac (8 August 1902 to 20 October 1984) laid the foundations of quantum electrodynamics, developed the theory of electrons, predicted the existence of the positron, and introduced the widely used bracket notation. His book *Principles of Quantum Mechanics* is still one of the best introductions to the subject. He was awarded the Nobel Prize for Physics in 1933.

■ Consider the Lorentz transformation $x'^{\mu} = \Lambda^{\mu}_{\nu}(v)x^{\nu}$. Which of the following set of equations respect the principle of relativity? Prove your answer by actual calculation where $x' = \gamma x - \gamma v t, t' = \gamma t - \gamma v x/c^2$ but $\psi'(x',t') = \psi(x,t)$, Lorentz scalar.

$$i\hbar\frac{\partial}{\partial t}\psi(x,t) = -\frac{\hbar^2}{2m}\frac{\partial^2}{\partial x^2}\psi(x,t) \tag{5.35}$$

$$i\hbar\frac{\partial}{\partial t}\psi(x,t) = -i\hbar c(\sigma\cdot\nabla)\psi(x,t) \tag{5.36}$$

$$\frac{\partial^2}{\partial x^2}A^{\mu}(x,t) = \frac{1}{c^2}\frac{\partial^2}{\partial t^2}A^{\mu}(x,t) \tag{5.37}$$

Here, $A^{\mu}$ is a four-vector.

To answer these equations, start with $x' = \gamma x - \gamma v t, t' = \gamma t - \gamma\frac{vx}{c^2}$. Thus,

$$\frac{\partial}{\partial x} = \frac{\partial x'}{\partial x}\frac{\partial}{\partial x'} + \frac{\partial t'}{\partial x}\frac{\partial}{\partial t'}\,;\ \frac{\partial}{\partial t} = \frac{\partial x'}{\partial t}\frac{\partial}{\partial x'} + \frac{\partial t'}{\partial t}\frac{\partial}{\partial t'} \tag{5.38}$$

$$\frac{\partial x'}{\partial x} = \gamma\,;\ \frac{\partial t'}{\partial x} = -\gamma\frac{v}{c^2}\,;\ \frac{\partial x'}{\partial t} = -\gamma v\,;\ \frac{\partial t'}{\partial t} = \gamma, \tag{5.39}$$

or

$$\frac{\partial}{\partial x} = \gamma\frac{\partial}{\partial x'} - \frac{\gamma v}{c^2}\frac{\partial}{\partial t'}\,;\ \frac{\partial}{\partial t} = -\gamma v\frac{\partial}{\partial x'} + \gamma\frac{\partial}{\partial t'}. \tag{5.40}$$

(a) If the first equation respects relativity, then we must assume also that

$$i\hbar\frac{\partial}{\partial t'}\psi'(x',t') = -\frac{\hbar^2}{2m}\frac{\partial^2}{\partial x'^2}\psi'(x',t'). \tag{5.41}$$

Substitute Eq. (5.40) into Eq. (5.35) to get (it is given that $\psi(x,t) = \psi'(x',t')$).

$$i\hbar\left(-\gamma v\frac{\partial}{\partial x'} + \gamma\frac{\partial}{\partial t'}\right)\psi'(x',t') = -\frac{\hbar^2}{2m}\left(\gamma\frac{\partial}{\partial x'} - \frac{\gamma v}{c^2}\frac{\partial}{\partial t'}\right)^2\psi'(x',t') \tag{5.42}$$

This is not the same equation as Eq. (5.41) as terms such as $\frac{\partial}{\partial x'}\frac{\partial}{\partial t'}$ are present in the transformed version of Eq. (5.35), whereas there should not be such terms if the equation respected the principle of relativity.

(b) Substitute Eq. (5.40) into Eq. (5.36) to get (again $\psi(x,t) = \psi'(x',t')$, and assume that $\psi$ does not depend on $y$ or $z$, thus $\frac{\partial}{\partial y} = \frac{\partial}{\partial z} = 0$),

$$i\hbar\left(-\gamma v\frac{\partial}{\partial x'}+\gamma\frac{\partial}{\partial t'}\right)\psi'(x',t')=-i\hbar c\,\sigma_x\left(\gamma\frac{\partial}{\partial x'}-\frac{\gamma v}{c^2}\frac{\partial}{\partial t'}\right)\psi'(x',t'). \quad (5.43)$$

This is consistent for all values of $v$ only if

$$i\hbar\left(\gamma\frac{\partial}{\partial t'}\right)\psi'(x',t')=-i\hbar c\,\sigma_x\left(\gamma\frac{\partial}{\partial x'}\right)\psi'(x',t') \quad (5.44)$$

and,

$$i\hbar\left(-\gamma v\frac{\partial}{\partial x'}\right)\psi'(x',t')=-i\hbar c\,\sigma_x\left(-\frac{\gamma v}{c^2}\frac{\partial}{\partial t'}\right)\psi'(x',t'). \quad (5.45)$$

In the first one we cancel $\gamma$ from both sides and this is nothing but Eq. (5.36) in primed coordinates. In the second one we cancel $-\gamma v$ from both sides, multiply by $c\,\sigma_x$, and use $\sigma_x^2=1$ to get the same equation.

(c) Inverting Eq. (5.40) is the same as choosing $-v$ instead of $v$ and choosing $t',x'$ instead of $t,x$ and vice versa.

$$\frac{\partial}{\partial x'}=\gamma\frac{\partial}{\partial x}+\frac{\gamma v}{c^2}\frac{\partial}{\partial t}\;;\;\frac{\partial}{\partial t'}=\gamma v\frac{\partial}{\partial x}+\gamma\frac{\partial}{\partial t} \quad (5.46)$$

Let us calculate the D'Almbertian.

$$\Delta'\equiv\frac{\partial^2}{\partial x'^2}-\frac{1}{c^2}\frac{\partial^2}{\partial t'^2}=\left(\gamma\frac{\partial}{\partial x}+\frac{\gamma v}{c^2}\frac{\partial}{\partial t}\right)^2-\frac{1}{c^2}\left(\gamma v\frac{\partial}{\partial x}+\gamma\frac{\partial}{\partial t}\right)^2$$

$$=\frac{\partial^2}{\partial x^2}-\frac{1}{c^2}\frac{\partial^2}{\partial t^2}\equiv\Delta \quad (5.47)$$

Thus, we have to show that

$$\Delta'A^{\mu'}(x',t')=0 \quad (5.48)$$

is the same as

$$\Delta A^{\mu}(x,t)=0. \quad (5.49)$$

We write,

$$A'^{\mu}(x',t')=\Lambda^{\mu}_{\nu}\,A^{\nu}(x,t). \quad (5.50)$$

Then,

$$\Delta'A^{\mu'}(x',t')=\Lambda^{\mu}_{\nu}\,\Delta A^{\nu}(x,t)=\Lambda^{\mu}_{\nu}\times 0=0. \quad (5.51)$$

Thus, $\Delta A^{\nu}(x,t)=0$ means $\Delta'A^{\mu'}(x',t')=0$.

## 5.3 Exercises

**Q.1** Discuss the symmetries and possible conserved quantities of the Lagrangians encountered in this chapter.

**Q.2** Find the Green function in momentum and frequency space for the case of free particles Eq. (5.28) and free fermions Eq. (5.33) and Eq. (5.34). Write

$$G(x,t) = \int_{-\infty}^{\infty} \frac{dk}{2\pi} \int_{-\infty}^{\infty} \frac{d\omega}{2\pi} e^{ikx} e^{-i\omega t} G(k,\omega), \tag{5.52}$$

and find $G(k,\omega)$ and also the spectral function $A(k,\omega) = -2Im(G(k,\omega - i\varepsilon))$ (here $\omega$ is real and $\varepsilon \to 0$).

**Q.3** Perturbatively evaluate the leading correction to the Green function in case of free fermions upon inclusion of quadratic corrections to the energy dispersion (i.e., set $E_{k_F+k} \approx E_F + v_F k + \frac{k^2}{2m}$ and expand in powers of $m^{-1}$).

**Q.4** Verify (or derive) that the formulas (Eq. (13.111)) for the right- and left-moving Green functions in case of a free fermion with energy $E_F$ in the presence of a delta function potential $V(x) = V_0 \delta(x)$ is correct.

# Chapter 6

# Concept of Functional Integration

So far we have pointed out that any dynamical equation may be thought of as the Lagrange equation of a suitable Lagrangian. The relevance of this is clear when we attempt to make a transition to quantum mechanics or when attempting to study thermal fluctuations of classical statistical systems. In both cases, a term proportional to the exponential of the action (time-integrated Lagrangian) serves as a 'weight' or probability distribution for the field in question as we shall see subsequently. Thus, all possible field configurations are allowed, but each comes with a weight which allows the computation of averages of the field (given by the solution of the classical equations) and the correlations between the values of the field at different points in space and time. To do this involves integrating over all possible field configurations with the weight mentioned above. Therefore, we have to now learn how to integrate over function spaces.

In this chapter, we introduce the reader to the concept of a functional integral or integration over spaces of functions. It must be stressed that the assertions of this chapter are far from rigorous and would not be satisfactory to a mathematician. The concept of functional integration has been made rigorous by mathematicians (see e.g., the series by Reed and Simon). The more ambitious reader may wish to consult these texts.

## 6.1 Integration over Functions

From a naive physicist's point of view, functional integration is just like ordinary integration, but the independent variable is a function of a real or complex number rather than itself being a real or complex number. An example may clarify this distinction. An ordinary Gaussian integral can be ($A > 0$),

$$I = \int_{-\infty}^{\infty} dx\, e^{-\frac{1}{2}Ax^2}. \tag{6.1}$$

The corresponding functional integral could be

$$I = \int_{-\infty}^{\infty} d[f]\, e^{-\frac{1}{2}A\int_a^b dx\, f(x)^2}. \tag{6.2}$$

In the above example, the integration is over all possible functions of $x$ defined in the interval $[a,b]$ rather than all possible real numbers. One is then called upon to make sense of this in the same sense in which mathematicians make sense of an integral as the limit of a sum. More interesting variations are possible with functional integration within the spirit of the Gaussian integral. For example, we could include derivatives of $f$ as well.

$$I = \int_{-\infty}^{\infty} d[f]\, e^{-\frac{1}{2}A\int_a^b dx\, (f'^2(x)+\lambda f(x)^2)} \tag{6.3}$$

In the subsequent few paragraphs, we try to make sense of the above identifications by relating them to ordinary integration. Imagine that we think of the interval $[a,b]$ as containing a finite number of points: $x_j = a + \frac{(b-a)}{N} j$ where $x_0 = a$ and $x_N = b$. Then we may define $f_j \equiv f(x_j)$. Thus, integrating over the function $f$ is the same as integrating over the numbers $f_j$, $j = 0,1,2,...,N-1,N$. Since $j = 0$ corresponds to $x = a$ and $j = N$ corresponds to $x = b$, we may choose to restrict the integration over functions that take on a predetermined value at these points. Alternatively, we could allow the derivatives of the functions that are to be integrated with respect to, to take on predetermined values. This latter case will be left to the exercises. Presently we focus on the situation where all functions $f$ obey $f(x_0 = a) = y_1$ and $f(x_N = b) = y_2$. We substitute the following identifications in Eq. (6.3)

$$\int_a^b dx\, (....) = \sum_{j=1}^{N-1} \Delta x\, (....) = \frac{(b-a)}{N}\sum_{j=1}^{N-1} (....). \tag{6.4}$$

Further we have,

$$f'(x_j) = \frac{(f(x_{j+1}) - f(x_j))}{\Delta x} \tag{6.5}$$

$$I = \int_{-\infty}^{\infty} d[f]\, e^{-\frac{1}{2}A\Delta x\sum_{j=0}^{N-1}\left(\frac{(f(x_{j+1})-f(x_j))^2}{(\Delta x)^2}+\lambda f(x_j)^2\right)}$$

$$= \int_{-\infty}^{\infty} df_1 \int_{-\infty}^{\infty} df_2\, ..... \int_{-\infty}^{\infty} df_{N-1}\, e^{-\frac{1}{2}A\Delta x\sum_{j=0}^{N-1}\left(\frac{(f(x_{j+1})-f(x_j))^2}{(\Delta x)^2}+\lambda f(x_j)^2\right)}. \tag{6.6}$$

In this way we can see that the functional integral is nothing but the product of ordinary Gaussian integrals. However, the integrals are all nested and it is hard to do them one by one. In order to evaluate this, we have to adopt a cleverer method. It involves first finding the function $f(x)$ that makes the integral $\int_a^b dx\,(f'^2(x)+\lambda f^2(x))$ an extremum. Consider two functions that are close to each other: $u(x)$ and $u(x)+\delta u(x)$ such that they both are subject to the condition that they are equal to $y_1$ at $x=a$ and $y_2$ at $x=b$. This means $\delta u(a)=\delta u(b)=0$. Thus,

$$0=\int_a^b dx\,(2u'(x)\delta u'(x)+2\lambda u(x)\delta u(x))$$

$$=\int_a^b dx\,(-2u''(x)\delta u(x)+2\lambda u(x)\delta u(x))+2u'(b)\delta u(b)-2u'(a)\delta u(a). \quad (6.7)$$

Since $\delta u(x)$ can be anything, this is obeyed only if at each point $x\in[a,b]$ we have

$$-u''(x)+\lambda u(x)=0. \quad (6.8)$$

This has to be solved subject to the condition that $u(a)=y_1$ and $u(b)=y_2$.

$$u(x)=C_1\,e^{\sqrt{\lambda}\,x}+C_2\,e^{-\sqrt{\lambda}\,x} \quad (6.9)$$

$$C_1=\frac{y_1\,e^{-\sqrt{\lambda}\,b}-y_2\,e^{-\sqrt{\lambda}\,a}}{(e^{\sqrt{\lambda}\,(a-b)}-e^{\sqrt{\lambda}\,(b-a)})} \quad (6.10)$$

$$C_2=\frac{y_1\,e^{\sqrt{\lambda}\,b}-y_2\,e^{\sqrt{\lambda}\,a}}{(e^{-\sqrt{\lambda}\,(a-b)}-e^{-\sqrt{\lambda}\,(b-a)})} \quad (6.11)$$

Now write $f(x)=u(x)+h(x)$, then $h(a)=h(b)=0$, and

$$I=e^{-\frac{1}{2}A\int_a^b dx\,(u'^2(x)+\lambda u(x)^2)}\int_{-\infty}^{\infty}d[h]\;e^{-\frac{1}{2}A\int_a^b dx\,(h'^2(x)+\lambda h(x)^2)}. \quad (6.12)$$

In order to proceed further, we Fourier transform $h(x)$.

$$h(x)=\sum_{n=1}^{\infty}c_n\,Sin(2\pi n\frac{(x-a)}{(b-a)}) \quad (6.13)$$

$$h'(x)=\sum_{n=1}^{\infty}c_n\frac{2\pi n}{(b-a)}\;Cos(2\pi n\frac{(x-a)}{(b-a)}) \quad (6.14)$$

$$\int_a^b dx\,(h'^2(x)+\lambda h^2(x))=\sum_{n=1}^{\infty}c_n^2\frac{(2\pi n)^2}{2(b-a)}+\lambda\sum_{n=1}^{\infty}c_n^2\frac{(b-a)}{2} \quad (6.15)$$

$$I = e^{-\frac{1}{2}A\int_a^b dx\,(u'^2(x)+\lambda u(x)^2)} \int_{-\infty}^{\infty} d[c]\; e^{-\frac{1}{2}A\sum_{n=1}^{\infty} c_n^2 [\frac{(2\pi n)^2}{2(b-a)}\; +\lambda\; \frac{(b-a)}{2}]} \tag{6.16}$$

Or (if $\lambda, A > 0$),

$$I = e^{-\frac{1}{2}A\int_a^b dx\,(u'^2(x)+\lambda u(x)^2)} \prod_{n=1}^{\infty} \sqrt{\frac{2\pi}{A[\frac{(2\pi n)^2}{2(b-a)} + \lambda\,\frac{(b-a)}{2}]}}. \tag{6.17}$$

The expression is formally divergent. In physics, however, we treat the integral in Eq. (6.3) as an integral over a random variable (function) $f(x)$ with the integrand being the probability distribution. Thus $P[f]\;df$ is the probability for finding the function between $f$ and $f+df$.

$$P[f]\;df = e^{-\frac{1}{2}A\int_a^b dx\,(f'^2(x)+\lambda f(x)^2)}\; d[f] \tag{6.18}$$

This interpretation is common, for example, in the path integral approach to quantum mechanics where $f[x]$ would be replaced by $X(t)$, the position of a particle at time $t$, and $P[X] \equiv e^{\frac{i}{\hbar}S[X]}$ would be the probability distribution with $S[X]$ as the action. Thus within this interpretation we should be thinking of calculating averages of various quantities. For example, $< f(y) >$ would mean

$$< f(y) > = \frac{e^{-\frac{1}{2}A\int_a^b dx\,(f'^2(x)+\lambda f(x)^2)} f(y)\; d[f]}{e^{-\frac{1}{2}A\int_a^b dx\,(f'^2(x)+\lambda f(x)^2)}\; d[f]}. \tag{6.19}$$

Similarly, we could also calculate $< f(y)f(z) >$ by inserting this instead of $f(y)$. In general, we could calculate the generating function $G[U]$, which is nothing but $< e^{\int_a^b f(x)U(x)dx} >$. For this we use the procedure already outlined, namely, write $f(x) = u(x) + h(x)$ and then,

$$G[U] \equiv < e^{\int_a^b f(x)U(x)dx} >$$

$$= e^{\int_a^b u(x)U(x)dx}\; \frac{e^{-\frac{1}{2}A\int_a^b dx\,(h'^2(x)+\lambda h(x)^2)} e^{\int_a^b h(x)U(x)dx}\; d[h]}{e^{-\frac{1}{2}A\int_a^b dx\,(h'^2(x)+\lambda h(x)^2)}\; d[h]}. \tag{6.20}$$

As usual we expand $h(x)$ in a Fourier series and get

$$G[U] = e^{\int_a^b u(x)U(x)dx}\; \frac{\int d[c]\; e^{-\frac{1}{2}A\sum_{n=1}^{\infty} c_n^2 [\frac{(2\pi n)^2}{2(b-a)}\; +\lambda\; \frac{(b-a)}{2}]} e^{\sum_{n=1}^{\infty} c_n U_n}}{\int d[c]\; e^{-\frac{1}{2}A\sum_{n=1}^{\infty} c_n^2 [\frac{(2\pi n)^2}{2(b-a)}\; +\lambda\; \frac{(b-a)}{2}]}}, \tag{6.21}$$

where $U_n = \int_a^b dx\, Sin(2\pi n\frac{(x-a)}{(b-a)})\; U(x)$. The numerator is nothing but a shifted Gaussian integral and is easily done to yield

$$G[U] = e^{\int_a^b u(x)U(x)dx}\; e^{\frac{1}{2}\sum_{n=1}^{\infty} \frac{U_n^2}{A[\frac{(2\pi n)^2}{2(b-a)}\; +\lambda\; \frac{(b-a)}{2}]}}. \tag{6.22}$$

Notice that in the above formula the formally divergent product over the index $n$, which appeared in the evaluation of $I$ earlier, has cancelled out since both the numerator and the denominator have the same terms. Thus, in order to calculate $< f(x) >$ we simply evaluate $\left(\frac{\delta}{\delta U(x)}G[U]\right)_{U\equiv 0}$ and in order to evaluate the product $< f(x)f(y) >= \left(\frac{\delta^2}{\delta U(x)\delta U(y)}G[U]\right)_{U\equiv 0}$ and so on. It is more convenient to evaluate the correlation functions: $C(x,y) =< f(x)f(y) > - < f(x) >< f(y) >$, $C(x,y,z) =< f(x)f(y)f(z) > - < f(x) > C(y,z) - < f(y) > C(x,z) - < f(z) > C(x,y) - < f(x) >< f(y) >< f(z) >$ and so on. This is most easily accomplished by first taking the natural logarithm and then differentiating.

$$C(x,y) = \frac{\delta^2}{\delta U(x)\delta U(y)} Log[G[U]] \tag{6.23}$$

$$C(x,y,z) = \frac{\delta^3}{\delta U(x)\delta U(y)\delta U(z)} Log[G[U]] \tag{6.24}$$

and so on.

## 6.2 Perturbation Theory

In most interesting applications in physics, we study interacting theories. In those theories, the integrand is not Gaussian but something more complicated. It could be a Gaussian plus a quartic term, for example. To keep things simple, we consider only those functions that vanish at the end points $a$ and $b$—in other words, the functions of the type, $h(x)$. Thus we wish to consider the probability distribution

$$P[h]\, dh = e^{-\frac{1}{2}A\int_a^b dx\, (h'^2(x)+\lambda h(x)^2+\frac{g}{4!}h^4(x))}\, d[h]. \tag{6.25}$$

Using this, we wish to calculate as before the generating function of correlations defined by

$$G[U] = \int e^{-\frac{1}{2}A\int_a^b dx\, (h'^2(x)+\lambda h(x)^2+\frac{g}{4!}h^4(x)+U(x)h(x))}\, d[h] = e^{F[U]}. \tag{6.26}$$

We wish to evaluate this by expanding $F[U]$ in powers of $g$.

$$F[U] = F_0[U] + gF_1[U] + \frac{g^2}{2!}F_2[U] + \ldots \tag{6.27}$$

Expanding both sides in a Taylor series we conclude,

$$\int e^{-\frac{1}{2}A\int_a^b dx\, (h'^2(x)+\lambda h(x)^2+U(x)h(x))}\, d[h] = e^{F_0[U]}$$

$$\frac{-\frac{1}{2}A\int_a^b dx \int \frac{1}{4!}h^4(x)e^{-\frac{1}{2}A\int_a^b dx\,(h'^2(x)+\lambda h(x)^2+U(x)h(x))}\,d[h]}{\int e^{-\frac{1}{2}A\int_a^b dx\,(h'^2(x)+\lambda h(x)^2+U(x)h(x))}\,d[h]}$$
$$= F_1[U] \quad (6.28)$$

and so on (the evaluation of $F_2$ is left for the exercises). Notice that,

$$\frac{\delta^4}{\delta U(x)^4}\int e^{-\frac{1}{2}A\int_a^b dx\,(h'^2(x)+\lambda h(x)^2+U(x)h(x))}\,d[h]$$
$$= (-\frac{1}{2}A)^4\int h^4(x)\,e^{-\frac{1}{2}A\int_a^b dx\,(h'^2(x)+\lambda h(x)^2+U(x)h(x))}\,d[h]. \quad (6.29)$$

Therefore,

$$F_1[U] = \frac{\frac{1}{4!}\int_a^b dx\frac{\delta^4}{\delta U(x)^4}e^{F_0[U]}}{(-\frac{1}{2}A)^3 e^{F_0[U]}}. \quad (6.30)$$

After some effort we may conclude,

$$F_1[U] = \frac{1}{4!(-\frac{1}{2}A)^3}\int_a^b dx\ \ (3(\frac{\delta^2 F_0[U]}{\delta U(x)\delta U(x)})^2 + 6(\frac{\delta F_0[U]}{\delta U(x)})^2(\frac{\delta^2 F_0[U]}{\delta U(x)\delta U(x)}))$$
$$+\frac{1}{4!(-\frac{1}{2}A)^3}\int_a^b dx\ \ (\frac{\delta F_0[U]}{\delta U(x)})^4, \quad (6.31)$$

from the earlier calculation we know,

$$F_0[U] = \frac{1}{2}\sum_{n=1}^{\infty}\frac{U_n^2}{A[\frac{(2\pi n)^2}{2(b-a)} + \lambda\,\frac{(b-a)}{2}]}. \quad (6.32)$$

Further, we may write,

$$U_n = \frac{2}{(b-a)}\int_a^b dx\ Sin(2\pi n\frac{(x-a)}{(b-a)})\ U(x) \quad (6.33)$$

$$F_0[U] = \frac{1}{(b-a)^2}\int_a^b dx\int_a^b dx'\ \ U(x)U(x')$$
$$\times \sum_{n=all}\frac{Sin(2\pi n\frac{(x-a)}{(b-a)})Sin(2\pi n\frac{(x'-a)}{(b-a)})}{A[\frac{(2\pi n)^2}{2(b-a)} + \lambda\,\frac{(b-a)}{2}]}. \quad (6.34)$$

To evaluate this we define,

$$W(x,x') = \sum_{n=all}\frac{Sin(2\pi n\frac{(x-a)}{(b-a)})Sin(2\pi n\frac{(x'-a)}{(b-a)})}{A[\frac{(2\pi n)^2}{2(b-a)} + \lambda\,\frac{(b-a)}{2}]}. \quad (6.35)$$

One may see that in the region $x,x^{'} \in [a,b]$ the function $W$ obeys $W(x,x^{'}) = W(x^{'},x)$ and,

$$(-\frac{d^2}{dx^2} + \lambda)W(x,x^{'}) = \frac{1}{A}\ \delta(x - x^{'}). \tag{6.36}$$

The solution to this that is consistent with $W(a,x^{'}) = W(b,x^{'}) = 0$ is

$$\begin{aligned} N_1(x,x^{'}) = &\, Sinh(\sqrt{\lambda}x)Sinh(\sqrt{\lambda}x^{'}) - Tanh(\sqrt{\lambda}a)\ Sinh(\sqrt{\lambda}x)Cosh(\sqrt{\lambda}x^{'}) \\ &- Tanh(\sqrt{\lambda}b)\ Cosh(\sqrt{\lambda}x)Sinh(\sqrt{\lambda}x^{'}) \\ &+ Tanh(\sqrt{\lambda}b)Tanh(\sqrt{\lambda}a)\ Cosh(\sqrt{\lambda}x)Cosh(\sqrt{\lambda}x^{'}) \end{aligned} \tag{6.37}$$

$$\begin{aligned} N_2(x,x^{'}) = &\, Sinh(\sqrt{\lambda}x)Sinh(\sqrt{\lambda}x^{'}) - Tanh(\sqrt{\lambda}a)\ Sinh(\sqrt{\lambda}x^{'})Cosh(\sqrt{\lambda}x) \\ &- Tanh(\sqrt{\lambda}b)\ Cosh(\sqrt{\lambda}x^{'})Sinh(\sqrt{\lambda}x) \\ &+ Tanh(\sqrt{\lambda}b)Tanh(\sqrt{\lambda}a)\ Cosh(\sqrt{\lambda}x^{'})Cosh(\sqrt{\lambda}x) \end{aligned} \tag{6.38}$$

$$\begin{aligned} W(x,x^{'}) = &\, \theta(x - x^{'})\ \frac{N_1(x,x^{'})}{\sqrt{\lambda}\,A\ (Tanh(\sqrt{\lambda}a) - Tanh(\sqrt{\lambda}b))} \\ &+ \theta(x^{'} - x)\ \frac{N_2(x,x^{'})}{\sqrt{\lambda}\,A\ (Tanh(\sqrt{\lambda}a) - Tanh(\sqrt{\lambda}b))}. \end{aligned} \tag{6.39}$$

Therefore,

$$F_0[U] = \frac{1}{(b-a)^2}\int_a^b dx \int_a^b dx^{'}\ \ U(x)U(x^{'})\ W(x,x^{'}) \tag{6.40}$$

and,

$$\begin{aligned} F_1[U] = &\, \frac{1}{4!(-\frac{1}{2}A)^3}\int_a^b dx\ \ (3(\frac{\delta^2 F_0[U]}{\delta U(x)\delta U(x)})^2 + 6(\frac{\delta F_0[U]}{\delta U(x)})^2(\frac{\delta^2 F_0[U]}{\delta U(x)\delta U(x)})) \\ &+ \frac{1}{4!(-\frac{1}{2}A)^3}\int_a^b dx\ \ ((\frac{\delta F_0[U]}{\delta U(x)})^4) \end{aligned} \tag{6.41}$$

and,

$$\frac{\delta^2 F_0[U]}{\delta U(x)\delta U(x)} = \frac{2}{(b-a)^2}\ \ W(x,x) \tag{6.42}$$

$$F_1[U] = \frac{1}{4!(-\frac{1}{2}A)^3}\int_a^b dx$$

$$(3(\frac{2}{(b-a)^2}\ W(x,x))^2+6(\frac{\delta F_0[U]}{\delta U(x)})^2\ \frac{2}{(b-a)^2}\ W(x,x)+(\frac{\delta F_0[U]}{\delta U(x)})^4)$$

$$+\frac{1}{4!(-\frac{1}{2}A)^3}\int_a^b dx\ 3(\frac{2}{(b-a)^2}\ W(x,x))^2$$

$$+\frac{1}{4!(-\frac{1}{2}A)^3}\int_a^b dx\ 6(\frac{\delta F_0[U]}{\delta U(x)})^2\ \frac{2}{(b-a)^2}\ W(x,x)$$

$$+\frac{1}{4!(-\frac{1}{2}A)^3}\int_a^b dx\ (\frac{\delta F_0[U]}{\delta U(x)})^4. \tag{6.43}$$

From the above assertions we conclude

(using $< h(x)h(x^{'}) >= \left(\frac{\delta^2 F[U]}{\delta U(x)\delta U(x^{'})}\right)_{U\equiv 0}$),

$$< h(x)h(x^{'}) >= \frac{2}{(b-a)^2}\ W(x,x^{'})$$

$$+g\frac{1}{4!(-\frac{1}{2}A)^3}\int_a^b dy\ 12\,W(x,y)\ W(x^{'},y)\left(\frac{2}{(b-a)^2}\right)^3\ W(y,y) \tag{6.44}$$

up to terms linear in $g$. The next order contribution may similarly be evaluated and is left to the exercises.

## 6.3 Exercises

**Q.1** Verify all the assertions made in this chapter by supplying all missing steps.

**Q.2** Find $< f(x) >$ and $< f(x)f(x^{'}) >$ with the Gaussian probability distribution as done in the main text, but with different boundary conditions. Use $f^{'}(a) = y_1^{'}$ and $f^{'}(b) = y_2^{'}$ this time.

**Q.3** Find $< h(x)h(x^{'}) >$ up to order $g^2$ for the case involving a non-Gaussian probability distribution (Eq. (6.27)). Do all the integrals explicitly and write the answer in terms of standard functions.

**Q.4** Why are we not choosing a cubic term, that is, what is wrong with

$$P[h]\ dh = e^{-\frac{1}{2}A\int_a^b dx\ (h^{'2}(x)+\lambda h(x)^2+\frac{g}{3!}h^3(x))}\ d[h]? \tag{6.45}$$

**Q.5** Consider the probability distribution

$$P[h]\,dh = e^{-\frac{1}{2}A\int_a^b dx\,(h'^2(x)-\lambda h(x)^2+\frac{g}{3!}h^4(x))}\,d[h], \tag{6.46}$$

where $g,\lambda,A>0$. Assuming that the minima of $G(y)=-\lambda y^2+\frac{g}{3!}y^4$ are deep (large and negative), find approximate expressions for $<h(x)>$ and $<h(x)h(x')>$ (bear in mind that $h(a)=h(b)=0$).

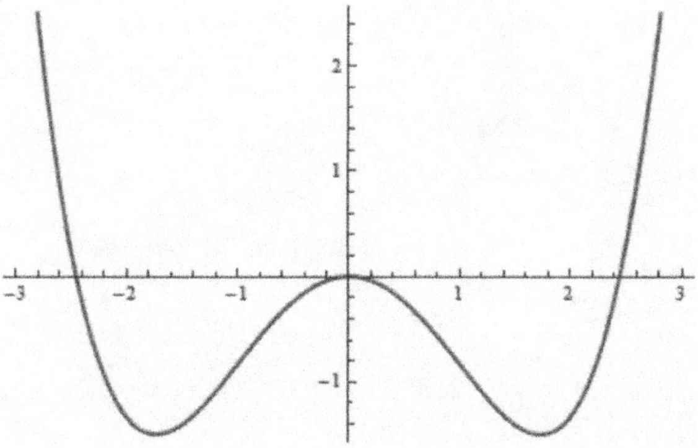


Figure 6.1: Two deep minima of Q.5 lead to two degenerate solutions and there can be tunneling between them (see later).

# Chapter 7

# Quantum Mechanics Using Lagrangians: Path Integrals

Quantum mechanics, as it is taught in various undergraduate courses, typically focuses on developing the formalism using the phase space or the Hamiltonian description of classical mechanics. This gives an impression to the student that Lagrangians cannot be used to develop a formalism of quantum mechanics. Of course, this is not true. Dirac is credited with originating the idea of the path integral in physics. However, it was Feynman who popularized this idea and made it widely accessible to physicists.

## 7.1 The Formalism

The physical motivation behind the path integral is as follows. Consider the problem of evaluating the matrix element of some operator in quantum mechanics: $< i,t_i|\hat{Q}(\mathbf{r},-i\hbar\nabla)|f,t_f >$. In order to evaluate this, we may either work in the Hamiltonian formalism of Schrodinger so that,

$$< i|\hat{Q}(\mathbf{r},-i\hbar\nabla)|f >= \int d^3r\ \psi^*(\mathbf{r},t_i)\hat{Q}(\mathbf{r},-i\hbar\nabla)\psi(\mathbf{r},t_f). \tag{7.1}$$

The evaluation of this requires the knowledge of the wavefunction obtained by solving the time-dependent Schrodinger equation

$$i\hbar\frac{\partial}{\partial t}\psi(\mathbf{r},t) = \hat{H}(\mathbf{r},-i\hbar\nabla)\psi(\mathbf{r},t), \tag{7.2}$$

with suitable initial and other conditions that make the solution unique. Notice that this approach explicitly invokes the Hamiltonian as distinct from the Lagrangian

Figure 7.1: Richard Feynman (11 May 1918 to 15 February 1988) was an American physicist who developed the path integral approach to quantum mechanics. He greatly influenced the development of science in late 20th century and pioneered the fields of nanotechnology and quantum computing. He was awarded the Nobel Prize for Physics in 1965.

of the system. Henceforth, for simplicity, we shall assume that the Hamiltonian is time independent. In this case, we may write,

$$|f>= e^{-\frac{i}{\hbar}(t_f-t_i)(\frac{p^2}{2m}+V(\mathbf{r}))}\ |i>$$

$$< i|\hat{Q}(\mathbf{r},-i\hbar\nabla)|f>=< i|\hat{Q}(\mathbf{r},-i\hbar\nabla)e^{-\frac{i}{\hbar}(t_f-t_i)(\frac{p^2}{2m}+V(\mathbf{r}))}\ |i> . \tag{7.3}$$

We now focus on the evolution operator,

$$e^{-\frac{i}{\hbar}(t_f-t_i)(\frac{p^2}{2m}+V(\mathbf{r}))} = (e^{-\frac{i}{\hbar}\frac{(t_f-t_i)}{N}(\frac{p^2}{2m}+V(\mathbf{r}))})^N. \tag{7.4}$$

The reason why we have written this in such a peculiar manner is because for $N$ large enough we may write,

$$e^{-\frac{i}{\hbar}\frac{(t_f-t_i)}{N}(\frac{p^2}{2m}+V(\mathbf{r}))} \approx e^{-\frac{i}{\hbar}\frac{(t_f-t_i)}{N}\frac{p^2}{2m}}e^{-\frac{i}{\hbar}\frac{(t_f-t_i)}{N}V(\mathbf{r})}. \tag{7.5}$$

The result follows from the observation that $e^{A+B} = Lim_{N\to\infty}(e^{\frac{A}{N}}e^{\frac{B}{N}})^N$, which is known as Trotter's product formula, which in turn follows from the Baker-Hausdorff formula (or Zassenhaus formula, see box description later), $e^{A+B} = e^Ae^Be^{-\frac{1}{2}[A,B]}$.... Setting $A = -\frac{i}{\hbar}\frac{(t_f-t_i)}{N}\frac{p^2}{2m}$ and $B = -\frac{i}{\hbar}\frac{(t_f-t_i)}{N}V(\mathbf{r})$, we conclude that

$[A,B] \sim \frac{1}{N^2}$ whereas $A$ and $B$ themselves are of order $\frac{1}{N}$. Hence the result. Set $\varepsilon = \frac{(t_f - t_i)}{N}$ and consider the matrix element,

$$< \mathbf{r}_k | e^{-\frac{i}{\hbar}\frac{(t_f-t_i)}{N}(\frac{p^2}{2m}+V(\mathbf{r}))} | \mathbf{r}_{k+1} >$$
$$\approx < \mathbf{r}_k | e^{-\frac{i}{\hbar}\frac{(t_f-t_i)}{N}\frac{p^2}{2m}} e^{-\frac{i}{\hbar}\frac{(t_f-t_i)}{N}V(\mathbf{r})} | \mathbf{r}_{k+1} > \tag{7.6}$$

between two eigenstates of position labeled $|\mathbf{r}_k>$ and $|\mathbf{r}_{k+1}>$ for reasons that will become clear shortly. Now insert a complete set of eigenstates of momentum using the resolution of the identity $1 = \int d^3p_k \, |\mathbf{p}_k>< \mathbf{p}_k|$.

$$< \mathbf{r}_k | e^{-\frac{i}{\hbar}\frac{(t_f-t_i)}{N}(\frac{p^2}{2m}+V(\mathbf{r}))} | \mathbf{r}_{k+1} >$$
$$\approx \int d^3p_k \, < \mathbf{r}_k | e^{-\frac{i}{\hbar}\frac{(t_f-t_i)}{N}\frac{p^2}{2m}} |\mathbf{p}_k >< \mathbf{p}_k | e^{-\frac{i}{\hbar}\frac{(t_f-t_i)}{N}V(\mathbf{r})} | \mathbf{r}_{k+1} > \tag{7.7}$$

Since $\frac{p^2}{2m}$ acts simply on an eigenstate of momentum and $V(\mathbf{r})$ acts simply on an eigenstate of position we get,

$$< \mathbf{r}_k | e^{-\frac{i}{\hbar}\frac{(t_f-t_i)}{N}(\frac{p^2}{2m}+V(\mathbf{r}))} | \mathbf{r}_{k+1} >$$
$$\approx \int d^3p_k \, < \mathbf{r}_k | \mathbf{p}_k >< \mathbf{p}_k | \mathbf{r}_{k+1} > \, e^{-\frac{i}{\hbar}\varepsilon\frac{p_k^2}{2m}} e^{-\frac{i}{\hbar}\varepsilon V(\mathbf{r}_{k+1})}. \tag{7.8}$$

We know that the momentum eigenstate in the position representation is

$$< \mathbf{r}_k | \mathbf{p}_k > = e^{\frac{i}{\hbar}\mathbf{r}_k \cdot \mathbf{p}_k}, \tag{7.9}$$

and its complex conjugate,

$$< \mathbf{p}_k | \mathbf{r}_{k+1} > = e^{-\frac{i}{\hbar}\mathbf{r}_{k+1} \cdot \mathbf{p}_k}. \tag{7.10}$$

This is easily verified by alternately using the position and momentum representation of the momentum operator

$$< \mathbf{r}_k | \hat{p} | \mathbf{p}_k > = \mathbf{p}_k < \mathbf{r}_k | \mathbf{p}_k > = -i\hbar \nabla_{\mathbf{r}_k} < \mathbf{r}_k | \mathbf{p}_k > . \tag{7.11}$$

Integrating this yields the desired result.

$$< \mathbf{r}_k | e^{-\frac{i}{\hbar}\frac{(t_f-t_i)}{N}(\frac{p^2}{2m}+V(\mathbf{r}))} | \mathbf{r}_{k+1} >$$
$$\approx \int d^3p_k \, e^{\frac{i}{\hbar}(\mathbf{r}_k - \mathbf{r}_{k+1})\cdot \mathbf{p}_k} \, e^{-\frac{i}{\hbar}\varepsilon\frac{p_k^2}{2m}} e^{-\frac{i}{\hbar}\varepsilon V(\mathbf{r}_{k+1})} \tag{7.12}$$

The integral over $\mathbf{p}_k$ is a shifted Gaussian integral which may be performed to give,

$$< \mathbf{r}_k | e^{-\frac{i}{\hbar}\frac{(t_f-t_i)}{N}(\frac{p^2}{2m}+V(\mathbf{r}))} | \mathbf{r}_{k+1} >$$

$$\approx e^{\frac{i}{\hbar}\frac{1}{2}m\frac{(\mathbf{r}_k-\mathbf{r}_{k+1})^2}{\varepsilon}}\ e^{-\frac{i}{\hbar}\varepsilon V(\mathbf{r}_{k+1})} \tag{7.13}$$

apart from an overall multiplicative constant. The terms in the exponent may be regrouped to give,

$$< \mathbf{r}_k|e^{-\frac{i}{\hbar}\frac{(t_f-t_i)}{N}(\frac{p^2}{2m}+V(\mathbf{r}))}|\mathbf{r}_{k+1} >$$

$$\approx e^{\frac{i}{\hbar}\varepsilon\ (\frac{1}{2}m\frac{(\mathbf{r}_{k+1}-\mathbf{r}_k)^2}{\varepsilon^2}-V(\mathbf{r}_{k+1}))}. \tag{7.14}$$

At this stage we interpret the index $k$ in the sequence $\mathbf{r}_{k=1,2,...}$ as the discrete version of a continuous parameter $s$. Thus, we identify $\mathbf{r}_k = \mathbf{r}(s_k)$. We may therefore regard $\frac{\mathbf{r}_{k+1}-\mathbf{r}_k}{\varepsilon} \approx \frac{d}{ds_{k+1}}\mathbf{r}(s_{k+1}) \equiv \dot{\mathbf{r}}(s_{k+1})$. Therefore, the term in the bracket is just the Lagrangian of the system. This allows us to write

$$< \mathbf{r}_k|e^{-\frac{i}{\hbar}\frac{(t_f-t_i)}{N}(\frac{p^2}{2m}+V(\mathbf{r}))}|\mathbf{r}_{k+1} >\approx e^{\frac{i}{\hbar}\varepsilon\ L(\mathbf{r}(s_{k+1}),\dot{\mathbf{r}}(s_{k+1}))}. \tag{7.15}$$

Now we go back to the Trotter product formula in Eq. (7.4). We wish to evaluate the matrix elements with respect to some position eigenstates which we denote as $|\mathbf{r}_0>$ and $|\mathbf{r}_N>$. We insert unity resolved using the position eigenstates as follows:

$$< \mathbf{r}_0|e^{-\frac{i}{\hbar}(t_f-t_i)(\frac{p^2}{2m}+V(\mathbf{r}))}|\mathbf{r}_N >=< \mathbf{r}_0|(e^{-\frac{i}{\hbar}\frac{(t_f-t_i)}{N}(\frac{p^2}{2m}+V(\mathbf{r}))})^N|\mathbf{r}_N >$$

$$=< \mathbf{r}_0|(e^{-\frac{i}{\hbar}\frac{(t_f-t_i)}{N}(\frac{p^2}{2m}+V(\mathbf{r}))})\int d^3r_1\ |\mathbf{r}_1 >< \mathbf{r}_1|(e^{-\frac{i}{\hbar}\frac{(t_f-t_i)}{N}(\frac{p^2}{2m}+V(\mathbf{r}))})|\mathbf{r}_2 > ....$$

$$< \mathbf{r}_{N-2}|(e^{-\frac{i}{\hbar}\frac{(t_f-t_i)}{N}(\frac{p^2}{2m}+V(\mathbf{r}))})\int d^3r_{N-1}\ |\mathbf{r}_{N-1} >$$

$$< \mathbf{r}_{N-1}|(e^{-\frac{i}{\hbar}\frac{(t_f-t_i)}{N}(\frac{p^2}{2m}+V(\mathbf{r}))})|\mathbf{r}_N > . \tag{7.16}$$

This may be written more compactly as

$$< \mathbf{r}_0|e^{-\frac{i}{\hbar}(t_f-t_i)(\frac{p^2}{2m}+V(\mathbf{r}))}|\mathbf{r}_N >$$

$$= \int d^3r_1 \int d^3r_2 .... \int d^3r_{N-1} \prod_{k=0}^{N-1} < \mathbf{r}_k|(e^{-\frac{i}{\hbar}\frac{(t_f-t_i)}{N}(\frac{p^2}{2m}+V(\mathbf{r}))})|\mathbf{r}_{k+1} >$$

$$= \int d^3r_1 \int d^3r_2 .... \int d^3r_{N-1} \prod_{k=0}^{N-1} e^{\frac{i}{\hbar}\varepsilon\ L(\mathbf{r}(s_{k+1}),\dot{\mathbf{r}}(s_{k+1}))}. \tag{7.17}$$

The last result follows from Eq. (7.15). But

$$\prod_{k=0}^{N-1} e^{\frac{i}{\hbar}\varepsilon\ L(\mathbf{r}(s_{k+1}),\dot{\mathbf{r}}(s_{k+1}))}$$

$$= e^{\frac{i}{\hbar}\varepsilon\ \sum_{k=0}^{N-1} L(\mathbf{r}(s_{k+1}),\dot{\mathbf{r}}(s_{k+1}))} \approx e^{\frac{i}{\hbar}\int_{s_1}^{s_N} ds\ L(\mathbf{r}(s),\dot{\mathbf{r}}(s))} \tag{7.18}$$

$$< \mathbf{r}_0 | e^{-\frac{i}{\hbar}(t_f - t_i)(\frac{p^2}{2m} + V(\mathbf{r}))} | \mathbf{r}_N >$$

$$= \int d^3r_1 \int d^3r_2 \, .... \int d^3r_{N-1} \prod_{k=0}^{N-1} < \mathbf{r}_k | (e^{-\frac{i}{\hbar}\frac{(t_f - t_i)}{N}(\frac{p^2}{2m} + V(\mathbf{r}))}) | \mathbf{r}_{k+1} >$$

$$= \int D[\mathbf{r}] \, e^{\frac{i}{\hbar} \int_{s_1}^{s_N} ds \, L(\mathbf{r}(s), \dot{\mathbf{r}}(s))}. \tag{7.19}$$

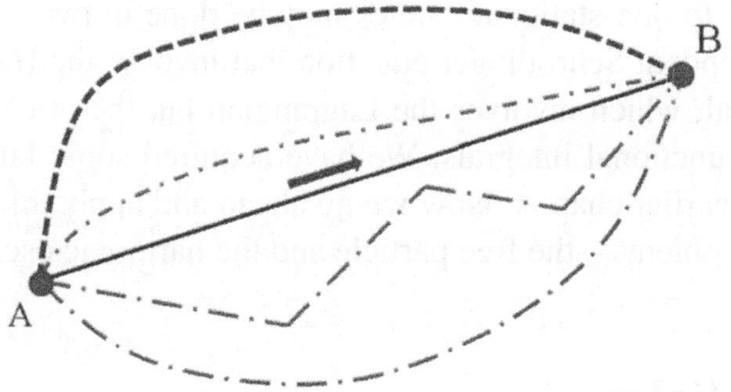


Figure 7.2: Various possible paths connecting the points $A$: $(\mathbf{r}_0, t_i)$ and $B$: $(\mathbf{r}_N, t_f)$. For a free particle, the solid straight line would be the classical path. In quantum mechanics, all paths are allowed (equally probable!) with different complex probability amplitudes (relative phases) associated with them.

Thus the path integral representation of this matrix element is

$$< \mathbf{r}_0 | e^{-\frac{i}{\hbar}(t_f - t_i)(\frac{p^2}{2m} + V(\mathbf{r}))} | \mathbf{r}_N > = \int D[\mathbf{r}] \, e^{\frac{i}{\hbar} \int_{s_1}^{s_N} ds \, L(\mathbf{r}(s), \dot{\mathbf{r}}(s))}. \tag{7.20}$$

The last identification is just a shorthand for the earlier step. The idea is, one has to evaluate a 'path integral' over all trajectories $\mathbf{r}(s)$ such that the end points are fixed $\mathbf{r}(s_0) \equiv \mathbf{r}_0$ and $\mathbf{r}(s_N) \equiv \mathbf{r}_N$. Now, going back to our earlier matrix element,

$$< i | \hat{Q}(\mathbf{r}, -i\hbar\nabla) | f > = < i | \hat{Q}(\mathbf{r}, -i\hbar\nabla) e^{-\frac{i}{\hbar}(t_f - t_i)(\frac{p^2}{2m} + V(\mathbf{r}))} \, | i >$$

$$= \int d^3r_0 \int d^3r_N \; < i | \hat{Q}(\mathbf{r}, -i\hbar\nabla) | \mathbf{r}_0 > < \mathbf{r}_0 | e^{-\frac{i}{\hbar}(t_f - t_i)(\frac{p^2}{2m} + V(\mathbf{r}))} | \mathbf{r}_N > < \mathbf{r}_N | i > \tag{7.21}$$

Using the path integral representation in Eq. (7.20) we get,

$$< i|\hat{Q}(\mathbf{r},-i\hbar\nabla)|f> = \int d^3r_0 \int d^3r_N \ < i|\hat{Q}(\mathbf{r},-i\hbar\nabla)|\mathbf{r}_0>$$

$$\int_{\mathbf{r}(s_0)\equiv\mathbf{r}_0;\mathbf{r}(s_N)=\mathbf{r}_N} D[\mathbf{r}]\ e^{\frac{i}{\hbar}\int_{s_0}^{s_N} ds\ L(\mathbf{r}(s),\dot{\mathbf{r}}(s))} < \mathbf{r}_N|i> . \tag{7.22}$$

In the above identification we have ignored the distinction between $s_0$ and $s_1$. From classical mechanics, we know that $S = \int_{s_0}^{s_N} ds\ L(\mathbf{r}(s),\dot{\mathbf{r}}(s))$ is the classical action of the particle. The above path integral is commonly written as

$$\int_{\mathbf{r}(s_0)\equiv\mathbf{r}_0;\mathbf{r}(s_N)=\mathbf{r}_N} D[\mathbf{r}]\ e^{\frac{i}{\hbar}S}. \tag{7.23}$$

Thus, the quantum mechanical problem of computing the matrix elements of an operator with respect to non-stationary states may be done in two ways-either by solving the time-dependent Schrodinger equation that invokes the Hamiltonian, or using the path integral, which involves the Lagrangian but the price to be paid is one has to evaluate functional integrals. We have acquired some familiarity with this technique in the earlier chapter. Now we go ahead and apply this technique to study two standard problems—the free particle and the harmonic oscillator.

## 7.2 Free Particles

First we study the free particle without any restrictions. Of interest is the Green function or propagator, which was expressed in the path integral language earlier (Eq. (7.20)).

$$< \mathbf{r}_0,t_i|\mathbf{r}_N,t_f > = \int_{\mathbf{r}(t_i)\equiv\mathbf{r}_0;\ \mathbf{r}(t_f)\equiv\mathbf{r}_N} D[\mathbf{r}]\ e^{\frac{i}{\hbar}\int_{t_i}^{t_f} ds\ L(\mathbf{r}(s),\dot{\mathbf{r}}(s))} \tag{7.24}$$

First we evaluate this for a free particle without any restrictions. Then we evaluate it for particles close to the Fermi surface of a metal. For free fermions, Pauli's exclusion principle is imposed by demanding that the energy of the particle always be larger or equal to the Fermi energy since all smaller energies are already occupied by pairs of fermions (with spin).

### 7.2.1 Free Particle

For free particles, the propagator is

$$< \mathbf{r}_0,t_i|\mathbf{r}_N,t_f > = \int_{\mathbf{r}(t_i)\equiv\mathbf{r}_0;\ \mathbf{r}(t_f)\equiv\mathbf{r}_N} D[\mathbf{r}]\ e^{\frac{i}{\hbar}\int_{t_i}^{t_f} ds\ \frac{1}{2}m\dot{\mathbf{r}}^2(s)}. \tag{7.25}$$

To facilitate the solution of this we make the substitution

$$\mathbf{r}(s) = \mathbf{r}_c(s) + \mathbf{q}(s), \tag{7.26}$$

where $\mathbf{r}_c(s)$ is the classical trajectory of the particle, which is a straight line connecting the points $(\mathbf{r}_0, t_i)$ and $(\mathbf{r}_N, t_f)$. Since the end points are fixed $\mathbf{r}(t_f) = \mathbf{r}_N$ and $\mathbf{r}(t_i) = \mathbf{r}_0$, we must also have $\mathbf{q}(t_i) = \mathbf{q}(t_f) = 0$.

$$\mathbf{r}_c(s) = \frac{(s-t_i)\mathbf{r}_N - (s-t_f)\mathbf{r}_0}{t_f - t_i} \tag{7.27}$$

The classical velocity is

$$\mathbf{v}_c = \frac{\mathbf{r}_N - \mathbf{r}_0}{t_f - t_i}$$

$$< \mathbf{r}_0, t_i | \mathbf{r}_N, t_f > = \int_{\mathbf{q}(t_i)\equiv 0;\ \mathbf{q}(t_f)\equiv 0} D[\mathbf{q}]\ e^{\frac{i}{\hbar}\int_{t_i}^{t_f} ds\ \frac{1}{2}m(\mathbf{v}_c + \dot{\mathbf{q}}(s))^2}$$

$$= \int_{\mathbf{q}(t_i)\equiv 0;\ \mathbf{q}(t_f)\equiv 0} D[\mathbf{q}]\ e^{\frac{i}{\hbar}\int_{t_i}^{t_f} ds\ \frac{1}{2}m(\mathbf{v}_c^2 + \dot{\mathbf{q}}^2(s))}\ e^{\frac{i}{\hbar}\int_{t_i}^{t_f} ds\ m(\mathbf{v}_c \cdot \dot{\mathbf{q}}(s))}. \tag{7.28}$$

The cross term drops out since $\int_{t_i}^{t_f} ds\ \dot{\mathbf{q}}(s) = \mathbf{q}(t_f) - \mathbf{q}(t_i) = 0 - 0 = 0$. Therefore,

$$< \mathbf{r}_0, t_i | \mathbf{r}_N, t_f >$$

$$= e^{\frac{i}{\hbar}\frac{1}{2}m\frac{(\mathbf{r}_N - \mathbf{r}_0)^2}{t_f - t_i}} \int_{\mathbf{q}(t_i)\equiv 0;\ \mathbf{q}(t_f)\equiv 0} D[\mathbf{q}]\ e^{\frac{i}{\hbar}\int_{t_i}^{t_f} ds\ \frac{1}{2}m\dot{\mathbf{q}}^2(s)}. \tag{7.29}$$

The integral over $\mathbf{q}(s)$ may be evaluated using the following prescription:

$$\mathbf{q}(s) = \sum_{n=1,2,...} \mathbf{q}_n\ sin(\pi n \frac{(s-t_i)}{(t_f - t_i)}). \tag{7.30}$$

This choice automatically respects the boundary conditions: $\mathbf{q}(t_i) = \mathbf{q}(t_f) = 0$.

$$\dot{\mathbf{q}}(s) = \sum_{n=1,2,...} \mathbf{q}_n\ cos(\pi n \frac{(s-t_i)}{(t_f - t_i)}) \frac{\pi n}{(t_f - t_i)}$$

$$\frac{i}{\hbar}\int_{t_i}^{t_f} ds\ \frac{1}{2}m\dot{\mathbf{q}}^2(s) = \frac{i}{\hbar} \sum_{n=1,2,...} \frac{m(\pi n)^2}{4(t_f - t_i)} \mathbf{q}_n^2 \tag{7.31}$$

However, the meaning of the term $D[\mathbf{q}]$ is not clear other than that it is proportional to the product of various $d^3q(t_1)d^3q(t_2).....d^3q(t_N)$. Since we do not know what that proportionality is, it is not useful to proceed to evaluate the integrals over $\mathbf{q}(s)$. Instead, we make the following observation that, at equal times, the propagator has to be a delta function since the eigenstates of position with different positions are orthogonal.

$$< \mathbf{r}_0, t_i | \mathbf{r}_N, t_f = t_i >= \delta(\mathbf{r}_N - \mathbf{r}_0) \tag{7.32}$$

This also means

$$Lim_{t_f \to t_i} \int d^3 r_N \ < \mathbf{r}_0, t_i | \mathbf{r}_N, t_f >= 1. \tag{7.33}$$

Now we observe that the integral over $\mathbf{q}$ is, at worst, some function of $t_f - t_i$. This allows us to write

$$< \mathbf{r}_0, t_i | \mathbf{r}_N, t_f >= e^{\frac{i}{\hbar} \frac{1}{2} m \frac{(\mathbf{r}_N - \mathbf{r}_0)^2}{t_f - t_i}} \ g(t_f - t_i) \tag{7.34}$$

To determine $g(t_f - t_i)$, we substitute the above ansatz into Eq. (7.33).

$$\int d^3 r_N \ < \mathbf{r}_0, t_i | \mathbf{r}_N, t_f >= (\frac{2i\pi\hbar}{m}(t_f - t_i))^{\frac{3}{2}} \ g(t_f - t_i) \tag{7.35}$$

$$g(t_f - t_i) = (\frac{2i\pi\hbar}{m}(t_f - t_i))^{-\frac{3}{2}} \tag{7.36}$$

Thus, the full expression for the propagator is

$$< \mathbf{r}_0, t_i | \mathbf{r}_N, t_f >= e^{\frac{i}{\hbar} \frac{1}{2} m \frac{(\mathbf{r}_N - \mathbf{r}_0)^2}{t_f - t_i}} \ (\frac{2i\pi\hbar}{m}(t_f - t_i))^{-\frac{3}{2}}. \tag{7.37}$$

Thus, the idea behind the path integral approach is to first find the classical trajectory connecting the two end points, then express the remaining in terms of an integral over fluctuations. This is typically indirectly evaluated by using conditions such as normalization. Now we study the situation typical in condensed matter physics where one has free fermions in a Fermi sea.

### 7.2.2 Free Fermions

For studying fermions, we prefer to focus only on one dimension. The higher dimensional case is somewhat more complicated and will not be done here. We have, at the back of our mind, the situation typical in condensed matter physics where all states with energy less than some energy $E_F$ are fully occupied so that fermions cannot scatter into those states. Thus, for studying the propagator of free fermions, we have to integrate over all paths subject to an additional constraint—the kinetic energy at all times has to be larger than or equal to the Fermi energy. This means that a state corresponding to a definite position will have some spread.

$$|\Psi_x >= \sum_{|p| > p_F} |p >< p|x > \tag{7.38}$$

In the position representation it is given by (we introduce a regularization factor $e^{-\varepsilon k}$ where $\varepsilon \to 0$ at the end),

$$\Psi_x(x') \equiv < x'|\Psi_x > = \sum_{|p|>p_F} < x'|p >< p|x > = \int_{|p|>p_F} \frac{dp}{2\pi\hbar} e^{\frac{i}{\hbar}p(x'-x)} e^{-\frac{\varepsilon}{\hbar}|p|}. \tag{7.39}$$

We may also write

$$\Psi_x(X) \equiv \int_{|k|>k_F} \frac{dk}{2\pi} e^{ik(X-x)} e^{-\varepsilon|k|}$$
$$= -\frac{1}{2\pi} \frac{e^{(i(X-x)-\varepsilon)k_F}}{(i(X-x)-\varepsilon)} + \frac{1}{2\pi} \frac{e^{-(i(X-x)+\varepsilon)k_F}}{(i(X-x)+\varepsilon)}. \tag{7.40}$$

From the diagram it is clear that the probability density of a fermion at a position $x$

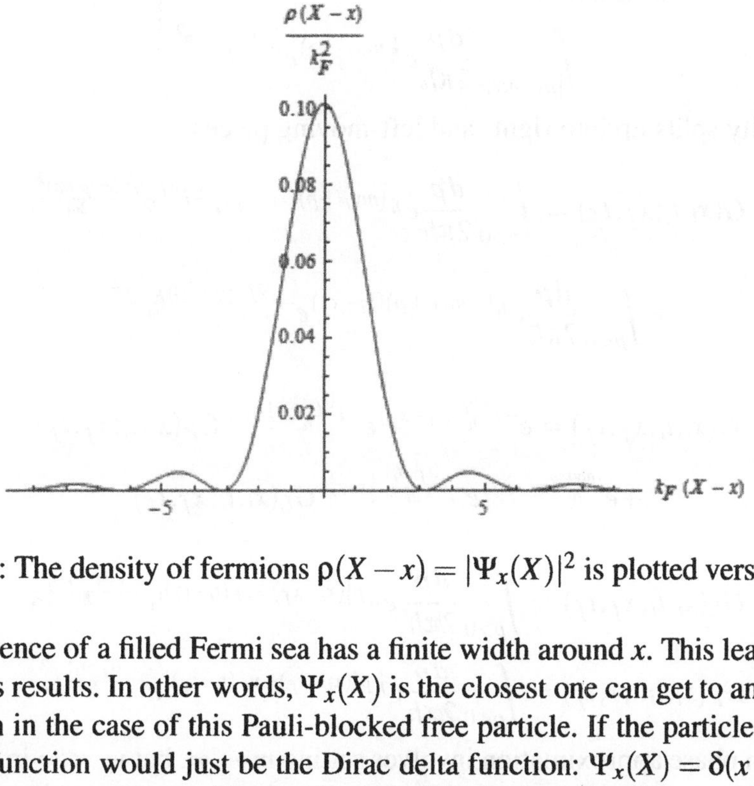


Figure 7.3: The density of fermions $\rho(X-x) = |\Psi_x(X)|^2$ is plotted versus $(X-x)$.

in the presence of a filled Fermi sea has a finite width around $x$. This leads to some anomalous results. In other words, $\Psi_x(X)$ is the closest one can get to an eigenstate of position in the case of this Pauli-blocked free particle. If the particle were truly free, this function would just be the Dirac delta function: $\Psi_x(X) = \delta(x-X)$. First we evaluate the average position and standard deviation. Consider,

$$< (X-x) > = \frac{\int_{-\infty}^{\infty} \Psi_0^*(Y) Y \Psi_0(Y) dY}{\int_{-\infty}^{\infty} \Psi_0^*(Y) \Psi_0(Y) dY} = 0. \tag{7.41}$$

This means the average position is $< X > = x$. The standard deviation diverges

$$\sigma^2 \equiv < (X-x)^2 > = \frac{\int_{-\infty}^{\infty} \Psi_0^*(Y) Y^2 \Psi_0(Y) dY}{\int_{-\infty}^{\infty} \Psi_0^*(Y) \Psi_0(Y) dY} = \infty. \tag{7.42}$$

This means that the notion of position is ambiguous for a fermion in the presence of a filled Fermi sea. Now we see if we can write down the path integral for such a fermion and rederive the propagator we derived earlier (Ch. 5) in Eq. (5.32). For this we write the propagator as usual. The Hamiltonian for such a fermion may be thought of as $H = \frac{p^2}{2m}C(p)$ where $C(p) = 1$ if $|p| > p_F$ and $C(p) = \infty$ if $|p| < p_F$. This ensures that the momentum of the particle close to the ground state will always be close to but larger than $p_F$. One may choose to evaluate the propagator below by resolving the identity using the momentum eigenstates. In the case the answer comes out immediately and naturally, but it is not a path integral.

$$G(x_i,t_i;x_f,t_f) = < x_i,t_i|x_f,t_f > \equiv < x_i,t_i|e^{-i\frac{(t_f-t_i)}{\hbar}C(p)\frac{p^2}{2m}}|x_f,t_i >$$
$$= \int_{-\infty}^{\infty}\frac{dp}{2\pi\hbar} < x_i,t_i|p >< p|x_f,t_i > e^{-i\frac{(t_f-t_i)}{\hbar}C(p)\frac{p^2}{2m}}$$
$$= \int_{|p|>mv_F}\frac{dp}{2\pi\hbar}e^{\frac{i}{\hbar}p(x_i-x_f)}e^{-i\frac{(t_f-t_i)}{\hbar}\frac{p^2}{2m}} \tag{7.43}$$

This naturally splits up into right- and left-moving pieces.

$$G(x_i,t_i;x_f,t_f) = \int_{p>0}\frac{dp}{2\pi\hbar}e^{\frac{i}{\hbar}(mv_F+p)(x_i-x_f)}e^{-i\frac{(t_f-t_i)}{\hbar}\frac{(mv_F+p)^2}{2m}}$$
$$+\int_{p<0}\frac{dp}{2\pi\hbar}e^{\frac{i}{\hbar}(-mv_F+p)(x_i-x_f)}e^{-i\frac{(t_f-t_i)}{\hbar}\frac{(-mv_F+p)^2}{2m}} \tag{7.44}$$

or,

$$G(x_i,t_i;x_f,t_f) = e^{-i\frac{mv_F}{\hbar}(x_i-x_f)}e^{-i\frac{(t_f-t_i)}{\hbar}\frac{1}{2}mv_F^2}G_R(x_i,t_i;x_f,t_f)$$
$$+e^{i\frac{mv_F}{\hbar}(x_i-x_f)}e^{-i\frac{(t_f-t_i)}{\hbar}\frac{1}{2}mv_F^2}G_L(x_i,t_i;x_f,t_f) \tag{7.45}$$

where,

$$G_L(x_i,t_i;x_f,t_f) = \int_{p>0}\frac{dp}{2\pi\hbar}e^{\frac{i}{\hbar}p[(x_i-x_f)+v_F(t_i-t_f)]}e^{-i\frac{(t_f-t_i)}{\hbar}\frac{p^2}{2m}} \tag{7.46}$$

$$G_R(x_i,t_i;x_f,t_f) = \int_{p<0}\frac{dp}{2\pi\hbar}e^{\frac{i}{\hbar}p[(x_i-x_f)-v_F(t_i-t_f)]}e^{-i\frac{(t_f-t_i)}{\hbar}\frac{p^2}{2m}}. \tag{7.47}$$

The random phase approximation involves setting $m \to \infty$, but $v_F < \infty$ in Eq. (7.46) and Eq. (7.47). It does not make sense to take this limit in the expression for the full Green function, since it contains oscillating terms that don't yield a definite limit. Alternatively, we may imagine that if either $|(x_f - x_i)| \to \infty$ or $|(t_f - t_i)| \to \infty$, the major contribution to the integral over $p$ comes from the small $p$ region. This means, we may ignore the $p^2$ in Eq. (7.46) and Eq. (7.47). The resulting expressions are known as the asymptotic parts of the Green functions.

$$G_L(x_i,t_i;x_f,t_f) = \int_{p>0}\frac{dp}{2\pi\hbar}e^{\frac{i}{\hbar}p[(x_i-x_f)+v_F(t_i-t_f)]}$$

$$= \frac{1}{2\pi i[(x_f - x_i) + v_F(t_f - t_i)]} \tag{7.48}$$

$$G_R(x_i,t_i;x_f,t_f) = \int_{p<0} \frac{dp}{2\pi\hbar} e^{\frac{i}{\hbar}p[(x_i-x_f)-v_F(t_i-t_f)]}$$

$$= \frac{1}{2\pi i[(x_i - x_f) - v_F(t_i - t_f)]} \tag{7.49}$$

The above analysis rests on the use of momentum eigenstates to resolve the identity. This effort leads to a simple closed formula for the propagator consistent with the description provided in Eq. (5.32) using the simple Schrodinger equation, but it does not lead to a path integral description. On the other hand, one may choose instead to resolve the identity by inserting position eigenstates. In this case the evolution operator has to be split up into slices that can be managed individually. This naturally leads, as we have seen, to the path integral.

$$G(x_i,t_i;x_f,t_f) = < x_i,t_i|x_f,t_f > \equiv < x_i,t_i|e^{-i\frac{(t_f-t_i)}{\hbar}C(p)\frac{p^2}{2m}}|x_f,t_i >$$

$$= < x_i,t_i|\underbrace{(e^{-i\frac{\varepsilon}{\hbar}C(p)\frac{p^2}{2m}})...(e^{-i\frac{\varepsilon}{\hbar}C(p)\frac{p^2}{2m}})}_{N}|x_f,t_i > \tag{7.50}$$

where $\varepsilon = \frac{(t_f-t_i)}{N}$. We now insert eigenstates of position using a resolution of the identity $1 = \int dx_k|x_k >< x_k|$.

$$G(x_i,t_i;x_f,t_f) = \int dx_1 ... \int dx_{N-1} \; < x_i,t_i|(e^{-i\frac{\varepsilon}{\hbar}C(p)\frac{p^2}{2m}})|x_1 >$$

$$\times < x_1|... < x_{N-1}|(e^{-i\frac{\varepsilon}{\hbar}C(p)\frac{p^2}{2m}})|x_f,t_i >$$

$$= \int dx_1 ... \int dx_{N-1} \prod_{k=0}^{N-1} < x_k|e^{-i\frac{\varepsilon}{\hbar}C(p)\frac{p^2}{2m}}|x_{k+1} >, \tag{7.51}$$

where we set $x_0 \equiv x_i$ and $x_N \equiv x_f$. Now,

$$< x_k|e^{-i\frac{\varepsilon}{\hbar}C(p)\frac{p^2}{2m}}|x_{k+1} > =$$

$$\int_{-\infty}^{\infty} dp_k \; < x_k|e^{-i\frac{\varepsilon}{\hbar}C(p)\frac{p^2}{2m}}|p_k >< p_k|x_{k+1} >$$

$$= \int_{-\infty}^{\infty} dp_k \; < x_k|p_k >< p_k|x_{k+1} > e^{-i\frac{\varepsilon}{\hbar}C(p_k)\frac{p_k^2}{2m}}$$

$$= \int_{-\infty}^{\infty} \frac{dp_k}{2\pi\hbar} e^{\frac{i}{\hbar}(x_k-x_{k+1})p_k} e^{-i\frac{\varepsilon}{\hbar}C(p_k)\frac{p_k^2}{2m}}. \tag{7.52}$$

We now make the following substitutions, $\hbar k_F = mv_F$ and $p_k = mv_F K$ and $E_F = \frac{1}{2}mv_F^2$. This renders the limits of integration constant of order unity. The presence of $C(p)$ means we have to avoid the region $-1 < K < 1$ since they are all fully occupied.

$$< x_k|e^{-i\frac{\varepsilon}{\hbar}C(p)\frac{p^2}{2m}}|x_{k+1} >$$
$$= \frac{mv_F}{h}\int_{-\infty}^{-1} dK\, e^{\frac{i}{\hbar}mv_F[(x_k-x_{k+1})K-\frac{\varepsilon v_F K^2}{2}]} + \frac{mv_F}{h}\int_{1}^{\infty} dK\, e^{\frac{i}{\hbar}mv_F[(x_k-x_{k+1})K-\frac{\varepsilon v_F K^2}{2}]}$$

$$< x_k|e^{-i\frac{\varepsilon}{\hbar}C(p)\frac{p^2}{2m}}|x_{k+1} >$$
$$= \frac{mv_F}{h}\int_{-\infty}^{-1} dK\, e^{\frac{i}{\hbar}mv_F[(x_k-x_{k+1})K-\frac{\varepsilon v_F K^2}{2}]} + \frac{mv_F}{h}\int_{1}^{\infty} dK\, e^{\frac{i}{\hbar}mv_F[(x_k-x_{k+1})K-\frac{\varepsilon v_F K^2}{2}]}$$
$$= \sqrt{\frac{m}{hi\varepsilon}}\; e^{\frac{im(x_k-x_{k+1})^2}{2\hbar\varepsilon}}\left(1+\frac{1}{2i}(Erfi(u_k(v_F))-Erfi(u_k(-v_F)))\right) \tag{7.53}$$

where,

$$u_k(v_F) = \frac{\left(\frac{1}{2}-\frac{i}{2}\right)\sqrt{m}(x_k - x_{k+1} + v_F\varepsilon)}{\sqrt{\hbar}\sqrt{\varepsilon}} \tag{7.54}$$

and $Erfi(z) = \frac{Erf(iz)}{i}$ is the imaginary error function. One may see that in the absence of a Fermi surface, $v_F \equiv 0$ and the present formula Eq. (7.53) reduces to the one derived earlier viz. Eq. (5.28). This means the Green function of a single electron in the presence of a Fermi surface is

$$G(x_i,t_i;x_f,t_f) = Lt_{N\to\infty}\int dx_1 ... \int dx_{N-1}\left(\sqrt{\frac{m}{hi\varepsilon}}\right)^N e^{\sum_{k=0}^{N-1}\frac{im(x_k-x_{k+1})^2}{2\hbar\varepsilon}}$$
$$\times \prod_{k=0}^{N-1}\left(1+\frac{1}{2i}(Erfi(u_k(v_F))-Erfi(u_k(-v_F)))\right). \tag{7.55}$$

If we take the straightforward approach, we are forced to conclude (with $v_k = \frac{x_{k+1}-x_k}{\varepsilon}$)

$$\prod_{k=0}^{N-1}\left(1+\frac{1}{2i}(Erfi(u_k(v_F))-Erfi(u_k(-v_F)))\right)$$
$$= e^{\sum_{k=0}^{N-1}Log\left(1+\frac{1}{2i}\left(Erfi((a+b)\sqrt{m\varepsilon})-Erfi((a-b)\sqrt{m\varepsilon})\right)\right)}$$
$$\approx e^{\sum_{k=0}^{N-1}\left(-\frac{2ib}{\sqrt{\pi}}\sqrt{m\varepsilon}+\frac{2b^2m\varepsilon}{\pi}-\frac{2i(-4b^3+3a^2b\pi+b^3\pi)(m\varepsilon)^{3/2}}{3\pi^{3/2}}\right)}, \tag{7.56}$$

where $a = -\frac{\left(\frac{1}{2}-\frac{i}{2}\right)v_k}{\sqrt{\hbar}}$ and $b = \frac{\left(\frac{1}{2}-\frac{i}{2}\right)v_F}{\sqrt{\hbar}}$. One may see that as $\varepsilon \to 0$ and $N \to \infty$ such that $\varepsilon N = (t_f - t_i) < \infty$ the nonconstant $v_k$ (or $a$) dependent terms cancel out. Thus, this approach does not lead to the correct propagator for a particle obeying Pauli's exclusion principle. Strictly speaking, this leads to a vanishing contribution since the divergent contribution from the first term leads to a rapidly oscillating contribution to the propagator due to the presence of the imaginary unit and this eventually cancels out. Thus this expansion is mathematically ill defined. It is indicative of an essential singularity. It is somewhat like Taylor expanding $f(x) = e^{-1/x^2}$ around $x = 0$—every term vanishes but the function clearly does not. We want the expansion of the Log to be proportional to $\varepsilon$ and contain $a$ so that a meaningful action may be written. The way to do this in the context of fermions is to invoke the random phase approximation at the outset viz. we set $m \to \infty$ keeping $v_F < \infty$. This means,

$$\prod_{k=0}^{N-1}\left(1+\frac{1}{2i}(Erfi(u_k(v_F)) - Erfi(u_k(-v_F)))\right)$$

$$= \prod_{k=0}^{N-1}\left(1+e^{(a-b)^2 m\varepsilon}\frac{i}{(a-b)\sqrt{4\pi\varepsilon m}} - e^{(a+b)^2 m\varepsilon}\frac{i}{(a+b)\sqrt{4\pi\varepsilon m}}\right). \tag{7.57}$$

This means we may write,

$$G_{RPA}(x_i,t_i;x_f,t_f) = Lt_{N\to\infty}\int dx_1 ... \int dx_{N-1}\left(\sqrt{\frac{m}{hi\varepsilon}}\right)^N e^{\sum_{k=0}^{N-1}\varepsilon\frac{imv_k^2}{2\hbar}}$$

$$\times \prod_{k=0}^{N-1}\left(1 - i\sqrt{\frac{ih}{\varepsilon m}}\frac{e^{-\frac{i(v_k+v_F)^2}{2\hbar}m\varepsilon}}{2\pi(v_k+v_F)} + i\sqrt{\frac{ih}{\varepsilon m}}\frac{e^{-\frac{i(v_k-v_F)^2}{2\hbar}m\varepsilon}}{2\pi(v_k-v_F)}\right). \tag{7.58}$$

Thus the RPA form of the path integral of a free particle obeying Pauli's exclusion principle in the presence of a filled Fermi sea may be written as (where $\varepsilon \equiv \frac{(t_f-t_i)}{N}$, $v(t) \equiv \frac{d}{dt}x(t)$),

$$G_{RPA}(x_i,t_i;x_f,t_f) = Lt_{N\to\infty}\int_{x(t_i)\equiv x_i}^{x(t_f)\equiv x_f} D[x(t)]\left(\sqrt{\frac{m}{hi\varepsilon}}\right)^N e^{\int_{t_i}^{t_f}dt\,\frac{imv^2(t)}{2\hbar}}$$

$$\times e^{\frac{1}{\varepsilon}\int_{t_i}^{t_f}dt\,Log\left(1 - i\sqrt{\frac{ih}{\varepsilon m}}\frac{e^{-\frac{i(v(t)+v_F)^2}{2\hbar}m\varepsilon}}{2\pi(v(t)+v_F)} + i\sqrt{\frac{ih}{\varepsilon m}}\frac{e^{-\frac{i(v(t)-v_F)^2}{2\hbar}m\varepsilon}}{2\pi(v(t)-v_F)}\right)}. \tag{7.59}$$

This continuum description is not particularly illuminating, however. It merely serves to highlight the complexities that one encounters while dealing with fermions in the context of path integrals. Specifically, the path integral is an approach closest in spirit to classical physics. Describing fermions using a formalism that is nearly classical in nature requires paying the price of introducing nonlocal actions (as in the present case) but with conventional commuting variables or, local actions but with anti-commuting (Grassmann) complex numbers (see later).

Reverting to the discrete RPA version in Eq. (7.58) we see that the most singular contribution comes from the term that has the largest number of potentially vanishing denominators. This means we may write,

$$G_{RPA}(x_i,t_i;x_f,t_f) \approx$$

$$Lt_{N\to\infty}\int dx_1 ... \int dx_{N-1}\left(\sqrt{\frac{m}{hi\varepsilon}}\right)^N$$

$$\left(\prod_{k=0}^{N-1}\left(-i\sqrt{\frac{ih}{\varepsilon m}}\frac{e^{\varepsilon\frac{imv_k^2}{2\hbar}}e^{-\frac{i(v_k+v_F)^2}{2\hbar}m\varepsilon}}{2\pi(v_k+v_F)}\right)+\prod_{k=0}^{N-1}\left(i\sqrt{\frac{ih}{\varepsilon m}}\frac{e^{\varepsilon\frac{imv_k^2}{2\hbar}}e^{-\frac{i(v_k-v_F)^2}{2\hbar}m\varepsilon}}{2\pi(v_k-v_F)}\right)\right) \tag{7.60}$$

$$\prod_{k=0}^{N-1} e^{\varepsilon\frac{imv_k^2}{2\hbar}}e^{-\frac{i(v_k+v_F)^2}{2\hbar}m\varepsilon} = e^{-\varepsilon\sum_{k=0}^{N-1}\frac{im}{2\hbar}(v_F^2+2v_Fv_k)} = e^{-(t_f-t_i)\frac{im}{2\hbar}v_F^2}e^{-\varepsilon\frac{im}{\hbar}v_F\sum_{k=0}^{N-1}v_k}. \tag{7.61}$$

Now $\sum_{k=0}^{N-1} v_k = \sum_{k=0}^{N-1}\frac{x_{k+1}-x_k}{\varepsilon} = \frac{x_f-x_i}{\varepsilon}$. Therefore, the propagator naturally splits up into two parts—right movers and left movers. For example, the left-moving piece may be written as,

$$G_L(x_i,t_i;x_f,t_f) \approx e^{-(t_f-t_i)\frac{im}{2\hbar}v_F^2}e^{-\frac{im}{\hbar}v_F(x_f-x_i)}$$

$$Lt_{N\to\infty}\int dx_1 ... \int dx_{N-1}\prod_{k=0}^{N-1}\frac{1}{2\pi i}\frac{1}{((x_{k+1}-x_k)+v_F\varepsilon)}. \tag{7.62}$$

For example, when $N = 3$ we are called upon to evaluate,

$$I = \int\int\frac{1}{(2\pi i)^3}\frac{dx_1dx_2}{((x_f-x_2)+v_F\varepsilon)((x_2-x_1)+v_F\varepsilon)((x_1-x_i)+v_F\varepsilon)}. \tag{7.63}$$

Interpreted as principal value these integrals vanish. However, one must imagine that there is a small imaginary part to $\varepsilon$ in which case the integrals exist. In this case this integral $I$ evaluates to,

$$I = \frac{1}{2\pi i}\frac{1}{(x_f-x_i+3v_F\varepsilon)}. \tag{7.64}$$

But $\varepsilon = \frac{(t_f-t_i)}{3}$ in this case, hence,

$$I = \frac{1}{2\pi i}\frac{1}{(x_f-x_i+v_F(t_f-t_i))}. \tag{7.65}$$

In general also, we get the same result. Thus, the left-moving Green function obtained using the path integral method is

$$G_L(x_i,t_i;x_f,t_f) \approx e^{-(t_f-t_i)\frac{im}{2\hbar}v_F^2} e^{-\frac{im}{\hbar}v_F(x_f-x_i)} \frac{1}{2\pi i} \frac{1}{(x_f - x_i + v_F(t_f-t_i))} \quad (7.66)$$

This result is identical to the Green function obtained earlier viz. Eq. (7.48). Now we go on to discuss the harmonic oscillator.

## 7.3 Harmonic Oscillator

The path integral for the harmonic oscillator is similar to the free particle except that now we have a potential energy function added in the Hamiltonian.

$$< x_i,t_i|x_f,t_f >= \int_{x(t_i)\equiv x_i}^{x(t_f)\equiv x_f} D[x]\ e^{\frac{i}{\hbar}\int_{t_i}^{t_f} dt(\frac{1}{2}m\dot{x}^2(t)-\frac{1}{2}m\omega^2x^2(t))} \quad (7.67)$$

The usual procedure for evaluating this is to write the path as the classical path connecting the events $(x_i,t_i)$ and $(x_f,t_f)$.

$$x_{cl}(t) = \frac{x_f\ sin(\omega(t-t_i)) - x_i\ sin(\omega(t-t_f))}{sin(\omega(t_f-t_i))} \quad (7.68)$$

This path obeys the equation of motion viz. $\ddot{x}_{cl}(t) = -\omega^2\ x_{cl}(t)$ and the initial and final conditions $x_{cl}(t_i) = x_i$ and $x_{cl}(t_f) = x_f$. The overall path is taken to be this path plus a quantum correction $\tilde{x}(t)$ so that $x(t) = x_{cl}(t) + \tilde{x}(t)$. This correction is now constrained to vanish at the end points: $\tilde{x}(t_i) = \tilde{x}(t_f) = 0$. This means we may write,

$$\tilde{x}(t) = \sum_{n=1}^{\infty} \tilde{x}_n\ sin(\frac{n\pi(t-t_i)}{t_f-t_i}). \quad (7.69)$$

At this state we introduce a dimensionless time $s$ defined through $t = t_f s + t_i(1-s)$ so that $t(s=0) = t_i$ and $t(s=1) = t_f$. Therefore, the action $S = \int_{t_i}^{t_f} dt\ L$ may be rewritten as,

$$\begin{aligned} S &= \int_{t_i}^{t_f} dt(\frac{1}{2}m\dot{x}^2(t) - \frac{1}{2}m\omega^2x^2(t)) \\ &= \int_{t_i}^{t_f} dt(\frac{1}{2}m\dot{x}_{cl}^2(t) + \frac{1}{2}m\dot{\tilde{x}}^2(t) - \frac{1}{2}m\omega^2x_{cl}^2(t) - \frac{1}{2}m\omega^2\tilde{x}^2(t)) \\ &+ \int_{t_i}^{t_f} dt\ m\dot{x}_{cl}(t)\dot{\tilde{x}}(t) - \int_{t_i}^{t_f} dt\ m\omega^2\tilde{x}(t)x_{cl}(t). \end{aligned} \quad (7.70)$$

But $\int_{t_i}^{t_f} dt\; m\dot{x}_{cl}(t)\dot{\tilde{x}}(t) = \int_{t_i}^{t_f} dt\; \frac{d}{dt}(m\dot{x}_{cl}(t)\tilde{x}(t)) - \int_{t_i}^{t_f} dt\; m\ddot{x}_{cl}(t)\tilde{x}(t)$. Since $\tilde{x}(t_f) = \tilde{x}(t_i) = 0$ and $\ddot{x}_{cl} = -\omega^2 x_{cl}$, the cross terms cancel out and we may write,

$$S = \int_{t_i}^{t_f} dt(\frac{1}{2}m\dot{x}^2(t) - \frac{1}{2}m\omega^2 x^2(t))$$

$$= \int_{t_i}^{t_f} dt(\frac{1}{2}m\dot{x}_{cl}^2(t) - \frac{1}{2}m\omega^2 x_{cl}^2(t)) + \int_{t_i}^{t_f} dt(\frac{1}{2}m\dot{\tilde{x}}^2(t) - \frac{1}{2}m\omega^2\tilde{x}^2(t)). \quad (7.71)$$

Set $dt = (t_f - t_i)ds$ and

$$X(s) = \tilde{x}(t(s)) = \sum_{n=1}^{\infty} \tilde{x}_n\; sin(n\pi s). \quad (7.72)$$

Thus,

$$\dot{x}_{cl}(t) = \omega\frac{x_f cos(\omega(t-t_i)) - x_i cos(\omega(t-t_f))}{sin(\omega(t_f-t_i))}. \quad (7.73)$$

We write, $S = S_{cl} + \tilde{S}$ where,

$$S_{cl} = \int_{t_i}^{t_f} dt(\frac{1}{2}m\dot{x}_{cl}^2(t) - \frac{1}{2}m\omega^2 x_{cl}^2(t))$$

$$= m\omega\frac{(x_f^2 + x_i^2)cos(\omega(t_f-t_i)) - 2x_f x_i}{2sin(\omega(t_f-t_i))} \quad (7.74)$$

and,

$$\tilde{S} = \int_{t_i}^{t_f} dt(\frac{1}{2}m\dot{\tilde{x}}^2(t) - \frac{1}{2}m\omega^2\tilde{x}^2(t))$$

$$= \int_{t_i}^{t_f} dt\; \frac{1}{2}m\sum_{n=1}^{\infty}\tilde{x}_n^2\,(\frac{(n\pi)^2}{(t_f-t_i)^2}\; cos^2(\frac{n\pi(t-t_i)}{t_f-t_i}) - \omega^2\; sin^2(\frac{n\pi(t-t_i)}{t_f-t_i}))$$

$$= \sum_{n=1}^{\infty}\frac{m\tilde{x}_n^2(t_f-t_i)}{4}\,(\frac{(n\pi)^2}{(t_f-t_i)^2} - \omega^2). \quad (7.75)$$

Thus we may write,

$$< x_i, t_i | x_f, t_f > = e^{\frac{i}{\hbar}S_{cl}}\; g(t_f - t_i)$$

$$g(t_f - t_i) = \int D[\tilde{x}_n]\; e^{\sum_{n=1}^{\infty}\frac{im\tilde{x}_n^2(t_f-t_i)}{4\hbar}\,(\frac{(n\pi)^2}{(t_f-t_i)^2} - \omega^2)} \quad (7.76)$$

$$\delta(x_i - x_f) \equiv Lim_{t_f\to t_i} < x_i, t_i | x_f, t_f > = Lim_{t_f\to t_i} e^{\frac{i}{\hbar}S_{cl}} g(t_f - t_i)$$

$$= Lim_{t_f\to t_i}\; e^{\frac{i}{\hbar}m\frac{(x_f-x_i)^2}{2(t_f-t_i)}}\; g(t_f - t_i). \quad (7.77)$$

This leads to the same expression as the one we obtained for the free particle,

$$g(t_f - t_i) = (\frac{2i\pi\hbar}{m}(t_f - t_i))^{-\frac{3}{2}}. \tag{7.78}$$

This means that the final expression for the propagator of a harmonic oscillator is

$$< x_i, t_i | x_f, t_f >= e^{\frac{i}{\hbar} m\omega \frac{(x_f^2 + x_i^2)cos(\omega(t_f - t_i)) - 2x_f x_i}{2sin(\omega(t_f - t_i))}} (\frac{2i\pi\hbar}{m}(t_f - t_i))^{-\frac{3}{2}}. \tag{7.79}$$

Later we shall derive this expression using what is known as the coherent state path integral where we resolve the identity using coherent states (eigenstates of the annihilation operator) rather than position eigenstates.

## 7.4 Two-Particle Green Functions

One may use the path integral to study the two-particle propagator. This means we start with a state such as $|x_{A0}t_{A0}; x_{B0}t_{B0}>$ that corresponds to one particle being at location $x_{A0}$ at time $t_{A0}$ and the other at $x_{B0}$ at time $t_{B0}$. The main point is, in quantum mechanics, the two particles are perfectly indistinguishable so that there can be an exchange of positions. We then wish to find the overlap between this and $|x_{AN}t_{AN}; x_{BN}t_{BN}>$. Firstly, since these particles are indistinguishable, the wavefunction must be either symmetric or antisymmetric. We choose the former now and the latter case is left to the exercises. Formally,

$$|x_{A0}t_{A0}; x_{B0}t_{B0}> = \frac{1}{\sqrt{2}}(|x_{A0}t_{A0}>_1 |x_{B0}t_{B0}>_2 + |x_{B0}t_{B0}>_1 |x_{A0}t_{A0}>_2) \tag{7.80}$$

$$< x_{AN}t_{AN}; x_{BN}t_{BN}| = \frac{1}{\sqrt{2}}(< x_{AN}t_{AN}|_1 < x_{BN}t_{BN}|_2 + < x_{BN}t_{BN}|_1 < x_{AN}t_{AN}|_2). \tag{7.81}$$

Thus the two-particle Green function may be written as,

$$< x_{A0}t_{A0}; x_{B0}t_{B0}|x_{AN}t_{AN}; x_{BN}t_{BN} >=< x_{A0}t_{A0}|x_{AN}t_{AN} >< x_{B0}t_{B0}|x_{BN}t_{BN} >$$
$$+ < x_{B0}t_{B0}|x_{AN}t_{AN} >< x_{A0}t_{A0}|x_{BN}t_{BN} > . \tag{7.82}$$

The above result, which involves decomposing two-particle Green functions into two one-particle Green functions, is valid only for free particles or free particles interacting with an external potential but not with each other. This result is known as Wick's theorem. Since the two particles are identical and interact with the same potential, we have dropped the subscripts since as functions the one-particle Green functions are the same for both particles. Wick's theorem is the statement that two-particle and higher correlations may be written as the product of several one-particle Green functions obtained by pairing of two points at a time. Clearly, there are many permutations possible. A simple sum yields an overall Green function that describes a set of bosons, whereas an sum with alternating signs, positive for an even permutation and negative for an odd permutation, yields a wavefunction that describes fermions. The two-particle Green function has both a direct as well as an exchange contribution (any one term can be called direct, there is nothing special about any of them). Now we wish to show how this exchange manifests itself in the path integral.

$$< x_{A0}t_{A0}; x_{B0}t_{B0}|x_{AN}t_{AN}; x_{BN}t_{BN} >=$$

$$\frac{1}{2} < x_{A0}t_{A0}; x_{B0}t_{B0}|e^{-\frac{i}{\hbar}(t_{AN}-t_{A0})H_1} e^{-\frac{i}{\hbar}(t_{BN}-t_{B0})H_2}|x_{AN}t_{A0} >_1 |x_{BN}t_{B0} >_2$$
$$+\frac{1}{2} < x_{A0}t_{A0}; x_{B0}t_{B0}|e^{-\frac{i}{\hbar}(t_{AN}-t_{A0})H_2} e^{-\frac{i}{\hbar}(t_{BN}-t_{B0})H_1}|x_{AN}t_{A0} >_2 |x_{BN}t_{B0} >_1 \quad (7.83)$$

As before, we want to insert a complete set of eigenstates of position. Consider ($a = 1$, $b = 2$ or $a = 2$,$b = 1$) where $\varepsilon_A = \frac{(t_{AN}-t_{A0})}{N}$ and $\varepsilon_B = (t_{BN} - t_{B0})/N$ and $\varepsilon_A^{'} = \frac{(t_{AN}-t_{B0})}{N}$ and $\varepsilon_B^{'} = (t_{BN} - t_{A0})/N$

$$_a< x_{A0}t_{A0}|\ _b< x_{B0}t_{B0}|e^{-\frac{i}{\hbar}(t_{AN}-t_{B0})H_b} e^{-\frac{i}{\hbar}(t_{BN}-t_{A0})H_a}|x_{AN}t_{B0} >_b |x_{BN}t_{A0} >_a$$
$$= \int dx_0 dx_1 ... dx_{N-1} dx_N \int dx_0^{'} dx_1^{'} ... dx_{N-1}^{'} dx_N^{'}$$
$$\times\ \delta(x_0^{'} - x_{A0})\delta(x_0 - x_{B0})\delta(x_N - x_{AN})\delta(x_N^{'} - x_{BN})$$
$$\times \prod_{k=0}^{N-1}\ _a< x_k^{'}t_{A0}|\ _b< x_k t_{B0}|e^{-\frac{i}{\hbar}\varepsilon_A^{'} H_b} e^{-\frac{i}{\hbar}\varepsilon_B^{'} H_a}|x_{k+1}t_{B0} >_b |x_{k+1}^{'}t_{A0} >_a . \quad (7.84)$$

Similarly,

$$_b< x_{A0}t_{A0}|\ _a< x_{B0}t_{B0}|e^{-\frac{i}{\hbar}(t_{AN}-t_{A0})H_b} e^{-\frac{i}{\hbar}(t_{BN}-t_{B0})H_a}|x_{AN}t_{A0} >_b |x_{BN}t_{B0} >_a$$
$$= \int dx_0 dx_1 ... dx_{N-1} dx_N \int dx_0^{'} dx_1^{'} ... dx_{N-1}^{'} dx_N^{'}$$
$$\times\ \delta(x_0^{'} - x_{A0})\delta(x_0 - x_{B0})\delta(x_N - x_{BN})\delta(x_N^{'} - x_{AN})$$
$$\times \prod_{k=0}^{N-1}\ _b< x_k^{'}t_{A0}|\ _a< x_k t_{B0}|e^{-\frac{i}{\hbar}\varepsilon_A H_b} e^{-\frac{i}{\hbar}\varepsilon_B H_a}|x_{k+1}^{'}t_{A0} >_b |x_{k+1}t_{B0} >_a . \quad (7.85)$$

As usual we write $H_a = \frac{p_a^2}{2m} + V(x_a)$ and use the Trotter product formula to get (the rest of the details are left to exercises),

$$< x_{A0}t_{A0}; x_{B0}t_{B0}|x_{AN}t_{AN}; x_{BN}t_{BN} > .$$

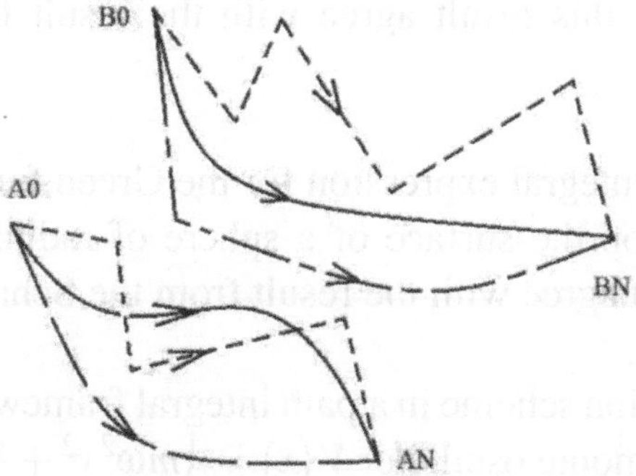


Figure 7.4: Path integral for two bosons showing one possible set of paths connecting the initial points $(A0, B0)$ to the final points $(AN, BN)$.

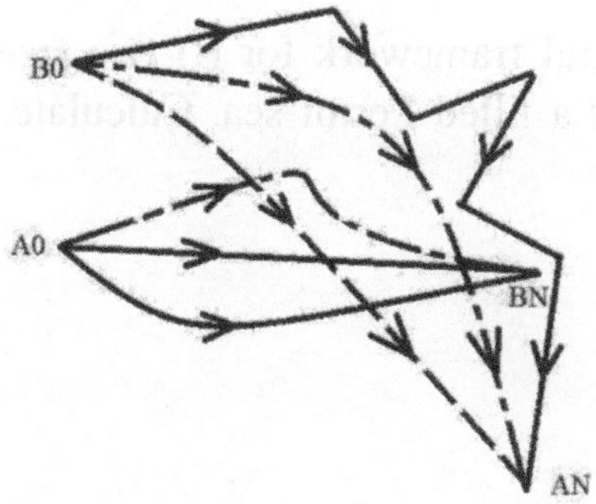


Figure 7.5: Path integral for two bosons showing another possible set of paths connecting the initial points $(A0, B0)$ to the final points $(AN, BN)$.

$$= \left( \int_{x(t_{A0}) \equiv x_{A0}}^{x(t_{AN}) = x_{AN}} D[x] \, e^{\frac{i}{\hbar} S[x,\dot{x}]} \right) \left( \int_{x(t_{B0}) \equiv x_{B0}}^{x(t_{BN}) = x_{BN}} D[x] \, e^{\frac{i}{\hbar} S[x,\dot{x}]} \right)$$
$$+ \left( \int_{x(t_{B0}) \equiv x_{B0}}^{x(t_{AN}) = x_{AN}} D[x] \, e^{\frac{i}{\hbar} S[x,\dot{x}]} \right) \left( \int_{x(t_{A0}) \equiv x_{A0}}^{x(t_{BN}) = x_{BN}} D[x] \, e^{\frac{i}{\hbar} S[x,\dot{x}]} \right) \tag{7.86}$$

Here, $S[x, \dot{x}] = \int_{t_i}^{t_f} dt (\frac{1}{2} m\dot{x}^2 - V(x))$ is the action between times specified in the path integral. In the exercises, one may encounter further examples.

## 7.5 Exercises

**Q.1** Write down a path integral expression for the Green function of an (otherwise) free particle confined to be present only at points on the positive x-axis. Evaluate this path integral. Does this result agree with the result from the Schrodinger's equation?

**Q.2** Write down a path integral expression for the Green function of a free particle confined to be present on the surface of a sphere of radius $R$. Evaluate this path integral. Does this result agree with the result from the Schrodinger's equation?

**Q.3** Develop a perturbation scheme in a path integral framework (see earlier chapter also) to study the anharmonic oscillator $V(x) = \frac{1}{2}m\omega^2x^2 + \frac{g}{4!}x^4$ where $m, \omega, g > 0$.

**Q.4** Develop the path integral framework close to one of the minima of $V(x) = -\frac{1}{2}\xi^2x^2 + \frac{g}{4!}x^4$ where $\xi, g > 0$. What is the tunneling probability from one minimum to another?

**Q.5** Develop the path integral framework for (i) two free fermions and (ii) two fermions in the presence of a filled Fermi sea. Calculate the two-particle Green function.

**Q.6** Verify Eq. (7.86).

# Chapter 8

# Creation and Annihilation Operators in Fock Space

In this chapter, we discuss the notion of creation and annihilation operators. These operators correspond to addition or removal of excitations of a system of particles or particles themselves. We show that these operators may be used to rewrite many-body Hamiltonians in a compact form where information about Bose or Fermi statistics is encoded in the Hamiltonian itself, which greatly reduces the effort required in studying its properties.

## 8.1 Introduction to Second Quantization

The older term for rewriting quantum formalisms using creation and annihilation operators in place of position and momentum operators is 'Second Quantization'. Second Quantization does not mean quantizing twice! We start by discussing the simple harmonic oscillator. Consider the Hamiltonian,

$$H = \frac{p^2}{2m} + \frac{1}{2}m\omega^2x^2 = E_0\left(\frac{p^2}{2mE_0} + \frac{m\omega^2x^2}{2E_0}\right). \tag{8.1}$$

This looks like $H = E_0(X^2+Y^2)$ $=E_0(Y+iX)(Y-iX) - iE_0[X,Y]$ where $Y = \frac{p}{\sqrt{2mE_0}}$ and $X = \sqrt{\frac{m}{2E_0}}\omega x$ for some appropriate choice of $E_0$. Set

$$a^{\dagger} = Y + iX;\ a = Y - iX \tag{8.2}$$

since X and Y are Hermitian. We are going to choose $E_0$ by demanding that, $[a,a^{\dagger}] = 1$. We see that $[X,Y] = \sqrt{\frac{m}{2E_0}}\omega\frac{i\hbar}{\sqrt{2mE_0}}$ but $[a,a^{\dagger}] = -2i[X,Y] = \frac{\hbar\omega}{E_0} = 1$

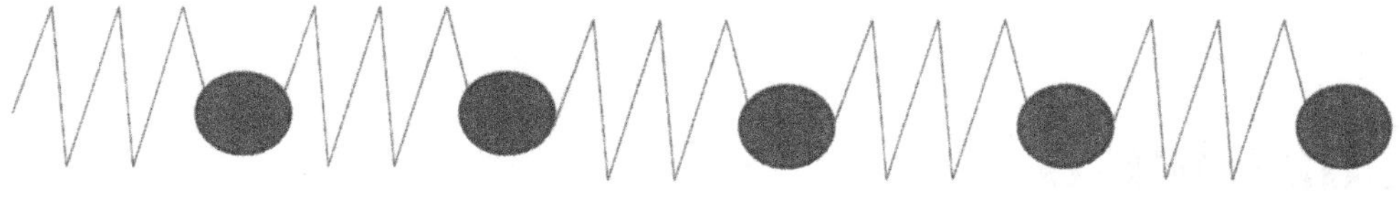

Figure 8.1: Shows a chain of masses and springs meant to illustrate the concept of a field.

or $E_0 = \hbar\omega$. Thus,

$$H = E_0(Y+iX)(Y-iX) - iE_0[X,Y] = \hbar\omega a^\dagger a + \frac{1}{2}\hbar\omega. \tag{8.3}$$

Here, $a^\dagger$ creates an excitation or a quantum of energy $\hbar\omega$. The quantity $N = a^\dagger a$ measures the number of quanta or excitations in a state this operator acts on.

Next we consider a chain of harmonic oscillators. It is described by,

$$H = \sum_{j=1}^{N} \frac{p_j^2}{2m} + \frac{1}{2}m\omega^2 \sum_{j=1}^{N-1} (x_{j+1} - x_j)^2. \tag{8.4}$$

We wish to rewrite this using creation and annihilation operators. Since the potential energy mixes up the position indices, we can expect the Hamiltonian in the second quantized language to be off diagonal in these indices. We make the ansatz

$$H = \sum_{i,j} \Omega(i,j) a^\dagger(i) a(j) + E_0. \tag{8.5}$$

Here, $E_0$ is the zero point energy, an ordinary number which in the case of one harmonic oscillator was $\frac{1}{2}\hbar\omega$. Here $\Omega(i,j)$ is some quantity (number) that depends on $i$ and $j$. We note that $a(i)$ is a linear combination of the 'p's and 'x's just as in the case of single harmonic oscillator. Thus we can say that $[a(i), p_k]$ and $[a(i), x_k]$ are ordinary numbers (called c number or commutating number). We use $[a(i), a(j)] = 0$ and $[a(i), a^\dagger(j)] = \delta_{i,j}$ and $[x_i, x_j] = 0$ and $[x_i, p_j] = i\hbar\delta_{i,j}$. To evaluate, consider the commutator, $[a(i), H]$

$$[a(i), H] = \sum_j \Omega(i,j) a(j) = \sum_{j=1}^{N} \frac{p_j [a(i), p_j]}{m}$$
$$+ m\omega^2 \sum_{j=1}^{N-1} (x_{j+1} - x_j)([a(i), x_{j+1}] - [a(i), x_j]). \tag{8.6}$$

Evaluate the commutator with $x_k$ to get

$$\sum_j \Omega(i,j)[x_k, a(j)] = i\hbar \frac{[a(i), p_k]}{m}. \tag{8.7}$$

Next, form the commutator with $p_k$,

$$\sum_j \Omega(i,j)[p_k,a(j)] = i\hbar m\omega^2([a(i),x_{k+1})] + [a(i),x_{k-1}] - 2[a(i),x_k]). \tag{8.8}$$

A sum such as $\sum_j \Omega(i,j)[p_k,a(j)]$ is a convolution, hence it suggests a Fourier transform solution. We write,

$$[a(j),p_k] = \sum_q A_p(q)e^{iqa(j-k)}; [a(j),x_k] = \sum_q A_x(q)e^{iqa(j-k)} \tag{8.9}$$

and

$$\Omega(i,j) = \sum_q \tilde{\Omega}_q e^{iqa(i-j)}. \tag{8.10}$$

We also employ the following identity, $\sum_j e^{i(q-q')aj} = N\delta_{q,q'}$ where N is the number of masses.

$$\sum_j \sum_{q'} \tilde{\Omega}_{q'} e^{iq'a(i-j)} \sum_q A_x(q)e^{iqa(j-k)} = -\frac{i\hbar}{m}\sum_q A_p(q)e^{iqa(i-k)}$$

$$\sum_j \sum_{q'} \tilde{\Omega}'_q e^{iq'a(i-j)} \sum_q A_p(q)e^{iqa(j-k)} = -i\hbar m\omega^2 \sum_q A_x(q)e^{iqa(i-k)}(e^{iqa} + e^{-iqa} - 2) \tag{8.11}$$

Performing the necessary summations we get

$$N\sum_q \tilde{\Omega}_q A_x(q)e^{iqa(i-k)} = -\frac{i\hbar}{m}\sum_q A_p(q)e^{iqa(i-k)}$$

$$\sum_q \tilde{\Omega}_q A_p(q)e^{iqa(i-k)} = i\hbar m\omega^2 \sum_q A_x(q)e^{iqa(i-k)} 4\sin^2(\frac{qa}{2}). \tag{8.12}$$

$A_s e^{iqa(i-k)}$ are linearly independent for each q, we may write,

$$N\tilde{\Omega}_q A_x(q) = -\frac{i\hbar}{m}A_p(q)$$

$$N\tilde{\Omega}_q A_p(q) = i\hbar m\omega^2 A_x(q) 4\sin^2(\frac{qa}{2}). \tag{8.13}$$

Multiplying these two equations together we get,

$$N^2\tilde{\Omega}_q^2 = \hbar^2\omega^2\, 4\sin^2(\frac{qa}{2}). \tag{8.14}$$

Define $\Omega_q = N\,\tilde{\Omega}_q$, since we require $\Omega(i,j) = \Omega(j,i)$. Thus,

$$H = \sum_{i,j}\Omega(i,j)a^\dagger(i)a(j) + E_0 = \frac{1}{N}\sum_{i,j,q}\Omega_q e^{iqa(i-j)}a^\dagger(i)a(j) + E_0$$

$$a_q = \frac{1}{\sqrt{N}} \sum_j e^{-iqaj}\, a(j) \tag{8.15}$$

$$H = \sum_q \Omega_q a_q^\dagger a_q + E_0 \tag{8.16}$$

where,

$$\Omega_q = \hbar\omega \left| 2sin(\frac{qa}{2}) \right|. \tag{8.17}$$

### 8.1.1 The Quantum Electromagnetic Field

A system made of masses and springs when quantized is not the only one that results in bosonic oscillators. One of the most important systems which also has this feature is the electromagnetic field. In an earlier chapter, we pointed out that the classical Hamiltonian of the electromagnetic field in free space may be written as

$$H = \int d^3r \frac{1}{8\pi}(\mathbf{E}^2(\mathbf{r}) + \mathbf{B}^2(\mathbf{r})). \tag{8.18}$$

Also we may make the following gauge choice $\mathbf{E} = -\frac{1}{c}\partial_t \mathbf{A}$ and $\mathbf{B} = \nabla \times \mathbf{A}$ with $\nabla \cdot \mathbf{A} \equiv 0$ so that the canonical variable becomes $\mathbf{A}$ and the canonical momentum would be as before,

$$\mathbf{P}_A = -\frac{1}{4\pi c}\mathbf{E} \tag{8.19}$$

Thus, the classical Hamiltonian for the electromagnetic field would have to be

$$H(\mathbf{A}, \mathbf{P}_A) = \int d^3r \frac{1}{8\pi}((4\pi c)^2 \mathbf{P}_A^2(\mathbf{r}) + (\nabla \times \mathbf{A}(\mathbf{r}))^2). \tag{8.20}$$

In order to quantize any Hamiltonian with a coordinate and a momentum, we are forced to impose the canonical commutation rules.

$$[A_j(\mathbf{r},t), A_k(\mathbf{r}',t)] = [P_{A,j}(\mathbf{r},t), P_{A,k}(\mathbf{r}',t)] = 0 \tag{8.21}$$

and the nontrivial one,

$$[A_j(\mathbf{r},t), P_{A,k}(\mathbf{r}',t)] = \delta_{j,k}\, i\hbar\, \delta^3(\mathbf{r} - \mathbf{r}') \tag{8.22}$$

where $j,k$ are spatial components. But since the transversality condition is operative, they only take two values (corresponding to two linearly independent polarization states). These commutation rules may be made simpler by the following substitutions. We demand that there exist bosonic oscillators $a_j(\mathbf{k})$ such that the Hamiltonian in Eq. (8.20) be purely diagonal in them. We further assume that there is an energy $\Omega(\mathbf{k})$ associated with each of these quanta so that apart from an additive constant we may write,

$$H = \sum_{j,\mathbf{k}} \Omega(\mathbf{k})\, a_j^{\dagger}(\mathbf{k}) a_j(\mathbf{k}). \tag{8.23}$$

These oscillators are required to obey the property that,

$$[a_j(\mathbf{k}), a_l(\mathbf{k}^{'})] = [a_j^{\dagger}(\mathbf{k}), a_l^{\dagger}(\mathbf{k}^{'})] = 0 \tag{8.24}$$

and,

$$[a_j(\mathbf{k}), a_l^{\dagger}(\mathbf{k}^{'})] = \delta_{j,l}\, \delta_{\mathbf{k},\mathbf{k}^{'}}. \tag{8.25}$$

We now observe that the commutator of Eq. (8.23) with $a_j(\mathbf{k})$ should be the same as the commutator of Eq. (8.20). Further, we know that since $P_A$ and $A_j$ are linear combinations of the oscillators, the commutator of these objects with the oscillators must be just numbers. Also since $\mathbf{B} = \nabla \times \mathbf{A}$ we have,

$$(\nabla \times \mathbf{A})^2 = \nabla \cdot (\mathbf{A} \times \mathbf{B}) + \mathbf{A} \cdot (\nabla \times \mathbf{B}). \tag{8.26}$$

Integrated over all space we get,

$$\int d^3r\, (\nabla \times \mathbf{A})^2 = \int d^3r\, \mathbf{A} \cdot (\nabla \times \mathbf{B}) = -\int d^3r\, \mathbf{A}(\mathbf{r}) \cdot \nabla^2 \mathbf{A}(\mathbf{r}). \tag{8.27}$$

The last result is from transversality. This means on the one hand,

$$[a_j(\mathbf{k}), H] = \int d^3r \frac{1}{4\pi}\, ((4\pi c)^2 \mathbf{P}_A(\mathbf{r}) \cdot [a_j(\mathbf{k}), \mathbf{P}_A(\mathbf{r})] - \sum_l [a_j(\mathbf{k}), A_l(\mathbf{r})] \nabla^2 A_l(\mathbf{r})). \tag{8.28}$$

On the other hand,

$$[a_j(\mathbf{k}), H] = \Omega(\mathbf{k})\, a_j(\mathbf{k}). \tag{8.29}$$

Keeping in mind $[a_j(\mathbf{k}), \mathbf{P}_A(\mathbf{r})]$ and $[a_j(\mathbf{k}), A_l(\mathbf{r})]$ are just numbers, we may successively take commutators of Eq. (8.28) and Eq. (8.29) with $A_m(\mathbf{r}^{'})$ and $P_{A,m}(\mathbf{r}^{'})$ to get on the one hand,

$$[[a_j(\mathbf{k}), H], A_m(\mathbf{r}^{'})] = \int d^3r \frac{1}{4\pi}\, (4\pi c)^2\, [\mathbf{P}_A(\mathbf{r}), A_m(\mathbf{r}^{'})] \cdot [a_j(\mathbf{k}), \mathbf{P}_A(\mathbf{r})]$$

$$= -i\hbar \int d^3r \frac{1}{4\pi}\, (4\pi c)^2\, \delta^3(\mathbf{r} - \mathbf{r}^{'}) [a_j(\mathbf{k}), P_{A,m}(\mathbf{r})] = -i\hbar \frac{1}{4\pi}\, (4\pi c)^2\, [a_j(\mathbf{k}), P_{A,m}(\mathbf{r}^{'})]. \tag{8.30}$$

On the other hand,

$$[[a_j(\mathbf{k}), H], A_m(\mathbf{r}^{'})] = \Omega(\mathbf{k})\, [a_j(\mathbf{k}), A_m(\mathbf{r}^{'})]. \tag{8.31}$$

Hence we have the first of the identities,

$$-i\hbar\frac{1}{4\pi}\,(4\pi c)^2\,[a_j(\mathbf{k}),P_{A,m}(\mathbf{r}^{'})] = \Omega(\mathbf{k})\,[a_j(\mathbf{k}),A_m(\mathbf{r}^{'})]. \tag{8.32}$$

Next we have,

$$[[a_j(\mathbf{k}),H],P_{A,m}(\mathbf{r}^{'})] = \int d^3r\frac{1}{4\pi}\,(-\sum_l[a_j(\mathbf{k}),A_l(\mathbf{r})]\nabla^2[A_l(\mathbf{r}),P_{A,m}(\mathbf{r}^{'})])$$
$$= -i\hbar\frac{1}{4\pi}\,\nabla^{'2}[a_j(\mathbf{k}),A_m(\mathbf{r}^{'})] \tag{8.33}$$

so that,

$$[[a_j(\mathbf{k}),H],P_{A,m}(\mathbf{r}^{'})] = \Omega(\mathbf{k})\,[a_j(\mathbf{k}),P_{A,m}(\mathbf{r}^{'})]. \tag{8.34}$$

Therefore the second relation is,

$$-i\hbar\frac{1}{4\pi}\,\nabla^{'2}[a_j(\mathbf{k}),A_m(\mathbf{r}^{'})] = \Omega(\mathbf{k})\,[a_j(\mathbf{k}),P_{A,m}(\mathbf{r}^{'})]. \tag{8.35}$$

Combining Eq. (8.32) and Eq. (8.35) we get,

$$\nabla^{'2}[a_j(\mathbf{k}),P_{A,m}(\mathbf{r}^{'})] = -(\frac{\Omega(\mathbf{k})}{c\hbar})^2\,[a_j(\mathbf{k}),P_{A,m}(\mathbf{r}^{'})]. \tag{8.36}$$

Therefore,

$$[a_j(\mathbf{k}),P_{A,m}(\mathbf{r}^{'})] = e^{i\hat{u}\cdot\mathbf{r}^{'}(\frac{\Omega(\mathbf{k})}{c\hbar})}\,[a_j(\mathbf{k}),P_{A,m}(0)] \tag{8.37}$$

and,

$$[a_j(\mathbf{k}),A_m(\mathbf{r}^{'})] = -i\hbar\frac{1}{4\pi\Omega(\mathbf{k})}\,(4\pi c)^2\,e^{i\hat{u}\cdot\mathbf{r}^{'}(\frac{\Omega(\mathbf{k})}{c\hbar})}\,[a_j(\mathbf{k}),P_{A,m}(0)] \tag{8.38}$$

for some unit vector $\hat{u}$ to be found now. Since $a_j(\mathbf{k})$ itself is a linear combination of $A_m$ and $P_{A,m}$, we must have (take commutators with $A_m$ and $P_{A,m}$ to verify),

$$a_j(\mathbf{k}) = \sum_m\frac{1}{i\hbar}\int d^3r^{'}\,[a_j(\mathbf{k}),P_{A,m}(\mathbf{r}^{'})]A_m(\mathbf{r}^{'}) + \sum_m\frac{1}{-i\hbar}\int d^3r^{'}\,[a_j(\mathbf{k}),A_m(\mathbf{r}^{'})]P_{A,m}(\mathbf{r}^{'}). \tag{8.39}$$

Thus,

$$a_j(\mathbf{k}) = \sum_m\frac{1}{i\hbar}\int d^3r^{'}\,e^{i\hat{u}\cdot\mathbf{r}^{'}(\frac{\Omega(\mathbf{k})}{c\hbar})}\,[a_j(\mathbf{k}),P_{A,m}(0)](A_m(\mathbf{r}^{'}) + \frac{i\hbar c}{\Omega(\mathbf{k})}\,(4\pi c)\,P_{A,m}(\mathbf{r}^{'})). \tag{8.40}$$

Now we examine the commutator of this with,

$$a^{\dagger}_{j^{'}}(\mathbf{k}^{'}) = \sum_{m^{'}}\frac{1}{-i\hbar}\int d^3r^{''}\,e^{-i\hat{u}^{'}\cdot\mathbf{r}^{''}(\frac{\Omega(\mathbf{k}^{'})}{c\hbar})}\,[a_{j^{'}}(\mathbf{k}^{'}),P_{A,m^{'}}(0)]^*(A_{m^{'}}(\mathbf{r}^{''})$$
$$-\frac{i\hbar c}{\Omega(\mathbf{k}^{'})}\,(4\pi c)\,P_{A,m^{'}}(\mathbf{r}^{''})). \tag{8.41}$$

But we must have

$$[a_j(\mathbf{k}), a^{\dagger}_{j'}(\mathbf{k}')] = \delta_{j,j'}\delta_{\mathbf{k},\mathbf{k}'}. \tag{8.42}$$

This is possible only if

$$\hat{u}(\frac{\Omega(\mathbf{k})}{c\hbar}) = \mathbf{k};\ \ \hat{u}'(\frac{\Omega(\mathbf{k}')}{c\hbar}) = \mathbf{k}'. \tag{8.43}$$

This means $\Omega(\mathbf{k}) = \hbar c|\mathbf{k}|$ and $\hat{u} = \hat{k}$ and,

$$\delta_{j,j'} = V\sum_{m}\frac{1}{\hbar^2}\ [a_j(\mathbf{k}), P_{A,m}(0)][a_{j'}(\mathbf{k}), P_{A,m}(0)]^{*}\frac{\hbar^2 c}{\Omega(\mathbf{k})}\ (2\pi c). \tag{8.44}$$

Since the definition of $a_j(\mathbf{k})$ is ambiguous up to an overall phase, we may look upon the commutators as real so that a possible set of explicit forms of these are,

$$[a_1(\mathbf{k}), P_{A,1}(0)] = [a_2(\mathbf{k}), P_{A,2}(0)] = \sqrt{\frac{\hbar|k|}{V(2\pi c)}}cos(\theta_k) \tag{8.45}$$

$$[a_1(\mathbf{k}), P_{A,2}(0)] = -[a_2(\mathbf{k}), P_{A,1}(0)] = sin(\theta_k)\sqrt{\frac{\hbar|k|}{V(2\pi c)}}. \tag{8.46}$$

This may be compactly written as

$$[a_1(\mathbf{k}), \mathbf{P}_A(0)] = \sqrt{\frac{\hbar|k|}{V(2\pi c)}}\hat{e}_{\mathbf{k},1}, \tag{8.47}$$

where

$$\hat{e}_{\mathbf{k},1} = cos(\theta_{\mathbf{k}})\hat{e}_1 + sin(\theta_{\mathbf{k}})\hat{e}_2 \tag{8.48}$$

$$[a_2(\mathbf{k}), \mathbf{P}_A(0)] = (-\hat{e}_1 sin(\theta_k) + \hat{e}_2 cos(\theta_k))\sqrt{\frac{\hbar|k|}{V(2\pi c)}}$$

$$= (\hat{k}\times\hat{e}_{\mathbf{k},1})\sqrt{\frac{\hbar|k|}{V(2\pi c)}} \equiv \hat{e}_{\mathbf{k},2}\sqrt{\frac{\hbar|k|}{V(2\pi c)}}. \tag{8.49}$$

Here $(\hat{e}_{\mathbf{k},1}, \hat{e}_{\mathbf{k},2}, \hat{k})$ form a right triad.

$$[a_j(\mathbf{k}), \mathbf{A}(\mathbf{r}')] = -\frac{i\hbar}{4\pi\Omega(\mathbf{k})}\ (4\pi c)^2\ e^{i\mathbf{k}\cdot\mathbf{r}'}\ [a_j(\mathbf{k}), \mathbf{P}_A(0)] \tag{8.50}$$

$$[\mathbf{A}(\mathbf{r}'), a^{\dagger}_j(\mathbf{k})] = \frac{i\hbar}{4\pi\Omega(\mathbf{k})}\ (4\pi c)^2\ e^{-i\mathbf{k}\cdot\mathbf{r}'}\ [a_j(\mathbf{k}), \mathbf{P}_A(0)] \tag{8.51}$$

so that,

$$\mathbf{A}(\mathbf{r}') = \sum_{j,\mathbf{k}}[\mathbf{A}(\mathbf{r}'),a_j^\dagger(\mathbf{k})]a_j(\mathbf{k}) - \sum_{j,\mathbf{k}}[\mathbf{A}(\mathbf{r}'),a_j(\mathbf{k})]a_j^\dagger(\mathbf{k})$$

$$= \sum_{j,\mathbf{k}} \frac{i\hbar}{4\pi\Omega(\mathbf{k})}(4\pi c)^2\, e^{-i\mathbf{k}\cdot\mathbf{r}'}\,[a_j(\mathbf{k}),\mathbf{P}_A(0)]a_j(\mathbf{k}) + h.c. \tag{8.52}$$

Here $h.c.$ stands for Hermitian conjugate. This means,

$$\mathbf{A}(\mathbf{r}) = \sum_{j,\mathbf{k}} \frac{i\hbar}{4\pi\Omega(\mathbf{k})}(4\pi c)^2\, e^{-i\mathbf{k}\cdot\mathbf{r}}\,\hat{e}_{\mathbf{k},j}\sqrt{\frac{\hbar|k|}{V(2\pi c)}}a_j(\mathbf{k}) + h.c. \tag{8.53}$$

So far we have described the creation and annihilation of excitations of a system of particles in terms of operators. Now we turn to a different problem of describing creation and annihilation of particles themselves. As we have alluded to earlier, this exercise facilitates the description of Bose and Fermi systems with quantum statistics built into the Hamiltonian itself. In relativistic frameworks, this approach enables the description of both matter and forces on an equal footing where material particles are thought of as excitations of matter fields and force quanta are thought of as excitations of force fields.

## 8.2 Creation and Annihilation Operators in Many-Body Physics

Consider the Hamiltonian

$$H = \sum_{i=1}^{N}\frac{p_i^2}{2m} + \frac{1}{2}\sum_{i\neq j}V(|\mathbf{r}_i - \mathbf{r}_j|). \tag{8.54}$$

This describes a system of $N$ particles, each of mass $m$, interacting mutually with a potential $V$. The conventional $\mathbf{r},\mathbf{p}$ description makes no mention of Bose or Fermi statistics. These have to be imposed on the wavefunctions of the system. We now wish to introduce a formalism that rewrites this Hamiltonian in such a way that quantum statistics are already in built into the Hamiltonian so that one may evaluate useful properties without having to first construct appropriately symmetrized wave-functions. The way this is done is to introduce creation and annihilation operators. For this we need several new concepts. First we note the allowed eigenfunction of H may be denoted as $\psi(\mathbf{r}_1,\mathbf{r}_2,...,\mathbf{r}_N)$. They have to have a permutation symmetry

$$\psi(\mathbf{r}_1,\mathbf{r}_2,...,\mathbf{r}_i,...,\mathbf{r}_j,...,\mathbf{r}_N) = s\psi(\mathbf{r}_1,\mathbf{r}_2,...,\mathbf{r}_j,...,\mathbf{r}_i,...,\mathbf{r}_N) \tag{8.55}$$

where $s = +1$ for bosons and $s = -1$ for fermions. We assume that the spin variable is subsumed into the notation $\mathbf{r}_1$. Therefore, when we write $c(\mathbf{r}_1)$ we mean $c(\mathbf{r}_1,\sigma_1)$

where $\sigma_1$ is the spin projection (for spin 1/2 fermions, $\sigma_1 = \uparrow, \downarrow$, for spin 1 bosons it is $\sigma_1 = 0, \pm 1$). In case a function does not possess these properties, we may enforce this through the action of symmetrizing operator (also called projection operator) $\mathfrak{P}_s$.

$$\mathfrak{P}_s \psi(\mathbf{r}_1, \mathbf{r}_2, ..., \mathbf{r}_i, ..., \mathbf{r}_j, ..., \mathbf{r}_N)$$
$$= \frac{1}{N!} \sum_P s^{|P|} \psi(\mathbf{r}_{P(1)}, \mathbf{r}_{P(2)}, ..., \mathbf{r}_{P(i)}, ..., \mathbf{r}_{P(j)}, ..., \mathbf{r}_{P(N)}) \tag{8.56}$$

Here P is a permutation of the numbers 1 to N. $|P|$ is the number of pairwise interchanges required to bring the existing permutation to the natural form, namely 1 to N. For example, we could have

$$P(1,2,3) = (2,1,3). \tag{8.57}$$

This means

$$P(1) = 2,\ P(2) = 1,\ P(3) = 3. \tag{8.58}$$

Here $|P| = 1$, since one change, namely interchange 2 and 1, brings back $(2,1,3)$ to its natural form $(1,2,3)$.

Now we define the creation and annihilation operators as follows. Consider any function (not necessarily a wavefunction of a system of particles) $\psi(\mathbf{r}_1, \mathbf{r}_2, .......\mathbf{r}_N)$. We define the following operator, known as the annihilation operator, at point $\mathbf{r}$. It is denoted by $a(\mathbf{r})$ and is defined through its action on the multivariable function as follows:

$$a(\mathbf{r})\psi(\mathbf{r}_1, \mathbf{r}_2, ..., \mathbf{r}_N) = \sqrt{N}\psi(\mathbf{r}_1, \mathbf{r}_2, ..., \mathbf{r}_{N-1}, \mathbf{r})$$
$$= \sqrt{N} \int \psi(\mathbf{r}_1, \mathbf{r}_2, ..., \mathbf{r}_N)\delta(\mathbf{r}_N - \mathbf{r}) d^d r_N. \tag{8.59}$$

That is, it makes the N-th particle disappear, the coordinate $\mathbf{r}_N$ no longer appears, and is replaced by a fixed (not dynamical) vector $\mathbf{r}$. We may also define a creation operator $a^\dagger(\mathbf{r})$.

$$a^\dagger(\mathbf{r})\psi(\mathbf{r}_1, \mathbf{r}_2, ..., \mathbf{r}_N) = \sqrt{N+1}\psi(\mathbf{r}_1, \mathbf{r}_2, ..., \mathbf{r}_N)\delta(\mathbf{r} - \mathbf{r}_{N+1}) \tag{8.60}$$

That is, it adds one more particle at position $\mathbf{r}$. The particle that annihilates a boson or a fermion is defined as

$$c(\mathbf{r}) = \mathfrak{P}_s a(\mathbf{r}) \mathfrak{P}_s. \tag{8.61}$$

The reason for this is intuitively obvious. This operator acts on some function of several variables and the action of the s-symmetrization operator $\mathfrak{P}_s$ makes the function properly symmetrized. When the annihilation operator acts on this new function, it does so democratically on all the variables even though $a(\mathbf{r})$ has a bias toward the variable $\mathbf{r}_N$ (the last one). The earlier step ensures that each of the variables gets a chance to be that 'last one' and hence is annihilated in turn by

$a(\mathbf{r})$. Upon annihilation by $a(\mathbf{r})$, the wavefunction is no longer properly symmetric. Hence a further symmetrization is needed to make the result a function respecting the statistics of the operator acting on it. Consider a symmetrized wavefunction, i.e., one that obeys, $\mathfrak{P}_s\psi_s = \psi_s$ then,

$$c(\mathbf{r})\psi_s(\mathbf{r}_1,\mathbf{r}_2,...,\mathbf{r}_N) = \psi_s(\mathbf{r}_1,\mathbf{r}_2,....,\mathbf{r}_{N-1},\mathbf{r})\sqrt{N} \tag{8.62}$$

and,

$$c^\dagger(\mathbf{r})\psi_s(\mathbf{r}_1,\mathbf{r}_2,...,\mathbf{r}_N) =$$

$$\frac{\sqrt{N+1}}{(N+1)!}\sum_P s^{|P|}\psi_s(\mathbf{r}_{P(1)},\mathbf{r}_{P(2)},....,\mathbf{r}_{P(N)})\delta(\mathbf{r}_{P(N+1)}-\mathbf{r}). \tag{8.63}$$

For example, for N = 2 we have

$$c^\dagger(\mathbf{r})\psi_s(\mathbf{r}_1,\mathbf{r}_2) = \frac{\sqrt{3}}{3!}\sum_P s^{|P|}\;\psi_s(\mathbf{r}_{P(1)},\mathbf{r}_{P(2)})\delta(\mathbf{r}_{P(3)}-\mathbf{r})$$

$$= \frac{\sqrt{3}}{3!}\psi_s(\mathbf{r}_1,\mathbf{r}_2)\delta(\mathbf{r}_3-\mathbf{r}) + \frac{\sqrt{3}}{3!}s\;\psi_s(\mathbf{r}_2,\mathbf{r}_1)\delta(\mathbf{r}_3-\mathbf{r}) + \frac{\sqrt{3}}{3!}s\;\psi_s(\mathbf{r}_1,\mathbf{r}_3)\delta(\mathbf{r}_2-\mathbf{r})$$

$$+\frac{\sqrt{3}}{3!}s\;\psi_s(\mathbf{r}_3,\mathbf{r}_2)\delta(\mathbf{r}_1-\mathbf{r}) + \frac{\sqrt{3}}{3!}\;\psi_s(\mathbf{r}_2,\mathbf{r}_3)\delta(\mathbf{r}_1-\mathbf{r}) + \frac{\sqrt{3}}{3!}\;\psi_s(\mathbf{r}_3,\mathbf{r}_1)\delta(\mathbf{r}_2-\mathbf{r})$$

$$= \frac{1}{\sqrt{3}}(\psi_s(\mathbf{r}_1,\mathbf{r}_2)\delta(\mathbf{r}_3-\mathbf{r}) + \psi_s(\mathbf{r}_2,\mathbf{r}_3)\delta(\mathbf{r}_1-\mathbf{r}) + \psi_s(\mathbf{r}_3,\mathbf{r}_1)\delta(\mathbf{r}_2-\mathbf{r})). \tag{8.64}$$

Thus we may convince ourselves of the validity of the general assertion as well. Now we prove the following important result viz. the commutation rules obeyed by the s-symmetric operators. Define $[A,B]_s = AB - s\;BA$, then we wish to prove the following:

**Theorem**:

$$[c(\mathbf{r}),c(\mathbf{r}')]_s = 0 \tag{8.65}$$

$$[c(\mathbf{r}),c^\dagger(\mathbf{r}')]_s = \delta(\mathbf{r}-\mathbf{r}') \tag{8.66}$$

This is proved by showing that

$$[c(\mathbf{r}),c(\mathbf{r}')]_s F_s(\mathbf{r}_1,\mathbf{r}_2,...,\mathbf{r}_N) = 0 \tag{8.67}$$

$$[c(\mathbf{r}),c^\dagger(\mathbf{r}')]_s F_s(\mathbf{r}_1,\mathbf{r}_2..,\mathbf{r}_N) = \delta(\mathbf{r}-\mathbf{r}')F_s(\mathbf{r}_1,\mathbf{r}_2..,\mathbf{r}_N) \tag{8.68}$$

for any function $F_s$ that obeys $\mathfrak{P}_sF_s = F_s$. First consider N = 2,

$$[c(\mathbf{r}),c(\mathbf{r}')]_s\psi_s(\mathbf{r}_1,\mathbf{r}_2) = [c(\mathbf{r})c(\mathbf{r}') - s\;c(\mathbf{r}')c(\mathbf{r})]\psi_s(\mathbf{r}_1,\mathbf{r}_2)$$

$$= [\psi_s(\mathbf{r},\mathbf{r}') - s\;\psi_s(\mathbf{r}',\mathbf{r})]\sqrt{2}\sqrt{1} = 0 \tag{8.69}$$

and,

$$
\begin{aligned}
[c(\mathbf{r}),c^\dagger(\mathbf{r}')]_s\psi_s(\mathbf{r}_1,\mathbf{r}_2) &= c(\mathbf{r})c^\dagger(\mathbf{r}')\psi_s(\mathbf{r}_1,\mathbf{r}_2) - s\, c^\dagger(\mathbf{r}')c(\mathbf{r})\psi_s(\mathbf{r}_1,\mathbf{r}_2) \\
&= [\psi_s(\mathbf{r}_1,\mathbf{r}_2)\delta(\mathbf{r}-\mathbf{r}') + \psi_s(\mathbf{r}_2,\mathbf{r})\delta(\mathbf{r}_1-\mathbf{r}') + \psi_s(\mathbf{r},\mathbf{r}_1)\delta(\mathbf{r}_2-\mathbf{r}')] \\
&-s\,\delta(\mathbf{r}_2-\mathbf{r}')\psi_s(\mathbf{r}_1,\mathbf{r}) - \delta(\mathbf{r}_1-\mathbf{r}')\psi_s(\mathbf{r}_2,\mathbf{r}) = \psi_s(\mathbf{r}_1,\mathbf{r}_2)\delta(\mathbf{r}-\mathbf{r}'). \qquad (8.70)
\end{aligned}
$$

Thus we have proved the commutation rules of the Fermi and Bose creation and annihilation operators when acting on two-particle wavefunctions. A general proof involves acting these operators on N-variable functions,

$$
\begin{aligned}
c(\mathbf{r}')c(\mathbf{r})\psi_s(\mathbf{r}_1,\mathbf{r}_2,...,\mathbf{r}_{N-1},\mathbf{r}_N) &= \sqrt{N}c(\mathbf{r}')\psi_s(\mathbf{r}_1,\mathbf{r}_2,...,\mathbf{r}_{N-1},\mathbf{r}) \\
&= \sqrt{N}\sqrt{N-1}\psi_s(\mathbf{r}_1,\mathbf{r}_2,...,\mathbf{r}_{N-2},\mathbf{r}',\mathbf{r}) \\
c(\mathbf{r})c(\mathbf{r}')\psi_s(\mathbf{r}_1,\mathbf{r}_2,...,\mathbf{r}_N) &= \sqrt{N}\sqrt{N-1}\,\psi_s(\mathbf{r}_1,\mathbf{r}_2,...,\mathbf{r}_{N-2},\mathbf{r},\mathbf{r}'). \qquad (8.71)
\end{aligned}
$$

Hence,

$$
\begin{aligned}
&[c(\mathbf{r}),c(\mathbf{r}')]_s\,\psi_s(\mathbf{r}_1,\mathbf{r}_2,...,\mathbf{r}_N) \\
&= \sqrt{N}\sqrt{N-1}[\psi_s(\mathbf{r}_1,\mathbf{r}_2,..,\mathbf{r}_{N-2},\mathbf{r},\mathbf{r}') - s\,\psi_s(\mathbf{r}_1,\mathbf{r}_2..,\mathbf{r}_{N-2},\mathbf{r}',\mathbf{r})] = 0. \qquad (8.72)
\end{aligned}
$$

Now we wish to prove the identity

$$
\begin{aligned}
&c^\dagger(\mathbf{r}')\psi_s(\mathbf{r}_1,\mathbf{r}_2,...,\mathbf{r}_N) \\
&= \frac{1}{\sqrt{N+1}}\sum_{P_{cyclic}} s^{|P_{cyclic}|}\psi_s(\mathbf{r}_{P_{cyclic}(1)},\mathbf{r}_{P_{cyclic}(2)},...,\mathbf{r}_{P_{cyclic}(N)}) \\
&\qquad \times\delta(\mathbf{r}_{P_{cyclic}(N+1)}-\mathbf{r}'), \qquad (8.73)
\end{aligned}
$$

where $P_{cyclic}$ refers to a cyclic permutation of $N+1$ vectors where a smaller index is mapped to the next larger one and the largest one is mapped to the first one. For example, for three number $1,2,3$, $P_{cyclic}(1)=2$, $P_{cyclic}(2)=3$ and $P_{cyclic}(3)=1$. The assertion in Eq. (8.73) rests crucially on $\psi_s$ being already s-symmetric. This redefinition (Eq. (8.73)) is made plausible, for example, by setting N = 1.

$$
\begin{aligned}
c^\dagger(\mathbf{r}')\psi_s(\mathbf{r}_1) &= \frac{1}{\sqrt{2}}\sum_{P_{cyclic}} s^{|P_{cyclic}|}\psi_s(\mathbf{r}_{P_{cyclic}(1)})\delta(\mathbf{r}_{P_{cyclic}(2)}-\mathbf{r}') \\
&= \frac{1}{\sqrt{2}}[\psi_s(\mathbf{r}_1)\delta(\mathbf{r}_2-\mathbf{r}') + s\,\psi_s(\mathbf{r}_2)\delta(\mathbf{r}_1-\mathbf{r}')]
\end{aligned}
$$

This is the same as what one would get, e.g., from the definition Eq. (8.63). We list the next one $N=2$ and leave the general proof to the exercises.

$$
\begin{aligned}
c^\dagger(\mathbf{r}')\psi_s(\mathbf{r}_1,\mathbf{r}_2) &= \frac{1}{\sqrt{3}}\sum_{P_{cyclic}} s^{|P_{cyclic}|}\psi_s(\mathbf{r}_{P_{cyclic}(1)},\mathbf{r}_{P_{cyclic}(2)})\delta(\mathbf{r}_{P_{cyclic}(3)}-\mathbf{r}') \\
= \frac{1}{\sqrt{3}}\psi_s(\mathbf{r}_1,\mathbf{r}_2)\delta(\mathbf{r}_3-\mathbf{r}') &+ \frac{1}{\sqrt{3}}\psi_s(\mathbf{r}_2,\mathbf{r}_3)\delta(\mathbf{r}_1-\mathbf{r}') + \frac{1}{\sqrt{3}}\psi_s(\mathbf{r}_3,\mathbf{r}_1)\delta(\mathbf{r}_2-\mathbf{r}')
\end{aligned}
$$

Reverting to the general case,

$$c(\mathbf{r})c^{\dagger}(\mathbf{r}')\psi_s(\mathbf{r}_1,\mathbf{r}_2,..,\mathbf{r}_N) = \int d^d r_{N+1}\delta(\mathbf{r}-\mathbf{r}_{N+1}) \sum_{P_{cyclic}} s^{|P_{cyclic}|}$$

$$\times\psi_s(\mathbf{r}_{P_{cyclic}(1)},\mathbf{r}_{P_{cyclic}(2)},...,\mathbf{r}_{P_{cyclic}(N)})\delta(\mathbf{r}_{P_{cyclic}(N+1)}-\mathbf{r}'), \tag{8.74}$$

and the operators in the reverse order give

$$c^{\dagger}(\mathbf{r}')c(\mathbf{r})\psi_s(\mathbf{r}_1,\mathbf{r}_2,...,\mathbf{r}_N) = \sqrt{N}c^{\dagger}(\mathbf{r}')\psi_s(\mathbf{r}_1,\mathbf{r}_2,..,\mathbf{r}_{N-1},\mathbf{r})$$

$$= \sum_{P_{cyclic}} s^{|P_{cyclic}|}\,\psi_s(\mathbf{r}_{P_{cyclic}(1)},\mathbf{r}_{P_{cyclic}(2)},...,\mathbf{r}_{P_{cyclic}(N-1)},\mathbf{r})\delta(\mathbf{r}_{P_{cyclic}(N)}-\mathbf{r}'). \tag{8.75}$$

Therefore,

$$[c(\mathbf{r}),c^{\dagger}(\mathbf{r}')]_s\psi_s(\mathbf{r}_1,\mathbf{r}_2,...,\mathbf{r}_N) =$$

$$\int d^d r_{N+1}\delta(\mathbf{r}-\mathbf{r}_{N+1}) \sum_{P_{cyclic}} s^{|P_{cyclic}|}\psi_s(\mathbf{r}_{P_{cyclic}(1)},\mathbf{r}_{P_{cyclic}(2)},..,\mathbf{r}_{P_{cyclic}(N)})$$

$$\delta(\mathbf{r}_{P_{cyclic}(N+1)}-\mathbf{r}')$$

$$-s\sum_{P_{cyclic}} s^{|P_{cyclic}|}\psi_s(\mathbf{r}_{P_{cyclic}(1)},\mathbf{r}_{P_{cyclic}(2)},..,\mathbf{r}_{P_{cyclic}(N-1)},\mathbf{r})\delta(\mathbf{r}_{P_{cyclic}(N)}-\mathbf{r}'). \tag{8.76}$$

We could evaluate the right-hand side of the above expression in general, but let us specialize to the case N = 2.

$$[c(\mathbf{r}),c^{\dagger}(\mathbf{r}')]_s\psi_s(\mathbf{r}_1,\mathbf{r}_2) =$$

$$\int d^d r_3\delta(\mathbf{r}-\mathbf{r}_3)\psi(\mathbf{r}_1,\mathbf{r}_2)\delta(\mathbf{r}_3-\mathbf{r}') + \int d^d r_3\delta(\mathbf{r}-\mathbf{r}_3)s^2\psi(\mathbf{r}_2,\mathbf{r}_3)\delta(\mathbf{r}_1-\mathbf{r}')$$

$$+\int d\mathbf{r}_3\delta(\mathbf{r}-\mathbf{r}_3)s^2\psi(\mathbf{r}_3,\mathbf{r}_1)\delta(\mathbf{r}_2-\mathbf{r}')$$

$$-s\psi_s(\mathbf{r}_1,\mathbf{r})\delta(\mathbf{r}_2-\mathbf{r}') - s^2\psi_s(\mathbf{r}_2,\mathbf{r})\delta(\mathbf{r}_1-\mathbf{r}') = \psi_s(\mathbf{r}_1,\mathbf{r}_2)\delta(\mathbf{r}-\mathbf{r}') \tag{8.77}$$

Thus we have,

$$[c(\mathbf{r}),c^{\dagger}(\mathbf{r}')]_s = \delta(\mathbf{r}-\mathbf{r}'). \tag{8.78}$$

Similarly, we can expect the other cases to work out too. The general proof is left to the exercises (prove by induction or some other method). Now we introduce the important concept of Green functions in many-body physics and how they may be expressed either in the conventional language of N-particle wavefunctions or in terms of creation and annihilation operators.

## 8.3 Green Functions in Many-Body Physics

The concept of a Green function in many-body physics is central to the understanding of the effect of correlations. While for special values of its arguments, it may be related to observables such as densities and currents, in general, it embodies a variety of information such as the lifetime and energy dispersion of quasiparticles. Green functions in many-body physics may be categorized as single-particle, two-particle, and so on. The single-particle Green function comes in two types. One is known as the hole propagator defined in the creation and annihilation operator language as,

$$G_{<}(\mathbf{r},t;\mathbf{r}',t') = < S|c^{\dagger}(\mathbf{r},t)c(\mathbf{r}',t')|S > \tag{8.79}$$

where $|S>$ is some state (not necessarily a stationary state) of the N-particle system at some time, which we designate at $time = 0$. The other is known as the particle propagator, which is defined in the operator language as

$$G_{>}(\mathbf{r},t;\mathbf{r}',t') = < S|c(\mathbf{r}',t')c^{\dagger}(\mathbf{r},t)|S > . \tag{8.80}$$

The symbol $c(\mathbf{r}',t')$ requires clarification since so far we have only introduced the meaning of $c(\mathbf{r}')$ - the annihilation operator. The meaning of this is the usual one except with one small modification,

$$c(\mathbf{r}',t') = e^{\frac{i}{\hbar}t' H_{N-1}} c(\mathbf{r}') e^{-\frac{i}{\hbar}t' H_N}. \tag{8.81}$$

Here $H_N$ is the Hamiltonian of an N-particle system and $H_{N-1}$ is the same Hamiltonian but with one less particle. For instance, if there are two body forces,

$$H_N = \sum_{i=1}^{N} \frac{p_i^2}{2m} + \frac{1}{2}\sum_{i\neq j=1}^{N} V(\mathbf{r}_i - \mathbf{r}_j) \tag{8.82}$$

whereas,

$$H_{N-1} = \sum_{i=1}^{N-1} \frac{p_i^2}{2m} + \frac{1}{2}\sum_{i\neq j=1}^{N-1} V(\mathbf{r}_i - \mathbf{r}_j), \tag{8.83}$$

so that for every s-symmetrized function of $N$ variables $\Psi(\mathbf{r}_1,...,\mathbf{r}_N)$ we may write,

$$\begin{aligned} c(\mathbf{r}',t')\Psi(\mathbf{r}_1,..,\mathbf{r}_{N-1},\mathbf{r}_N) &= e^{\frac{i}{\hbar}t' H_{N-1}} c(\mathbf{r}') e^{-\frac{i}{\hbar}t' H_N}\Psi(\mathbf{r}_1,..,\mathbf{r}_{N-1},\mathbf{r}_N) \\ = e^{\frac{i}{\hbar}t' H_{N-1}} c(\mathbf{r}')\Psi(\mathbf{r}_1,..,\mathbf{r}_{N-1},\mathbf{r}_N;t') &= \sqrt{N}\, e^{\frac{i}{\hbar}t' H_{N-1}}\Psi(\mathbf{r}_1,..,\mathbf{r}_{N-1},\mathbf{r}';t') \\ &\equiv \sqrt{N}\,\Psi_{hole,t'}(\mathbf{r}_1,..,\mathbf{r}_{N-1},\mathbf{r}';0) \end{aligned} \tag{8.84}$$

where $\Psi_{hole,t'}(\mathbf{r}_1,..,\mathbf{r}_{N-1},\mathbf{r}';0)$ is that wavefunction at time $T = 0$ of $N-1$ variables $\mathbf{r}_1,\mathbf{r}_2,....,\mathbf{r}_{N-1}$, which when evolved from this time $T = 0$ to time $T = t'$ using $H_{N-1}$ becomes a function identical to the function $\psi(\mathbf{r}_1,...,\mathbf{r}_{N-1},\mathbf{r}';t')$ (which is nothing but the original $\psi(\mathbf{r}_1,..,\mathbf{r}_{N-1},\mathbf{r}_N)$ time evolved from time $T=0$ to $T=t^{'}$ and then replacing $\mathbf{r}_N$ by $\mathbf{r}^{'}$ (annihilation)). If the starting state is represented in position space as $|S>\to \Psi_S(\mathbf{r}_1,..,\mathbf{r}_N)$, then the Green function may be reexpressed in a mixed language of operators and wavefunctions as follows ($d^d[r] \equiv d^d r_1 d^d r_2 ... d^d r_N$).

$$G_<(\mathbf{r},t;\mathbf{r}^{'},t^{'}) = < S|c^{\dagger}(\mathbf{r},t)c(\mathbf{r}^{'},t^{'})|S>$$

$$= \int d^d[r]\ \Psi^*_S(\mathbf{r}_1,..,\mathbf{r}_N)c^{\dagger}(\mathbf{r},t)c(\mathbf{r}^{'},t^{'})\Psi_S(\mathbf{r}_1,..,\mathbf{r}_N). \tag{8.85}$$

We may also rewrite this as

$$G_<(\mathbf{r},t;\mathbf{r}^{'},t^{'}) = \int d^d[r]\ \Psi^*_S(\mathbf{r}_1,..,\mathbf{r}_N)c^{\dagger}(\mathbf{r},t)c(\mathbf{r}^{'},t^{'})\Psi_S(\mathbf{r}_1,..,\mathbf{r}_N)$$

$$= \int d^d r_1 d^d r_2 ... d^d r_{N-1}\ (c(\mathbf{r},t)\Psi_S(\mathbf{r}_1,..,\mathbf{r}_N))^{\dagger}(c(\mathbf{r}^{'},t^{'})\Psi_S(\mathbf{r}_1,..,\mathbf{r}_N))$$

$$= N\int d^d r_1 d^d r_2 ... d^d r_{N-1}\ \Psi^*_{hole,t}(\mathbf{r}_1,..,\mathbf{r}_{N-1},\mathbf{r};0)\ \Psi_{hole,t^{'}}(\mathbf{r}_1,..,\mathbf{r}_{N-1},\mathbf{r}^{'};0). \tag{8.86}$$

The last step follows when we insert the last step of Eq. (8.84) into the penultimate step of the above equation. Notice that we have converted the effect of the creation operator into the effect of an annihilation operator followed by a Hermitian conjugate. This automatically makes one variable, namely $\mathbf{r}_N$, disappear. However, it is possible to get the same answer more directly, but this is left to the exercises. Thus the important identity relating to the hole propagator is,

$$< S|c^{\dagger}(\mathbf{r},t)c(\mathbf{r}^{'},t^{'})|S>$$

$$= N\int d^d r_1 d^d r_2 ... d^d r_{N-1}\ \Psi^*_{hole,t}(\mathbf{r}_1,..,\mathbf{r}_{N-1},\mathbf{r};0)\ \Psi_{hole,t^{'}}(\mathbf{r}_1,..,\mathbf{r}_{N-1},\mathbf{r}^{'};0). \tag{8.87}$$

The reason for the extra factor $N$ (that follows rigorously from the definitions) may be understood by realizing that the quantity $\int d^d r\ < S|c^{\dagger}(\mathbf{r},t)c(\mathbf{r},t)|S> \equiv N$ is the total number of particles. Succinctly, Eq. (8.87) says that the operator definition of the hole propagator is equivalent in the wavefunction language to evaluating the overlap of two functions obtained using the following procedure. (i) First evolve the initial (s-symmetric) wavefunction to some time, then (ii) replace one of the variables (annihilation) with either $\mathbf{r}$ or $\mathbf{r}^{'}$ so that the wavefunction has one fewer dynamical variables, and then (iii) de-evolve it back to the initial time now using the Hamiltonian with one less variable than earlier. This procedure makes it clear that the result is not the same as annihilating at the initial time itself. (iv) Finally, evaluate the overlap between two such wavefunctions at different times and positions.

One may similarly try and find a meaning of the particle propagator in terms of wavefunctions. Define,

$$\Psi_{t,prtcl}(\mathbf{r}_1,...,\mathbf{r}_N,\mathbf{r}_{N+1};\mathbf{r},0) =$$

$$e^{\frac{i}{\hbar}tH_{N+1}}\mathfrak{P}_s\left(\delta(\mathbf{r}-\mathbf{r}_{N+1})e^{-\frac{i}{\hbar}tH_N}\Psi_S(\mathbf{r}_1,...,\mathbf{r}_N)\right). \tag{8.88}$$

This means, (i) first evolve the initial wavefunction with $N$ positions up to time $T = t$ from $T = 0$, (ii) multiply by a delta function (creation) that introduces a new position variable $\mathbf{r}_{N+1}$ at location $\mathbf{r}$, (iii) s-symmetrize the result, and finally (iv) de-evolve using the Hamiltonian $H_{N+1}$ back to time $T = 0$. The particle Green function is then the overlap between two such functions.

$$G_>(\mathbf{r},t;\mathbf{r}',t') = < S|c(\mathbf{r}',t')c^\dagger(\mathbf{r},t)|S> =$$

$$(N+1)\int d^dr_1...d^dr_{N+1}\ \Psi^*_{t',prtcl}(\mathbf{r}_1,..,\mathbf{r}_{N+1};\mathbf{r}',0)\Psi_{t,prtcl}(\mathbf{r}_1,..,\mathbf{r}_{N+1};\mathbf{r},0). \tag{8.89}$$

Now we examine how we may reexpress Hamiltonians that are originally in terms of positions and momenta using these operators.

## 8.4 Hamiltonians Using Creation and Annihilation Operators

As before, consider the Hamiltonian,

$$H = \sum_{i=1}^{N}\frac{p_i^2}{2m} + \frac{1}{2}\sum_{i\neq j}V(|\mathbf{r}_i - \mathbf{r}_j|). \tag{8.90}$$

We wish to show that this can also be written as (in this section $d\mathbf{r}$ is the volume element (d-dimensional) and $\delta(\mathbf{r}-\mathbf{r}')$ is the d-dimensional Dirac delta function),

$$H = \int d\mathbf{r}\ c^\dagger(\mathbf{r})\frac{p^2}{2m}c(\mathbf{r}) + \frac{1}{2}\int d\mathbf{r}\int d\mathbf{r}'c^\dagger(\mathbf{r})c^\dagger(\mathbf{r}')c(\mathbf{r}')c(\mathbf{r})V(|\mathbf{r}-\mathbf{r}'|). \tag{8.91}$$

To do this we have to show that $H_1F_s(\mathbf{r}_1,...,\mathbf{r}_N) = H_2F_s(\mathbf{r}_1,...,\mathbf{r}_N)$ for any $F_s = \mathfrak{P}_sF$ and $H_1$ is the expression in Eq. (8.90) and $H_2$ is the expression in Eq. (8.91). Consider

$$H\psi(\mathbf{r}_1,\mathbf{r}_2) = \int d\mathbf{r}\ c^\dagger(\mathbf{r})\frac{p^2}{2m}c(\mathbf{r})\psi(\mathbf{r}_1,\mathbf{r}_2)$$

$$+\frac{1}{2}\int d\mathbf{r}\int d\mathbf{r}'\ c^\dagger(\mathbf{r})c^\dagger(\mathbf{r}')c(\mathbf{r}')c(\mathbf{r})V(|\mathbf{r}-\mathbf{r}'|)\psi(\mathbf{r}_1,\mathbf{r}_2). \tag{8.92}$$

We may evaluate the action of the kinetic energy as

$$\int d\mathbf{r}\, c^{\dagger}(\mathbf{r})\frac{p^2}{2m}c(\mathbf{r})\psi_s(\mathbf{r}_1,\mathbf{r}_2) = \sqrt{2}\int d\mathbf{r}\, c^{\dagger}(\mathbf{r})\frac{-\hbar^2\nabla^2}{2m}c(\mathbf{r})\psi_s(\mathbf{r}_1,\mathbf{r})$$

$$= \int d\mathbf{r}\, \delta(\mathbf{r}-\mathbf{r}_2)\frac{-\hbar^2\nabla^2}{2m}\psi_s(\mathbf{r}_1,\mathbf{r}) + s\int d\mathbf{r}\, \delta(\mathbf{r}-\mathbf{r}_1)\frac{-\hbar^2\nabla^2}{2m}\psi_s(\mathbf{r}_2,\mathbf{r})$$

$$= \left(\frac{-\hbar^2\nabla_1^2}{2m} + \frac{-\hbar^2\nabla_2^2}{2m}\right)\psi_s(\mathbf{r}_1,\mathbf{r}_2), \tag{8.93}$$

which is what we expect from ordinary quantum mechanics written in the position and momentum notation. The action of the potential energy in the 'second quantized' notation yields,

$$\frac{1}{2}\int d\mathbf{r}\int d\mathbf{r}'\, c^{\dagger}(\mathbf{r})c^{\dagger}(\mathbf{r}')c(\mathbf{r}')c(\mathbf{r})V(|\mathbf{r}-\mathbf{r}'|)\psi_s(\mathbf{r}_1,\mathbf{r}_2)$$

$$= \frac{1}{\sqrt{2}}\int d\mathbf{r}\int d\mathbf{r}'\, c^{\dagger}(\mathbf{r})c^{\dagger}(\mathbf{r}')V(|\mathbf{r}-\mathbf{r}'|)\psi_s(\mathbf{r}',\mathbf{r})$$

$$= \frac{1}{2}\int d\mathbf{r}\int d\mathbf{r}'[\delta(\mathbf{r}-\mathbf{r}_2)\delta(\mathbf{r}_1-\mathbf{r}') + s\delta(\mathbf{r}-\mathbf{r}_1)\delta(\mathbf{r}_2-\mathbf{r}')]V(|\mathbf{r}-\mathbf{r}'|)\psi_s(\mathbf{r}',\mathbf{r})$$

$$= V(|\mathbf{r}_1-\mathbf{r}_2|)\psi_s(\mathbf{r}_1,\mathbf{r}_2). \tag{8.94}$$

Again, this is as it should be. Therefore the action of the second quantized Hamiltonian on the wavefunction is,

$$H\psi_s(\mathbf{r}_1,\mathbf{r}_2) = \left(\frac{-\hbar^2\nabla_1^2}{2m} + \frac{-\hbar^2\nabla_2^2}{2m}\right)\psi_s(\mathbf{r}_1,\mathbf{r}_2) + V(|\mathbf{r}_1-\mathbf{r}_2|)\psi_s(\mathbf{r}_1,\mathbf{r}_2), \tag{8.95}$$

which is same as Eq. (8.90) with N = 2.

■ Find the second quantized version of the Hamiltonian

$$H = \sum_{i=1}^{N}\frac{p_i^2}{2m} + \frac{1}{3!}\sum_{i,j,k} V(|\mathbf{r}_i-\mathbf{r}_j|,|\mathbf{r}_i-\mathbf{r}_k|,|\mathbf{r}_j-\mathbf{r}_k|). \tag{8.96}$$

This contains a so-called three-body potential that occurs commonly in nuclear physics. The answer is,

$$H = \int d\mathbf{r}\, c^{\dagger}(\mathbf{r})\frac{p^2}{2m}c(\mathbf{r})$$

$$+\frac{1}{3!}\int d\mathbf{r}\, d\mathbf{r}'\, d\mathbf{r}'\, c^{\dagger}(\mathbf{r})c^{\dagger}(\mathbf{r}')c^{\dagger}(\mathbf{r}'')c(\mathbf{r}'')c(\mathbf{r}')c(\mathbf{r})V(|\mathbf{r}-\mathbf{r}'|,|\mathbf{r}-\mathbf{r}''|,|\mathbf{r}'-\mathbf{r}''|). \tag{8.97}$$

In case there are vector potentials it is equally starightforward. If the original version is

$$H = \sum_{i=1}^{N} \frac{1}{2m}(\mathbf{p}_i - \frac{e}{c}\mathbf{A}(\mathbf{r}_i))^2, \tag{8.98}$$

the second quantized version is

$$H = \int d\mathbf{r} \frac{1}{2m} c^{\dagger}(\mathbf{r})(\mathbf{p} - \frac{e}{c}\mathbf{A}(\mathbf{r}))^2\, c(\mathbf{r}). \tag{8.99}$$

Now we discuss an important topic, namely the current algebra in quantum field theory.

## 8.5 Current Algebra

We may define current density and particle density operators as follows.

$$\rho(\mathbf{r}) = \sum_{i=1}^{N} \delta(\mathbf{r} - \mathbf{r}_i) \tag{8.100}$$

$$J(\mathbf{r}) = \sum_{i=1}^{N} \frac{\mathbf{p}_i}{2m}\delta(\mathbf{r} - \mathbf{r}_i) + \sum_{i=1}^{N} \delta(\mathbf{r} - \mathbf{r}_i)\frac{\mathbf{p}_i}{2m} \tag{8.101}$$

The second quantized versions are (show this),

$$\rho(\mathbf{r}) = c^{\dagger}(\mathbf{r})c(\mathbf{r}) \tag{8.102}$$

$$J(\mathbf{r}) = -\frac{i\hbar}{2}c^{\dagger}(\mathbf{r})[\nabla c(\mathbf{r})] + \frac{i\hbar}{2}[\nabla c^{\dagger}(\mathbf{r})]c(\mathbf{r}). \tag{8.103}$$

These obey a closed set of commutation rules known as current algebra (we already encountered this while studying classical fluids). The important point is that these are valid independent of the nature of the underlying statistics of $c$ (i.e., valid for bosons as well as fermions).

$$[\rho(\mathbf{r}), \rho(\mathbf{r}^{'})] = 0 \tag{8.104}$$

$$[\rho(\mathbf{r}), J_a(\mathbf{r}^{'})] = i\hbar\, \rho(\mathbf{r}^{'})\nabla^{'}_a\delta(\mathbf{r}^{'} - \mathbf{r}) \tag{8.105}$$

$$[J_a(\mathbf{r},t), J_b(\mathbf{r}^{'},t)] = -i\hbar\, J_a(\mathbf{r}^{'},t)\ (\nabla_{r_b}\delta(\mathbf{r} - \mathbf{r}^{'})) + i\hbar\, J_b(\mathbf{r},t)\ (\nabla_{r^{'}_a}\delta(\mathbf{r}^{'} - \mathbf{r})) \tag{8.106}$$

While discussing classical fluids, we had occasion to introduce an irrotational velocity field so that the current density would have the following expression $\mathbf{J}(\mathbf{r}) = -\rho(\mathbf{r})\nabla\Pi(\mathbf{r})$ where $[\Pi(\mathbf{r}), \rho(\mathbf{r}^{'})] = i\hbar\, \delta(\mathbf{r} - \mathbf{r}^{'})$ and $[\Pi(\mathbf{r}), \Pi(\mathbf{r}^{'})] = 0$. While this choice is certainly consistent with current algebra (Eq. (8.104), Eq. (8.105), and Eq. (8.106)), it is by no means clear that this is the only choice or the most general choice. In particular, one may propose the seemingly more general possibility where we write $\mathbf{J}(\mathbf{r}) = -\rho(\mathbf{r})\nabla_{\mathbf{r}}\Pi(\mathbf{r}) + \mathbf{V}(\mathbf{r};[\rho])$ and $\mathbf{V}$ depends on $\rho$ and not $\Pi$. This choice clearly obeys Eq. (8.105), but when one tries to impose Eq. (8.106) on it, we obtain,

$$[J_a(\mathbf{r}), J_b(\mathbf{r}')] =$$
$$[-\rho(\mathbf{r})\nabla_a\Pi(\mathbf{r}) + V_a(\mathbf{r};[\rho]), -\rho(\mathbf{r}')\nabla'_b\Pi(\mathbf{r}') + V_b(\mathbf{r}';[\rho])]$$
$$= i\hbar\rho(\mathbf{r})\nabla'_b\Pi(\mathbf{r}')\ \nabla_a\delta(\mathbf{r}-\mathbf{r}') - i\hbar\rho(\mathbf{r}')\nabla_a\Pi(\mathbf{r})\ \nabla'_b\delta(\mathbf{r}'-\mathbf{r})$$
$$-\rho(\mathbf{r})\nabla_a[\Pi(\mathbf{r}), V_b(\mathbf{r}';[\rho])] - \rho(\mathbf{r}')\nabla'_b[V_a(\mathbf{r};[\rho]), \Pi(\mathbf{r}')]. \tag{8.107}$$

Now,

$$i\hbar\rho(\mathbf{r})\nabla'_b\Pi(\mathbf{r}')\ \nabla_a\delta(\mathbf{r}-\mathbf{r}') = i\hbar\rho(\mathbf{r})\nabla_a(\delta(\mathbf{r}-\mathbf{r}')\nabla'_b\Pi(\mathbf{r}'))$$
$$= i\hbar\rho(\mathbf{r})\nabla_a(\delta(\mathbf{r}-\mathbf{r}')\nabla_b\Pi(\mathbf{r}))$$
$$= i\hbar\rho(\mathbf{r})\nabla_b\Pi(\mathbf{r})\nabla_a\delta(\mathbf{r}-\mathbf{r}') + i\hbar\delta(\mathbf{r}-\mathbf{r}')\rho(\mathbf{r})\nabla_a\nabla_b\Pi(\mathbf{r})$$
$$= i\hbar(J_b(\mathbf{r}) - V_b([\rho];\mathbf{r}))\nabla'_a\delta(\mathbf{r}-\mathbf{r}') + i\hbar\delta(\mathbf{r}-\mathbf{r}')\rho(\mathbf{r})\nabla_a\nabla_b\Pi(\mathbf{r}). \tag{8.108}$$

Similarly,

$$-i\hbar\rho(\mathbf{r}')\nabla_a\Pi(\mathbf{r})\ \nabla'_b\delta(\mathbf{r}'-\mathbf{r}) =$$
$$-i\hbar\rho(\mathbf{r}')\ \delta(\mathbf{r}'-\mathbf{r})(\nabla'_b\nabla'_a\Pi(\mathbf{r}')) - i\hbar(J_a(\mathbf{r}') - V_a([\rho];\mathbf{r}'))\ \nabla_b\delta(\mathbf{r}'-\mathbf{r}). \tag{8.109}$$

Therefore,

$$[J_a(\mathbf{r}), J_b(\mathbf{r}')] =$$
$$= i\hbar J_b(\mathbf{r})\nabla'_a\delta(\mathbf{r}-\mathbf{r}') - i\hbar J_a(\mathbf{r}')\ \nabla_b\delta(\mathbf{r}'-\mathbf{r})$$
$$-i\hbar V_b([\rho];\mathbf{r})\nabla'_a\delta(\mathbf{r}-\mathbf{r}') + i\hbar V_a([\rho];\mathbf{r}')\ \nabla_b\delta(\mathbf{r}'-\mathbf{r})$$
$$-\rho(\mathbf{r})\nabla_a[\Pi(\mathbf{r}), V_b(\mathbf{r}';[\rho])] - \rho(\mathbf{r}')\nabla'_b[V_a(\mathbf{r};[\rho]), \Pi(\mathbf{r}')]. \tag{8.110}$$

This means that the terms involving $V$ should add up to zero. Thus after some minor changes we get,

$$i\hbar V_b([\rho];\mathbf{r})\nabla_a\delta(\mathbf{r}-\mathbf{r}') - i\hbar V_a([\rho];\mathbf{r}')\ \nabla'_b\delta(\mathbf{r}'-\mathbf{r})$$
$$-\rho(\mathbf{r})\nabla_a[\Pi(\mathbf{r}), V_b(\mathbf{r}';[\rho])] - \rho(\mathbf{r}')\nabla'_b[V_a(\mathbf{r};[\rho]), \Pi(\mathbf{r}')] = 0. \tag{8.111}$$

To obtain $V$ we make the substitution $V_a(\mathbf{r};[\rho]) = \rho(\mathbf{r})\ \tilde{V}_a(\mathbf{r};[\rho])$ so that,

$$i\hbar\ \delta(\mathbf{r}-\mathbf{r}')(\nabla_b\tilde{V}_a(\mathbf{r}) - \nabla_a\tilde{V}_b(\mathbf{r}))$$
$$= -\rho(\mathbf{r})[\nabla'_b\Pi(\mathbf{r}'), \tilde{V}_a(\mathbf{r})] + \rho(\mathbf{r})[\nabla_a\Pi(\mathbf{r}), \tilde{V}_b(\mathbf{r}')]. \tag{8.112}$$

One may compute $\tilde{V}$ by making an ansatz of the form

$$\tilde{V}_a(\mathbf{r}) = \tilde{V}_{a,0}(\mathbf{r}) + \sum_{N=1}^{\infty} \int d\mathbf{r}_1...d\mathbf{r}_N \frac{\tilde{V}_a^N(\mathbf{r}_1,\mathbf{r}_2,...,\mathbf{r}_N;\mathbf{r})}{N!} \rho(\mathbf{r}_1)\rho(\mathbf{r}_2)....\rho(\mathbf{r}_N). \tag{8.113}$$

Substituting into the earlier equation we get,

$$[\nabla_b^{'}\Pi(\mathbf{r}^{'}),\tilde{V}_a(\mathbf{r})] = i\hbar \sum_{N=1}^{\infty} \int d\mathbf{r}_2...d\mathbf{r}_N \frac{\nabla_b^{'}\tilde{V}_a^N(\mathbf{r}^{'},\mathbf{r}_2,...,\mathbf{r}_N;\mathbf{r})}{(N-1)!} \rho(\mathbf{r}_2)....\rho(\mathbf{r}_N). \tag{8.114}$$

Hence we may conclude after inserting Eq. (8.113) and Eq. (8.114) into Eq. (8.112),

$$(\nabla_b\tilde{V}_{a,0}(\mathbf{r}) - \nabla_a\tilde{V}_{b,0}(\mathbf{r})) = 0 \tag{8.115}$$

and,

$$(\nabla_b\tilde{V}_a^N(\mathbf{r}_1,\mathbf{r}_2,...,\mathbf{r}_N;\mathbf{r}) - \nabla_a\tilde{V}_b^N(\mathbf{r}_1,\mathbf{r}_2,...,\mathbf{r}_N;\mathbf{r})) = \nabla_a\tilde{V}_b^N(\mathbf{r},\mathbf{r}_2,...,\mathbf{r}_N;.)\ \delta(\mathbf{r}-\mathbf{r}_1). \tag{8.116}$$

The notation $f(.)$ means $f(.) \equiv \int d^d r f(\mathbf{r})$. Note that crucial use has been made of the assumption that the densities are non-zero since the last few assertions are true only upon dividing by the densities. Now integrate Eq. (8.116) with respect to $\mathbf{r}$ (i.e., perform $\int d^d r$ on both sides) to conclude,

$$\nabla_a^1\tilde{V}_b^N(\mathbf{r}_1,\mathbf{r}_2,...,\mathbf{r}_N;.) = 0. \tag{8.117}$$

Since $\mathbf{r}_1$ could be anything including $\mathbf{r}$, it implies that the right-hand side of Eq. (8.116), and therefore the left-hand side, are zero. Hence,

$$(\nabla_b\tilde{V}_a^N(\mathbf{r}_1,\mathbf{r}_2,...,\mathbf{r}_N;\mathbf{r}) - \nabla_a\tilde{V}_b^N(\mathbf{r}_1,\mathbf{r}_2,...,\mathbf{r}_N;\mathbf{r})) = 0, \tag{8.118}$$

and therefore $\tilde{V}_a^N(\mathbf{r}_1,\mathbf{r}_2,...,\mathbf{r}_N;\mathbf{r}) \equiv \nabla_a\tilde{\omega}^N(\mathbf{r}_1,\mathbf{r}_2,...,\mathbf{r}_N;\mathbf{r})$ is the gradient of some function making the velocity field irrotational in general, in regions where $\rho(\mathbf{r}) \neq 0$. This means $\mathbf{J}(\mathbf{r}) = -\rho(\mathbf{r})\nabla\Pi(\mathbf{r})$ wherever $\rho \neq 0$ and current algebra ensures that $[\Pi(\mathbf{r}),\Pi(\mathbf{r}^{'})] = 0$ and $[\Pi(\mathbf{r}),\rho(\mathbf{r}^{'})] = i\hbar\ \delta^d(\mathbf{r}-\mathbf{r}^{'})$. This has important ramifications to an effort where one attempts to express single-particle properties in terms of correlations between observables. 'Bosonization' may be thought of as the act of expressing the density and velocities (ratio of current to density) in terms of the quantum fields as in Eq. (8.102) and Eq. (8.103). Its reverse is known as 'refermionization' (in some circles). This involves inverting Eq. (8.102) and Eq. (8.103) and expressing $c(\mathbf{r})$ and $c^{\dagger}(\mathbf{r})$ in terms of $\mathbf{J}(\mathbf{r})$ (equivalently, $\mathbf{v}(\mathbf{r})$ and $\rho(\mathbf{r})$). This enables the computation of the single-particle properties in terms of the correlation function between currents and densities. To what extent this is possible or meaningful will be addressed in the last few chapters.

## 8.6 Exercises

**Q.1** Consider a function $f(\mathbf{r}_1, \mathbf{r}_2, \mathbf{r}_3) = \mathbf{r}_1.\mathbf{r}_2 + \mathbf{r}_3.\mathbf{a}$. Make this a wavefunction that describes three bosons by acting the projection operator on it.

**Q.2** Consider a function $f(\mathbf{r}_1, \mathbf{r}_2, \mathbf{r}_3) = \mathbf{r}_1^2(|\mathbf{r}_2| - |\mathbf{r}_3|)$. Make this a wavefunction that describes three fermions by acting the projection operator on it.

**Q.3** Show that $\mathfrak{P}_s^2 F(\mathbf{r}_1, \mathbf{r}_2, .......\mathbf{r}_N) = \mathfrak{P}_s F(\mathbf{r}_1, \mathbf{r}_2, .......\mathbf{r}_N)$.

**Q.4** Prove the general cyclic permutation reduction of the creation operator (viz. Eq.(8.73)) using mathematical induction or otherwise.

**Q.5** Provide a general proof that the right-hand side of Eq. (8.76) is nothing but the Dirac delta function times the wavefunction on the left-hand side.

**Q.6** Prove that ($[A,B] \equiv AB - BA$ and $[A,B]_s \equiv AB - sBA$, $s = \pm 1$),

$$[a, bc] = [a, b]_s c + s\, b[a, c]_s \tag{8.119}$$

Also show that

$$[ab, cd] = [ab, c]d + c[ab, d]. \tag{8.120}$$

Combine these two and show that the simplified expression for $[c^\dagger(\mathbf{r}_a)c(\mathbf{r}_a^{'}), c^\dagger(\mathbf{r}_b)c(\mathbf{r}_b^{'})]$ is independent of the statistical parameter "$s$" (same for bosons as well as fermions).

**Q.7** Explain the meaning of $< S|c^\dagger(\mathbf{r}_1, t_1)c^\dagger(\mathbf{r}_2, t_2)c(\mathbf{r}_3, t_3)c(\mathbf{r}_4, t_4)|S >$ in the wave-function language.

**Q.8** Explain how the current density and particle density operate on a multiparticle wavefunction.

# Chapter 9

# Quantum Fields on a Lattice

So far we have discussed particles moving in space that is a continuum. In solid-state physics it is more useful to employ what is known as a tight-binding picture where an electron is assumed to be tied to a particular atom, occasionally hopping to a neighboring atom. In this picture, the electron's position is a discrete quantity—it is either on one atom or on its neighbor and not anywhere in between. This tight-binding picture may be derived using appropriate basis functions. For our purposes, we postulate that the kinetic motion of the electron is brought about by hopping. Therefore the kinetic energy is

$$K = -t \sum_{<ij>} c^{\dagger}_{i\sigma} c_{j\sigma}, \tag{9.1}$$

where $< ij >$ signifies that sites $i$ and $j$ are nearest neighbors. The negative sign implies that hopping lowers the energy of the system by an amount $t$. But there is a price to be paid for hopping, in the form of a potential energy. The potential energy is assumed to be short-ranged, namely it is present only if a site has two electrons so that they repel with some energy $U$. But Pauli's exclusion principle forbids two electrons with the same spin from residing at the same site. Thus while hopping lowers energy, hopping onto a site already occupied by an electron is either forbidden or has an energy cost $U$. The potential energy may be written as

$$V = U \sum_{i} n_{i\uparrow} n_{i\downarrow}, \tag{9.2}$$

where $n_{i\sigma} = c^{\dagger}_{i\sigma} c_{i\sigma}$ is the number of electrons in site $i$. These two put together, form the famous Hubbard model of condensed matter physics. It is mathematically well defined given that short distances have a lower bound, namely the lattice spacing. The phases of this model depend on the ratio $U/t$ and also the number of electrons per site (and temperature, if present). One can have variants of this model as well.

The extended Hubbard model includes interaction between the nearest neighbors. Then we have a term such as

$$V' = U' \sum_{<ij>} n_i n_j, \tag{9.3}$$

where $n_i = n_{i,\uparrow} + n_{i,\downarrow}$. The fundamental problem involves computing the Green function of these models.

## 9.1 Derivation of the Tight Binding Picture

A derivation of this picture may be motivated by the following set of assumptions. We postulate that in a crystal, the Hamiltonian of an electron in the position and momentum representation may be written as

$$H(\mathbf{r}, -i\hbar\nabla_{\mathbf{r}}) \equiv \sum_{\mathbf{n}} H_{at}(\mathbf{r} - \mathbf{R_n}, -i\hbar\nabla_{\mathbf{r}}) + \Delta U(\mathbf{r}), \tag{9.4}$$

where $H_{at}(\mathbf{r} - \mathbf{R_n}, -i\hbar\nabla_{\mathbf{r}})$ is the Hamiltonian of an electron in the vicinity of a single atom located at lattice position $\mathbf{R_n}$ and $\Delta U(\mathbf{r})$ ensures that the full Hamiltonian is not merely a sum of disjoint contributions from isolated atoms. In order to obtain the second quantized Hamiltonian, we follow the procedure of the earlier chapters and write,

$$H \equiv \sum_{\mathbf{n}} \int d^d r\, c^\dagger(\mathbf{r}) H_{at}(\mathbf{r} - \mathbf{R_n}, -i\hbar\nabla_{\mathbf{r}}) c(\mathbf{r}) + \int d^d r\, c^\dagger(\mathbf{r}) \Delta U(\mathbf{r}) c(\mathbf{r}). \tag{9.5}$$

Here $c(\mathbf{r})$ is the annihilation operator of an electron at position $\mathbf{r}$. The tight-binding approximation involves expanding $c(\mathbf{r})$ is a basis of the eigenstates of $H_{at}(\mathbf{r} - \mathbf{R_n}, -i\hbar\nabla_{\mathbf{r}})$. Let $\varphi_m(\mathbf{r})$ be the wavefunction of an electron in the atom, where $m$ labels the eigenstates.

$$H_{at}(\mathbf{r}, -i\hbar\nabla_{\mathbf{r}}) \varphi_l(\mathbf{r}) = E_l\, \varphi_l(\mathbf{r}) \tag{9.6}$$

For the Hamiltonian,

$$H_{tot}(\mathbf{r}, -i\hbar\nabla_{\mathbf{r}}) = \sum_{\mathbf{n}} H_{at}(\mathbf{r} - \mathbf{R_n}, -i\hbar\nabla_{\mathbf{r}}). \tag{9.7}$$

The wavefunctions of $H_{tot}$ have to obey Bloch's theorem. Thus, if

$$H_{tot}(\mathbf{r}, -i\hbar\nabla_{\mathbf{r}}) \Psi_{\mathbf{k}}(\mathbf{r}) = \mathcal{E}(\mathbf{k})\, \Psi_{\mathbf{k}}(\mathbf{r}), \tag{9.8}$$

thus

$$\Psi_{\mathbf{k}}(\mathbf{r}) = e^{i\mathbf{k}\cdot\mathbf{r}} u_{\mathbf{k}}(\mathbf{r}). \tag{9.9}$$

The function $u_{\mathbf{k}}(\mathbf{r}+\mathbf{R_n}) = u_{\mathbf{k}}(\mathbf{r})$ is periodic.

Bloch's theorem, which leads to Eq. (9.9) may be understood as follows. Since $H_{tot}(\mathbf{r}+\mathbf{R}_l) = H_{tot}(\mathbf{r})$ is periodic, it follows that $\Psi_{\mathbf{k}}(\mathbf{r})$ and $\Psi_{\mathbf{k}}(\mathbf{r}+\mathbf{R}_l)$ are two eigenfunctions of the same Hamiltonian with the same energy. Thus if this energy is nondegenerate, these two wavefunctions must be proportional to each other.

$$\Psi_{\mathbf{k}}(\mathbf{r}+\mathbf{R}_l) = C(\mathbf{R}_l)\ \Psi_{\mathbf{k}}(\mathbf{r}) \tag{9.10}$$

Now replace $\mathbf{r}$ by $\mathbf{r}+\mathbf{R}_{l'}$ so that,

$$\Psi_{\mathbf{k}}(\mathbf{r}+\mathbf{R}_l+\mathbf{R}_{l'}) = C(\mathbf{R}_l)\ \Psi_{\mathbf{k}}(\mathbf{r}+\mathbf{R}_{l'})$$
$$= C(\mathbf{R}_l)C(\mathbf{R}_{l'})\ \Psi_{\mathbf{k}}(\mathbf{r}). \tag{9.11}$$

But $\mathbf{R}_l+\mathbf{R}_{l'}$ is also a lattice vector. Hence,

$$\Psi_{\mathbf{k}}(\mathbf{r}+\mathbf{R}_l+\mathbf{R}_{l'}) = C(\mathbf{R}_l+\mathbf{R}_{l'})\ \Psi_{\mathbf{k}}(\mathbf{r}). \tag{9.12}$$

Hence, $C(\mathbf{R}_l+\mathbf{R}_{l'}) = C(\mathbf{R}_l)C(\mathbf{R}_{l'})$. Therefore,

$$C(\mathbf{R}_l) = e^{i\mathbf{k}\cdot\mathbf{R}_l} \tag{9.13}$$

for some $\mathbf{k}$. Here $\mathbf{k}$ has to be real since $|\Psi(\mathbf{r}+\mathbf{R}_l)|^2 = |\Psi(\mathbf{r})|^2$. Now define $u_{\mathbf{k}}(\mathbf{r}) = e^{-i\mathbf{k}\cdot\mathbf{r}}\Psi(\mathbf{r})$. It follows that $u_{\mathbf{k}}(\mathbf{r}+\mathbf{R}_l) = u_{\mathbf{k}}(\mathbf{r})$ is periodic. Inverting this allows us to write Eq. (9.9).

This periodic function may be expressed as a linear combination,

$$u_{\mathbf{k}}(\mathbf{r}) = \sum_{\mathbf{n},l} \varphi_l(\mathbf{r}-\mathbf{R_n})b_l(\mathbf{k}). \tag{9.14}$$

While the above assertion is quite general, the tight-binding approximation consists of exploiting the observation that the atomic orbitals $\varphi_l(\mathbf{r}-\mathbf{R_n})$ are sharply peaked at $\mathbf{r} = \mathbf{R_n}$—the location of the $\mathbf{n}^{th}$ atom, so that it is legitimate to write the following simplified expression for the Bloch wavefunction,

$$\Psi_{\mathbf{k}}(\mathbf{r}) = \sum_{\mathbf{n},l} e^{i\mathbf{k}\cdot\mathbf{r}}\varphi_l(\mathbf{r}-\mathbf{R_n})b_l(\mathbf{k}) \approx \sum_{\mathbf{n},l} e^{i\mathbf{k}\cdot\mathbf{R}_n}\varphi_l(\mathbf{r}-\mathbf{R_n})b_l(\mathbf{k}). \tag{9.15}$$

Using this as a basis set (as opposed to, say, a plane wave basis), the field operator takes the form (since we are talking about electrons in a solid, we now explicitly display a discrete spin index $\sigma = \{\uparrow, \downarrow\}$),

$$c_\sigma(\mathbf{r}) = \sum_{\mathbf{k}} \Psi_{\mathbf{k}}(\mathbf{r}) c_{\mathbf{k},\sigma} = \sum_{\mathbf{n},l} \left( \sum_{\mathbf{k}} e^{i\mathbf{k}\cdot\mathbf{R}_n} c_{\mathbf{k},\sigma} b_l(\mathbf{k}) \right) \varphi_l(\mathbf{r} - \mathbf{R}_\mathbf{n}). \tag{9.16}$$

At this stage we choose to make the identification

$$c_{\mathbf{n},\sigma}(l) \equiv \sum_{\mathbf{k}} e^{i\mathbf{k}\cdot\mathbf{R}_\mathbf{n}} c_{\mathbf{k},\sigma} b_l(\mathbf{k}), \tag{9.17}$$

and its inverse,

$$\sum_{\mathbf{n}} e^{-i\mathbf{k}\cdot\mathbf{R}_\mathbf{n}} c_{\mathbf{n},\sigma}(l) \equiv N\, c_{\mathbf{k},\sigma} b_l(\mathbf{k}), \tag{9.18}$$

where $N$ is the total number of lattice points, so that,

$$c_\sigma(\mathbf{r}) = \sum_{\mathbf{n},l} c_{\mathbf{n},\sigma}(l) \varphi_l(\mathbf{r} - \mathbf{R}_\mathbf{n}). \tag{9.19}$$

We impose (of course here, $s = -1$ for fermions),

$$[c_{\mathbf{n},\sigma}(l), c_{\mathbf{n}',\sigma'}(l')]_s = 0 \tag{9.20}$$

and,

$$[c_{\mathbf{n},\sigma}(l), c^\dagger_{\mathbf{n}',\sigma'}(l')]_s = \delta_{\mathbf{n},\mathbf{n}'} \delta_{\sigma,\sigma'} \delta_{l,l'} \tag{9.21}$$

so that the expectation involving the total number of particles is verified,

$$\int d^d r\, c^\dagger_\sigma(\mathbf{r}) c_\sigma(\mathbf{r}) = \sum_{\mathbf{n},l} c^\dagger_{\mathbf{n},\sigma}(l) c_{\mathbf{n},\sigma}(l) \tag{9.22}$$

provided the orthogonality condition between the orbitals is obeyed.

$$\int d^d r\, \varphi^*_{l'}(\mathbf{r} - \mathbf{R}_{\mathbf{n}'}) \varphi_l(\mathbf{r} - \mathbf{R}_\mathbf{n}) = \delta_{l',l} \delta_{\mathbf{n}',\mathbf{n}} \tag{9.23}$$

Furthermore, using the completeness condition of atomic orbitals allows us to write,

$$[c_\sigma(\mathbf{r}), c^\dagger_{\sigma'}(\mathbf{r}')]_s = \sum_{\mathbf{n},l} \varphi_l(\mathbf{r} - \mathbf{R}_\mathbf{n}) \varphi^*_l(\mathbf{r}' - \mathbf{R}_\mathbf{n}) \delta_{\sigma,\sigma'} =$$

$$= N\, \delta(\mathbf{r} - \mathbf{r}') \delta_{\sigma,\sigma'}. \tag{9.24}$$

Now on the one hand,

$$H_{tot}(\mathbf{r}, -i\hbar\nabla_\mathbf{r}) c_\sigma(\mathbf{r}) = H_{tot}(\mathbf{r}, -i\hbar\nabla_\mathbf{r}) \sum_{\mathbf{k}} \Psi_{\mathbf{k}}(\mathbf{r}) c_{\mathbf{k},\sigma} = \sum_{\mathbf{k}} \mathcal{E}(\mathbf{k})\, \Psi_{\mathbf{k}}(\mathbf{r}) c_{\mathbf{k},\sigma}$$

$$= \sum_{\mathbf{n},l} \varphi_l(\mathbf{r} - \mathbf{R}_\mathbf{n}) \sum_{\mathbf{k}} \mathcal{E}(\mathbf{k}) e^{i\mathbf{k}\cdot\mathbf{R}_n} b_l(\mathbf{k}) c_{\mathbf{k},\sigma}$$

$$= \sum_{\mathbf{n},\mathbf{n}',l} \varphi_l(\mathbf{r}-\mathbf{R_n})\, c_{\mathbf{n}',\sigma}(l)\, W(\mathbf{R_n}-\mathbf{R_{n'}}), \tag{9.25}$$

where

$$W(\mathbf{R_n}-\mathbf{R_{n'}}) = \left(\frac{1}{N}\sum_{\mathbf{k}} \mathcal{E}(\mathbf{k}) e^{i\mathbf{k}\cdot(\mathbf{R_n}-\mathbf{R_{n'}})}\right). \tag{9.26}$$

This means,

$$\sum_{\sigma}\int d^d r\, c^{\dagger}_{\sigma}(\mathbf{r}) H_{tot}(\mathbf{r}, -i\hbar\nabla_{\mathbf{r}}) c_{\sigma}(\mathbf{r}) =$$

$$\sum_{\sigma}\sum_{\mathbf{n},\mathbf{n}',l} c^{\dagger}_{\mathbf{n},\sigma}(l) c_{\mathbf{n}',\sigma}(l)\; W(\mathbf{R_n}-\mathbf{R_{n'}}). \tag{9.27}$$

In the above summation, if we include only $\mathbf{n}' = \mathbf{n}$ we get just the total number of particles times a constant, which is a conserved quantity. Hence we should be including $\mathbf{n}' \neq \mathbf{n}$. In the tight-binding approximation, one includes only the nearest neighbors, implying that $W(\mathbf{R})$ is sharply peaked at $\mathbf{R} = 0$. Set $\mathbf{n} = \mathbf{n}' + \delta$ where $\delta$ is the nearest neighbor vector. Further we assume that $W(\mathbf{R}_\delta) = -t$ is the same for all nearest neighbors. This means,

$$\sum_{\sigma}\int d^d r\, c^{\dagger}_{\sigma}(\mathbf{r}) H_{tot}(\mathbf{r}, -i\hbar\nabla_{\mathbf{r}}) c_{\sigma}(\mathbf{r}) = -t \sum_{\delta,\mathbf{n}',l,\sigma} c^{\dagger}_{\mathbf{n}'+\delta,\sigma}(l) c_{\mathbf{n}',\sigma}(l). \tag{9.28}$$

A similar argument is applicable when we look at the remaining term. First we exploit the periodicity of $\Delta U(\mathbf{r})$ to write,

$$\Delta U(\mathbf{r}) = \sum_{\mathbf{m}} w(\mathbf{r}-\mathbf{R_m}) \tag{9.29}$$

so that,

$$\int d^d r\, c^{\dagger}_{\sigma}(\mathbf{r}) c_{\sigma}(\mathbf{r}) \Delta U(\mathbf{r}) = \sum_{\mathbf{m}}\int d^d r\, c^{\dagger}_{\sigma}(\mathbf{r}) c_{\sigma}(\mathbf{r})\; w(\mathbf{r}-\mathbf{R_m})$$

$$= \sum_{\mathbf{m}}\sum_{\mathbf{n}',l'}\sum_{\mathbf{n},l} c^{\dagger}_{\mathbf{n}',\sigma}(l') c_{\mathbf{n},\sigma}(l)\; \left(\int d^d r\; \varphi^*_{l'}(\mathbf{r}-\mathbf{R_{n'}})\varphi_l(\mathbf{r}-\mathbf{R_n})\; w(\mathbf{r}-\mathbf{R_m})\right). \tag{9.30}$$

If $w(\mathbf{r}-\mathbf{R_m})$ peaks at $\mathbf{r} = \mathbf{R_m}$, the above integral may be approximated as,

$$\int d^d r\, c^{\dagger}_{\sigma}(\mathbf{r}) c_{\sigma}(\mathbf{r}) \Delta U(\mathbf{r}) =$$

$$\approx \sum_{\mathbf{m}}\sum_{\mathbf{n}',l'}\sum_{\mathbf{n},l} c^{\dagger}_{\mathbf{n}',\sigma}(l') c_{\mathbf{n},\sigma}(l)\; \left(\int d^d r\; \varphi^*_{l'}(\mathbf{r}+\mathbf{R_{m-n'}})\varphi_l(\mathbf{r}+\mathbf{R_{m-n}})\right) w_{peak}. \tag{9.31}$$

Since the orbitals around atoms are highly localized, $\varphi^*_{l'}(\mathbf{r}+\mathbf{R}_{\mathbf{m}-\mathbf{n}'})\varphi_l(\mathbf{r}+\mathbf{R}_{\mathbf{m}-\mathbf{n}})) \approx 0$ when $\mathbf{n} \neq \mathbf{n}'$. When $\mathbf{n} = \mathbf{n}'$, however, we get,

$$\int d^dr\, c^\dagger_\sigma(\mathbf{r})c_\sigma(\mathbf{r})\Delta U(\mathbf{r}) =$$

$$\approx \sum_{\mathbf{m},\mathbf{n},l} c^\dagger_{\mathbf{n}',\sigma}(l')c_{\mathbf{n},\sigma}(l)\;\delta_{l,l'}\;w_{peak} \propto N_{tot} \times N_{ele}, \tag{9.32}$$

where $N_{tot}$ is the total number of sites and $N_{ele}$ is the total number of electrons, which makes this term a constant. So this case is uninteresting. Now consider the case when $w(\mathbf{r}-\mathbf{R_m})$ has a node at $\mathbf{r}=\mathbf{R_m}$. This may be captured by saying that perhaps $w(\mathbf{r}-\mathbf{R_m})$ peaks midway between $\mathbf{R_m}$ and its nearest neighbor viz. at, $\mathbf{r}=\frac{1}{2}(\mathbf{R_m}+\mathbf{R}_{\mathbf{m}'})$ where $\mathbf{R}_{\mathbf{m}'}$ is the nearest neighbor. In this case,

$$\int d^dr\, c^\dagger_\sigma(\mathbf{r})c_\sigma(\mathbf{r})\Delta U(\mathbf{r}) = \sum_{\mathbf{m}}\int d^dr\, c^\dagger_\sigma(\mathbf{r})c_\sigma(\mathbf{r})\; w(\mathbf{r}-\mathbf{R_m})$$

$$= \sum_{\mathbf{m}}\sum_{\mathbf{n}',l'}\sum_{\mathbf{n},l} c^\dagger_{\mathbf{n}',\sigma}(l')c_{\mathbf{n},\sigma}(l)\;\Big(\int d^dr\; \varphi^*_{l'}(\mathbf{r}-\mathbf{R}_{\mathbf{n}'})\varphi_l(\mathbf{r}-\mathbf{R_n})\; w(\mathbf{r}-\mathbf{R_m})\Big). \tag{9.33}$$

It is easy to argue that when $\mathbf{n} \neq \mathbf{n}'$ the above term is significant only when $\mathbf{n}$ and $\mathbf{n}'$ are nearest neighbors and $\mathbf{R_m} = \frac{1}{2}(\mathbf{R_n}+\mathbf{R}_{\mathbf{n}'})$.

$$\int d^dr\, c^\dagger_\sigma(\mathbf{r})c_\sigma(\mathbf{r})\Delta U(\mathbf{r})$$

$$\approx \sum_{\mathbf{n}',l'}\sum_{\mathbf{n},l} c^\dagger_{\mathbf{n}',\sigma}(l')c_{\mathbf{n},\sigma}(l)\;\Big(\int d^dr\; \varphi^*_{l'}(\mathbf{r}+\frac{1}{2}(\mathbf{R_n}-\mathbf{R}_{\mathbf{n}'}))\varphi_l(\mathbf{r}-\frac{1}{2}(\mathbf{R_n}-\mathbf{R}_{\mathbf{n}'}))\; w(\mathbf{r})\Big). \tag{9.34}$$

For nearest neighbors, $\mathbf{R_n}-\mathbf{R}_{\mathbf{n}'} \sim \vec{a}_{\mathbf{n}}$, where $\vec{a}_{\mathbf{n}}$ is the basis vector in the appropriate direction. Thus the overlap integral will be small but nonzero since the orbitals are now evaluated midway between two lattice points. The main point is that in such cases we may always write the tight-binding Hamiltonian as,

$$H_{hop} = -t\sum_{\delta,\mathbf{n}',l,\sigma} c^\dagger_{\mathbf{n}'+\delta,\sigma}(l)c_{\mathbf{n}',\sigma}(l). \tag{9.35}$$

Observe the large number of simplifying assumptions involved in deriving this picture. In a similar vein, we may derive the tight-binding picture when mutual interactions are present. In this case we have to evaluate,

$$H_I = \frac{1}{2}\sum_{\sigma,\sigma'}\int d^dr\int d^dr'\; V(\mathbf{r}-\mathbf{r}')c^\dagger_\sigma(\mathbf{r})c^\dagger_{\sigma'}(\mathbf{r}')c_{\sigma'}(\mathbf{r}')c_\sigma(\mathbf{r}). \tag{9.36}$$

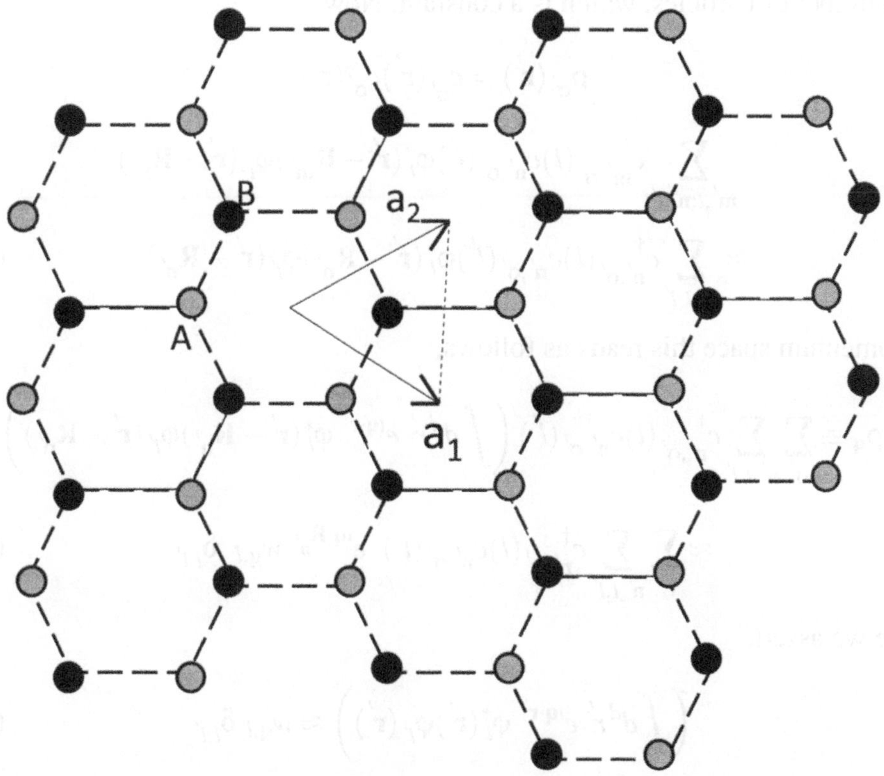


Figure 9.1: Graphene is a two-dimensional lattice made of just carbon atoms arranged in a honeycomb pattern. The lighter circles are carbon atoms that form a triangular lattice known as the A sublattice. The darker circles, also carbon atoms, form another triangular lattice known as the B sublattice. The centers of three hexagons that share edges, form an equilateral triangle.

Now we evaluate the density field in the lattice picture. For this we first write,

$$H_I = \frac{1}{2}\sum_{\sigma,\sigma'}\int d^d r \int d^d r' \; V(\mathbf{r}-\mathbf{r}')c^\dagger_\sigma(\mathbf{r})c^\dagger_{\sigma'}(\mathbf{r}')c_{\sigma'}(\mathbf{r}')c_\sigma(\mathbf{r})$$

$$= \sum_{\mathbf{q}} \frac{v_q}{2\Omega}\rho_{\mathbf{q}}\rho_{-\mathbf{q}} + const. \qquad (9.37)$$

where $V(\mathbf{r}-\mathbf{r}') = \frac{1}{\Omega}\sum_{\mathbf{q}} e^{i\mathbf{q}\cdot(\mathbf{r}-\mathbf{r}')}v_q$ and,

$$\rho_{\mathbf{q}} \equiv \sum_{\sigma'}\int d^d r' \, e^{i\mathbf{q}\cdot\mathbf{r}'} \; c^\dagger_{\sigma'}(\mathbf{r}')c_{\sigma'}(\mathbf{r}'). \qquad (9.38)$$

The new form comes from writing $H_I$ as the product of two densities after passing one of the annihilation operators across, which picks up a term proportional to the total number of particles, which is a constant. Now,

$$\rho_{\sigma'}(\mathbf{r}') = c^{\dagger}_{\sigma'}(\mathbf{r}')c_{\sigma'}(\mathbf{r}')$$

$$= \sum_{\mathbf{m}',l;\mathbf{n}',l'} c^{\dagger}_{\mathbf{m}',\sigma'}(l)c_{\mathbf{n}',\sigma'}(l')\varphi^*_l(\mathbf{r}'-\mathbf{R}_{\mathbf{m}'})\varphi_{l'}(\mathbf{r}'-\mathbf{R}_{\mathbf{n}'})$$

$$\approx \sum_{\mathbf{n}',l,l'} c^{\dagger}_{\mathbf{n}',\sigma'}(l)c_{\mathbf{n}',\sigma'}(l')\varphi^*_l(\mathbf{r}'-\mathbf{R}_{\mathbf{n}'})\varphi_{l'}(\mathbf{r}'-\mathbf{R}_{\mathbf{n}'}). \tag{9.39}$$

In momentum space this reads as follows,

$$\rho_{\mathbf{q}} \equiv \sum_{\sigma'}\sum_{\mathbf{n}',l,l'} c^{\dagger}_{\mathbf{n}',\sigma'}(l)c_{\mathbf{n}',\sigma'}(l')\left(\int d^dr'\, e^{i\mathbf{q}\cdot\mathbf{r}'}\, \varphi^*_l(\mathbf{r}'-\mathbf{R}_{\mathbf{n}'})\varphi_{l'}(\mathbf{r}'-\mathbf{R}_{\mathbf{n}'})\right)$$

$$\approx \sum_{\sigma'}\sum_{\mathbf{n}',l,l'} c^{\dagger}_{\mathbf{n}',\sigma'}(l)c_{\mathbf{n}',\sigma'}(l')\; e^{i\mathbf{q}\cdot\mathbf{R}_{\mathbf{n}'}}\; w_{\mathbf{q},l}\;\delta_{l,l'} \tag{9.40}$$

where we assert,

$$\left(\int d^dr'\, e^{i\mathbf{q}\cdot\mathbf{r}'}\, \varphi^*_l(\mathbf{r}')\varphi_{l'}(\mathbf{r}')\right) \approx w_{\mathbf{q},l}\;\delta_{l,l'} \tag{9.41}$$

since the orbitals are orthogonal and for small $q$ this is the leading contribution. Therefore,

$$H_I = \sum_{\mathbf{q}} \frac{v_q}{2\Omega}\rho_{\mathbf{q}}\rho_{-\mathbf{q}}$$

$$= \sum_{\sigma',\sigma,\mathbf{n}',\mathbf{n}'',l,l'} c^{\dagger}_{\mathbf{n}',\sigma'}(l)c_{\mathbf{n}',\sigma'}(l)\; c^{\dagger}_{\mathbf{n}'',\sigma''}(l')c_{\mathbf{n}'',\sigma''}(l')\; Q_{l,l'}(\mathbf{R}_{\mathbf{n}'}-\mathbf{R}_{\mathbf{n}''}) \tag{9.42}$$

where,

$$Q_{l,l'}(\mathbf{R}_{\mathbf{n}'}-\mathbf{R}_{\mathbf{n}''}) = \sum_{\mathbf{q}} \frac{v_q}{2\Omega} e^{i\mathbf{q}\cdot(\mathbf{R}_{\mathbf{n}'}-\mathbf{R}_{\mathbf{n}''})}\; w_{-\mathbf{q},l'}\; w_{\mathbf{q},l}. \tag{9.43}$$

In the usual case of Coulomb interactions, $v_q$ is appreciable only for small $q$ and $w_{0,l} = 1$, hence the above $Q$ factor peaks at $\mathbf{R}_{\mathbf{n}'} = \mathbf{R}_{\mathbf{n}''}$. In case there is only one band, the interaction term reads as follows.

$$H_I = \frac{U}{2}\sum_{\sigma,\sigma'} n_{i\sigma}n_{i\sigma'}, \tag{9.44}$$

where $U$ is the energy scale deduced from the $Q$ factor above.

$$U = \sum_{\mathbf{q}} \frac{v_{\mathbf{q}}}{\Omega}\; w_{-\mathbf{q},l}w_{\mathbf{q},l} \tag{9.45}$$

When $\sigma = \sigma'$ we use idempotence to conclude that this term is proportional to the total number, which is just a constant. Hence we should only include,

$$H_I = U \sum_i n_{i\uparrow} n_{i\downarrow}. \tag{9.46}$$

This potential energy and the hopping term Eq. (9.35) together make the one-band Hubbard model. It is usually written as,

$$H = -t \sum_{<ij>\sigma} c^{\dagger}_{i\sigma} c_{j\sigma} + U \sum_i n_{i\uparrow} n_{i\downarrow}. \tag{9.47}$$

Many variants of this exist. One may include next-nearest neighbor hopping and even more through a hopping term such as $t_{ij}$. One may also consider next-nearest neighbor interactions in the case of interactions.

$$H = -\sum_{<ij>\sigma} t_{ij}\, c^{\dagger}_{i\sigma} c_{j\sigma} + U \sum_i n_{i\uparrow} n_{i\downarrow} + V \sum_{<ij>} n_i n_j \tag{9.48}$$

where $n_i = \sum_{a=\{\uparrow,\downarrow\}} n_{i,a}$. Another example involves mobile electrons hybridizing with localized electrons. This is called the Anderson (lattice) model.

$$H = -t \sum_{<ij>\sigma} c^{\dagger}_{i\sigma} c_{j\sigma} + \xi \sum_{i\sigma} (c^{\dagger}_{i\sigma} d_{i\sigma} + d^{\dagger}_{i\sigma} c_{i\sigma}) + U \sum_i (d^{\dagger}_{i\uparrow} d_{i\uparrow})(d^{\dagger}_{i\downarrow} d_{i\downarrow}) \tag{9.49}$$

Having derived the lattice version of the Hamiltonian of mutually interacting particles, we are now faced with the prospect of deducing its properties. By this, one means computing what is known as the phase diagram. The phase diagram of the system consists of the following ingredients. Firstly, there are three mutually perpendicular axes, one each for temperature, number of particles per site (total number of particles divided by total number of sites), and the ratio $U/t$. Secondly, different regions in this space are identified, separated from each other by surfaces. Each of these regions corresponds to a particular phase. The surface represents a boundary across which a phase transition happens. A phase is characterized by a set of non-vanishing 'order parameters'. An order parameter is an operator that has a nonvanishing expectation value in that phase. It is more convenient to use the following definition. If $\mathfrak{O}(i)$ is an order parameter, then the average of the product of two such operators (at equal times) has the property,

$$Lim_{|i-j|\to\infty} < \mathfrak{O}^{\dagger}(i)\mathfrak{O}(j) > = < \mathfrak{O}^{\dagger}(i) > < \mathfrak{O}(j) > . \tag{9.50}$$

In other words, the average of two $\mathfrak{O}$'s factor into two independent nonvanishing quantities (which, in a translationally invariant system, would be independent of the location) as shown when the spatial separation between the two operators becomes large. When this condition is true we say that the system possesses long-range order and the state of the system with respect to which the expectation value has been taken is in a phase characterized by the order parameter $\mathfrak{O}$. When a chosen $\mathfrak{O}$ does not obey this property, then the above correlation function typically has the property,

$$Lim_{|i-j|\to\infty} < \mathfrak{O}^{\dagger}(i)\mathfrak{O}(j) > \sim e^{-\frac{|i-j|}{\xi}}. \tag{9.51}$$

In one spatial dimension however, there can never be long-range order (except at absolute zero temperature) due to what is known as the Mermin-Wagner theorem. Instead, the typical situation which replaces the assertion in Eq. (9.50) in one dimension is,

$$Lim_{|i-j|\to\infty} < \mathfrak{O}^{\dagger}(i)\mathfrak{O}(j) > \sim \frac{1}{|i-j|^{\delta}}. \tag{9.52}$$

Instead of converging to a constant, the above average slowly decays to zero as a power law. This is the closest one can get to long-range order in one dimension—still acceptable considering that the absence of long-range order means an exponential decay (Eq. (9.51)). Of course, all these assertions are in hindsight. One is still faced with the formidable task of evaluating these averages and verifying these expectations. In this book we do not embark upon this calculation as this is the subject matter of many-body theory. However, it is still important to list the sort of operators $\mathfrak{O}(i)$ that are used to describe various phases. An incomplete list (for electrons in a solid, for example) is given in the box (here $\sigma = \uparrow$ has a numerical value of $+1$ and $\sigma = \downarrow$ has a numerical value of $-1$).

Charge density wave (CDW):
$\mathfrak{O}_{CDW}(i) = \sum_{\sigma=\uparrow,\downarrow} \psi^{\dagger}_{\sigma}(i)\psi_{\sigma}(i)$

Spin density wave (SDW):
$\vec{\mathfrak{O}}_{SDW}(i) = \sum_{\sigma,\sigma'=\uparrow,\downarrow} \psi^{\dagger}_{\sigma}(i)\vec{\sigma}_{\sigma,\sigma'}\psi_{\sigma'}(i)$

Singlet superconductivity (SS):
$\mathfrak{O}_{SS}(i) = \sum_{\sigma,\sigma'=\uparrow,\downarrow} \sigma\psi_{\sigma}(i)\delta_{\sigma,\sigma'}\psi_{-\sigma'}(i)$

Triplet superconductivity (TS):
$\vec{\mathfrak{O}}_{TS}(i) = \sum_{\sigma,\sigma'=\uparrow,\downarrow} \sigma\psi_{\sigma}(i)\vec{\sigma}_{\sigma,\sigma'}\psi_{-\sigma'}(i)$

Thus one is now faced with the task of evaluating the averages $< \mathfrak{O}^{\dagger}(i)\mathfrak{O}(j) >$. Various approximation schemes are used in the literature for this purpose. It would take us too far afield to describe these approaches. We conclude our discussion of lattice models by examining the large U limit of the Hubbard model.

## 9.2 Schrieffer-Wolff Transformation

Even in the oversimplified one-band Hubbard model, there lurks a rich variety of phenomena, most of which are poorly understood in more than one dimension. In one dimension, a variety of methods starting from the rigorous Bethe ansatz to bosonization (to be discussed in the last chapter) provide a satisfactory description of the basic physics (by this one means the phase diagram). In more than one dimension, one of the most important phenomena is known as the Mott-Hubbard transition. This is a metal insulator transition at absolute zero temperature that is applicable only when there is exactly one electron per site, is driven by strong correlations. This means the ratio $U/t$ determines the nature of the phase at absolute zero—below a critical value, the system is gapless (a gap is the difference in energy between the ground state and the 'first' excited state. In infinite systems, in a metallic state, this quantity is zero, as the 'first' excited state together with the ground state form a continuum) and above this value, it is gapped. Establishing this is one of the most important goals of the physics of strong correlations. Of course, as we have pointed out several times, this book simply discusses the framework or the mathematical language in which meaningful and interesting questions such as these, may be posed. Answering them is still largely the subject matter of ongoing research—though many results are available, very few of them are universally accepted. We now turn to the description of the simple Hubbard model when $U/t$ is much larger than unity. Intuitively, it is easy to see what might happen. Imagine a situation where there are exactly as many electrons as there are sites (this is quite possible and common when each atom contributes one electron to the conduction process). In such a situation, when $U/t$ is large, the ground state is one where each electron stays put at an atom so that hopping is suppressed, since hopping would entail one of the sites having two electrons (allowed by Pauli's principle if they have opposite spins), which makes the energy of the system large. Thus, the only degree of freedom left to the electron after hopping is suppressed is spin flipping. The spins of the electrons on neighboring sites interact with one another leading to a magnetic insulator. To understand this phenomenon, one notes that Hamiltonians that differ from each other by unitary transformations describe the same physics. By this, one means the following. Suppose one wants to compute a correlation function of the type

$$C(t,t') \equiv < \Phi|A(t)B(t')|\Phi> = < \Phi|e^{i\frac{t}{\hbar}H}A(0)e^{-i\frac{t}{\hbar}H}e^{i\frac{t'}{\hbar}H}B(0)e^{-i\frac{t'}{\hbar}H}|\Phi> . \quad (9.53)$$

Upon a unitary transformation, the same correlation function may be written as,[1]

$$C(t,t') = < \Phi(\lambda)|e^{i\frac{t}{\hbar}H(\lambda)}A_\lambda(0)e^{-i\frac{t}{\hbar}H(\lambda)}e^{i\frac{t'}{\hbar}H(\lambda)}B_\lambda(0)e^{-i\frac{t'}{\hbar}H(\lambda)}|\Phi(\lambda)> \quad (9.54)$$

[1]Not wanting to confuse the time parameter with the hopping parameter, both of which are denoted by $t$, we rename the hopping parameter $\lambda$.

where the $\lambda$-dependent quantities are unitarily transformed versions,

$$H(\lambda) = e^{i\lambda G} H e^{-i\lambda G};\ A_\lambda(0) = e^{i\lambda G} A(0) e^{-i\lambda G};\ B_\lambda(0) = e^{i\lambda G} B(0) e^{-i\lambda G} \tag{9.55}$$

and,

$$|\Phi(\lambda)> = e^{i\lambda G}|\Phi> \tag{9.56}$$

(keeping in mind that $e^{i\frac{t}{\hbar}H(\lambda)} = e^{i\lambda G} e^{i\frac{t}{\hbar}H} e^{-i\lambda G}$). Now we wish to take advantage of this observation to recast the Hubbard model in the situation when $U/\lambda \gg 1$. This transformation is known as the Schrieffer-Wolff transformation. To achieve this, we first introduce various projection operators. The single occupation projection operator is,

$$\mathfrak{P}_i^s = 1 - n_{i\uparrow} n_{i\downarrow}. \tag{9.57}$$

This operator, when acting on a state, prevents the resulting state from having two electrons at site labeled $i$, since if it did, $n_{i\uparrow} = n_{i\downarrow} = 1$ and $\mathfrak{P}_i^s = 0$ and the state would not exist. Similarly, the operator that projects onto doubly occupied sites is,

$$\mathfrak{P}_i^d = n_{i\uparrow} n_{i\downarrow}. \tag{9.58}$$

The operator that ensures that no site is doubly occupied $\mathfrak{P}^s = \prod_i \mathfrak{P}_i^s$. We note that $\mathfrak{P}_i^s + \mathfrak{P}_i^d = 1$. We now rewrite the Hamiltonian by projecting out states that have more than one site doubly occupied using the above resolution of the identity,

$$H = 1.H.1 = (\mathfrak{P}^s + (1-\mathfrak{P}^s))H(\mathfrak{P}^s + (1-\mathfrak{P}^s)). \tag{9.59}$$

This means the full Hamiltonian is a sum of four pieces:

$$H_{11} = \mathfrak{P}^s H \mathfrak{P}^s \tag{9.60}$$

$$H_{12} = \mathfrak{P}^s H (1-\mathfrak{P}^s) \tag{9.61}$$

$$H_{21} = (1-\mathfrak{P}^s) H \mathfrak{P}^s \tag{9.62}$$

$$H_{22} = (1-\mathfrak{P}^s) H (1-\mathfrak{P}^s). \tag{9.63}$$

Of the four terms, only $H_{22}$ contains the $U$-dependent contribution since the others contain the projection $\mathfrak{P}_s$, which annihilates the potential energy term in $H$. Since $U$ is large, and $H_{22}$ contains at least one doubly occupied site, the eigenvalues of $H_{22}$ are such that, $H_{22} > U \gg H_{11}, H_{12}, H_{21}$. Now we perform a unitary transformation on this Hamiltonian so that the terms that mix the sectors containing no doubly occupied states to the sector that does are eliminated. The form that does not involve mixing of the sectors is $H_d \equiv T^{-1} H T$ so that,

$$H_d \approx \mathfrak{P}^s (H_{11} - \frac{H_{12} H_{21}}{U}) \mathfrak{P}^s + (1-\mathfrak{P}^s)(H_{22} + \frac{H_{21} H_{12}}{U})(1-\mathfrak{P}^s). \tag{9.64}$$

These terms are nothing but the eigenvalues of the matrix form of $H_{ij}$ (we set $H_{22} \sim U$ when it appears in the denominator, since this sector is assumed to contain states with exactly one doubly occupied site). One may determine the needed $T$ matrix by using the modified condition $TH_d = HT$ so that,

$$T\mathfrak{P}^s(H_{11} - \frac{H_{12}H_{21}}{U})\mathfrak{P}^s + T(1-\mathfrak{P}^s)(H_{22} + \frac{H_{21}H_{12}}{U})(1-\mathfrak{P}^s) = HT. \quad (9.65)$$

Now we post-multiply and pre-multiply by one of $\mathfrak{P}^s$ and $1-\mathfrak{P}^s$ and conclude that the following (non-unique) choice suffices,

$$T = \mathfrak{P}^s 1 \mathfrak{P}^s + \mathfrak{P}^s \frac{H_{12}}{U}(1-\mathfrak{P}^s) + (1-\mathfrak{P}^s)(-\frac{H_{21}}{U})\mathfrak{P}^s + (1-\mathfrak{P}^s)1(1-\mathfrak{P}^s). \quad (9.66)$$

One may see from Eq. (9.64) that the double occupancy sector is at a scale $\sim U$ much larger than the no-double-occupancy sector. Thus for studying low-energy phenomena, we focus on the first term. When double occupancy is not allowed, the $U$ term in the Hamiltonian drops out so that,

$$H_{11} = -t \sum_{<ij>\sigma} \mathfrak{P}^s c^\dagger_{i\sigma} c_{j\sigma} \mathfrak{P}^s \quad (9.67)$$

$$H_{12} = -t \sum_{<ij>\sigma} \mathfrak{P}^s c^\dagger_{i\sigma} c_{j\sigma}(1-\mathfrak{P}^s) \quad (9.68)$$

$$H_{21} = -t \sum_{<ij>\sigma} (1-\mathfrak{P}^s) c^\dagger_{i\sigma} c_{j\sigma} \mathfrak{P}^s. \quad (9.69)$$

Now we make use of the assertion that $1-\mathfrak{P}^s$ contains at most one doubly occupied state so that,

$$1 = \prod_i (\mathfrak{P}^s_i + \mathfrak{P}^d_i) \approx \prod_i \mathfrak{P}^s_i + \sum_k \left( \prod_{i \neq k} \mathfrak{P}^s_i \right) \mathfrak{P}^d_k + \ldots \quad (9.70)$$

Therefore,

$$H_{12}H_{21} = t^2 \sum_{<ij>\sigma} \sum_{<i^{\prime}j^{\prime}>\sigma^{\prime}} \mathfrak{P}^s c^\dagger_{i\sigma} c_{j\sigma}(1-\mathfrak{P}^s) c^\dagger_{i^{\prime}\sigma^{\prime}} c_{j^{\prime}\sigma^{\prime}} \mathfrak{P}^s$$

$$= t^2 \sum_{<ij>\sigma} \sum_{<i^{\prime}j^{\prime}>\sigma^{\prime}} \mathfrak{P}^s c^\dagger_{i\sigma} c_{j\sigma} \sum_k \left( \prod_{p \neq k} \mathfrak{P}^s_p \right) \mathfrak{P}^d_k c^\dagger_{i^{\prime}\sigma^{\prime}} c_{j^{\prime}\sigma^{\prime}} \mathfrak{P}^s. \quad (9.71)$$

Now we note,

$$\mathfrak{P}^d_k c^\dagger_{i^{\prime}\sigma^{\prime}} c_{j^{\prime}\sigma^{\prime}} = c^\dagger_{i^{\prime}\sigma^{\prime}} c_{j^{\prime}\sigma^{\prime}} \mathfrak{P}^d_k + c^\dagger_{i^{\prime},\sigma^{\prime}} c_{j^{\prime}\sigma^{\prime}} \, (\delta_{k,i^{\prime}} - \delta_{j^{\prime},k}) n_{k,\bar{\sigma^{\prime}}}. \quad (9.72)$$

Using similar ideas we may pass the remaining projection operators in the middle to the extreme right to obtain,

$$H_{12}H_{21} =$$
$$t^2 \sum_{<ij>\sigma} \sum_{<i^{'}j^{'}>\sigma^{'}} \mathfrak{P}^s c^{\dagger}_{i\sigma} c_{j\sigma} \mathfrak{P}^s_{j^{'}} c^{\dagger}_{i^{'},\sigma^{'}} c_{j^{'}\sigma^{'}} n_{i^{'},\bar{\sigma}^{'}} \mathfrak{P}^s$$
$$-t^2 \sum_{<ij>\sigma} \sum_{<i^{'}j^{'}>\sigma^{'}} \mathfrak{P}^s c^{\dagger}_{i\sigma} c_{j\sigma} \mathfrak{P}^s_{i^{'}} c^{\dagger}_{i^{'},\sigma^{'}} c_{j^{'}\sigma^{'}} n_{j^{'},\bar{\sigma}^{'}} \mathfrak{P}^s. \tag{9.73}$$

In the second term of the above equation, there is the product $c_{j^{'}\sigma^{'}} n_{j^{'},\bar{\sigma}^{'}} \mathfrak{P}^s$. The projection operator together with the occupation number ensures that $n_{j^{'},\sigma^{'}} = 0$; in other words, the $j^{'}\sigma^{'}$ state is empty. When annihilated, this gives zero (see exercises also). Thus only the first term remains. Using the definition of the projection operator in the middle we obtain,

$$H_{12}H_{21} = t^2 \sum_{<ij>\sigma} \sum_{<i^{'}j^{'}>\sigma^{'}} \mathfrak{P}^s c^{\dagger}_{i\sigma} c_{j\sigma} c^{\dagger}_{i^{'},\sigma^{'}} c_{j^{'}\sigma^{'}} n_{i^{'},\bar{\sigma}^{'}} \mathfrak{P}^s. \tag{9.74}$$

There are many terms in this. One may single out the 'coherent contributions' wherein $(i,j) \equiv (i^{'},j^{'})$ or $(i,j) \equiv (j^{'},i^{'})$. One then includes $\sigma^{'} = \sigma$ followed by $\sigma^{'} = \bar{\sigma}$.

$$H_{12}H_{21} = -t^2 \sum_{<ij>\sigma} \mathfrak{P}^s (c^{\dagger}_{i\sigma} c^{\dagger}_{i,\bar{\sigma}} c_{j\bar{\sigma}} c_{j\sigma} - n_{i\sigma} n_{j,\bar{\sigma}}) \mathfrak{P}^s \tag{9.75}$$

It is possible to relate this to a term that corresponds to interaction between spins at sites $i$ and $j$. Define (we have set $\hbar = 1$)

$$\vec{S}_i = \sum_{a,b=\uparrow,\downarrow} c^{\dagger}_{ia} \frac{1}{2} \vec{\sigma}_{ab}\, c_{ib}, \tag{9.76}$$

and $\vec{\sigma}$ are the three Pauli matrices. Also, $n_{i\sigma} = c^{\dagger}_{i\sigma} c_{i\sigma}$ and $n_i = \sum_{\sigma} n_{i\sigma}$. It is left to the exercises to show that,

$$(\vec{S}_i \cdot \vec{S}_j - \frac{1}{4} n_i n_j) = \frac{1}{2} \sum_{\sigma} (c^{\dagger}_{i\sigma} c^{\dagger}_{i,\bar{\sigma}} c_{j\bar{\sigma}} c_{j\sigma} - n_{i\sigma} n_{j,\bar{\sigma}}). \tag{9.77}$$

Therefore, the low-energy sector of the diagonal form of the Hamiltonian is

$$H_d \approx \mathfrak{P}^s (H_t + \frac{2t^2}{U} \sum_{<ij>} (\vec{S}_i \cdot \vec{S}_j - \frac{1}{4} n_i n_j)) \mathfrak{P}^s + ..., \tag{9.78}$$

where $H_t = -t \sum_{<ij>\sigma} c^{\dagger}_{i\sigma} c_{j\sigma}$ is the original hopping term. Several points are apparent in this calculation. The first is that the above is only the leading contribution; we have ignored many terms that couple distant sites. When there is exactly one electron per site (half-filling), the hopping is suppressed and the dominant physics is the anti-ferromagnetic coupling between the spins of the electrons on neighboring sites. Thus we obtain a magnetic insulator.

One may introduce variants of the above. Of particular importance is the Anderson impurity model. In this model, otherwise free electrons on a lattice hybridize with a localized orbital. Including these two terms makes the model solvable. However, important physics is contained, when in addition, two electrons of opposite spins on the orbital repel with an energy $U \gg t, \xi$.

$$H = -t \sum_{<ij>\sigma} c^\dagger_{i\sigma} c_{j\sigma} + \xi (c^\dagger_{0,\sigma} d_\sigma + d^\dagger_\sigma c_{0,\sigma}) + U d^\dagger_\uparrow d_\uparrow d^\dagger_\downarrow d_\downarrow \tag{9.79}$$

The term with $\xi$ is known as hybridization. For large $U$, the model above may be transformed by a Schrieffer-Wolff transformation to the Kondo model,

$$H = -t \sum_{<ij>\sigma} c^\dagger_{i\sigma} c_{j\sigma} + g \sum_{\alpha,\alpha'} c^\dagger_{0,\alpha} \vec{\sigma}_{\alpha\alpha'} c_{0,\alpha'} \cdot \vec{S}, \tag{9.80}$$

where $\vec{S} = \sum_{a,b} d^\dagger_a \vec{\sigma}_{ab} d_b$ is the local spin. Implicit is the no double occupancy constraint viz. $(d^\dagger_\uparrow d_\uparrow)(d^\dagger_\downarrow d_\downarrow) = 0$. Some more examples are found in the exercises.

## 9.3 Exercises

**Q.1** Verify the identity in Eq. (9.77).

**Q.2** Verify Eq. (9.73) and Eq. (9.74).

**Q.3** What kind of continuum picture would lead to the Anderson lattice model? (Hint: Think in terms of each atom contributing two kinds of electrons, one whose wavefunctions overlap with neighboring atoms and the other does not. The repulsion is only for the localized electrons).

**Q.4** Derive the Kondo model Eq. (9.80) from the Anderson model Eq. (9.79).

**Q.5** What kind of Kondo model emerges by applying the Schrieffer-Wolff transformation to the 'extended hybridization' model?

$$H = -t \sum_{<ij>\sigma} c^\dagger_{i\sigma} c_{j\sigma} + \sum_{k=0,\pm 1,\sigma} \xi_k (c^\dagger_{k,\sigma} d_\sigma + d^\dagger_\sigma c_{k,\sigma}) + U d^\dagger_\uparrow d_\uparrow d^\dagger_\downarrow d_\downarrow \tag{9.81}$$

where $\xi_1 = \xi_{-1} \neq \xi_0$.

# Chapter 10

# Green Functions: Matsubara and Nonequilibrium

In this chapter, we discuss the central concept in many-body physics, namely the single-particle Green function. It is shown that many physical observables, such as current and number density, are related to this object. However, this quantity is more important than that since it provides information of the nature of 'quasiparticles'. If we choose to characterize the atomic constituents of the model being studied minus the mutual interactions among the constituents—'particles', then 'quasiparticles' would be the effective description of such particles in the presence of such mutual interaction between the constituents. The single-particle Green function contains information about the energy momentum relation of these quasiparticles and also their lifetime. We consider this concept both for systems in thermodynamic equilibrium and for systems out of equilibrium.

## 10.1 Matsubara Green Functions

In this section, we define the finite temperature, or Matsubara Green function, of a system of particles for systems that statistical mechanics would classify as grand canonical. This means that we imagine the system exchanging energy and particles with a reservoir where only the average energy and average number of particles are fixed. We imagine the system to be described by a Hamiltonian that may be written in the form $H = H_0 + V$, where $H_0$ is that part of the Hamiltonian which may be handled exactly. The remaining is denoted by $V$. Consider the following operator,

$$U(s) = e^{s(H_0+V)}. \tag{10.1}$$

If $H_0V = VH_0$, then we can trivially factorize this as,

$$U(s) = e^{sH_0}e^{sV}. \tag{10.2}$$

In general, this is not true. However, for practical applications we need such a factorization. Thus we write

$$U(s) = e^{sH_0}S(s), \tag{10.3}$$

where

$$S(s) = e^{-sH_0}e^{s(H_0+V)}. \tag{10.4}$$

We want to express S(s) in some simple way in terms of V. To this end, let us examine some of its properties. In particular let us compute,

$$\frac{dS(s)}{ds} = -H_0e^{-sH_0}e^{s(H_0+V)} + e^{-sH_0}(H_0+V)e^{s(H_0+V)}$$

$$= -H_0e^{-sH_0}e^{s(H_0+V)} + e^{-sH_0}H_0e^{s(H_0+V)} + e^{-sH_0}Ve^{s(H_0+V)} = e^{-sH_0}Ve^{s(H_0+V)}$$

$$= e^{-sH_0}Ve^{sH_0}e^{-sH_0}e^{s(H_0+V)}. \tag{10.5}$$

Define

$$\hat{V}(s) = e^{-sH_0}Ve^{sH_0}. \tag{10.6}$$

Therefore,

$$\frac{dS(s)}{ds} = \hat{V}(s)S(s). \tag{10.7}$$

If $[H_0, V] = 0$, then $\hat{V}(s) = V$ and the solution would simply be,

$$S(s) = e^{sV} \equiv e^{\int_0^s dt\ V}. \tag{10.8}$$

Since trivially, $s = \int_0^s dt$. In general, $S(s)$ is not this simple, hence the solution is symbolically denoted as,

$$S(s) = T[e^{\int_0^s dt\ \hat{V}(t)}]. \tag{10.9}$$

The meaning of the symbol T[...] will be made clear soon. It may be regarded as a symbolic way of writing the iterative solution to Eq. (10.7). The solution to Eq. (10.7) can be generated as follows,

$$S(s) = 1 + \int_0^s \hat{V}(t)dt + \int_0^s dt \int_0^t dt' \hat{V}(t)\hat{V}(t')$$

$$+ \int_0^s dt \int_0^t dt' \int_0^{t'} dt'' \hat{V}(t)\hat{V}(t')\hat{V}(t'') + .. \tag{10.10}$$

Now we wish to express all of this in a compact form. For this we introduce the notion of time ordering, which involves the use of Heaviside's unit step function.

But first let us examine the following identity obtained by simply interchanging t and $t^{'}$.

$$\int_0^s dt \int_0^t dt^{'}\theta(t-t')\,\hat{V}(t)\hat{V}(t^{'}) = \int_0^s dt^{'} \int_0^s dt\,\theta(t'-t)\hat{V}(t^{'})\hat{V}(t) \tag{10.11}$$

Here, $\theta(s)$ is Heaviside's step function. If $A = B$, then $A = \frac{1}{2}(A+B)$, hence we may add these two equal quantities and then take half the resultant

$$\int_0^s dt \int_0^s dt^{'}\,\theta(t-t')\hat{V}(t)\hat{V}(t^{'})$$

$$= \frac{1}{2}\int_0^s dt \int_0^s dt^{'}[\theta(t-t')\hat{V}(t)\hat{V}(t^{'}) + \theta(t'-t)\hat{V}(t^{'})\hat{V}(t)]$$

$$= \frac{1}{2}\int_0^s dt \int_0^s dt^{'}T[\hat{V}(t)\hat{V}(t^{'})]. \tag{10.12}$$

By a similar method we may conclude,

$$\int_0^s dt \int_0^s dt^{'} \int_0^s dt^{''}\theta(t-t')\theta(t^{'}-t'')\hat{V}(t)\hat{V}(t^{'})\hat{V}(t^{''})$$

$$= \frac{1}{3!}\int_0^s dt \int_0^s dt^{'} \int_0^s dt^{''}T[\hat{V}(t)\hat{V}(t^{'})\hat{V}(t^{''})]. \tag{10.13}$$

Therefore we may write,

$$S(s) = 1 + \int_0^s \hat{V}(t)dt + \frac{1}{2!}\int_0^s dt \int_0^s dt^{'}T[\hat{V}(t)\hat{V}(t^{'})]$$

$$+\frac{1}{3!}\int_0^s dt \int_0^s dt^{'} \int_0^s dt^{''}T[\hat{V}(t)\hat{V}(t^{'})\hat{V}(t^{''})] + ...$$

$$= T[e^{\int_0^s \hat{V}(t)dt}]. \tag{10.14}$$

Time ordering is an idempotent operation. This means if one tries to time order an already time-ordered expression, there is no effect on the expression. Mathematically, $T[S(s)] = S(s)$. Now we wish to define a certain quantity known as the Green function. Imagine that a N-particle system at time t = 0 is in a eigenstate of the Hamiltonian

$$H = -\int d^3r\ c^{\dagger}(\mathbf{r})\frac{\hbar^2\nabla^2}{2m}c(\mathbf{r}) + \frac{1}{2}\int d^3r \int d^3r^{'}V(|\mathbf{r}-\mathbf{r}^{'}|)c^{\dagger}(\mathbf{r})c^{\dagger}(\mathbf{r}^{'})c(\mathbf{r}^{'})c(\mathbf{r}). \tag{10.15}$$

Let us denote this eigenstate as $|I>$. The subscript I denotes the I-th excited state, and I = 0 corresponds to the ground state. We now consider the time evolution of operators with respect to the time-independent part of the full hamiltonian,

$$\hat{c}(\mathbf{r},t) = e^{\frac{itH}{\hbar}}c(\mathbf{r},0)e^{\frac{-itH}{\hbar}}. \tag{10.16}$$

In the presence of the external time-dependent potential, we have to define the time evolution differently. We formally introduce a time-dependent external potential that is also defined in imaginary time.

$$c(\mathbf{r},t) = U^{\dagger}(t)c(\mathbf{r},0)U(t) \tag{10.17}$$

$$i\hbar\frac{\partial}{\partial t}U(t) = \left(H + \int d^3r\, W(\mathbf{r},t)c^{\dagger}(\mathbf{r})c(\mathbf{r})\right)U(t)$$

$$-i\hbar\frac{\partial}{\partial t}U^{\dagger}(t) = \left(H + \int d^3r\, W(\mathbf{r},t)c^{\dagger}(\mathbf{r})c(\mathbf{r})\right)\hat{U}^{\dagger}(t) \tag{10.18}$$

$U(0) = 1$. Using the time-ordering decomposition we may write,

$$U(t) = e^{\frac{-itH}{\hbar}}\; S(t) \tag{10.19}$$

$$S(t) = T\left(e^{-\frac{i}{\hbar}\int_0^t ds \int d^3r\; W(\mathbf{r},s)\; \hat{c}^{\dagger}(\mathbf{r},s)c(\mathbf{r},s)}\right). \tag{10.20}$$

First we show that $S(t)$ is unitary. This is the same as proving that $U(t)$ is unitary since

$$U^{\dagger}(t)U(t) = S^{\dagger}(t)e^{i\frac{tH}{\hbar}}e^{-i\frac{tH}{\hbar}}S(t) = S^{\dagger}(t)S(t). \tag{10.21}$$

For this we show that $U^{\dagger}(t)U(t)$ is a constant. Then we show that this constant is unity. Consider

$$i\hbar\frac{\partial}{\partial t}(U^{\dagger}(t)U(t)) = i\hbar\frac{\partial U^{\dagger}(t)}{\partial t}U(t) + U^{\dagger}(t)i\hbar\frac{\partial}{\partial t}U(t)$$

$$= -U^{\dagger}(t)\left(H + \int d^3r\, W(\mathbf{r},t)c^{\dagger}(\mathbf{r})c(\mathbf{r})\right)U(t)$$

$$+U^{\dagger}(t)\left(H + \int d^3r\, W(\mathbf{r},t)c^{\dagger}(\mathbf{r})c(\mathbf{r})\right)U(t) = 0. \tag{10.22}$$

But $U^{\dagger}(0)U(0) = \hat{1}$. This means $U^{\dagger}(t)U(t) = \hat{1}$ and $U(t)$ and therefore $U(t)$ and $S(t)$ are unitary. Continuing in this vein we have,

$$c(\mathbf{r},t) = S^{\dagger}(t)e^{i\frac{tH}{\hbar}}c(\mathbf{r},0)e^{-i\frac{tH}{\hbar}}S(t)$$

$$= S^{\dagger}(t)\hat{c}(\mathbf{r},t)S(t). \tag{10.23}$$

We define the following quantities, which are known as the Green functions of the system. This makes use of the grand canonical ensemble of statistical mechanics in the interaction picture,

$$G(\mathbf{r},t;\mathbf{r}',t') = -i\frac{Tr\left(e^{-\beta(H-\mu N)}\;T\;(S(-i\beta\hbar)\hat{c}(\mathbf{r},t)\hat{c}^{\dagger}(\mathbf{r}',t'))\right)}{Tr\left(e^{-\beta(H-\mu N)}S(-i\beta\hbar)\right)}; \tag{10.24}$$

here $Tr[...] = \sum_{I,N=0}^{\infty} < I,N|[...]|I,N >$

$$G(\mathbf{r},t;\mathbf{r}',t')$$

$$= -i\,\frac{\sum_{I,N=0}^{\infty} < I,N|e^{-\beta(\varepsilon_{I,N}-\mu N)}\,T\left(S(-i\beta\hbar)\,\hat{c}(\mathbf{r},t)\hat{c}^{\dagger}(\mathbf{r}',t')\right)|I,N>}{\sum_{I,N=0}^{\infty} e^{-\beta(\varepsilon_{I,N}-\mu N)} < I,N|S(-i\beta\hbar)|I,N>}. \quad (10.25)$$

Here $\varepsilon_{I,N}$ is the energy of the I-th eigenstate of an N-particle system. That is, $H|I,N> = \varepsilon_{I,N}|I,N>$ and N is the total number of particles. We show in the next section that the complicated-looking expression in Eq. (10.24) that at present is not properly motivated is actually a simple and intuitive concept. The time-ordering notion in the expression required clarification. We imagine that all times $t,t'$ lie in the interval $[0,-i\beta\hbar]$. This means we consider the times to be imaginary. This purely mathematical device enables us to exploit a certain kind of periodicity that occurs only in imaginary time. In order to define time ordering we have to determine that some point in this interval is 'greater' or 'less' than other points. We postulate $-i\beta\hbar \geq t,t' \geq 0$. That is, the times $t,t'$ etc. go from the 'smallest' possible value 0 to the largest possible value $-i\beta\hbar$ along the vertical imaginary axis. For particles with statistics '$\sigma$' ($\sigma = +1$ for bosons and $\sigma = -1$ for fermions), we may define the following time-ordering prescriptions:

$$T[c(\mathbf{r},t)c^{\dagger}(\mathbf{r}',t')] = c(\mathbf{r},t)c^{\dagger}(\mathbf{r}',t');\ t > t' \quad (10.26)$$

$$T[c(\mathbf{r},t)c^{\dagger}(\mathbf{r}',t')] = \sigma\, c^{\dagger}(\mathbf{r}',t')c(\mathbf{r},t);\ t' > t. \quad (10.27)$$

Here $t > t'$ means t is closer to $-i\beta\hbar$ than $t'$ and both lie on the imaginary axis between the points 0 and $-i\beta\hbar$.

**Some Simple Cases**: Before studying the difficult problem of mutually interacting particles in an external field, which is what evaluation of the Green function in Eq. (10.24) is meant to do, we first focus on the simple case of non-interacting particles (V = 0) with no external field (W = 0). In this case the Green function is simply,

$$G_0(\mathbf{r},t;\mathbf{r}',t') = -i\frac{[Tre^{-\beta(H_0-\mu N)}T[c(\mathbf{r},t)c^{\dagger}(\mathbf{r}',t')]]}{Tr[e^{-\beta(H_0-\mu N)}]} \quad (10.28)$$

$$c(\mathbf{r},t) = e^{\frac{i}{\hbar}H_0 t}c(\mathbf{r},0)e^{-\frac{i}{\hbar}H_0 t} \quad (10.29)$$

$$H_0 = -\int d^3r\, c^{\dagger}(\mathbf{r})\frac{\hbar^2\nabla^2}{2m}c(\mathbf{r}). \quad (10.30)$$

In order to facilitate progress, we use the momentum state representation (the spatial Fourier transform)

$$c(\mathbf{r}) = \frac{1}{\sqrt{N}}\sum_{\mathbf{k}} e^{i\mathbf{k}\cdot\mathbf{r}}c_{\mathbf{k}}. \quad (10.31)$$

We make a distinction between real space objects, where the position variable is in the parenthesis, and momentum space objects, where the momentum labels are in the subscripts. In three dimensions,

$$\sum_k [....] = \frac{V}{(2\pi)^3} \int d^3r\, [...] \tag{10.32}$$

where V is the volume of the system. Firstly, since $[c(\mathbf{r}), c(\mathbf{r}')]_\sigma = 0$ and $[c(\mathbf{r}), c^\dagger(\mathbf{r}')]_\sigma = \delta(\mathbf{r}-\mathbf{r}')$, we must have $[c_\mathbf{k}, c_{\mathbf{k}'}]_\sigma = 0$ and $[c_\mathbf{k}, c^\dagger_{\mathbf{k}'}]_\sigma = \delta_{\mathbf{k},\mathbf{k}'}$. If we substitute Eq. (10.31) into Eq. (10.30) we get

$$H_0 = \sum_\mathbf{k} \varepsilon_k\, c^\dagger(\mathbf{k}) c(\mathbf{k}), \tag{10.33}$$

where $\varepsilon_k \equiv \frac{\hbar^2 k^2}{2m}$. Also, the time-evolved annihilation operators are

$$c(\mathbf{r},t) = \frac{1}{\sqrt{V}} \sum_\mathbf{k} e^{i\mathbf{k}\cdot\mathbf{r}} e^{\frac{i}{\hbar}H_0 t} c_\mathbf{k} e^{-\frac{i}{\hbar}H_0 t} = \frac{1}{\sqrt{V}} \sum_\mathbf{k} e^{i\mathbf{k}\cdot\mathbf{r}} c_\mathbf{k} e^{-\frac{i}{\hbar}\varepsilon_k t}. \tag{10.34}$$

Let us first assume that in Eq. (10.28) $t > t'$, this means $t'$ is closer to $-i\beta\hbar$ than $t$ and both lie between 0 and $-i\beta\hbar$ on the negative imaginary axis. Further, we set Z $= Tr[e^{-\beta(H_0-\mu N)}]$.

$$G_0^<(\mathbf{r},t;\mathbf{r}',t') = -\sigma \frac{i}{Z} Tr[e^{-\beta(H_0-\mu N)} c^\dagger(\mathbf{r}',t') c(\mathbf{r},t)] \tag{10.35}$$

In other words,

$$G_0^<(\mathbf{r},t;\mathbf{r}',t') = -\sigma\, \frac{i}{Z}\frac{1}{V} \sum_{\mathbf{k},\mathbf{k}'} e^{i(\mathbf{k}\cdot\mathbf{r}-\mathbf{k}'\cdot\mathbf{r}')} e^{-\frac{i}{\hbar}(\varepsilon_k t - \varepsilon_{k'} t')}$$

$$\times \sum_{I,N} < I,N| \left( e^{-\beta\sum_\mathbf{p}(\varepsilon_p-\mu)c^\dagger_\mathbf{p} c_\mathbf{p}} c^\dagger_{\mathbf{k}'} c_\mathbf{k} \right) |I,N> . \tag{10.36}$$

The many-particle states of a non-interacting system are obtained by creating particles with well-defined momenta. For bosons we can have an arbitrary number of particles in each momentum state, but for fermions we can have only one.

$$\frac{1}{Z}\sum_{I,N} < I,N| \left( e^{-\beta\sum_\mathbf{p}(\varepsilon_p-\mu)c^\dagger_\mathbf{p} c_\mathbf{p}} c^\dagger_{\mathbf{k}'} c_\mathbf{k} \right) |I,N>$$

$$= \frac{1}{Z}\sum_{I,N} e^{-\beta\sum_\mathbf{p}(\varepsilon_p-\mu)n_\mathbf{p}} < I,N| c^\dagger_{\mathbf{k}'} c_\mathbf{k} |I,N> \tag{10.37}$$

The states $c_{\mathbf{k}}|I,N>$ and $c_{\mathbf{k}'}|I,N>$ are orthogonal for $\mathbf{k}' \neq \mathbf{k}$ and $n_{\mathbf{p}}|I,N>= N_{\mathbf{p}}|I,N>$. If $\sigma = -1$ (fermions) the above expression becomes

$$\frac{1}{Z}\sum_{I,N} e^{-\beta\sum_p(\varepsilon-\mu)n_p} < I,N|c^{\dagger}_{\mathbf{k}'}c_{\mathbf{k}}|I,N>=$$

$$\delta_{\mathbf{k},\mathbf{k}'}\frac{\sum_{\{N_{\mathbf{p}}\}=0,1} e^{-\beta\sum_{\mathbf{p}\neq\mathbf{k}}(\varepsilon_p-\mu)N_{\mathbf{p}}}\sum_{\{N_{\mathbf{k}}\}=0,1} e^{-\beta(\varepsilon_k-\mu)N_{\mathbf{k}}}N_{\mathbf{k}}}{\sum_{\{N_p\}=0,1} e^{-\beta\sum_{\mathbf{p}\neq\mathbf{k}}(\varepsilon_p-\mu)N_{\mathbf{p}}}\sum_{\{N_{\mathbf{k}}\}=0,1} e^{-\beta(\varepsilon_k-\mu)N_{\mathbf{k}}}}. \tag{10.38}$$

If $\sigma = +1$ (bosons) the above expression becomes

$$\frac{1}{Z}\sum_{I,N} e^{-\beta\sum_{\mathbf{p}}(\varepsilon_{\mathbf{p}}-\mu)n_{\mathbf{p}}} < I,N|\left(c^{\dagger}_{\mathbf{k}}c_{\mathbf{k}}\right)|I,N>=$$

$$\delta_{\mathbf{k},\mathbf{k}'}\frac{\sum_{\{N_{\mathbf{p}}\}=0,1,2,3,..} e^{-\beta\sum_{\mathbf{p}\neq k}(\varepsilon_{\mathbf{p}}-\mu)N_{\mathbf{p}}}\sum_{\{N_{\mathbf{k}}\}=0,1,2,3,..} e^{-\beta(\varepsilon_{\mathbf{k}}-\mu)N_{\mathbf{k}}}N_{\mathbf{k}}}{\sum_{\{N_{\mathbf{p}}\}=0,1,2,3...} e^{-\beta\sum_{\mathbf{p}\neq\mathbf{k}}(\varepsilon_{\mathbf{p}}-\mu)N_{\mathbf{p}}}\sum_{\{N_{\mathbf{k}}\}=0,1,2,3,..} e^{-\beta(\varepsilon_{\mathbf{k}}-\mu)N_{\mathbf{k}}}}. \tag{10.39}$$

Define the following quantities

$$f_B(\lambda) = \sum_{N=0,1,2,3,..} e^{-\lambda N} = \frac{1}{1-e^{-\lambda}} \tag{10.40}$$

$$f_F(\lambda) = \sum_{N=0,1} e^{-\lambda N} = 1 + e^{-\lambda}. \tag{10.41}$$

Then,

$$-\frac{d}{d\lambda}Ln[f_B(\lambda)] = \frac{\sum_{N=0,1,2,3,..} N\, e^{-\lambda N}}{\sum_{N=0,1,2,3,..} e^{-\lambda N}} = \frac{1}{e^{\lambda}-1}$$

$$-\frac{d}{d\lambda}Ln[f_F(\lambda)] = \frac{\sum_{N=0,1} N\, e^{-\lambda N}}{\sum_{N=0,1} e^{-\lambda N}} = \frac{1}{e^{\lambda}+1}. \tag{10.42}$$

If we set $\lambda = \beta(\varepsilon_{\mathbf{k}} - \mu)$, then we can say

$$G_0^<(\mathbf{r},t;\mathbf{r}',t') = -\sigma\, i\frac{1}{V}\sum_{\mathbf{k}} e^{i\mathbf{k}.(\mathbf{r}-\mathbf{r}')}e^{-\frac{i}{\hbar}\varepsilon_k(t-t')}n_\sigma(\mathbf{k}), \tag{10.43}$$

where

$$n_\sigma(k) = \frac{1}{e^{\beta(\varepsilon_k-\mu)} - \sigma}. \tag{10.44}$$

Similarly, we can show that when $t > t'$,

$$G_0^>(\mathbf{r},t;\mathbf{r}',t') = -\frac{i}{V}\sum_{\mathbf{k}} e^{i\mathbf{k}.(\mathbf{r}-\mathbf{r}')}e^{-\frac{i}{\hbar}\varepsilon_k(t-t')}(1+\sigma\, n_\sigma(k)). \tag{10.45}$$

We may now relate these two quantities using a clever observation.

$$G_0^{>}(\mathbf{r},t=-i\beta\hbar;\mathbf{r}^{'},t^{'}) = -i\frac{1}{V}\sum_{\mathbf{k}} e^{i\mathbf{k}.(\mathbf{r}-\mathbf{r}^{'})}e^{\frac{i}{\hbar}\epsilon_k t^{'}}e^{-\beta\epsilon_k}(1+\sigma\, n_{\sigma}(k)) \tag{10.46}$$

Now consider,

$$e^{-\beta\epsilon_k}(1+\sigma\, n_{\sigma}(k)) = e^{-\beta\mu}n_{\sigma}(k). \tag{10.47}$$

Therefore,

$$G_0^{>}(\mathbf{r},t=-i\beta\hbar;\mathbf{r}^{'},t^{'}) = -i\frac{1}{V}\sum_{\mathbf{k}} e^{i\mathbf{k}.(\mathbf{r}-\mathbf{r}^{'})}e^{\frac{i}{\hbar}\epsilon_k t^{'}}e^{-\beta\mu}n_{\sigma}(k)$$

$$= \sigma\, e^{-\beta\mu}G_0^{<}(\mathbf{r},t=0;\mathbf{r}^{'},t^{'}). \tag{10.48}$$

Therefore we have

$$G_0(\mathbf{r},t=-i\beta\hbar;\mathbf{r}^{'},t^{'}) = \sigma\, e^{-\beta\mu}G_0(\mathbf{r},t=0;\mathbf{r}^{'},t^{'}). \tag{10.49}$$

Here, $G_0$ is the time-ordered Green function. The above equation is known as the Kubo-Martin-Schwinger (KMS) boundary condition. We may now rewrite the Green function of the noninteracting system as $G_0(\mathbf{r}-\mathbf{r}^{'},t-t^{'};\mathbf{0},0)$ since we know that it depends only on the differences between the time and position coordinates. A quantity that obeys such a periodicity property may be discrete Fourier transformed. These discrete frequencies are known as Matsubara frequencies. We may write

$$G_0(\mathbf{r}-\mathbf{r}^{'},t-t^{'};\mathbf{0},0)$$
$$= \frac{1}{V}\sum_{\mathbf{k}}\frac{1}{-i\beta\hbar}\sum_{n} e^{i\mathbf{k}.(\mathbf{r}-\mathbf{r}^{'})}e^{-\frac{i}{\hbar}\mu(t-t^{'})}e^{z_n(t-t^{'})}G_0(\mathbf{k},z_n) \tag{10.50}$$

Since $G_0$ has to obey the KMS boundary condition we must have,

$$e^{z_n(-i\beta\hbar-t^{'})} = \sigma e^{z_n(0-t^{'})} \tag{10.51}$$

$$e^{-iz_n\beta\hbar} = \sigma. \tag{10.52}$$

Thus $z_n = \frac{(2n+1)\pi}{\beta\hbar}$ if $\sigma = -1$ and $z_n = \frac{(2n)\pi}{\beta\hbar}$ if $\sigma = 1$. We make some observations about these Matsubara frequencies,

$$\frac{1}{-i\beta\hbar}\sum_{n} e^{z_n(t-t^{'})} = \delta_P(t-t^{'}). \tag{10.53}$$

Here $\delta_P(t-t^{'})$ is the periodic delta function. It has the property $\delta_P(-i\beta\hbar-t^{'}) = \sigma\,\delta_P(0-t^{'})$. For any function f(t) defined in the interval $[0,-i\beta\hbar]$ and obeying the property, $f(-i\beta\hbar) = \sigma\, f(0)$, we have,

$$\int_0^{-i\beta\hbar}\delta_P(t-t^{'})f(t^{'})dt^{'} = f(t). \tag{10.54}$$

We can similarly define a periodic step function

$$\frac{\partial}{\partial t}\theta_P(t-t^{'}) = \delta_P(t-t^{'}) \tag{10.55}$$

or

$$\theta_P(t-t^{'}) = \frac{1}{-i\beta\hbar}\sum_n \frac{e^{z_n(t-t^{'})}}{z_n+i\delta}. \tag{10.56}$$

From Eq. (10.11) and Eq. (10.12) we find,

$$i\hbar\frac{\partial}{\partial t}G_0^{>/<}(\mathbf{r},t;\mathbf{r}^{'},t) = -\frac{\hbar^2\nabla^2}{2m}G_0^{>/<}(\mathbf{r},t;\mathbf{r}^{'},t^{'}). \tag{10.57}$$

But we know that

$$G_0(\mathbf{r},t;\mathbf{r}^{'},t^{'}) = \theta_P(t-t^{'})G_0^{>}(\mathbf{r},t;\mathbf{r}^{'},t^{'}) + \theta_P(t-t^{'})G_0^{<}(\mathbf{r},t;\mathbf{r}^{'},t^{'}). \tag{10.58}$$

Therefore,

$$\begin{aligned} i\hbar\frac{\partial}{\partial t}G_0(\mathbf{r},t;\mathbf{r}^{'},t) &= i\hbar\frac{\partial\theta_P(t-t^{'})}{\partial t}G_0^{>}(\mathbf{r},t;\mathbf{r}^{'},t^{'}) + i\hbar\frac{\partial\theta_P(t^{'}-t)}{\partial t}G_0^{<}(\mathbf{r},t;\mathbf{r}^{'},t^{'}) \\ &+ i\hbar\theta_P(t-t^{'})\frac{\partial G_0^{>}(\mathbf{r},t;\mathbf{r}^{'},t^{'})}{\partial t}) + i\hbar\theta_P(t^{'}-t)\frac{\partial G_0^{<}(\mathbf{r},t;\mathbf{r}^{'},t^{'})}{\partial t} \\ &= i\hbar\delta_P(t-t^{'})[G_0^{>}(\mathbf{r},t;\mathbf{r}^{'},t^{'}) - G_0^{<}(\mathbf{r},t;\mathbf{r}^{'},t^{'})] - \frac{\hbar^2\nabla^2}{2m}G_0(\mathbf{r},t;\mathbf{r}^{'},t^{'}). \end{aligned} \tag{10.59}$$

From Eq.(10.11) and Eq.(10.12) we have

$$\begin{aligned} &[G_0^{>}(\mathbf{r},t;\mathbf{r}^{'},t) - G_0^{<}(\mathbf{r},t;\mathbf{r}^{'},t^{'})] \\ &= -i\frac{1}{V}\sum_{\mathbf{k}} e^{i\mathbf{k}.(\mathbf{r}-\mathbf{r}^{'})}(1+\sigma\, n_\sigma(k)) + \sigma\, i\frac{1}{V}\sum_k e^{i\mathbf{k}.(\mathbf{r}-\mathbf{r}^{'})}n_\sigma(k) \\ &= -i\frac{1}{V}\sum_{\mathbf{k}} e^{i\mathbf{k}.(\mathbf{r}-\mathbf{r}^{'})} = -i\delta(\mathbf{r}-\mathbf{r}^{'}), \end{aligned} \tag{10.60}$$

and

$$\left(i\hbar\frac{\partial}{\partial t} + \frac{\hbar^2\nabla^2}{2m}\right)G_0(\mathbf{r}-\mathbf{r}^{'},t-t^{'};\mathbf{0},0) = \hbar\,\delta_P(t-t^{'})\delta(r-r^{'}). \tag{10.61}$$

Eq. (10.16) into Eq. (10.17), we get a formula for $G(\mathbf{k},z_n)$.

$$\frac{1}{V}\sum_{\mathbf{k}}\frac{1}{-i\beta\hbar}\sum_n e^{i\mathbf{k}.(\mathbf{r}-\mathbf{r}^{'})}e^{-\frac{i}{\hbar}\mu(t-t^{'})}e^{z_n(t-t^{'})}(i\hbar z_n - \varepsilon_k + \mu)G_0(\mathbf{k},z_n)$$

$$= \hbar\, \delta_P(t-t')\delta_P(\mathbf{r}-\mathbf{r}') \tag{10.62}$$

If we choose $(i\hbar z_n - \varepsilon_k + \mu)G_0(\mathbf{k}, z_n) = \hbar$, then the left-hand side becomes,

$$LHS = \frac{1}{V}\sum_k \frac{1}{-i\beta\hbar}\sum_n e^{i\mathbf{k}.(\mathbf{r}-\mathbf{r}')}e^{-\frac{i}{\hbar}\mu(t-t')}e^{z_n(t-t')}\hbar$$

$$= \hbar \int \frac{d^3k}{(2\pi)^3} e^{i\mathbf{k}.(\mathbf{r}-\mathbf{r}')} \frac{1}{-i\beta\hbar}\sum_n e^{z_n(t-t')}e^{-\frac{i}{\hbar}\mu(t-t')}$$

$$= \hbar\delta_P(t-t')\delta_P(\mathbf{r}-\mathbf{r}') = RHS. \tag{10.63}$$

Therefore, the Green function in Fourier space has a particularly simple form,

$$G_0(\mathbf{k}, z_n) = \frac{1}{(i\hbar z_n - \varepsilon_k + \mu)}. \tag{10.64}$$

In the literature it is customary to work in natural units where $\hbar = 1$, which makes the above Green function take the familiar form

$$G_0(\mathbf{k}, z_n) = \frac{1}{(iz_n - \varepsilon_k + \mu)} \tag{10.65}$$

where $z_n = \frac{(2n+1)\pi}{\beta}$ for fermions and $z_n = \frac{2n\pi}{\beta}$ for bosons. When mutual interactions between particles are present, we shall see subsequently that the Green function may always be written as (assuming that the system is translationally invariant both in space and time)

$$G(\mathbf{k}, z_n) = \frac{1}{(iz_n - \varepsilon_k + \mu - \Sigma(\mathbf{k}, iz_n))}. \tag{10.66}$$

Here $\Sigma(\mathbf{k}, iz_n)$ is called the 'self-energy' of the system. In the absence of the self-energy (i.e., for free particles), we may see that the spectral function

$$A_0(\mathbf{k}, \omega) = -2\ Im(G_0(\mathbf{k}, -i\omega + \delta)) \tag{10.67}$$

encodes both the dispersion relation and the lifetime of quasiparticles. In this case,

$$A_0(\mathbf{k}, \omega) = -2Im(G_0(\mathbf{k}, -i\omega + \delta)) = 2\pi\,\delta(-\omega + \varepsilon_k - \mu). \tag{10.68}$$

This says that the energy of quasiparticles is $\varepsilon_k - \mu$, which is nothing but the free particle dispersion. Furthermore, the delta function says that the spectral function is peaked at this energy, which means that the lifetime is infinite at each value of $\omega, \mathbf{k}$. The general relation when self-energy is present is given by

$$A(\mathbf{k}, \omega) = -2Im(G(\mathbf{k}, -i\omega + \delta)). \tag{10.69}$$

One may obtain the momentum distribution of particles from the spectral function using the relation

$$n_\sigma(\mathbf{k}) = \int_{-\infty}^{\infty} \frac{d\omega}{2\pi} \frac{A(\mathbf{k},\omega)}{e^{\beta\omega} - \sigma}, \tag{10.70}$$

where $\sigma = +1$ is for bosons and $\sigma = -1$ is for fermions (here $\beta = (k_B T)^{-1}$ is the inverse temperature). When self energy is present, we may write,

$$G(\mathbf{k}, -i\omega + \delta) = \frac{1}{(\omega - \varepsilon_k + \mu - \Sigma(\mathbf{k}, \omega + i\delta))}. \tag{10.71}$$

Now we write, $\Sigma(\mathbf{k}, \omega + i\delta) = Re(\Sigma(\mathbf{k}, \omega + i\delta)) + i\, Im(\Sigma(\mathbf{k}, \omega + i\delta))$. This means,

$$A(\mathbf{k},\omega) = \frac{-2\, Im(\Sigma(\mathbf{k}, \omega + i\delta))}{(\omega - \varepsilon_k + \mu - Re(\Sigma(\mathbf{k}, \omega + i\delta)))^2 + (Im(\Sigma(\mathbf{k}, \omega + i\delta)))^2}. \tag{10.72}$$

This says that the energy $\omega$ of quasiparticles versus $k$ is given by the solution to $\omega - \varepsilon_k + \mu - Re(\Sigma(\mathbf{k}, \omega + i\delta)) = 0$. These quasiparticles have a lifetime given by $\frac{\pi}{|Im(\Sigma(\mathbf{k}, \omega + i\delta))|}$. The quasiparticles that have $|Im(\Sigma(\mathbf{k}, \omega + i\delta))| = 0$ are infinitely long-lived. The rest are short-lived. There is another illuminating way of writing Eq. (10.66). Consider the zeros of the denominator of that equation, viz. the solutions of $(\omega - \varepsilon_k + \mu - \Sigma(\mathbf{k}, \omega)) = 0$. We denote them as $\omega = \omega_1(\mathbf{k}) + i\omega_2(\mathbf{k})$ where $\omega_{1,2}$ are real. In that case we have a pole (typically). In the vicinity of the pole we may write

$$G(\mathbf{k}, -i\omega) \approx \frac{Z(\mathbf{k})}{(\omega - \omega_1(\mathbf{k}) - i\omega_2(\mathbf{k}))} + G_{reg}(\mathbf{k}, -i\omega), \tag{10.73}$$

where $G_{reg}$ is non-singular at the pole. Here the quantity $Z(\mathbf{k})$ is called the quasiparticle residue and measures the jump in the momentum distribution across the Fermi surface for fermions and the condensate fraction for bosons.

## 10.2 Nonequilibrium Green Functions

The discussion of the previous section was limited to time-independent situations since the evolution operator was assumed to be $e^{-i\frac{t}{\hbar}H}$. When the Hamiltonian depends on time, the evolution operator is $U(t, t_r) \equiv T\left(e^{-i\int_{t_r}^{t} \frac{ds}{\hbar} H(s)}\right)$. Here the time ordering is the natural ordering of real numbers since the time-dependent part may be defined only for real times and may not admit an analytic continuation to imaginary times. Besides, there is no concept of temperature in nonequilibrium systems, so there is no incentive to think of the imaginary time domain. Since temperature has no meaning now, imagine instead, a state $|\Phi>$ with respect to which the expectation value of the product of two fields at two different times is to be evaluated,

$$G(\mathbf{x},t;\mathbf{x}^{'},t^{'}) \equiv <\Phi|T[\psi(\mathbf{x},t)\psi^{\dagger}(\mathbf{x}^{'},t^{'})]|\Phi> . \tag{10.74}$$

More generally, one could imagine an ensemble of such states weighted by some predesignated weight $p(\Phi)$ so that,

$$G(\mathbf{x},t;\mathbf{x}^{'},t^{'}) \equiv \sum_{\Phi} p(\Phi) \ <\Phi|T[\psi(\mathbf{x},t)\psi^{\dagger}(\mathbf{x}^{'},t^{'})]|\Phi> . \tag{10.75}$$

Using the definition of $U(t,t_r)$, the time-evolved operator relative to some reference time $t_r$ may be written as,

$$\psi(\mathbf{x},t) = U^{\dagger}(t,t_r)\psi(\mathbf{x},t_r)U(t,t_r). \tag{10.76}$$

Since we have at the back of our mind some sort of perturbation expansion in the external time-dependent potential, we now introduce the so-called interaction picture where the time-independent part of the Hamiltonian is left out of the evolution operator and absorbed into the field operator on the right-hand side. Let $H(s) = H + V(s)$ be the full time-dependent Hamiltonian and $H$ be the time-independent part. We define

$$U(t,t_r) = T\left(e^{-i\int_{t_r}^{t}\frac{ds}{\hbar}H(s)}\right) = T\left(e^{-i\int_{t_r}^{t}\frac{ds}{\hbar}(H+V(s))}\right)$$

$$= e^{-i\frac{(t-t_r)}{\hbar}H}\ S(t,t_r) \tag{10.77}$$

where,

$$S(t,t_r) = e^{i\frac{(t-t_r)}{\hbar}H}\ T\left(e^{-i\int_{t_r}^{t}\frac{ds}{\hbar}(H+V(s))}\right) \equiv e^{i\frac{(t-t_r)}{\hbar}H}\ U(t,t_r). \tag{10.78}$$

The quantity $S(t,t_r)$ is called the S-matrix. The above is a trivial identity. It also follows that this object is unitary $S^{\dagger}(t,t_r)S(t,t_r) = S(t,t_r)S^{\dagger}(t,t_r) = 1$ since $U$ is unitary also. We may derive a compact albeit formal expression for $S(t,t_r)$ by examining its equation of motion,

$$i\hbar\frac{\partial}{\partial t}S(t,t_r) = -HS(t,t_r) + e^{i\frac{(t-t_r)}{\hbar}H}\ i\hbar\frac{\partial}{\partial t}U(t,t_r)$$

$$= -HS(t,t_r) + e^{i\frac{(t-t_r)}{\hbar}H}\ (H+V(t))U(t,t_r), \tag{10.79}$$

using the evolution equation for $U(t,t_r)$ namely, $i\hbar\frac{\partial}{\partial t}U(t,t_r) = (H+V(t))U(t,t_r)$. Now substitute the inverse relation $U(t,t_r) = e^{-i\frac{(t-t_r)}{\hbar}H}\ S(t,t_r)$ into the above equation to get,

$$i\hbar\frac{\partial}{\partial t}S(t,t_r) = -HS(t,t_r) + e^{i\frac{(t-t_r)}{\hbar}H}\ (H+V(t))e^{-i\frac{(t-t_r)}{\hbar}H}\ S(t,t_r)$$

$$= e^{i\frac{(t-t_r)}{\hbar}H}\, V(t) e^{-i\frac{(t-t_r)}{\hbar}H}\, S(t,t_r). \tag{10.80}$$

Now call $\hat{V}(t) \equiv e^{i\frac{(t-t_r)}{\hbar}H}\, V(t) e^{-i\frac{(t-t_r)}{\hbar}H}$. Then,

$$i\hbar\frac{\partial}{\partial t}S(t,t_r) = \hat{V}(t)\, S(t,t_r), \tag{10.81}$$

which may then be formally solved as

$$S(t,t_r) = T\left(e^{-\frac{i}{\hbar}\int_{t_r}^{t}\hat{V}(s)\, ds}\right), \tag{10.82}$$

since $S(t_r,t_r) \equiv 1$. Going back to Eq. (10.76), we get,

$$\psi(\mathbf{x},t) = S^{\dagger}(t,t_r) e^{i\frac{(t-t_r)}{\hbar}H} \psi(\mathbf{x},t_r) e^{-i\frac{(t-t_r)}{\hbar}H}\, S(t,t_r)$$

$$= S^{\dagger}(t,t_r)\hat{\psi}(\mathbf{x},t)S(t,t_r), \tag{10.83}$$

where

$$\hat{\psi}(\mathbf{x},t) = e^{i\frac{(t-t_r)}{\hbar}H}\psi(\mathbf{x},t_r)e^{-i\frac{(t-t_r)}{\hbar}H}. \tag{10.84}$$

In general,

$$S(t,t') = T\left(e^{-\frac{i}{\hbar}\int_{t'}^{t}\hat{V}(s)\, ds}\right). \tag{10.85}$$

Clearly, $S(t,t) = 1$. The above function obeys the equation of motion,

$$i\hbar\frac{\partial}{\partial t}S(t,t') = \hat{V}(t)\, S(t,t') \tag{10.86}$$

$$i\hbar\frac{\partial}{\partial t'}S(t,t') = -S(t,t')\hat{V}(t'), \tag{10.87}$$

and its Hermitian conjugate obeys

$$-i\hbar\frac{\partial}{\partial t}S^{\dagger}(t,t') = S^{\dagger}(t,t')\, \hat{V}(t). \tag{10.88}$$

This is unitary $S^{\dagger}(t,t')S(t,t') = 1$. To prove this, we first show that $S^{\dagger}(t,t')S(t,t')$ is independent of $t$ by differentiating with respect to $t$. Then we choose $t = t'$ since it is independent of $t$ and the result follows.

$$i\hbar\frac{\partial}{\partial t}(S^{\dagger}(t,t')S(t,t'))$$

$$= (S^{\dagger}(t,t')i\hbar\frac{\partial}{\partial t}S(t,t')) + (i\hbar\frac{\partial}{\partial t}S^{\dagger}(t,t')S(t,t'))$$

$$= (S^{\dagger}(t,t')\hat{V}(t)\, S(t,t')) - (S^{\dagger}(t,t')\, \hat{V}(t)S(t,t')) = 0 \tag{10.89}$$

$S^{\dagger}(t,t')S(t,t') = S^{\dagger}(t',t')S(t',t') = 1$. This object also obeys group properties,

$$S(t,t')S(t',t'') = S(t,t''). \tag{10.90}$$

To show this, differentiate the left-hand side of the above identity with respect to $t'$.

$$\begin{aligned}
&\frac{\partial}{\partial t'}S(t,t')S(t',t'') \\
&= \left(\frac{\partial S(t,t')}{\partial t'}\right) S(t',t'') + S(t,t') \left(\frac{\partial S(t',t'')}{\partial t'}\right) \\
&= -S(t,t')\,\hat{V}(t')\,S(t',t'') + S(t,t')\,\hat{V}(t')\,S(t',t'') = 0
\end{aligned} \tag{10.91}$$

This means we may set $t' = t$ with impunity, so that $S(t,t')S(t',t'') = S(t,t'')$. Lastly $S^{\dagger}(t,t') = S(t',t)$. In order to do perturbation theory we have to rewrite Eq. (10.74) in the interaction picture. This means replace the fields in this equation by fields evolving according to the time-independent part of the Hamiltonian, namely using Eq. (10.84). The simplest way to do this is by proving the following result.

Lemma:

$$T(\psi(\mathbf{x},t)\psi(\mathbf{x}',t')) = S(-\infty,\infty)\; T(\hat{\psi}(\mathbf{x},t)\hat{\psi}^{\dagger}(\mathbf{x}',t')S(\infty,-\infty)). \tag{10.92}$$

The proof involves starting from the right-hand side and reproducing the left-hand side. Let $t > t'$. In this case (even otherwise) we may always write $S(\infty,-\infty) = S(\infty,t)S(t,t')S(t',-\infty)$. Inserting this into Eq. (10.92) we get

$$\begin{aligned}
&S(-\infty,\infty)\; T(\hat{\psi}(\mathbf{x},t)\hat{\psi}^{\dagger}(\mathbf{x}',t')S(\infty,-\infty)) \\
&= S(-\infty,\infty)\; T(\hat{\psi}(\mathbf{x},t)\hat{\psi}^{\dagger}(\mathbf{x}',t')S(\infty,t)S(t,t')S(t',-\infty)).
\end{aligned} \tag{10.93}$$

As per the diagram in Figure 10.1, the term $S(\infty,t)$ is on the extreme left, followed by $\hat{\psi}(\mathbf{x},t)$ then $S(t,t')$, followed by $\hat{\psi}^{\dagger}(\mathbf{x}',t')$ and lastly $S(t',-\infty)$. Also, since $\hat{\psi}(\mathbf{x},t)$ and $\hat{\psi}^{\dagger}(\mathbf{x}',t')$ have not exchanged places, there is no need for the statistical permutation parameter we have been calling $\sigma$ ($\sigma = +1$ for bosons and $\sigma = -1$ for fermions).

$$\begin{aligned}
&S(-\infty,\infty)\; T(\hat{\psi}(\mathbf{x},t)\hat{\psi}^{\dagger}(\mathbf{x}',t')S(\infty,-\infty)) \\
&= S(-\infty,\infty)\; S(\infty,t)\hat{\psi}(\mathbf{x},t)S(t,t')\hat{\psi}^{\dagger}(\mathbf{x}',t')S(t',-\infty) \\
&= S(-\infty,t)\hat{\psi}(\mathbf{x},t)S(t,t')\hat{\psi}^{\dagger}(\mathbf{x}',t')S(t',-\infty)
\end{aligned} \tag{10.94}$$

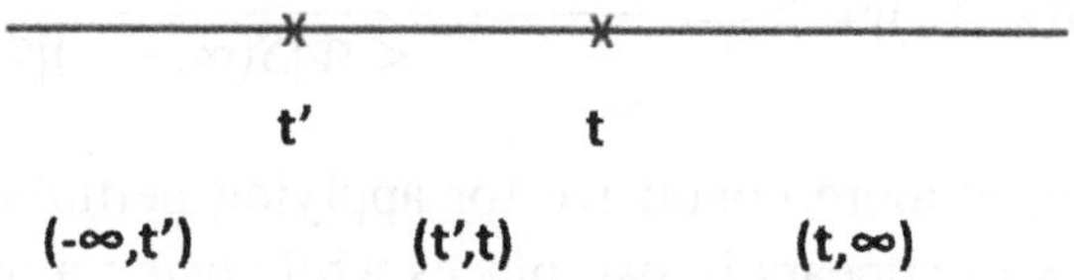


Figure 10.1: The intervals involved when $t > t'$ are shown.

Now we may infer from the earlier discussion that

$$\hat{\psi}(\mathbf{x},t) = S(t,t_r)\psi(\mathbf{x},t)S(t_r,t). \tag{10.95}$$

Now we set $t_r = -\infty$ since it is up to us to choose the reference time $t_r$. This expression is now substituted into Eq. (10.94) to give

$$\begin{aligned} &S(-\infty,\infty)\, T(\hat{\psi}(\mathbf{x},t)\hat{\psi}^{\dagger}(\mathbf{x}',t')S(\infty,-\infty)) \\ &= S(-\infty,t)S(t,-\infty)\psi(\mathbf{x},t)S(-\infty,t)S(t,t')S(t',-\infty) \\ &\quad \times \psi^{\dagger}(\mathbf{x}',t')S(-\infty,t')S(t',-\infty) = \psi(\mathbf{x},t)\psi^{\dagger}(\mathbf{x}',t') \end{aligned} \tag{10.96}$$

which proves the Lemma in Eq. (10.92) for $t > t'$. The proof for $t < t'$ is similar and left to the reader as an exercise. Now the average with respect to some two states may be written as

$$\begin{aligned} &< \Phi|T(\psi(\mathbf{x},t)\psi(\mathbf{x}',t'))|\Phi > \\ &=< \Phi|S(-\infty,\infty)\, T(\hat{\psi}(\mathbf{x},t)\hat{\psi}^{\dagger}(\mathbf{x}',t')S(\infty,-\infty))|\Phi > . \end{aligned} \tag{10.97}$$

As it stands, the above expression is not particularly convenient since the S-matrix appears in two different places. Two different perturbation series will have to be performed. There is one special form of the time-dependent potential that allows a simpler interpretation. This is the adiabatic equilibrium assumption. In this discussion we set the time-dependent potential to be weakly time dependent for most times except in the remote past and the remote future where the potential exponentially decays to zero. Mathematically, this means, $V(t) = e^{-\varepsilon|t|}V_0$, where $\varepsilon \to 0$ and $V_0$ is time independent. In this case we may assert that if the state in the remote past viz. $|\Phi>$ was non-degenerate, then the state in the remote future, namely, $S(\infty,-\infty)|\Phi>$ has to be essentially the same as $|\Phi>$ since the potential has been switched on and off adiabatically. Since both states are normalized, this means

$$S(\infty,-\infty)|\Phi>= e^{i\theta}|\Phi> . \tag{10.98}$$

Therefore,

$$< \Phi|S(-\infty,\infty) = e^{-i\theta} < \Phi| \;;\; e^{i\theta} =< \Phi|S(\infty,-\infty)|\Phi >$$

$$< \Phi|T(\psi(\mathbf{x},t)\psi(\mathbf{x}',t'))|\Phi >_{eq} = \frac{< \Phi|T(\hat{\psi}(\mathbf{x},t)\hat{\psi}^{\dagger}(\mathbf{x}',t')S(\infty,-\infty))|\Phi >}{< \Phi|S(\infty,-\infty)|\Phi >}. \tag{10.99}$$

The above expression is more conducive for applying perturbation theory. Even though the S-matrix still appears in two places while using a perturbation expansion, the denominator of the above expression cancels with terms in the numerator. To see this, write

$$S(\infty,-\infty) = 1 + \lambda S_1 + \lambda^2 S_2 + \ldots \tag{10.100}$$

where $S_1$ contains terms linear in the time-dependent potential, $S_2$ contains terms quadratic in this potential, and so on. Finally we set $\lambda = 1$ as this is just a book-keeping device.

$$< \Phi|T(\psi(\mathbf{x},t)\psi(\mathbf{x}',t'))|\Phi >_{eq}$$

$$= \frac{< \Phi|T(\hat{\psi}(\mathbf{x},t)\hat{\psi}^{\dagger}(\mathbf{x}',t')(1+\lambda S_1 + \lambda^2 S_2 + ..))|\Phi >}{< \Phi|(1+\lambda S_1 + \lambda^2 S_2 + ..)|\Phi >}$$

$$=< \Phi|T(\hat{\psi}(\mathbf{x},t)\hat{\psi}^{\dagger}(\mathbf{x}',t')|\Phi >$$

$$+\lambda(< T(\hat{\psi}(\mathbf{x},t)\hat{\psi}^{\dagger}(\mathbf{x}',t')S_1) > - < T(\hat{\psi}(\mathbf{x},t)\hat{\psi}^{\dagger}(\mathbf{x}',t')) >< S_1 >) + \ldots \tag{10.101}$$

One may see that powers of $\lambda$ in this expansion only involve connected terms. This means that unless $S_1, S_2, ..$ are linked to the fields, the terms vanish. In the general case of time-dependent non-adiabatic switching, the Lemma in Eq. (10.92) involves two S-matrices.

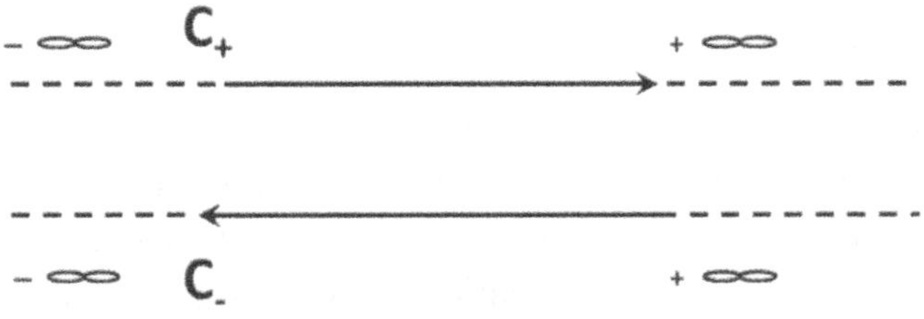


Figure 10.2: This shows the the time loop used in the definition of the nonequilibrium Green function.

This is not such a big deal of course, since in the end, equilibrium or otherwise, all one has to calculate are the particle propagator $< \psi(\mathbf{x},t)\psi^{\dagger}(\mathbf{x}',t') >$ and the hole propagator $< \psi^{\dagger}(\mathbf{x}',t')\psi(\mathbf{x},t) >$. The time evolution differential equations for each one will involve the other as well and when interactions are present, two-particle and higher-order Green functions are also present in conjunction. These have to be solved in a coupled manner. The insistence of having just one S-matrix in the interaction picture stems from the need to do perturbation more cleanly and as we shall soon see, doing this also enables the derivation of a function differential equation for the Green function with a source known as the Schwinger-Dyson equation, which is a closed equation for just the one-particle Green function without involving higher-order Green functions. At least for this reason we wish to recast Eq. (10.92) in a form that involves only one S-matrix. The way to do this is to somehow insert the S-matrix on the extreme left of Eq. (10.92) into the time ordering. This cannot be done unless we take the point of view that time ordering is along two branches—an upper branch and a lower branch—so that each time on the lower branch is postulated to be larger than each time on the upper branch. Thus, the two branches put together form a loop called $C \equiv C_{+} \bigcup C_{-}$ and $C_{+} : -\infty \rightarrow +\infty$ and $C_{-} : +\infty \rightarrow -\infty$. If $t \in C_{-}$ and $t^{'} \in C_{+}$, then $t > t^{'}$ always. However for $t, t^{'} \in C_{+}$, the numerically larger one is also larger in the sense of loop ordering, but for $t, t^{'} \in C_{-}$ the numerically smaller one is larger in the sense of loop ordering. This state of affairs is depicted in the diagram (Fig. 10.2). In this case we may regard the S-matrix on the extreme left as containing times on the lower branch and the times inside the time ordering in Eq. (10.92) as being on the upper branch so that the former are always larger in the C-loop sense than the latter. This allows us to take the S-matrix on the extreme left inside the time ordering to get,

$$
\begin{aligned}
& T_C(\psi(\mathbf{x},t)\psi(\mathbf{x}^{'},t^{'})) \\
&= T(S_{C_-}(-\infty,\infty)S_{C_+}(\infty,-\infty)\ \hat{\psi}(\mathbf{x},t)\hat{\psi}^{\dagger}(\mathbf{x}^{'},t^{'})) \\
&\equiv T(S(C)\ \hat{\psi}(\mathbf{x},t)\hat{\psi}^{\dagger}(\mathbf{x}^{'},t^{'})). \qquad (10.102)
\end{aligned}
$$

The S-matrix on the combined loop is $S(C) \equiv S_{C_-}(-\infty,\infty)S_{C_+}(\infty,-\infty)$, which is traversed in the clockwise direction. Typically, one is interested in weighted averages. For the purposes of the next section, we use the definition involving the grand canonical ensemble,

$$
G(\mathbf{x},t;\mathbf{x}^{'},t^{'}) = \frac{Tr\left(e^{-\beta(H-\mu N)}\ T(S(C)\ \hat{\psi}(\mathbf{x},t)\hat{\psi}^{\dagger}(\mathbf{x}^{'},t^{'}))\right)}{Tr\left(e^{-\beta(H-\mu N)}\ S(C)\right)}. \qquad (10.103)
$$

The reason for the denominator will be made clear in the next section.

## 10.3 Schwinger-Dyson Equations

In this section, we consider a collection of particles that are mutually interacting via a two-body potential. We also assume that an external potential is present. This external potential defined in imaginary time in the interval $[0, -i\beta\hbar]$ may be set to zero at the end to make the whole system translationally invariant in space and time so that we may then extract the self-energy function of the quasi particles, which is the real intention of this section. Alternatively, one could regard the external potential as present in real time and use the closed time loop approach discussed earlier.

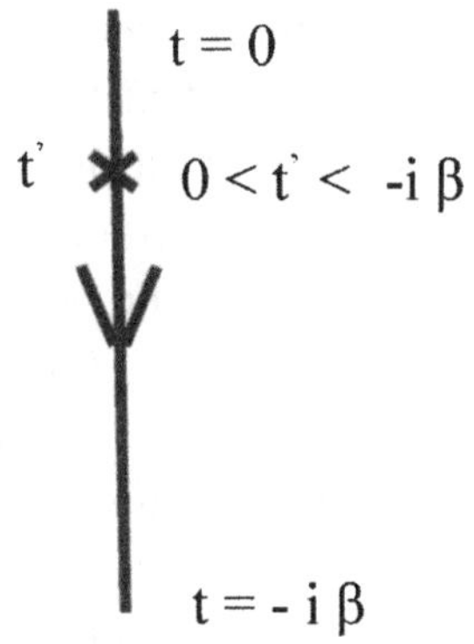


Figure 10.3: In the Matsubara formalism, the times are assumed to lie in the interval shown on the imaginary axis. For time ordering, we regard times closer to $t = 0$ as smaller and those closer to $t = -i\beta$ as being larger.

We shall see that the device of introducing an external potential leads to an exact closed functional differential equation for the Green function. A necessarily approximate scheme has to be invoked to solve this, which will yield the necessary self-energy function. The Hamiltonian of such a collection is written as $H = H_0 + H_{ext}(t)$

$$H_0 = \int d^3r\, c^{\dagger}(\mathbf{r}) \left( \frac{-\hbar^2\nabla^2}{2m} + \frac{1}{2}\int d^3r'\, V(\mathbf{r}-\mathbf{r}')c^{\dagger}(\mathbf{r}')c(\mathbf{r}') \right) c(\mathbf{r})$$

$$H_{ext}(t) = \int d^3r\, c^{\dagger}(\mathbf{r})c(\mathbf{r})U(\mathbf{r},t). \tag{10.104}$$

The external potential could be time dependent. One could take the point of view that this external potential is simply a device introduced to write down a closed equation for a Green function of an interacting system. In this case we are free to dictate its properties since in the end it will be set to zero. In such a case we shall assume that it is legitimate to change the real times to imaginary times and $U(\mathbf{r},t)$ is defined in imaginary time in the interval $[0, -i\beta\hbar]$. On the other hand, it is possible that there is a nonequilibrium external potential defined at real times like we did in the earlier section, in this case, the integral from $[0, -i\beta\hbar]$ shall be replaced by the contour described in the earlier section. In this section, we focus on the former case and refer the reader to the references for the latter case (e.g., see the paper on Nonequilibrium Dynamical Screening). (in this section we set $\hbar = 1$). We wish to derive a differential equation for the one-particle Green function. This quantity is to be calculated by solving the said equation by supplying appropriate boundary conditions, namely the KMS boundary conditions (in the present context). To this end, let us study first the time evolution of the annihilation operator in the Heisenberg picture.

$$i\frac{\partial}{\partial t}c(\mathbf{r},t) = [c(\mathbf{r},t),H] =$$

$$\left(\frac{-\hbar^2\nabla^2}{2m} + \int d^3r'\, V(\mathbf{r}-\mathbf{r}')c^\dagger(\mathbf{r}',t)c(\mathbf{r}',t)\right)\, c(\mathbf{r},t) + c(\mathbf{r},t)U(\mathbf{r},t) \quad (10.105)$$

We may define the following new time-evolved operators with respect to only $H_0$.

$$i\frac{\partial}{\partial t}\hat{c}(\mathbf{r},t) = [\hat{c}(\mathbf{r},t),H_0] =$$

$$\left(\frac{-\hbar^2\nabla^2}{2m} + \int d^3r' V(\mathbf{r}-\mathbf{r}')\hat{c}^\dagger(\mathbf{r}',t)\hat{c}(\mathbf{r}',t)\right)\hat{c}(\mathbf{r},t) \quad (10.106)$$

We now define the following Green functions (in equilibrium the time ordering is on the imaginary interval, but for nonequilibrium systems it is on the closed time loop. Here we restrict ourselves to the former case):

$$G_{full}(\mathbf{r}_1,t_1;\mathbf{r}_2,t_2;U) \equiv G_{full}(1,2;U) = -i < T(c(\mathbf{r}_1,t_1)c^\dagger(\mathbf{r}_2,t_2)) > \quad (10.107)$$

where,

$$< T(...) >= \frac{Tr(T(e^{i\int_0^{-i\beta} dt(\mu N - H(t))})(...))}{Tr(T(e^{i\int_0^{-i\beta} dt(\mu N - H(t))}))}. \quad (10.108)$$

The above definition is the intuitive definition of a particle or a hole propagator. In the earlier sections we had also introduced the Green function

$$G'_{full}(\mathbf{r}_1,t_1;\mathbf{r}_2,t_2;U) \equiv G'_{full}(1,2;U)$$

$$= -i\frac{< T\ S(-i\beta,0)\ \hat{c}(\mathbf{r}_1,t_1)\hat{c}^\dagger(\mathbf{r}_2,t_2) >_0}{< T\ S(-i\beta,0) >_0}, \quad (10.109)$$

where

$$S(-i\beta,0) = e^{-i\int_0^{-i\beta} dt \int d^3r\ U(\mathbf{r},t)\ \hat{c}^\dagger(\mathbf{r},t)\hat{c}(\mathbf{r},t)} \quad (10.110)$$

and

$$< T(...) >_0 = \frac{Tr(T(e^{i\int_0^{-i\beta} dt\ (\mu N - H_0)})(...))}{Tr(T(e^{i\int_0^{-i\beta} dt\ (\mu N - H_0)}))} = \frac{Tr(e^{-\beta(H_0-\mu N)}T(...))}{Tr(e^{-\beta(H_0-\mu N)})}. \quad (10.111)$$

Note the compact notation $1 \equiv (\mathbf{r}_1,t_1)$ etc. We remarked that this latter definition is not well motivated. We remedy that here by proving the equivalence of the natural definition in Eq. (10.107) and the version from the interaction picture in Eq. (10.109).

**Theorem** : $G'_{full}(1,2;U) \equiv G_{full}(1,2;U)$.

To prove this we first observe that when $-i\beta > t_1 > t_2 > 0$,

$$G'_{full,>}(1,2;U) = -i\frac{< T\ S(-i\beta,0)\ \hat{c}(\mathbf{r}_1,t_1)\hat{c}^\dagger(\mathbf{r}_2,t_2) >_0}{< T\ S(-i\beta,0) >_0}$$

$$= -i\frac{< T\ S(-i\beta,t_1)S(t_1,t_2)S(t_2,0)\ \hat{c}(\mathbf{r}_1,t_1)\hat{c}^\dagger(\mathbf{r}_2,t_2) >_0}{< T\ S(-i\beta,0) >_0}$$

$$= -i\frac{< S(-i\beta,t_1)\hat{c}(\mathbf{r}_1,t_1)S(t_1,t_2)\hat{c}^\dagger(\mathbf{r}_2,t_2)S(t_2,0)\ >_0}{< T\ S(-i\beta,0) >_0}. \tag{10.112}$$

But,

$$\hat{c}(\mathbf{r}_1,t_1) = S(t_1,0)c(\mathbf{r}_1,t_1)S(0,t_1) \tag{10.113}$$

etc. Therefore,

$$G'_{full,>}(1,2;U) = -i\frac{< T\ S(-i\beta,0)\ \hat{c}(\mathbf{r}_1,t_1)\hat{c}^\dagger(\mathbf{r}_2,t_2) >_0}{< T\ S(-i\beta,0) >_0}$$

$$= -i\frac{< S(-i\beta,t_1)S(t_1,0)c(\mathbf{r}_1,t_1)S(0,t_1)S(t_1,t_2)S(t_2,0)c^\dagger(\mathbf{r}_2,t_2)S(0,t_2)S(t_2,0)\ >_0}{< T\ S(-i\beta,0) >_0}$$

$$= -i\frac{< S(-i\beta,0)c(\mathbf{r}_1,t_1)c^\dagger(\mathbf{r}_2,t_2) >_0}{< T\ S(-i\beta,0) >_0}$$

$$= -i\frac{Tr(e^{-\beta(H_0-\mu N)}\ S(-i\beta,0)c(\mathbf{r}_1,t_1)c^\dagger(\mathbf{r}_2,t_2))}{Tr(e^{-\beta(H_0-\mu N)}\ S(-i\beta,0))}. \tag{10.114}$$

But,

$$e^{-\beta(H_0-\mu N)}\ S(-i\beta,0) = Te^{i\int_0^{-i\beta}dt(\mu N - H(t))}. \tag{10.115}$$

Hence,

$$G'_{full,>}(1,2;U) = -i\frac{Tr(Te^{i\int_0^{-i\beta}dt(\mu N-H(t))}\ c(\mathbf{r}_1,t_1)c^\dagger(\mathbf{r}_2,t_2))}{Tr(Te^{i\int_0^{-i\beta}dt(\mu N-H(t))})} = G_{full,>}(1,2;U). \tag{10.116}$$

Similarly, when $t_1 < t_2$ we have,

$$G'_{full,<}(1,2;U) = -i\sigma\ \frac{< S(-i\beta,0)c^\dagger(\mathbf{r}_2,t_2)c(\mathbf{r}_1,t_1) >_0}{< T\ S(-i\beta,0) >_0}. \tag{10.117}$$

This is same as $G_{full}(1,2;U)$ for reasons just explained and the proof is complete. Now consider the equation of motion,

$$i\frac{\partial}{\partial t_1}c(\mathbf{r}_1,t_1) = \left(-\frac{\nabla_1^2}{2m} + \int d^3r'\,V(\mathbf{r}_1 - \mathbf{r}')c^\dagger(\mathbf{r}',t_1)c(\mathbf{r}',t_1)\right)c(\mathbf{r}_1,t_1)$$

$$+c(\mathbf{r}_1,t_1)U(\mathbf{r}_1,t_1). \tag{10.118}$$

This means,

$$i\frac{\partial}{\partial t_1} < -ic(\mathbf{r}_1,t_1)c^\dagger(\mathbf{r}_2,t_2) >= -\frac{\nabla_1^2}{2m} < -ic(\mathbf{r}_1,t_1)c^\dagger(\mathbf{r}_2,t_2) >$$
$$+\int d^3r' V(\mathbf{r}_1-\mathbf{r}') < -ic^\dagger(\mathbf{r}',t_{1+})c(\mathbf{r}',t_{1+})c(\mathbf{r}_1,t_1)c^\dagger(\mathbf{r}_2,t_2) >$$
$$+ < -ic(\mathbf{r}_1,t_1)c^\dagger(\mathbf{r}_2,t_2) > U(\mathbf{r}_1,t_1) \tag{10.119}$$

and

$$i\frac{\partial}{\partial t_1} < -isc^\dagger(\mathbf{r}_2,t_2)c(\mathbf{r}_1,t_1) >= -\frac{\nabla_1^2}{2m} < -isc^\dagger(\mathbf{r}_2,t_2)c(\mathbf{r}_1,t_1) >$$
$$+\int d^3r' V(\mathbf{r}_1-\mathbf{r}') < -isc^\dagger(\mathbf{r}_2,t_2)c^\dagger(\mathbf{r}',t_1)c(\mathbf{r}',t_1)c(\mathbf{r}_1,t_1) >$$
$$+ < -isc^\dagger(\mathbf{r}_2,t_2)c(\mathbf{r}_1,t_1) > U(\mathbf{r}_1,t_1). \tag{10.120}$$

But we know,

$$G_{full}(1,2;U) = \theta(t_1-t_2) < -ic(\mathbf{r}_1,t_1)c^\dagger(\mathbf{r}_2,t_2) >$$
$$+\theta(t_2-t_1) < -isc^\dagger(\mathbf{r}_2,t_2)c(\mathbf{r}_1,t_1) >$$
$$= \theta(t_1-t_2)G^>_{full}(1,2;U) + \theta(t_2-t_1)G^<_{full}(1,2;U). \tag{10.121}$$

Thus,

$$i\frac{\partial}{\partial t_1}G_{full}(1,2;U) = \delta(t_1-t_2)\ (c(\mathbf{r}_1,t_1)c^\dagger(\mathbf{r}_2,t_1) + sc^\dagger(\mathbf{r}_2,t_1)c(\mathbf{r}_1,t_1))$$
$$+\theta(t_1-t_2)i\frac{\partial}{\partial t_1}G^>_{full}(1,2;U) + \theta(t_2-t_1)\ i\frac{\partial}{\partial t_1}G^<_{full}(1,2;U). \tag{10.122}$$

Combining all this we may write (here $\delta(1-2) = \delta(t_1-t_2)\delta(\mathbf{r}_1-\mathbf{r}_2)$),

$$\left(i\frac{\partial}{\partial t_1} + \frac{\nabla_1^2}{2m} - U(\mathbf{r}_1,t_1)\right) G_{full}(1,2;U)$$
$$= \delta(1-2) + \int d^3r' V(\mathbf{r}_1-\mathbf{r}')[< -iTc^\dagger(\mathbf{r}',t_{1+})c(\mathbf{r}',t_{1+})c(\mathbf{r}_1,t_1)c^\dagger(\mathbf{r}_2,t_2) >]. \tag{10.123}$$

Since $G_{full}(1,2;U) = G'_{full}(1,2;U)$, we must have,

$$\left(i\frac{\partial}{\partial t_1} + \frac{\nabla_1^2}{2m} - U(\mathbf{r}_1,t_1)\right) G_{full}(1,2;U)$$

$$= \delta(1-2) + \int d^3r' V(\mathbf{r}_1 - \mathbf{r}') \frac{< -iT\, S\, \hat{c}^\dagger(\mathbf{r}', t_1)\hat{c}(\mathbf{r}', t_1)\hat{c}(\mathbf{r}_1, t_1)\hat{c}^\dagger(\mathbf{r}_2, t_2) >_0}{< TS >_0}. \tag{10.124}$$

Consider the following identity,

$$i\frac{\delta}{\delta U(\mathbf{r}', t_{1+})} G_{full}(1,2;U) = \frac{< -iT\, S\, \hat{c}^\dagger(\mathbf{r}', t_{1+})\hat{c}(\mathbf{r}', t_{1+})\hat{c}(\mathbf{r}_1, t_1)\hat{c}^\dagger(\mathbf{r}_2, t_2) >_0}{< TS >_0}$$

$$- \frac{< -iT\, S\, \hat{c}(\mathbf{r}_1, t_1)\hat{c}^\dagger(\mathbf{r}_2, t_2) >_0}{< TS >_0} \frac{< T\, S\, \hat{c}^\dagger(\mathbf{r}', t_1)\hat{c}(\mathbf{r}', t_1) >_0}{< TS >_0}. \tag{10.125}$$

We may substitute this into the equation for $G_{full}$ to arrive at the Schwinger-Dyson equation for $G_{full}$ (henceforth simply called $G$):

$$\left( i\frac{\partial}{\partial t_1} + \frac{\nabla_1^2}{2m} - U_{eff}(\mathbf{r}_1, t_{1+}) \right) G(1,2;U)$$

$$= \delta(1-2) + \int d^3r' V(\mathbf{r}_1 - \mathbf{r}') i \frac{\delta}{\delta U(\mathbf{r}', t_1)} G(1,2;U), \tag{10.126}$$

where

$$U_{eff}(\mathbf{r}_1, t_1) = U(\mathbf{r}_1, t_1) + \int d^3r'\ V(\mathbf{r}_1 - \mathbf{r}')\ \frac{< TS\hat{c}^\dagger(\mathbf{r}', t_{1+})\hat{c}(\mathbf{r}', t_1) >_0}{< TS >_0}. \tag{10.127}$$

The Schwinger-Dyson equation is a functional differential equation. The mathematics of such equations is ill-developed. The usual method for solving such equations is by introducing the concept of self-energies. First, some notation. By $\int d1$ we mean, $\int_0^{-i\beta} dt_1 \int d^3r_1$ . We first introduce the concept of inverse of F:

$$\int d3\, F(1,3) F^{-1}(3,2) = \delta(1-2). \tag{10.128}$$

We make a useful observation from this identity. Just by differentiating with respect to $U(4)$ we get,

$$\int d3\ \frac{\delta F(1,3)}{\delta U(4)} F^{-1}(3,2) + \int d3\, F(1,3) \frac{\delta F^{-1}(3,2)}{\delta U(4)} = 0. \tag{10.129}$$

Multiply by $F(2,1')$ and integrate with respect to 2, we get,

$$\frac{\delta F(1,1')}{\delta U(4)} = -\int d3 \int d2\, F(1,3) \frac{\delta F^{-1}(3,2)}{\delta U(4)} F(2,1'). \tag{10.130}$$

Define $G_0$ to be the solution to,

$$\left( i\frac{\partial}{\partial t_1} + \frac{\nabla_1^2}{2m} - U_{eff}(\mathbf{r}_1, t_1) G_0(1,2,U) \right) = \delta(1-2). \tag{10.131}$$

Then the self-energy function $\Sigma(1,2;U)$ is defined as

$$\Sigma(1,2;U) = G_0^{-1}(1,2;U) - G^{-1}(1,2;U), \tag{10.132}$$

where the inverses are in the matrix sense. Postmultiply Eq. (10.126) by $G^{-1}(2,1';U)$ and integrate over 2 to get,

$$\left(i\frac{\partial}{\partial t_1} + \frac{\nabla_1^2}{2m} - U_{eff}(\mathbf{r}_1,t_1)\right)\delta(1-1')$$

$$= G^{-1}(1,1';U) + \int d^3r' V(\mathbf{r}_1 - \mathbf{r}') \int d2 \left( i\frac{G(1,2;U)}{\delta U(\mathbf{r}',t_{1+})} G^{-1}(2,1';U)\right). \tag{10.133}$$

We know that,

$$\left(i\frac{\partial}{\partial t_1} + \frac{\nabla_1^2}{2m} - U_{eff}(\mathbf{r}_1,t_1)\right)\delta(1-1') = G_0^{-1}(1,1',U). \tag{10.134}$$

Now we write,

$$G^{-1}(2,1;U) = G_0^{-1}(2,1;U) - \Sigma(1,2;U). \tag{10.135}$$

This can also be written in the form below, which is called the Dyson equation:

$$G(1,2;U) = G_0(1,2;U) + \int d3 \int d2\; G_0(1,2;U)\Sigma(2,3;U)G(3,1;U). \tag{10.136}$$

By making these substitutions we conclude that,

$$\Sigma(1,1';U) = \int d^3r' V(\mathbf{r}_1 - \mathbf{r}') \int d2 \left( i\frac{G(1,2;U)}{\delta U(\mathbf{r}',t_{1+})} G^{-1}(2,1';U)\right). \tag{10.137}$$

To evaluate this, we observe that since $\int d2\; G(1,2;U)G^{-1}(2,1';U) = 1$, we may differentiate this to get,

$$\int d2 \frac{\delta G(1,2;U)}{\delta U(\mathbf{r}',t_{1+})} G^{-1}(2,1';U) = -\int d2\; G(1,2;U) \frac{\delta G^{-1}(2,1';U)}{\delta U(\mathbf{r},t_{1+})}. \tag{10.138}$$

Therefore,

$$\Sigma(1,1';U)$$

$$= -\int d^3r' V(\mathbf{r}_1 - \mathbf{r}') \int d2 \left( i\, G(1,2;U) \frac{\delta(G_0^{-1}(2,1';U) - \Sigma(2,1';U))}{\delta U(\mathbf{r}',t_{1+})}\right). \tag{10.139}$$

So far everything has been exact. Eq. (10.136) is the Dyson equation, an integral equation for the Green function in terms of the self-energy. This self-energy has to be computed using Kadanoff and Baym's expression Eq. (10.139) for the self-energy (the formalism is due to Schwinger). Clearly, the self-energy involves the

Green function we are trying to compute, making this a coupled integral-differential equation. Thus an exact approach is out of the question. The so-called GW approximation involves neglecting the derivative of the self-energy on the right side of the above equation. This means,

$$\Sigma(1,1';U)$$
$$\approx -\int d^3r' V(\mathbf{r}_1-\mathbf{r}') \int d2 \left( i\, G(1,2;U) \frac{\delta}{\delta U(\mathbf{r}',t_{1+})} G_0^{-1}(2,1';U) \right). \quad (10.140)$$

Substituting for $G_0^{-1}$ from Eq. (10.134) we get,

$$\Sigma(1,1';U) \approx G(1,1';U) \int d^3r'\; V(\mathbf{r}_1-\mathbf{r}') \int d2\, i \frac{\delta U_{eff}(1')}{\delta U(\mathbf{r}',t_{1+})}$$
$$= i\, v_{eff}(1,1';U) G(1,1';U). \quad (10.141)$$

The effective Coulomb interaction $v_{eff}$ is defined as,

$$v_{eff}(1,1';U) = \int d^3r'\; V(\mathbf{r}_1-\mathbf{r}_1') \frac{\delta U_{eff}(1')}{\delta U(\mathbf{r}',t_1)}. \quad (10.142)$$

Using Eq. (10.127) we get,

$$v_{eff}(1,1';U) = V(\mathbf{r}_1-\mathbf{r}_1')\delta(t_1-t_{1'})$$
$$-i \int d^3r' V(\mathbf{r}_1-\mathbf{r}') \int d^3r''\; V(\mathbf{r}_1'-\mathbf{r}'') \ll \rho(\mathbf{r}',t_1), \rho(\mathbf{r}'',t_{1'}) \gg \quad (10.143)$$

and

$$\ll \rho(\mathbf{r}',t_1), \rho(\mathbf{r}'',t_{1'}) \gg = \frac{\left\langle TS\, \hat{\rho}(\mathbf{r}',t_1)\hat{\rho}(\mathbf{r}'',t_{1'}) \right\rangle_0}{<TS>_0}$$
$$-\frac{\left\langle TS\hat{\rho}(\mathbf{r}',t_1) \right\rangle_0 \left\langle \hat{\rho}(\mathbf{r}'',t_{1'}) \right\rangle}{<TS>_0 \quad <TS>_0}. \quad (10.144)$$

Therefore, the formal solution for the Green function is

$$G(1,1';U) = (G_0^{-1}(1,1';U) - \Sigma(1,1';U))^{-1}, \quad (10.145)$$

where the inverse is in the matrix sense and $\Sigma$ is given by the GW approximation

$$\Sigma(1,1';U) \approx i\, v_{eff}(1,1';U) G(1,1';U) \quad (10.146)$$

to be determined self-consistently. All these ideas also apply to systems out of equilibrium, provided the time integration from 0 to $-i\beta$ is replaced by integration along the contour described earlier. While the GW approximation is no doubt a popular approximation in the literature, the jury is still out on whether, and if so in what sense, is this a 'controlled approximation'. One must be wary of making approximations that cannot be thought of as a conscious expansion in powers of a dimensionless quantity that is demonstrably small compared to unity. In this strict sense, the above and nearly all approximations used in physics, are not 'controlled approximations'. The computations using this formalism are highly nontrivial even with present-day computing resources. For nonequilibrium systems, it is even more difficult. However, the screened potential in case of a highly nonequilibrium laser-excited semiconductor has been evaluated using what may be described as a controlled approximation (see references).

■ In this example, we consider the application of the creation annihilation operators to study stimulated and spontaneous emission. We have chosen to present a complementary description of these phenomena to those found in standard texts on quantum mechanics. There, one uses time-dependent perturbation theory and Fermi's golden rule to evaluate the emission rate. In order to extract a meaningful answer, one has to posit a continuum of states. Here we choose instead to illustrate the phenomenon in a setting more typically found in quantum optics viz. that of a two-level system interacting with a single-mode radiation field. Depending upon whether the radiation is treated classically or quantum mechanically, one is able to investigate the phenomena of stimulated and spontaneous emission.

Imagine a collection of atoms, each of which can either be in the ground state with energy $E_0$ or an excited state with energy $E_1$. In the absence of any other fields, the Hamiltonian may be written as,

$$H = E_0 a_0^\dagger a_0 + E_1 a_1^\dagger a_1. \tag{10.147}$$

Obviously, $N_0 = a_0^\dagger a_0$ is the number of atoms in the ground state and $N_1 = a_1^\dagger a_1$ is the number of atoms in the excited state. We assume that there is no restriction on the number of such atoms in each state, so that $N_0, N_1 = 0, 1, 2, ....$ We now couple this to a photon described by creation operator $b^\dagger$. We ignore all other labels such as momentum and polarization, anticipating that when two levels such as $E_0$ and $E_1$ are involved, photons that have maximum influence on the processes between these levels are the ones with energy $\hbar\omega \sim E_1 - E_0$. Thus we only consider these photons and suppress all additional labels. The interaction of this radiation with the atoms is included a coupling term,

$$H_{coupl} = \xi \, (a_1^\dagger a_0 b + a_0^\dagger a_1 b^\dagger). \tag{10.148}$$

A term such as $a_1^\dagger a_0 b$ says that in order to create an atom in an excited state, we not only have to annihilate one from the ground state but also annihilate a photon from the surroundings, for that is what supplies the energy difference ($\xi$ is the strength of the coupling). Then there is the energy of the photon itself,

$$H_{photon} = \hbar\omega \, b^\dagger b. \tag{10.149}$$

One last term needed to be added, namely the source of the electromagnetic field

$$H_{source} = (bf(t) + b^{\dagger} f^{*}(t)), \tag{10.150}$$

where $f(t)$ is the source. The reason why a term linear in $b, b^{\dagger}$ enters the Hamiltonian may be seen by realizing that in the presence of a radiation field, the interaction term in elementary quantum mechanics has the form $-\frac{e}{mc}\mathbf{A}\cdot\mathbf{p}$ and noting that the vector potential $\mathbf{A}$ is nothing but a linear combination of $b$ and $b^{\dagger}$. Thus the overall Hamiltonian is,

$$H_{tot} = E_0 a_0^{\dagger} a_0 + E_1 a_1^{\dagger} a_1 + \xi\ (a_1^{\dagger} a_0 b + a_0^{\dagger} a_1 b^{\dagger}) + \hbar\omega\ b^{\dagger} b + (bf(t) + b^{\dagger} f^{*}(t)). \tag{10.151}$$

Now we write the light field as the sum of two pieces—a classical average and a quantum fluctuation

$$b(t) =< b(t) > +c(t), \tag{10.152}$$

where $c(t) = b(t) - < b(t) >$. The average $< b(t) >$ is chosen to be the average when only the source of the radiation is present, but the atoms are absent. The condition is, $\frac{\delta}{\delta b^{\dagger}}(\hbar\omega\ b^{\dagger} b + (bf(t) + b^{\dagger} f^{*}(t))) = 0$. This means,

$$\hbar\omega\ < b(t) > + f^{*}(t) = 0. \tag{10.153}$$

Thus we may rewrite Eq. (10.151) as

$$H_{tot} = E_0 a_0^{\dagger} a_0 + E_1 a_1^{\dagger} a_1 + \xi\ (a_1^{\dagger} a_0 < b(t) > + a_0^{\dagger} a_1 < b(t) >^{*}),$$

$$+ \xi\ (a_1^{\dagger} a_0 c + a_0^{\dagger} a_1 c^{\dagger}) + \hbar\omega\ c^{\dagger} c + E_s(t) \tag{10.154}$$

where $E_s$ is a time-dependent commuting quantity (classical average of the photon and source energies).

**Rabi Oscillation**: In this example we ignore the quantum nature of light (i.e., ignore the $c, c^{\dagger}$ operators in Eq. (10.154)). This allows us to write the semiclassical Hamiltonian as,

$$H_{semi} = E_0 a_0^{\dagger} a_0 + E_1 a_1^{\dagger} a_1 + \xi\ (a_1^{\dagger} a_0 < b(t) > + a_0^{\dagger} a_1 < b(t) >^{*}). \tag{10.155}$$

Now imagine an initial state that contains $N$ atoms in the $E_1$ state and none in the $E_0$ state. This is the usual situation in case of lasers where there is population inversion. Let $|N_0, N_1 >$ be the simultaneous eigenstate of the number of atoms in $E_0$ and $E_1$ levels. The initial state we have in mind is,

$$< N_0, N_1 | \psi(t = 0) > = < N_0, N_1 | 0, N > = \delta_{N_0,0} \delta_{N_1,N}. \tag{10.156}$$

We wish to ascertain the nature of the state after some time $t$.

$$i\hbar \frac{d}{dt} |\psi(t) > = H |\psi(t) > \tag{10.157}$$

Now imagine the matrix element $< N_0, N_1|\psi(t) >$. Keeping in mind that

$$< N_0, N_1|\xi\, a_1^\dagger a_0 < b(t) > = \xi\,\sqrt{(N_0+1)N_1} < b(t) >< N_0+1, N_1-1|, \quad (10.158)$$

this also obeys

$$
\begin{aligned}
i\hbar\frac{d}{dt} < N_0, N_1|\psi(t) > &= (N_0E_0 + N_1E_1) < N_0, N_1|\psi(t) > \\
&+ \xi\,\sqrt{(N_0+1)N_1} < b(t) >< N_0+1, N_1-1|\psi(t) > \\
&+ \xi\,\sqrt{(N_1+1)N_0} < b(t) >^* < N_0-1, N_1+1|\psi(t) > . \quad (10.159)
\end{aligned}
$$

Rabi oscillations are periodic oscillations in the distribution functions brought about by energy being exchanged by radiation and the atoms. This frequency is determined by the intensity of the incident classical radiation and is typically much smaller than the frequency of the applied field. This is seen clearly by the following analysis.

In order to simplify proceedings, we make the substitution,

$$< N_0, N_1|\psi(t) > = e^{-\frac{i}{\hbar}t(N_0E_0+N_1E_1)}\, g(N_0, N_1; t). \quad (10.160)$$

In addition to substituting Eq. (10.160) into Eq. (10.159), we also set $< b(t) > = < b(0) > e^{-i\omega_s t}$. This leads to the following simplifications.

$$
\begin{aligned}
i\hbar\frac{d}{dt}g(N_0, N_1; t) &= \xi\,\sqrt{(N_0+1)N_1}\, e^{-i\,\omega_s\,t} < b(0) > e^{\frac{i}{\hbar}t(-E_0+E_1)}\, g(N_0+1, N_1-1; t) \\
&+ \xi\,\sqrt{(N_1+1)N_0}\, e^{i\,\omega_s\,t} < b(0) >^*\, e^{-\frac{i}{\hbar}t(-E_0+E_1)}\, g(N_0-1, N_1+1; t) \quad (10.161)
\end{aligned}
$$

It is easy to understand the meaning of the two terms on the right-hand side. The first corresponds to an atom in the excited state leaving and reaching the ground state accompanied with the emission of light. The second term does the reverse and therefore corresponds to absorption. Now we have to use the initial condition in Eq. (10.156) to solve the above equation. To simplify the solution as much as possible we consider a situation where there is precisely one atom in the higher state and inquire about the nature of the time evolution of this state. This means,

$$g(N_0, N_1; 0) = \delta_{N_0,0}\delta_{N_1,1}. \quad (10.162)$$

Thus we need be concerned about only two functions $g(0,1;t)$ and $g(1,0;t)$.

$$i\hbar\frac{d}{dt}g(0,1;t) = \xi < b(0) > e^{-i\omega_s t} e^{\frac{i}{\hbar}t(-E_0+E_1)}\, g(1,0;t) \quad (10.163)$$

$$i\hbar\frac{d}{dt}g(1,0;t) = \xi < b(0) >^* e^{i\omega_s t}e^{-\frac{i}{\hbar}t(-E_0+E_1)}\, g(0,1;t) \tag{10.164}$$

It is easy to see that the solutions of these equations are periodic. This means a state that initially has one atom in the excited state and none in the ground state, will, after a half a time period, completely convert to a state where there is one atom in the ground state and none in the excited state. This will go on indefinitely. Thus the 'lifetime' in this case is not due to decay but due to an oscillation. These oscillations are known as Rabi oscillations, first studied by I.I. Rabi in he context of nuclear magnetic resonance. We define detuning $\Delta \equiv \omega - \frac{1}{\hbar}(E_1 - E_0)$, then the Rabi frequency is,

$$\Omega_R = \frac{\sqrt{\hbar^2\Delta^2 + 4| < b(0) > |^2\xi^2}}{2\hbar} \tag{10.165}$$

and,

$$g(0,1;t) = e^{-\frac{i}{2}\Delta t}(cos(\Omega_R t) + \frac{i\Delta}{2\Omega_R}sin(\Omega_R t)) \tag{10.166}$$

$$g(1,0;t) = -\frac{2i < b(0) >^* e^{\frac{i}{2}t\Delta}\xi sin(\Omega_R t)}{2\hbar\Omega_R}. \tag{10.167}$$

This means,

$$< 0,1|\psi(t) >= e^{-\frac{i}{\hbar}tE_1}\, e^{-\frac{i}{2}\Delta t}(cos(\Omega_R t) + \frac{i\Delta}{2\Omega_R}sin(\Omega_R t))$$

$$< 1,0|\psi(t) >= -e^{-\frac{i}{\hbar}tE_0}\, \frac{2i < b(0) >^* e^{\frac{i}{2}t\Delta}\xi sin(\Omega_R t)}{2\hbar\Omega_R}. \tag{10.168}$$

In Rabi oscillation, emission and absorption are both stimulated by external radiation. In this phenomenon, a quantity such as $| < N_0,N_1|\psi(t) > |^2$ oscillates with a frequency $\Omega_R$ determined by the intensity of the light, which is much slower than the applied frequency $\omega_s$. We have seen in this example that emission is assisted or stimulated by the external radiation. In particular, if coupling to radiation was not there, the excited state would be stable. However, this is true only because we have disregarded the quantum nature of light. We see in the next example spontaneous emission, viz. emission of light of an excited state on its own. Even though an external source of radiation is not present to stimulate the transition, it nevertheless occurs due to quantum fluctuations of the radiation field.

**Spontaneous Emission**: The earlier result implies that an excited state is stable unless stimulated to undergo a transition (which then reverses, setting up a periodic oscillation, due to the closed nature of the system under consideration). However, this is not true since quantum fluctuations can induce a similar effect even in the absence of a source, as we have mentioned already. To study this we start with the following Hamiltonian.

$$H_{tot} = E_0 a_0^\dagger a_0 + E_1 a_1^\dagger a_1 + \xi \, (a_1^\dagger a_0 c + a_0^\dagger a_1 c^\dagger) + \hbar\omega \, c^\dagger c \tag{10.169}$$

Now the eigenstates of the Hamiltonian in Eq. (10.169) are labeled by $|N_0, N_1, N_\nu>$ where $N_\nu$ represents now the number of light quanta ($c^\dagger c|N_0, N_1, N_\nu> = N_\nu|N_0, N_1, N_\nu>$). As before, we wish to study the timeevolution of the amplitude $< N_0, N_1, N_\nu|\psi(t)>$ subject to the initial condition, $< N_0, N_1, N_\nu|\psi(0)> = \delta_{N_0,0}\delta_{N_\nu,0}\delta_{N_1,1}$. Schrodinger's equation for this amplitude is,

$$i\hbar\frac{d}{dt} < N_0, N_1, N_\nu|\psi(t)> = < N_0, N_1, N_\nu|H|\psi(t)> . \tag{10.170}$$

This means,

$$\begin{aligned} i\hbar\frac{d}{dt} &< N_0, N_1, N_\nu|\psi(t)> \\ = &< N_0, N_1, N_\nu|(E_0 a_0^\dagger a_0 + E_1 a_1^\dagger a_1 + \xi \, (a_1^\dagger a_0 c + a_0^\dagger a_1 c^\dagger) + \hbar\omega \, c^\dagger c)|\psi(t)> \\ &= (E_0 N_0 + E_1 N_1 + \hbar\omega \, N_\nu) < N_0, N_1, N_\nu|\psi(t)> \\ &+ \xi \, \sqrt{N_1(N_0+1)(N_\nu+1)} < N_0+1, N_1-1, N_\nu+1|\psi(t)> \\ &+ \xi \, \sqrt{N_0 N_\nu(N_1+1)} < N_0-1, N_1+1, N_\nu-1|\psi(t)> . \end{aligned} \tag{10.171}$$

This follows from the observation,

$$\begin{aligned} &< N_0, N_1, N_\nu|\xi \, a_1^\dagger a_0 c \\ &= (c^\dagger a_0^\dagger a_1|N_0, N_1, N_\nu > \xi \,)^\dagger = \sqrt{N_1(N_0+1)(N_\nu+1)} < N_0+1, N_1-1, N_\nu+1|\xi \end{aligned} \tag{10.172}$$

and,

$$\begin{aligned} &< N_0, N_1, N_\nu|\xi \, a_0^\dagger a_1 c^\dagger = \\ &(\xi \, c a_1^\dagger a_0|N_0, N_1, N_\nu >)^\dagger = \xi \, \sqrt{N_0 N_\nu(N_1+1)} < N_0-1, N_1+1, N_\nu-1|. \end{aligned} \tag{10.173}$$

As usual we make the substitution,

$$< N_0, N_1, N_\nu|\psi(t)> = e^{-\frac{i}{\hbar}t(E_0 N_0 + E_1 N_1 + \hbar\omega \, N_\nu)} \, g(N_0, N_1, N_\nu; t). \tag{10.174}$$

Finally we get,

$$\begin{aligned} i\hbar\frac{d}{dt} \, & g(N_0, N_1, N_\nu; t) \\ = \xi \, & \sqrt{N_1(N_0+1)(N_\nu+1)} e^{-\frac{i}{\hbar}t(E_0 - E_1 + \hbar\omega)} \, g(N_0+1, N_1-1, N_\nu+1; t) \\ + \xi \, & \sqrt{N_0 N_\nu(N_1+1)} e^{-\frac{i}{\hbar}t(-E_0 + E_1 - \hbar\omega)} \, g(N_0-1, N_1+1, N_\nu-1; t). \end{aligned} \tag{10.175}$$

As before, we consider the situation where a $t = 0$, there is one atom in the $E_1$ level and none in the $E_0$ level and there are no photons to begin with. This means the above rate equation has only these non-vanishing components,

$$i\hbar\frac{d}{dt}\,g(0,1,0;t) = \xi\, e^{-i\Delta t}\ \ g(1,0,1;t) \tag{10.176}$$

$$i\hbar\frac{d}{dt}\,g(1,0,1;t) = \xi\, e^{i\Delta t}\ \ g(0,1,0;t) \tag{10.177}$$

whereas before $\hbar\Delta = \hbar\omega - (E_1 - E_0)$ was the detuning. The important point of the above formulas is that these equations do not contain an external field and yet have nontrivial solutions as we shall see below.

$$g(0,1,0,t) = e^{-\frac{i}{2}t\Delta}(cos(\Omega_R t) + \frac{i\Delta}{2\Omega_R} sin(\Omega_R t)) \tag{10.178}$$

$$g(1,0,1,t) = -2i\, e^{\frac{i}{2}t\Delta}\, \xi\, \frac{sin(\Omega_R t)}{2\hbar\Omega_R} \tag{10.179}$$

The square of the amplitudes oscillate with the vacuum Rabi frequency,

$$\Omega_R = \frac{\sqrt{\hbar^2\Delta^2 + 4\xi^2}}{2\hbar}. \tag{10.180}$$

The figure depicts the probability that the higher state is occupied viz. $|g(0,1,0,t)|^2$ as a function of $\Omega_R t$ for two different values of detuning $\Delta = 0, \Omega_R$. One sees that there is an initial decay of the state followed by revival. We may define the lifetime of the excited state as half the period of oscillation ($\tau = \frac{\pi}{2\Omega_R}$).

## 10.4 Exercises

**Q.1** Verify Eq. (10.13).

**Q.2** Reconcile Eq. (10.66) and Eq. (10.73) with Eq. (10.70) using Lehmann's representation,

$$G(\mathbf{k},\omega) = \int_{-\infty}^{\infty} d\omega' \frac{A(\mathbf{k},\omega')}{\omega - \omega' + i\eta\ (\hbar\omega' - \mu)}. \tag{10.181}$$

**Q.3** Consider fermions interacting mutually with a delta function potential $V(\mathbf{r} - \mathbf{r}') = \lambda\, \delta^3(\mathbf{r} - \mathbf{r}')$. By expanding all the formulas in the chapter in powers of $\lambda$, find the Green function up to second order in $\lambda$.

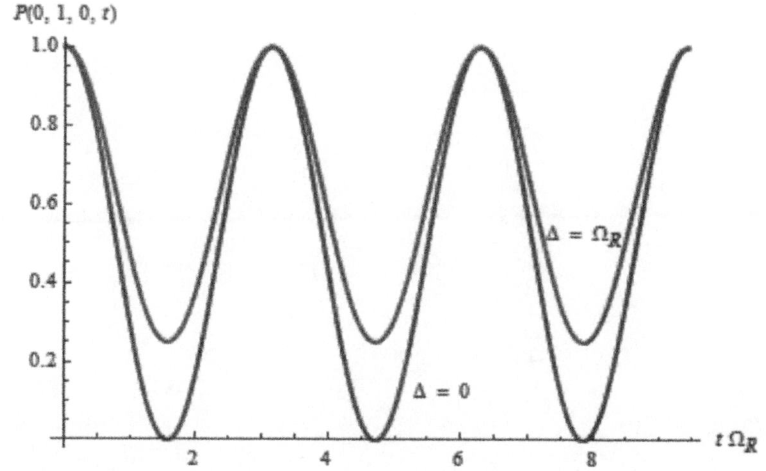


Figure 10.4: Periodic oscillations in the probability that the higher level is occupied is seen even in the absence of an external field. The excited state decays but eventually revives in this model.

**Q.4** This chapter discussed fermions as a continuum. Imagine the reverse where there is only one quantum particle in the presence of a nonequilibrium external potential. What would be the analogs of contour ordering and the Schwinger-Dyson equation?

**Q.5** In the earlier question, imagine there are two quantum particles interacting with each other via a potential $V(\mathbf{r}-\mathbf{r}')$ in the presence of a nonequilibrium external potential. What would be the analogs of contour ordering and the Schwinger-Dyson equation?

**Q.6** Try to solve the rate equations of stimulated and spontaneous emission numerically (e.g., try solving Eq. (10.175) with $N_0 = 0, N_1 = 10, N_v = 0$ at $t = 0$.)

# Chapter 11

# Coherent State Path Integrals

In the earlier sections, we introduced creation and annihilation operators as alternatives to position and momenta in recasting quantum mechanics of many-particle systems. The description using creation and annihilation operators has been so far tied to the Hamiltonian framework. Just as we did earlier, it is natural to ask whether the creation and annihilation operator approach can be introduced through a Lagrangian framework. It indeed can, and it is the subject of this section known as the coherent state path integral (CSPI). We first introduce the concept in the context of a single degree of freedom.

## 11.1 CSPI for a Harmonic Oscillator

Consider the familiar harmonic oscillator,

$$H = \hbar\omega(a^\dagger a + \frac{1}{2}). \tag{11.1}$$

First, one introduces the notion of a coherent state. A coherent state, simply put, is an eigenstate of the creation operator.

$$a^\dagger|\bar{z}>= \bar{z}|\bar{z}> \tag{11.2}$$

Since the creation operator $a^\dagger$ is non-Hermitian, the eigenvalue $\bar{z}$ is complex. The annihilation operator $a$ may be thought of as the derivative with respect to this variable $\bar{z}$.

$$a|\bar{z}>\equiv \frac{\partial}{\partial \bar{z}}|\bar{z}> \tag{11.3}$$

We may see that this is consistent with the commutation rule $[a, a^\dagger] = 1$ since,

$$a(a^\dagger|\bar{z}>) = a\bar{z}|\bar{z}> \tag{11.4}$$

$$a^\dagger(a|\bar{z}>) = \bar{z}a|\bar{z}> . \tag{11.5}$$

Subtracting one from the other gives,

$$|\bar{z}>= [a,\bar{z}]|\bar{z}> . \tag{11.6}$$

This means $[a,\bar{z}] = 1$ when acting on $|\bar{z}>$. This means we may set $a \equiv \frac{\partial}{\partial \bar{z}}$ since $[a,\bar{z}] = \frac{\partial}{\partial \bar{z}}\bar{z} - \bar{z}\frac{\partial}{\partial \bar{z}} = 1 + \bar{z}\frac{\partial}{\partial \bar{z}} - \bar{z}\frac{\partial}{\partial \bar{z}} = 1$. Just as one may form eigenstates of momentum in the position representation as an exponential $< x|p >= e^{\frac{i}{\hbar}px}$, here too we may enquire as to the nature of the eigenstates of the annihilation operator (conjugate to the creation operator) in the $\bar{z}$ space. We set,

$$a|z>= z|z> \tag{11.7}$$

as an eigenstate of $a$. If we wish to determine the meaning of $a^\dagger|z>$, it is useful to use Bargmann's representation.

$$|z>= e^{za^\dagger}|0> \tag{11.8}$$

where $|0>$ is the vacuum state, $a|0>= 0$. It is easy to verify that the Bargmann representation obeys the definition of a coherent state namely, Eq. (11.7).

$$a|z>= a\, e^{za^\dagger}|0>= e^{za^\dagger}\left(e^{-za^\dagger}\, a\, e^{za^\dagger}\right)|0>$$

$$= e^{za^\dagger}\left(a - z[a^\dagger,a]\right)|0>= z\, e^{za^\dagger}|0>= z\,|z> \tag{11.9}$$

Hadamard's formula states,

$$e^X Y e^{-X} = Y + [X,Y] + \frac{1}{2!}[X,[X,Y]] + \frac{1}{3!}[X,[X,[X,Y]]] + ... \quad (11.10)$$

This is proved by first defining $f(s) = e^{sX} Y e^{-sX}$ and expressing $f(s) = f(0) + sf'(0) + \frac{s^2}{2!}f''(0) + \frac{s^3}{3!}f'''(0) + ...$ and evaluating $f(0) = Y$, $f'(0) = [X,Y]$ and so on. Baker Hausdorff theorem states,

$$e^X e^Y = e^{X+Y+\frac{1}{2}[X,Y]+\frac{1}{12}[X,[X,Y]]-\frac{1}{12}[Y,[X,Y]]-\frac{1}{24}[Y,[X,[X,Y]]]-...} \quad (11.11)$$

The dual of this is the Zassenhaus formula,

$$e^{X+Y} = e^X e^Y e^{-\frac{1}{2}[X,Y]} e^{\frac{1}{6}(2[Y,[X,Y]]+[X,[X,Y]])} ... \quad (11.12)$$

In most practical applications in physics, $[X,Y]$ is proportional to the identity operator so this series truncates.

$$e^{X+Y} \approx e^X e^Y e^{-\frac{1}{2}[X,Y]} \quad (11.13)$$

$$e^X e^Y \approx e^{X+Y} e^{\frac{1}{2}[X,Y]} \quad (11.14)$$

$$e^X Y e^{-X} \approx Y + [X,Y] \quad (11.15)$$

Here we have used Hadamard's formula $e^A B e^{-A} = B + [A,B] + \frac{1}{2!}[A,[A,B]] + ...$ (see inset). From this it is clear that,

$$a^{\dagger}|z> = \frac{\partial}{\partial z}|z> . \quad (11.16)$$

Now we wish to evaluate $<\bar{z}|z'>$. Consider the matrix element $<\bar{z}|a|z'>$. On the one hand,

$$<\bar{z}|a|z'> = z' <\bar{z}|z'>; \quad (11.17)$$

on the other hand,

$$<\bar{z}|a|z'> = \frac{\partial}{\partial \bar{z}} <\bar{z}|z'> . \quad (11.18)$$

Equating these two we get,

$$\frac{\partial}{\partial \bar{z}} <\bar{z}|z'> = z' <\bar{z}|z'> \quad (11.19)$$

or,

$$<\bar{z}|z'> = e^{z'\bar{z}} . \quad (11.20)$$

It is also true that $< \bar{z}^{'}|w >= e^{w\bar{z}^{'}}$. Now keeping in mind that $z^{'} = x^{'} + iy^{'}$ and $\bar{z}^{'} = x^{'} - iy^{'}$ and the Jacobian relation,

$$d\bar{z}^{'}dz^{'} = \begin{vmatrix} \frac{\partial \bar{z}^{'}}{\partial x^{'}} & \frac{\partial z^{'}}{\partial x^{'}} \\ \frac{\partial \bar{z}^{'}}{\partial y^{'}} & \frac{\partial z^{'}}{\partial y^{'}} \end{vmatrix} dx^{'}dy^{'}$$

$$= \begin{vmatrix} 1 & 1 \\ -i & i \end{vmatrix} dx^{'}dy^{'} = 2i\, dx^{'}dy^{'}, \tag{11.21}$$

we may write,

$$\int \frac{dz^{'}d\bar{z}^{'}}{2\pi i} e^{-\bar{z}^{'}z^{'}} < \bar{z}|z^{'} >< \bar{z}^{'}|w > \equiv \int \frac{dz^{'}d\bar{z}^{'}}{2\pi i} e^{-\bar{z}^{'}z^{'}} e^{z^{'}\bar{z}} e^{w\bar{z}^{'}} =$$

$$= \iint \frac{dx^{'}dy^{'}}{\pi} e^{-(x^{'2}+y^{'2})} e^{(x^{'}+iy^{'})\bar{z}} e^{w(x^{'}-iy^{'})}$$

$$= \int_{-\infty}^{\infty} \frac{dx^{'}}{\pi} e^{-x^{'2}} e^{x^{'}(\bar{z}+w)} \int_{-\infty}^{\infty} dy^{'} e^{-y^{'2}} e^{iy^{'}(\bar{z}-w)} = e^{\frac{(\bar{z}+w)^2-(\bar{z}-w)^2}{4}}$$

$$= e^{\bar{z}w} =< \bar{z}|w > . \tag{11.22}$$

This means,

$$\int \frac{dz^{'}d\bar{z}^{'}}{2\pi i} e^{-\bar{z}^{'}z^{'}} < \bar{z}|z^{'} >< \bar{z}^{'}|w > \equiv < \bar{z}|1|w > \tag{11.23}$$

or,

$$\int \frac{dz^{'}d\bar{z}^{'}}{2\pi i} e^{-\bar{z}^{'}z^{'}} |z^{'} >< \bar{z}^{'}| \equiv 1. \tag{11.24}$$

This is the so-called resolution of identity and it means that we may insert the left-hand side into any expression with impunity since it is equal to unity. As before, we are interested in the propagator of the harmonic oscillator.

$$G(x_i,t_i;x_f,t_f) \equiv < x_i,t_i|x_f,t_f > \tag{11.25}$$

This may also be written using the evolution operator as ($\varepsilon = \frac{(t_f-t_i)}{N}$, $h = \hbar\omega\, a^{\dagger}a$ and $H = h + \frac{1}{2}\hbar\omega$),

$$G(x_i,t_i;x_f,t_f) \equiv < x_i,t_i|e^{-\frac{i}{\hbar}H(t_f-t_i)}|x_f,t_i >$$

$$= e^{-\frac{i}{\hbar}\frac{1}{2}\hbar\omega(t_f-t_i)} < x_i,t_i|\underbrace{e^{-\frac{i}{\hbar}h\varepsilon}...e^{-\frac{i}{\hbar}h\varepsilon}}_{N}|x_f,t_i > \tag{11.26}$$

We now insert the resolution of identity in Eq. (11.24) in between each fragment of the evolution factors,

$$G(x_i,t_i;x_f,t_f)$$

$$= e^{-\frac{i}{2}\omega(t_f - t_i)} \int \frac{[d\bar{z}dz]}{2\pi i} e^{-\sum_{k=0}^{N} \bar{z}_k z_k}$$

$$< x_i, t_i | z_0 > \prod_{k=0}^{N-1} < \bar{z}_k | e^{-\frac{i}{\hbar} h\varepsilon} | z_{k+1} > < \bar{z}_N | x_f, t_i > . \tag{11.27}$$

Now we have to evaluate the matrix element,

$$< \bar{z}_k | e^{-\frac{i}{\hbar} h\varepsilon} | z_{k+1} > \approx < \bar{z}_k | (1 - \frac{i}{\hbar} h\varepsilon) | z_{k+1} >$$

$$= < \bar{z}_k | z_{k+1} > - \frac{i}{\hbar} \varepsilon < \bar{z}_k | h | z_{k+1} >$$

$$= < \bar{z}_k | z_{k+1} > - i\varepsilon \, \omega < \bar{z}_k | a^\dagger a | z_{k+1} >$$

$$= (1 - i\varepsilon \, \omega \, z_{k+1} \bar{z}_k) e^{\bar{z}_k z_{k+1}} \approx e^{-i\varepsilon\omega z_{k+1}\bar{z}_k} e^{\bar{z}_k z_{k+1}} . \tag{11.28}$$

The above assertions are valid up to order $\varepsilon$.

$$G(x_i, t_i; x_f, t_f) = e^{-\frac{i}{2}\omega(t_f - t_i)} \int \frac{[d\bar{z}dz]}{2\pi i}$$

$$< x_i, t_i | z_0 > \left( \prod_{k=0}^{N-1} e^{-i\varepsilon\omega z_{k+1}\bar{z}_k} e^{\bar{z}_k z_{k+1}} e^{-\bar{z}_k z_k} \right) e^{-\bar{z}_N z_N} < \bar{z}_N | x_f, t_i > \tag{11.29}$$

We now invoke a time sequence $t_k = t_i + \frac{k}{N}(t_f - t_i) = t_i + k\varepsilon$. This means we may use the assertion, $\sum_{k=0}^{N} \varepsilon g_k \approx \int_{t_i}^{t_f} dt \; g(t)$ for any $g_k$. Also $z_{k+1} - z_k = \varepsilon \frac{z_{k+1} - z_k}{\varepsilon} = \varepsilon \; \dot{z}(t_k)$. We now write $e^{-\sum_{k=0}^{N-1} \bar{z}_k z_k} = e^{-\sum_{k=0}^{N-1} \frac{1}{2}\bar{z}_k z_k} e^{-\sum_{k=0}^{N-1} \frac{1}{2}\bar{z}_{k+1} z_{k+1}} e^{\frac{1}{2}(\bar{z}_N z_N - \bar{z}_0 z_0)}$. Therefore,

$$G(x_i, t_i; x_f, t_f) = e^{-\frac{i}{2}\omega(t_f - t_i)} \int \frac{[d\bar{z}dz]}{2\pi i} e^{-\frac{1}{2}(\bar{z}_N z_N + \bar{z}_0 z_0)}$$

$$< x_i, t_i | z_0 > \left( e^{i \int_{t_i}^{t_f} dt \; (\frac{i}{2}\dot{\bar{z}}(t) z(t) - \frac{i}{2}\bar{z}(t)\dot{z}(t) - \omega \bar{z}(t) z(t))} \right) < \bar{z}_N | x_f, t_i > . \tag{11.30}$$

We may think of $L[Z, \dot{Z}] = (\frac{i}{2}\dot{\bar{z}}(t) z(t) - \frac{i}{2}\bar{z}(t)\dot{z}(t) - \omega\bar{z}(t) z(t))$ where $Z = (z, \bar{z})$ as the Lagrangian of the system. This path integral has to be evaluated using the boundary conditions, $z(t_i) \equiv z_0$ and $z(t_f) = z_N$. Finally, an integration over $z_0$ and $z_N$ completes the calculation. Here $\Psi_{z_0}(x_i) \equiv < x_i, t_i | z_0 >$ is the $|z_0 >$ in the position representation. It is obtained as a solution to $a\Psi_{z_0}(x_i) = z_0 \Psi_{z_0}(x_i)$.

### 11.1.1 Evaluation of the Path Integral

In order to evaluate the path integral in Eq. (11.30), we use the methods we have already introduced earlier. The path is written as the sum of two terms. The first is the classical path and the other is the quantum correction. The classical path obeys the variational principle,

$$0 = \delta S = \int_{t_i}^{t_f} dt\ \delta L(Z, \dot{Z}) =$$

$$\int_{t_i}^{t_f} dt\ (\frac{i}{2}(\delta\dot{\bar{z}}(t))z(t) - \frac{i}{2}(\delta\bar{z}(t))\dot{z}(t) - \omega(\delta\bar{z}(t))z(t))$$
$$+ \int_{t_i}^{t_f} dt\ (\frac{i}{2}\dot{\bar{z}}(t)(\delta z(t)) - \frac{i}{2}\bar{z}(t)(\delta\dot{z}(t)) - \omega\bar{z}(t)(\delta z(t)))$$
$$= \int_{t_i}^{t_f} dt\ (-i(\delta\bar{z}(t))\dot{z}(t) - \omega(\delta\bar{z}(t))z(t))$$
$$+ \int_{t_i}^{t_f} dt\ (i\dot{\bar{z}}(t)(\delta z(t)) - \omega\bar{z}(t)(\delta z(t))). \tag{11.31}$$

This means,

$$-i\dot{z}_c(t) - \omega z_c(t) = 0;\ i\dot{\bar{z}}_c(t) - \omega\bar{z}_c(t) = 0. \tag{11.32}$$

The solution is,

$$z_c(t) = z_0\ e^{i\omega(t-t_i)}. \tag{11.33}$$

The constraint that the end points be fixed leads to the following relation between the starting and end coherent state eigenvalues.

$$z_N = z_c(t_f) = z_0\ e^{i\omega(t_f-t_i)} \tag{11.34}$$

We now define $z(t) = z_c(t) + \tilde{z}(t)$. Imposition of $\tilde{z}(t_i) = \tilde{z}(t_f) = 0$ ensures that $z(t_i) = z_0$ and $z(t_f) = z_N$. This means we may write the action as

$$S = \int_{t_i}^{t_f} dt\ (\frac{i}{2}\dot{\bar{\tilde{z}}}(t)\tilde{z}(t) - \frac{i}{2}\bar{\tilde{z}}(t)\dot{\tilde{z}}(t) - \omega\bar{\tilde{z}}(t)\tilde{z}(t)), \tag{11.35}$$

since the classical action vanishes identically. The Green function may then be written as,

$$G(x_i, t_i; x_f, t_f) = e^{-\frac{i}{2}\omega(t_f-t_i)} \int \frac{[d\bar{z}dz]}{2\pi i}\ e^{-\frac{1}{2}(\bar{z}_N z_N + \bar{z}_0 z_0)} \tag{11.36}$$

$$< x_i, t_i | z_0 > \left( e^{i\int_{t_i}^{t_f} dt\ (\frac{i}{2}\dot{\bar{\tilde{z}}}(t)\tilde{z}(t) - \frac{i}{2}\bar{\tilde{z}}(t)\dot{\tilde{z}}(t) - \omega\bar{\tilde{z}}(t)\tilde{z}(t))} \right) < \bar{z}_N | x_f, t_i > . \tag{11.37}$$

Due to the periodicity, we may write $\tilde{z}(t) = \sum_{n=1}^{\infty} sin(\frac{2n\pi(t-t_i)}{(t_f-t_i)})\ c_n$. Now we evaluate $< x|z >$. Keeping in mind that,

$$a^{\dagger} = \frac{p}{\sqrt{2m\hbar\omega}} + i\sqrt{\frac{m}{2\hbar\omega}}\omega x;\ a = \frac{p}{\sqrt{2m\hbar\omega}} - i\sqrt{\frac{m}{2\hbar\omega}}\omega x \tag{11.38}$$

and,

$$a < x|z >= z < x|z > \tag{11.39}$$

we see that,

$$(\frac{d}{dx} + \frac{m\omega}{\hbar}x) < x|z >= iz\sqrt{\frac{2m\omega}{\hbar}} < x|z > . \tag{11.40}$$

The solution may be written as,

$$< x|z >= e^{ix\sqrt{\frac{2m\omega}{\hbar}}z} e^{-\frac{m\omega}{2\hbar}x^2} C_1. \tag{11.41}$$

The constant $C_1$ is evaluated below. We may also write,

$$< \bar{z}|x' >= e^{-ix'\sqrt{\frac{2m\omega}{\hbar}}\bar{z}} e^{-\frac{m\omega}{2\hbar}x'^2} C_1^* \tag{11.42}$$

so that,

$$\int \frac{dzd\bar{z}}{2\pi i} e^{-\bar{z}z} < x|z >< \bar{z}|x' >=$$

$$\int \frac{dzd\bar{z}}{2\pi i} e^{-\bar{z}z} e^{i\sqrt{\frac{2m\omega}{\hbar}}(xz - x'\bar{z})} e^{-\frac{m\omega}{2\hbar}x^2} e^{-\frac{m\omega}{2\hbar}x'^2} |C_1|^2. \tag{11.43}$$

A choice,

$$|C_1|^2 = \sqrt{\frac{2m\omega}{h}} e^{\frac{1}{2}(z^2+\bar{z}^2)} \tag{11.44}$$

ensures that,

$$\int \frac{dzd\bar{z}}{2\pi i} e^{-\bar{z}z} < x|z >< \bar{z}|x' >= \delta(x - x'). \tag{11.45}$$

Going back to Eq. (11.37)

$$G(x_i,t_i;x_f,t_f) = e^{-\frac{i}{2}\omega(t_f-t_i)} e^{-\bar{z}_0 z_0}$$

$$e^{ix_i\sqrt{\frac{2m\omega}{\hbar}}z_0} e^{-ix_f\sqrt{\frac{2m\omega}{\hbar}}\bar{z}_0} e^{-i\omega(t_f-t_i)} e^{-\frac{m\omega}{2\hbar}(x_i^2+x_f^2)} \sqrt{\frac{2m\omega}{h}} e^{\frac{1}{4}(z_0^2+\bar{z}_0^2)} e^{\frac{1}{4}(z_N^2+\bar{z}_N^2)}$$

$$\int \frac{[d\bar{z}dz]}{2\pi i} \left( e^{i\int_{t_i}^{t_f} dt\, (\frac{i}{2}\dot{\bar{z}}(t)z(t) - \frac{i}{2}\bar{z}(t)\dot{z}(t) - \omega\bar{z}(t)z(t))} \right). \tag{11.46}$$

We may now evaluate the same quantity using conventional Hamiltonian methods. For this we invoke the occupation number basis.

$$G(x_i,t_i;x_f,t_f) \equiv < x_i,t_i|e^{-\frac{i}{\hbar}H(t_f-t_i)}|x_f,t_i >$$

$$= \sum_{n=0}^{\infty} < x_i,t_i|n >< n|x_f,t_i > e^{-\frac{i}{\hbar}(n+\frac{1}{2})\hbar\omega(t_f-t_i)}$$

$$< x_i,t_i|n >= \frac{(-\frac{i\hbar}{\sqrt{2m\hbar\omega}}\frac{\partial}{\partial x_i} + i\sqrt{\frac{m}{2\hbar\omega}}\omega x_i)^n}{\sqrt{n!}} < x_i|0 >$$

$$< n|x_f,t_i >= \frac{(\frac{i\hbar}{\sqrt{2m\hbar\omega}}\frac{\partial}{\partial x_f} - i\sqrt{\frac{m}{2\hbar\omega}}\omega x_f)^n}{\sqrt{n!}} < 0|x_f > \tag{11.47}$$

so that,

$$G(x_i,t_i;x_f,t_f)$$
$$= e^{-\frac{i}{2}\omega(t_f-t_i)} \sum_{n=0}^{\infty} \frac{(Q_f^\dagger Q_i)^n}{n!} < x_i|0 >< 0|x_f >$$
$$= e^{-\frac{i}{2}\omega(t_f-t_i)}\ exp[Q_f^\dagger Q_i]\ < x_i|0 >< 0|x_f >$$
$$= e^{-\frac{i}{2}\omega(t_f-t_i)}\ exp[Q_f^\dagger Q_i]\ < x_i|0 >< 0|x_f > \tag{11.48}$$

where,

$$Q_i = e^{i\omega t_i}(-\frac{i\hbar}{\sqrt{2m\hbar\omega}}\frac{\partial}{\partial x_i} + i\sqrt{\frac{m}{2\hbar\omega}}\omega x_i) \tag{11.49}$$

$$Q_f^\dagger = e^{-i\omega t_f}(\frac{i\hbar}{\sqrt{2m\hbar\omega}}\frac{\partial}{\partial x_f} - i\sqrt{\frac{m}{2\hbar\omega}}\omega x_f). \tag{11.50}$$

We may use the so-called Hubbard-Stratanovich transformation to write,

$$e^{Q_f^\dagger Q_i} = \int \frac{d\alpha d\bar{\alpha}}{2\pi i}\ e^{-\bar{\alpha}\alpha} e^{\alpha Q_f^\dagger} e^{\bar{\alpha} Q_i} \tag{11.51}$$

$$G(x_i,t_i;x_f,t_f) =$$
$$e^{-\frac{i}{2}\omega(t_f-t_i)} \int \frac{d\alpha d\bar{\alpha}}{2\pi i}\ e^{-\bar{\alpha}\alpha}$$
$$e^{\bar{\alpha}e^{i\omega t_i} i\sqrt{\frac{m}{2\hbar\omega}}\omega x_i - \bar{\alpha}e^{i\omega t_i}\frac{i\hbar}{\sqrt{2m\hbar\omega}}\frac{\partial}{\partial x_i}}\ \Psi_0(x_i)$$
$$e^{-\alpha e^{-i\omega t_f} i\sqrt{\frac{m}{2\hbar\omega}}\omega x_f + \alpha e^{-i\omega t_f}\frac{i\hbar}{\sqrt{2m\hbar\omega}}\frac{\partial}{\partial x_f}}\ \Psi_0^*(x_f). \tag{11.52}$$

Here, $\Psi_0(x) \equiv < x|0 >$ is the ground state wavefunction of the harmonic oscillator in the position space. We now use the Zassenhaus formula to write $e^{cx+c'\frac{d}{dx}} = e^{cx}e^{c'\frac{d}{dx}}e^{\frac{1}{2}cc'}$. This means, $e^{cx+c'\frac{d}{dx}}f(x) = e^{cx}f(x+c')e^{\frac{1}{2}cc'}$. Thus we may write,

$$G(x_i,t_i;x_f,t_f) =$$
$$e^{-\frac{i}{2}\omega(t_f-t_i)} \int \frac{d\alpha d\bar{\alpha}}{2\pi i}\ e^{-\bar{\alpha}\alpha}$$

$$e^{\bar{\alpha}e^{i\omega t_i} i\sqrt{\frac{m}{2\hbar\omega}}\omega x_i} e^{\frac{1}{4}\bar{\alpha}^2 e^{2i\omega t_i}} \Psi_0(x_i - \bar{\alpha}e^{i\omega t_i}\frac{i\hbar}{\sqrt{2m\hbar\omega}})$$

$$e^{-\alpha e^{-i\omega t_f} i\sqrt{\frac{m}{2\hbar\omega}}\omega x_f} e^{\frac{1}{4}\alpha^2 e^{-2i\omega t_f}} \Psi_0(x_f + \alpha e^{-i\omega t_f}\frac{i\hbar}{\sqrt{2m\hbar\omega}}). \tag{11.53}$$

The ground state wavefunction is given by,

$$\Psi_0(x) = \left(\frac{m\omega}{\pi\hbar}\right)^{\frac{1}{4}} e^{-\frac{m\omega x^2}{2\hbar}}. \tag{11.54}$$

The simplified expression reads as follows,

$$G(x_i,t_i;x_f,t_f) =$$

$$e^{-\frac{i}{2}\omega(t_f-t_i)} e^{-\frac{m\omega}{2\hbar}(x_f^2+x_i^2)} \left(\frac{m\omega}{\pi\hbar}\right)^{\frac{1}{2}}$$

$$\int \frac{d\alpha d\bar{\alpha}}{2\pi i} e^{-\bar{\alpha}\alpha} e^{2i\bar{\alpha}e^{i\omega t_i}\sqrt{\frac{m\omega}{2\hbar}}x_i} e^{\frac{1}{2}\bar{\alpha}^2 e^{2i\omega t_i}} e^{-2i\alpha e^{-i\omega t_f}\sqrt{\frac{m\omega}{2\hbar}}x_f} e^{\frac{1}{2}\alpha^2 e^{-2i\omega t_f}}. \tag{11.55}$$

This may be evaluated using the substitutions $\alpha = X + iY$ and $\bar{\alpha} = X - iY$ and

$$\frac{d\alpha d\bar{\alpha}}{2\pi i} \equiv \frac{dXdY}{\pi} \tag{11.56}$$

$$G(x_i,t_i;x_f,t_f) = e^{-\frac{m\omega}{2\hbar}(x_f^2+x_i^2)} \left(\frac{m\omega}{\pi\hbar}\right)^{\frac{1}{2}} \frac{Exp[\frac{\frac{m\omega}{\hbar}(x_f^2+x_i^2)e^{i\omega(t_i-t_f)} - \frac{2m\omega}{\hbar}x_f x_i}{2i\ sin(\omega(t_i-t_f))}]}{(2i\ sin(\omega(t_f-t_i)))^{\frac{1}{2}}}. \tag{11.57}$$

This expression in Eq. (11.57) is of course identical to Eq. (7.79) obtained using the conventional path integral. This is the so-called coherent state path integral for the harmonic oscillator. Of course, there is no particular advantage to writing this expression since the harmonic oscillator, being simple, its Green function can be obtained using several methods—many of them simpler than this approach. Its true usefulness lies when applied to systems with infinitely many degrees of freedom where one encounters fields. The noncommuting quantum fields in the Hamiltonian framework are replaced by simple functions in the Lagrangian framework, which is when this approach becomes useful. Now we discuss the same idea in the context of fermions.

## 11.2 CSPI for a Fermionic Oscillator

A fermionic oscillator is analogous to the harmonic oscillator except that the number of possibilities for the occupation number are limited to only two, namely zero or one. Thus we postulate that the full Hilbert space is spanned by two vectors $|0>$ and $|1>$. Now we write,

$$c|0>=0\,;\ |1>=c^{\dagger}|0>;\ c|1>=|0>;\ c^{\dagger}|1>=0. \tag{11.58}$$

Thus within this Hilbert space $c^2 = c^{\dagger 2} = 0$ and $cc^{\dagger} + c^{\dagger}c = 1$. In this case the coherent states involve objects that are a generalization of complex numbers called Grassmann numbers. For example we write,

$$c|\eta>=\eta|\eta>. \tag{11.59}$$

Acting again with $c$ and using the identity $c^2 = 0$ leads us to conclude that $\eta^2 = 0$. This is one of the properties of a Grassmann number. Similarly, we could define another coherent state $|\bar{\xi}>$ such that,

$$c^{\dagger}|\bar{\xi}>=\bar{\xi}|\bar{\xi}>. \tag{11.60}$$

Here too we find $\bar{\xi}^2 = 0$. Now we make the following observation. Since all second and higher powers of a Grassmann variable are zero, a function can be at most linear in such a variable. Thus $f(\eta) = f(0) + \eta f^{'}(0)$ for any $f$. If it is a function of two such variables $f(\eta,\omega) = f(0,0) + \omega f^{(0,1)}(0,0) + \eta f^{(1,0)}(0,0) + \eta\omega f^{(1,1)}(0,0)$ and so on. This means,

$$|\eta>=|0>+\eta|1>, \tag{11.61}$$

where $c|0>=0$ and $|1>=c^{\dagger}|0>$. One can see that this obeys the defining equation for a coherent state. Similarly,

$$|\bar{\xi}>=\bar{\xi}|0>+|1>. \tag{11.62}$$

Differentiation of Grassmann variables is defined as $\frac{d}{d\eta}f(\eta) \equiv f^{'}(0)$. We will define integration later. As usual we wish to evaluate the overlap. For this we observe that on the one hand,

$$<\bar{\xi}|c|\eta>=\eta\ <\bar{\xi}|\eta>. \tag{11.63}$$

We may choose to write $c \equiv \frac{\partial}{\partial\bar{\xi}}$ in the same representation in which $c^{\dagger} \equiv \bar{\xi}$. From this we may see that $cc^{\dagger} + c^{\dagger}c = \frac{\partial}{\partial\bar{\xi}}\bar{\xi} + \bar{\xi}\frac{\partial}{\partial\bar{\xi}} = 1$, since now $\{\bar{\xi}, \frac{\partial}{\partial\bar{\xi}}\} = 1$ where $\{A,B\} = 1$. This means,

$$\frac{\partial}{\partial\bar{\xi}} <\bar{\xi}|\eta>=\eta\ <\bar{\xi}|\eta>. \tag{11.64}$$

This means,

$$<\bar{\xi}|\eta>=e^{\bar{\xi}\eta}=1+\bar{\xi}\eta. \tag{11.65}$$

In order to understand how to resolve the identity, we have to introduce the meaning of integration over Grassmann numbers. One way to do this is to introduce the notion of a 'Fourier transform'. The Fourier transform of a constant is a delta function. Hence we first focus on trying to represent a delta function using Grassmann variables. We would like to know what $\delta(\eta-\eta')$ means. One property is $\eta\delta(\eta)=0$. Keeping in mind that $\delta(\eta)=c+\eta\, d$ for some constants, we conclude that $c=0$ and $d\neq 0$, which we set to unity $d=1$ in anticipation. Therefore,

$$\delta(\eta)\equiv\eta. \tag{11.66}$$

Note that this obeys the expectation that for any function $f(\eta)$, $\delta(\eta)f(\eta)=\eta(f(0)+\eta f'(0))=\eta f(0)$, since the delta function forces the argument of any function multiplying it to be zero. Now we invoke the notion of integration. We know that the delta function integrated over all values of an argument yields unity.

$$\int d\eta\ \delta(\eta)=1 \tag{11.67}$$

This means,

$$\int d\eta\ \eta=1. \tag{11.68}$$

We also know that the delta function admits a Fourier representation. In particular, just as $\int\frac{dp}{2\pi\hbar}e^{i\frac{p}{\hbar}(x-x')}=\delta(x-x')$ we expect ($[x,\frac{p}{i\hbar}]=1$, $\{\xi,\bar{\xi}\}=1$),

$$\int d\bar{\xi}\ e^{\bar{\xi}(\eta-\eta')}=\delta(\eta-\eta'). \tag{11.69}$$

In other words,

$$\int d\bar{\xi}\ (1+\bar{\xi}(\eta-\eta'))=(\eta-\eta'). \tag{11.70}$$

Therefore, we must also have,

$$\int d\bar{\xi}\ 1=0. \tag{11.71}$$

Now we prove a rather amusing result that this integration is the same as differentiation with Grassmann variables. Notice that the integration of Grassmann variables we have used so far implies that the range of integration is over all values of that variable, i.e., it is a definite integral rather than an indefinite integral. Now consider,

$$\int d\theta\ f(\theta)=\int d\theta\ (f(0)+\theta f'(0))=0+1f'(0)=f'(0)=\frac{d}{d\theta}f(\theta). \tag{11.72}$$

Therefore, integrating out a Grassmann variable is the same as differentiating with that variable. Since the functions are at most linear in these variables, both operations get rid of that variable. Normally, a definite integral can never be written as a derivative of some function, unless the range of integration is finite and the function is linear. In the present case, the functions are always linear, so it works and Grassmann numbers are in some sense bounded since $\eta^2=0$. Coming back to the question of resolving the identity, we assert that for some weight function $W(\theta,\bar{\theta})$ to be computed,

$$\int d\bar{\theta}d\theta\, W(\theta,\bar{\theta})\ \ |\theta><\bar{\theta}| \equiv 1. \qquad (11.73)$$

Imagine that we insert this resolved identity as shown $<\alpha|\gamma>=<\alpha|1|\gamma>$. This means,

$$<\alpha|\gamma>= \int d\bar{\theta}d\theta\ \ W(\theta,\bar{\theta})\ \ <\alpha|\theta><\bar{\theta}|\gamma>. \qquad (11.74)$$

We write, $<\alpha|\gamma>= 1+\alpha\gamma$, $<\alpha|\theta>= 1+\alpha\theta$ and $<\bar{\theta}|\gamma>= 1+\bar{\theta}\gamma$. We assert that the choice

$$W(\theta,\bar{\theta}) = e^{-\bar{\theta}\theta} = 1-\bar{\theta}\theta \qquad (11.75)$$

is able to reproduce Eq. (11.73). To see this, consider,

$$<\alpha|\gamma>= 1+\alpha\gamma = \int d\bar{\theta}d\theta\ (1-\bar{\theta}\theta)(1+\alpha\theta)(1+\bar{\theta}\gamma) \qquad (11.76)$$

or,

$$1+\alpha\gamma = \int d\bar{\theta}d\theta\ (1-\bar{\theta}\theta)(1+\bar{\theta}\gamma+\alpha\theta+\alpha\theta\bar{\theta}\gamma). \qquad (11.77)$$

This means,

$$1+\alpha\gamma =$$
$$\int d\bar{\theta}d\theta\ (1+\bar{\theta}\gamma+\alpha\theta+\alpha\theta\bar{\theta}\gamma) - \int d\bar{\theta}d\theta\ (\bar{\theta}\theta)$$
$$= \int d\bar{\theta}d\theta\ \alpha\theta\bar{\theta}\gamma - \int d\bar{\theta}d\theta\ \bar{\theta}\theta. \qquad (11.78)$$

since all other terms drop out. Now we make use of the idea that every Grassmann object commutes with every other. Therefore, $d\theta\ \bar{\theta} = -\bar{\theta}\ d\theta$ and this makes the second term become equal to unity. On the other hand, $\alpha\theta\bar{\theta} = \theta\bar{\theta}\alpha$ so that the net result is $1+\alpha\gamma$ as it should be. Therefore, we resolve the identity in Grassmann variables as,

$$\int d\bar{\theta}d\theta\, e^{-\bar{\theta}\theta}\ \ |\theta><\bar{\theta}| \equiv 1. \qquad (11.79)$$

We now wish to use this to write down the Lagrangian approach for a fermionic oscillator. We wish to compute the fermion propagator defined as,

$$<G|c^{\dagger}(t)c(t^{'})|G>= e^{\frac{i}{\hbar}(t-t^{'})E_G}\ \ <G|c^{\dagger}(0)e^{-\frac{i}{\hbar}(t-t^{'})H}c(0)|G> \qquad (11.80)$$

where $H|G>= E_G|G>$ is an eigenstate of the Hamiltonian. For a fermionic oscillator, this means either $E_G = 0$ or $E_G = \hbar\omega$. As usual we rewrite this as,

$$<G|c^{\dagger}(t)c(t^{'})|G>= e^{\frac{i}{\hbar}(t-t^{'})E_G}\ \ <G|c^{\dagger}(0)\underbrace{e^{-\frac{i}{\hbar}\varepsilon H}...e^{-\frac{i}{\hbar}\varepsilon H}}_{N}c(0)|G>. \qquad (11.81)$$

Now we imagine inserting the resolution of the identity between each of the infinitesimal propagators. The matrix elements of the infinitesimal evolution operator (keep in mind that $\varepsilon = \frac{t-t'}{N}$) between coherent states is,

$$< \bar{\theta}_k | e^{-i\varepsilon\omega c^\dagger c} | \theta_{k+1} > = < \bar{\theta}_k | \theta_{k+1} > \; e^{-i\varepsilon\omega\bar{\theta}_k\theta_{k+1}} \tag{11.82}$$

$$< G | c^\dagger(t) c(t') | G > = e^{\frac{i}{\hbar}(t-t')E_G} \; < G | c^\dagger(0) \underbrace{e^{-\frac{i}{\hbar}\varepsilon H} ... e^{-\frac{i}{\hbar}\varepsilon H}}_{N} c(0) | G >$$

$$= e^{\frac{i}{\hbar}(t-t')E_G} \; < G | c^\dagger(0) | \theta_0 >$$

$$\times \int [d\bar{\theta} d\theta] \prod_{k=0}^{N-1} e^{-\bar{\theta}_k\theta_k} \; < \bar{\theta}_k | e^{-i\varepsilon\omega c^\dagger c} | \theta_{k+1} > < \bar{\theta}_N | c(0) | G > . \tag{11.83}$$

The matrix elements at the extremities are nothing but,

$$< \bar{\theta}_N | c(0) | G > = \frac{\partial < \bar{\theta}_N | G >}{\partial \bar{\theta}_N} ; < G | c^\dagger(0) | \theta_0 > = \frac{\partial < G | \theta_0 >}{\partial \theta_0} . \tag{11.84}$$

Therefore,

$$< G | c^\dagger(t) c(t') | G > = e^{\frac{i}{\hbar}(t-t')E_G} \; \int [d\bar{\theta} d\theta] \, \frac{\partial < G | \theta_0 >}{\partial \theta_0}$$

$$e^{i\int_{t'}^{t} ds \, (-\frac{i}{2}\bar{\theta}(s)\dot{\theta}(s) + \frac{i}{2}\dot{\bar{\theta}}(s)\theta(s) - \omega\bar{\theta}(s)\theta(s))} e^{-\frac{1}{2}(\bar{\theta}_0\theta_0 + \bar{\theta}_N\theta_N)} \frac{\partial < \bar{\theta}_N | G >}{\partial \bar{\theta}_N} . \tag{11.85}$$

This path integral has to be performed keeping in mind that $\theta(t') = \theta_0$ and $\theta(t) = \theta_N$ and finally one integrates over $\theta_0$ and $\theta_N$ as well.

### 11.2.1 Evaluating the Path Integral

In order to evaluate the path integral, we proceed as usual. First we write the Grassmann path as the classical solution plus a quantum correction

$$\theta(s) = \theta_{cl}(s) + \tilde{\theta}(s) \tag{11.86}$$

where $\theta_{cl}(s)$ obeys,

$$\frac{d}{ds} \frac{\partial L}{\partial \dot{\bar{\theta}}(s)} = \frac{\partial L}{\partial \bar{\theta}(s)} \tag{11.87}$$

with,

$$L[\bar{\theta}; \dot{\bar{\theta}}] = i \, \dot{\bar{\theta}}(s)\theta(s) - \omega\bar{\theta}(s)\theta(s) \tag{11.88}$$

$$i\,\dot{\theta}_{cl}(s) = -\omega\,\theta_{cl}(s). \tag{11.89}$$

Since this is a first-order equation, the end points are not independent.

$$\theta_{cl}(s) = e^{i\omega(s-t')}\theta_0 \tag{11.90}$$

But,

$$\theta_N \equiv \theta_{cl}(t) = e^{i\omega(t-t')}\theta_0. \tag{11.91}$$

Also,

$$\tilde{\theta}(s) = \sum_{n=1}^{\infty} sin(n\pi\frac{(s-t')}{t-t'})\,\tilde{\theta}_n. \tag{11.92}$$

Therefore,

$$< G|c^{\dagger}(t)c(t')|G >= e^{\frac{i}{\hbar}(t-t')E_G}\int[d\bar{\theta}d\theta]$$
$$\times\frac{\partial < G|\theta_0 >}{\partial\theta_0}\,e^{i\int_{t'}^{t} ds\,L[\bar{\theta};\dot{\theta}]}\,e^{-\bar{\theta}_0\theta_0}\frac{\partial < \bar{\theta}_N|G >}{\partial\bar{\theta}_N}. \tag{11.93}$$

The action may be written as,

$$S = \int_{t'}^{t} ds\,L[\bar{\theta};\dot{\theta}] = \int_{t'}^{t} ds\,(i\dot{\bar{\theta}}(s)\theta(s) - \omega\,\bar{\theta}(s)\theta(s))$$
$$= \sum_{n\neq l=1}^{\infty}\frac{iln(-1+cos(l\pi)cos(n\pi))}{l^2-n^2}\bar{\tilde{\theta}}_l\tilde{\theta}_n, \tag{11.94}$$

which is independent of $t$ and $t'$. Also since $\theta_N$ and $\theta_0$ are proportional we may write (after ignoring terms that do not depend on $t$ and $t'$),

$$< G|c^{\dagger}(t)c(t')|G >= e^{\frac{i}{\hbar}(t-t')E_G}\int[d\bar{\theta}_0 d\theta_0]\,\Psi'^{*}_G(\theta_0)\,e^{-\bar{\theta}_0\theta_0}\Psi'_G(\bar{\theta}_0) \tag{11.95}$$

where, $\Psi_G(\theta) \equiv < \theta|G >$.

$$\Psi^{*}_G(\theta) \equiv < G|\theta > = < G|e^{\theta c^{\dagger}}|0 > = < G|0 > + \theta < G|1 > . \tag{11.96}$$

Therefore,

$$< G|c^{\dagger}(t)c(t')|G >= e^{\frac{i}{\hbar}(t-t')E_G}\int[d\bar{\theta}_0 d\theta_0]\;< G|1 >\;e^{-\bar{\theta}_0\theta_0} < 1|G >$$
$$= e^{\frac{i}{\hbar}(t-t')E_G}\;< G|1 >< 1|G > . \tag{11.97}$$

If $|G > = |0 >$, the hole propagator above vanishes since the annihilation operator annihilates the vacuum. When $|G > = |1 >$ we get,

$$< G|c^{\dagger}(t)c(t')|G >= e^{\frac{i}{\hbar}(t-t')E_G}. \tag{11.98}$$

Thus fermions may also be studied in the path integral language by invoking Grassmann variables.

## 11.3 Generalization to Fields

So far, we have studied the CSPI method for systems with one degree of freedom. It is possible, of course, to generalize the formalism to include infinitely many degrees of freedom. Using a method similar to what we saw in the earlier section, we could easily accept that the correct way of handling this situation would be to first ensure that the Hamiltonian is normal ordered. This means all the creation operators are to the left of the annihilation operators. This is important because in the CSPI approach, the identity is resolved using the eigenstates of the annihilation operator rather than the creation operator, as a result, matrix elements such as $< \bar{z}_k|a^{\dagger}a|z_{k+1} >$ found in Eq. (11.28) are again proportional to $< \bar{z}_k|z_{k+1} >$, which is why this method works. If instead we chose the Hamiltonian as $H = \hbar\omega\, aa^{\dagger}$, the matrix element $< \bar{z}_k|aa^{\dagger}|z_{k+1} >$ would involve derivatives and would not lead to anything useful. Therefore, for a Hamiltonian such as the one in Eq. (8.91) (we assume it describes bosons for simplicity), the way to formulate CSPI would be to first write the action that appears in Eq. (11.30) as,

$$S_{i\to f} = \int_{t_i}^{t_f} dt\, \Big( \int d\mathbf{r}\, (\frac{i}{2}\dot{\bar{z}}(\mathbf{r},t)z(\mathbf{r},t) - \frac{i}{2}\bar{z}(\mathbf{r},t)\dot{z}(\mathbf{r},t)) - H[z,\bar{z}] \Big) \tag{11.99}$$

where,

$$H[z,\bar{z}] = \int d\mathbf{r}\, \bar{z}(\mathbf{r},t)\frac{p^2}{2m}z(\mathbf{r},t) + \frac{1}{2}\int d\mathbf{r} \int d\mathbf{r}'\bar{z}(\mathbf{r},t)\bar{z}(\mathbf{r}',t)z(\mathbf{r}',t)z(\mathbf{r},t)V(|\mathbf{r}-\mathbf{r}'|). \tag{11.100}$$

Therefore, if the aim is to evaluate an overlap such as $< \Psi_i,t_i|\Psi_f,t_f >$, we would write,

$$< \Psi_i,t_i|\Psi_f,t_f >= \int_{z(\mathbf{r},t_i)\equiv z_0(\mathbf{r});z(\mathbf{r},t_f)\equiv z_N(\mathbf{r})} \frac{[d\bar{z}dz]}{2\pi i}\, e^{-\frac{1}{2}\int d\mathbf{r}\, (\bar{z}_N(\mathbf{r})z_N(\mathbf{r})+\bar{z}_0(\mathbf{r})z_0(\mathbf{r}))}$$
$$< \Psi_i,t_i|\{z_0\} > e^{iS_{i\to f}} < \{\bar{z}_N\}|\Psi_f,t_i > . \tag{11.101}$$

If the initial and final states are both position eigenstates of a single particle, then we would write

$$|\Psi_f,t_f >\equiv c^{\dagger}(\mathbf{r}_f,t_f)|G >; < \Psi_i,t_i| \equiv< G|c(\mathbf{r}_i,t_i), \tag{11.102}$$

where $|G >$ is the ground state of the system. The coherent states,

$$|\{z_0\} >= e^{\int d\mathbf{r}\, z_0(\mathbf{r})c^{\dagger}(\mathbf{r},t_i)}|vac >; \;\; < \{\bar{z}_N\}| =< vac|e^{\int d\mathbf{r}\, \bar{z}_N(\mathbf{r})c(\mathbf{r},t_i)} \tag{11.103}$$

where $|vac >$ is the state annihilated by each $c(\mathbf{r},t_i)$. Therefore,

$$< \{\bar{z}_N\}|\Psi_f,t_i >= \bar{z}_N(\mathbf{r}_f) \;\; < vac|e^{\int d\mathbf{r}\, \bar{z}_N(\mathbf{r})c(\mathbf{r},t_i)}|G >= \bar{z}_N(\mathbf{r}_f) \;\; < \{\bar{z}_N\}|G > . \tag{11.104}$$

Similarly,

$$< \Psi_i, t_i | \{z_0\} >= z_0(\mathbf{r}_i) \ < G | \{z_0\} > . \tag{11.105}$$

Hence the propagator is,

$$< \mathbf{r}_i, t_i | \mathbf{r}_f, t_f >=$$

$$\int_{z(\mathbf{r},t_i) \equiv z_0(\mathbf{r}); z(\mathbf{r},t_f) \equiv z_N(\mathbf{r})} \frac{[d\bar{z}\, dz]}{2\pi i} e^{-\frac{1}{2}\int d\mathbf{r}\ (\bar{z}_N(\mathbf{r}) z_N(\mathbf{r}) + \bar{z}_0(\mathbf{r}) z_0(\mathbf{r}))} \tag{11.106}$$

$$z_0(\mathbf{r}_i) \ < G | \{z_0\} > e^{iS_{i\to f}} \ \bar{z}_N(\mathbf{r}_f) \ < \{\bar{z}_N\} | G > . \tag{11.107}$$

In order to complete the calculation, one has to provide the ground state description of the many-body system in the coherent state basis.

## 11.4 Exercises

**Q.1** A Hamiltonian with bosonic oscillators is given by $H = \hbar\omega(a^\dagger a + \frac{1}{2}) + \lambda(a^\dagger a^\dagger + aa)$. Develop the coherent state path integral approach for this Hamiltonian. This means write down the coherent state path integral for the Green function $< Ta(t)a^\dagger(t') >$.

**Q.2** A Hamiltonian with two types of fermionic oscillators is given by $H = \hbar\omega(a^\dagger a + b^\dagger b) + \lambda(b^\dagger a^\dagger + ab)$. Develop the coherent state path integral approach for this Hamiltonian.

**Q.3** An unperturbed Hamiltonian with two bosonic oscillators is given by $H_0 = \hbar\omega(a^\dagger a + b^\dagger b)$. Imagine the perturbation to be $V = \lambda(a^\dagger ab + b^\dagger a^\dagger a)$. Develop a perturbative scheme using the coherent state path integral approach for this Hamiltonian.

**Q.4** Do the same if the oscillators in the above question were fermions.

# Chapter 12

# Nonlocal Operators

The description of many-particle quantum mechanics given till now does not give us much insight into how one may go about computing the Green function of quantum systems with infinitely many coupled degrees of freedom. Apart from the free particle case and perhaps the harmonic oscillator, there are precious few exact computations of Green functions of many-particle systems. Even the approximate methods are found wanting since we have already alluded to the 'uncontrolled' nature of most of the approximations that have been proposed to date. There is one method that offers some hope in this regard. This method is (wrongly) called 'bosonization'. This method will be the main focus of much of this and the next chapter. The main mathematical tool used is the introduction of operators that are 'non-local' in a sense to be made precise later, which enables the exact computation of the asymptotic properties of the Green function ($G(\mathbf{x}-\mathbf{x}^{'},t-t^{'})$ in the regime $|\mathbf{x}-\mathbf{x}^{'}| \to \infty$ and/or $|t-t^{'}| \to \infty$) under some further restrictive assumptions. We wish to ease into this subject through the study of quantum vortices in charged bosons where the notion of nonlocality makes its presence felt in a relatively more familiar setting.

## 12.1 Quantum Vortices in a Charged Boson Fluid

We wish to introduce the concept of quantum vortices in a charged boson fluid. Vortex strength is defined as the circulation of the velocity field around a closed loop. The velocity field is the ratio of the current density and the number density. Current density and number density in turn are defined in terms of the boson field. Define a complex scalar $\psi(x)$ where $x = (\mathbf{x},t)$. Then we have $\rho(x) = \psi^{\dagger}(\mathbf{x},t)\psi(\mathbf{x},t)$ and $\mathbf{J}(x) = Im[\psi^{\dagger}(\mathbf{x},t)\nabla\psi(\mathbf{x},t)]$. We choose to polar decompose $\psi(x)$ as follows: $\psi(x) = e^{-i\Pi(x)}\sqrt{\rho(x)}$. In the CSPI approach (generalized to incorporate fields),

$\psi(x)$ is a simple complex number and therefore admits such a decomposition. Therefore the velocity field becomes $\mathbf{v}(x) = \frac{1}{\rho(x)}\mathbf{J}(x) \equiv -\nabla\Pi(x)$. This definition is meaningful only in regions where $\rho(x) \neq 0$. Thus, multiplying connected regions require special care which we shall address later. It is important to stress that in this work, $\psi(x)$ is not a complex order parameter, rather it is the field variable, namely the de-quantized version of the field operator that is used in the coherent state path integral approaches. For fermions, which we shall study later on, this would be a Grassmann variable. This is in contrast with the usual approach for dealing with vortices in, say, superfluids or superconductors where $\psi(x)$ is a complex order parameter. The action for free bosons is written in the coherent state path integral approach as $S_{0,B} = \int_0^{-i\beta} dt \int d^3x\, \psi^\dagger(x)(i\hbar\partial_t + \frac{\hbar^2\nabla^2}{2m})\psi(x)$. Thus,

$$S_{0,B} = \int d^3x \int_0^{-i\beta\hbar} dt\ [\hbar\ \rho\partial_t\Pi - \frac{\rho\hbar^2(\nabla\Pi)^2 + \frac{\hbar^2(\nabla\rho)^2}{4\rho}}{2m}]. \tag{12.1}$$

Now we wish to introduce a coupling to $U(1)$ gauge field. For this we have to follow the minimal coupling prescription, $p_\mu \to p_\mu - \frac{q_c}{c}A_\mu$ or, $i\hbar\partial_t \to i\hbar\partial_t - q\phi$ and $-i\hbar\nabla \to -i\hbar\nabla - \frac{q_c}{c}\mathbf{A}$.

$$S_{full,B} = \int_0^{-i\beta\hbar} dt \int d^3x\, \psi^\dagger(x)(i\hbar\partial_t - q\phi - \frac{(-i\hbar\nabla - \frac{q_c}{c}\mathbf{A})^2}{2m})\psi(x)$$

$$+ \int_0^{-i\beta\hbar} dt \int d^3x\, \frac{1}{8\pi}(\mathbf{E}^2 - \mathbf{B}^2) \tag{12.2}$$

where $\mathbf{E} = -\nabla\phi - \frac{1}{c}\frac{\partial}{\partial t}\mathbf{A}$ and $\mathbf{B} = \nabla \times \mathbf{A}$. The above action is invariant under gauge transformations: $\psi(x) \to e^{i\frac{q}{\hbar c}\lambda(x)}\psi(x)$, $\mathbf{A} \to \mathbf{A} + \nabla\lambda$, $\phi \to \phi - \frac{1}{c}\frac{\partial}{\partial t}\lambda$. We now wish to polar decompose the field and rewrite this in terms of the phase variable $\Pi$ and the density $\rho$ and the gauge fields. Since $\Pi \to \Pi - \frac{q}{\hbar c}\lambda$, we may define gauge-invariant quantities $\mathbf{v}(x) = -\nabla\Pi - \frac{q}{\hbar c}\mathbf{A}$ and $v_0(x) = \partial_t\Pi - \frac{q}{\hbar}\phi$. We note that,

$$\mathbf{E} = \frac{\hbar}{q}(\partial_t\mathbf{v}(x) + \nabla v_0(x))\ ;\ \mathbf{B} = -\frac{\hbar c}{q}(\nabla \times \mathbf{v}). \tag{12.3}$$

Thus the full gauge invariant action is given by (with an external scalar potential $U(x)$ assumed time independent; in condensed matter physics terminology it would be called disorder),

$$S_{full,B} = \int d^3x \int_0^{-i\beta\hbar} dt\ [\rho\ (\hbar v_0 - U) - \frac{\rho\hbar^2\mathbf{v}^2 + \frac{\hbar^2(\nabla\rho)^2}{4\rho}}{2m}]$$

$$+ \int d^3x \int_0^{-i\beta\hbar} dt\ \frac{\hbar^2}{8\pi q^2}((\partial_t\mathbf{v}(x) + \nabla v_0(x))^2 - c^2(\nabla \times \mathbf{v})^2). \tag{12.4}$$

If we so wish, we could evaluate the Green function of the charged bosons using the formal functional integral using the non-local but gauge-invariant form of the field,

$$\psi(x) = e^{i\int^{\mathbf{x}} d\mathbf{r}\cdot\mathbf{v}(x)}\ \sqrt{\rho(x)} \equiv \Psi[\rho,\mathbf{v};x] \tag{12.5}$$

$$G(x-x^{'}) \equiv \int D[v^{\mu}] \int D[\rho]\Psi[\rho,\mathbf{v};x]\Psi^{*}[\rho,\mathbf{v};x^{'}]e^{iS_{full,B}}. \tag{12.6}$$

Note that any scheme for evaluating the above functional integral is going to be non-perturbative in the coupling $q^2$ that appears now in the denominator of the action. However, we wish to evaluate something equally interesting viz. the correlation function between vortices. For this we have to start making some approximations. We assume that $\rho(x) = \rho_0 + \tilde{\rho}(x)$ where $\rho_0$ is the assumed uniform average density of particles and that it is legitimate to retain only the harmonic terms in the density fluctuations in the action. This approximation is likely to be valid in the $\rho_0 \to \infty$ limit. We may integrate out $v_0$ and $\tilde{\rho}$ and write down only an effective action for $\mathbf{v}$. Start with (harmonic approximation),

$$S_{harm,B} \approx \int d^3x \int_0^{-i\beta\hbar} dt\ [\tilde{\rho}\ (\hbar v_0 - U) - \frac{\rho_0\hbar^2\mathbf{v}^2 + \frac{\hbar^2(\nabla\rho)^2}{4\rho_0}}{2m}]$$

$$+ \int d^3x \int_0^{-i\beta\hbar} dt\ \frac{\hbar^2}{8\pi q^2}((\partial_t\mathbf{v}(x) + \nabla v_0(x))^2 - c^2(\nabla\times\mathbf{v})^2). \tag{12.7}$$

First we may integrate over $\tilde{\rho}$ (that is, $e^{i\frac{S_{eff,v_0}}{\hbar}} = \int D[\tilde{\rho}]e^{i\frac{S_{harm,B}}{\hbar}}$ ) to obtain the effective action,

$$S_{eff,v_0} \approx \int d^3x \int_0^{-i\beta\hbar} dt\ \frac{8m\rho_0}{\hbar^2} \int d^3x^{'} G(x-x^{'})(\hbar v_0(x^{'})$$
$$-U(x^{'}))(\hbar v_0(x) - U(x)) - \int d^3x \int_0^{-i\beta\hbar} dt\ \frac{\rho_0\hbar^2\mathbf{v}(x)^2}{2m}$$
$$+ \int d^3x \int_0^{-i\beta\hbar} dt\ \frac{\hbar^2}{8\pi q^2}((\partial_t\mathbf{v}(x) + \nabla v_0(x))^2 - c^2(\nabla\times\mathbf{v})^2), \tag{12.8}$$

where $\nabla^2 G(x-x^{'}) = \delta(\mathbf{x}-\mathbf{x}^{'})\delta(t-t^{'})$. In order to further integrate over $v_0$ we have to keep in mind that we are working in the $\rho_0 \to \infty$ limit. In this limit, the integrand oscillates rapidly and cancels out unless $\hbar v_0(x) = U(x)$. Thus,

$$S_{eff} \approx -\int d^3x \int_0^{-i\beta\hbar} dt\ \frac{\rho_0\hbar^2\mathbf{v}^2}{2m}$$

$$+ \int d^3x \int_0^{-i\beta\hbar} dt\ \frac{\hbar^2}{8\pi q^2}((\partial_t\mathbf{v}(x) + \frac{\nabla U(x)}{\hbar})^2 - c^2(\nabla\times\mathbf{v})^2). \tag{12.9}$$

Lastly, we have to keep in mind that $U(x)$ is time independent. This means that the cross term with $U$ and $\partial_t \mathbf{v}$ drops out and the term quadratic in $U$ is irrelevant (a constant). Thus,

$$S_{\mathbf{v}} \approx -\int_0^{-i\beta\hbar} dt \int d^3x \, \frac{\hbar^2 \rho_0}{2m} \mathbf{v}^2(x)$$
$$+\frac{\hbar^2 c^2}{8\pi q^2} \int_0^{-i\beta\hbar} dt \int d^3x \, ((-\frac{1}{c}\partial_t \mathbf{v})^2 - (\nabla \times \mathbf{v})^2). \quad (12.10)$$

This means that external scalar potentials that are time independent (time-independent electric fields) do not affect the properties of vortices. Thus we may suspect that the properties of vortices are insensitive to disorder. We may now couple this to sources and calculate the velocity–velocity correlation function (see exercises). Set $\omega_0 = \sqrt{\frac{\rho_0}{m} 4\pi q^2}$, which is the plasma frequency. Then we conclude,

$$< v_i(\mathbf{x},t) v_j(\mathbf{x}',t') > = \frac{4\pi q^2}{V} \sum_{\mathbf{k}} e^{i\mathbf{k}\cdot(\mathbf{x}-\mathbf{x}')} e^{-i\omega_0(t-t')} \frac{k_i k_j}{k^2(2\hbar\omega_0)}$$
$$+\frac{4\pi q^2}{V} \sum_{\mathbf{k}} e^{i\mathbf{k}\cdot(\mathbf{x}-\mathbf{x}')} e^{-i\sqrt{\omega_0^2+c^2k^2}(t-t')} \frac{k_i k_j + \delta_{i,j}(-k^2)}{(2\hbar\sqrt{\omega_0^2+c^2k^2})(-k^2)}. \quad (12.11)$$

Vortex strength is defined as $\phi_\Gamma(t) = \oint_\Gamma dx_i \, v_i(x)$. Taking into account the fact that $\oint dx_i k_i e^{i\mathbf{k}\cdot\mathbf{x}} = -i\oint (d\mathbf{x}\cdot\nabla) e^{i\mathbf{k}\cdot\mathbf{x}} = 0$ we obtain,

$$< \phi_\Gamma(t) \phi_{\Gamma'}(t') > = \frac{4\pi q^2}{V} \sum_{\mathbf{k}} e^{-i\sqrt{\omega_0^2+c^2k^2}(t-t')} \frac{\oint d\mathbf{x} \cdot \oint d\mathbf{x}' e^{i\mathbf{k}\cdot(\mathbf{x}-\mathbf{x}')}}{(2\hbar\sqrt{\omega_0^2+c^2k^2})}. \quad (12.12)$$

We focus on the equal-time correlation only, and taking into account that we are working in the $\rho_0 \to \infty$ limit, we obtain (it is more convincing to perform the summation over $\mathbf{k}$ in Eq. (12.12) first, which is left to the exercises),

$$< \phi_\Gamma(t) \phi_{\Gamma'}(t) > = \frac{4\pi q^2}{V} \sum_{\mathbf{k}} \frac{\oint d\mathbf{x} \cdot \oint d\mathbf{x}' e^{i\mathbf{k}\cdot(\mathbf{x}-\mathbf{x}')}}{(2\hbar\sqrt{\omega_0^2+c^2k^2})}$$
$$\approx \frac{4\pi q^2}{2\hbar\omega_0 V} \oint d\mathbf{x} \cdot \oint d\mathbf{x}' \sum_{\mathbf{k}} e^{i\mathbf{k}\cdot(\mathbf{x}-\mathbf{x}')} e^{-\frac{c^2k^2}{2\omega_0^2}} \quad (12.13)$$

or,

$$< \phi_\Gamma(t) \phi_{\Gamma'}(t) > = \frac{q^2}{\hbar\omega_0} \frac{1}{\sqrt{2\pi}} \frac{\omega_0^3}{c^3} \oint_\Gamma d\mathbf{x} \cdot \oint_{\Gamma'} d\mathbf{x}' e^{-\frac{\omega_0^2}{c^2}(\mathbf{x}-\mathbf{x}')^2}. \quad (12.14)$$

This clearly shows that as $\omega_0 \to \infty$, the only contribution to the vortex–vortex correlation that survives are the terms in the vicinity of the points where the two loops

$\Gamma$ and $\Gamma'$ intersect. We parameterize the loops as $\mathbf{x}(s)$ and $\mathbf{x}'(s')$ and define tangent vectors $\mathbf{t}(s) = \frac{d\mathbf{x}(s)}{ds}$ etc. Further, let us assume that $\mathbf{x}(s_i) = \mathbf{x}'(s'_i) = \mathbf{p}_i$ are the points at which the loops intersect. Thus, $\mathbf{x}(s) \approx \mathbf{p}_i + (s - s_i)\mathbf{t}_i$ and $\mathbf{x}'(s') \approx \mathbf{p}_i + (s' - s'_i)\mathbf{t}'_i$. Therefore, if $n_c$ is the number of points at which the loops cross and $\lambda = s - s_i$ and $\lambda' = s' - s'_i$,

$$< \phi_\Gamma(t)\phi_{\Gamma'}(t) > \approx \frac{q^2}{\hbar\omega_0}\frac{1}{\sqrt{2\pi}}\frac{\omega_0^3}{c^3}\sum_{i=1}^{n_c}\int_{-\infty}^{\infty} d\lambda \int_{-\infty}^{\infty} d\lambda' \; (\mathbf{t}_i \cdot \mathbf{t}'_i)\, e^{-\frac{\omega_0^2}{c^2}(\mathbf{t}_i\lambda - \mathbf{t}'_i\lambda')^2}. \quad (12.15)$$

This means,

$$< \phi_\Gamma(t)\phi_{\Gamma'}(t) > = \frac{q^2}{(2\hbar c)}\sqrt{2\pi}\sum_{i=1}^{n_c}\frac{(\mathbf{t}_i \cdot \mathbf{t}'_i)}{\sqrt{(\mathbf{t}_i)^2(\mathbf{t}'_i)^2 - (\mathbf{t}_i \cdot \mathbf{t}'_i)^2}}$$

$$= \alpha\sqrt{\frac{\pi}{2}}\sum_i cot(\theta_i). \quad (12.16)$$

Thus the vortex–vortex correlation function is proportional to the sum of the cotangent of the angles between the tangent vectors of the two loops at each point where they intersect. The coefficient of proportionality is the universal fine structure constant and is not related to material parameters. From the Aharonov-Bohm argument we know that the circulation of the velocity field, which is nothing but the circulation of the vector potential, is related by Stokes theorem to the magnetic flux through the loops. Thus $\phi_\Gamma(t)$ is also the magnetic flux through the loop $\Gamma$ (times $-\frac{q}{\hbar c}$). Thus the vortex–vortex correlation function is nothing but the flux–flux correlation. These are quantum vortices in the sense that the average of the vortex strength is zero but the fluctuation is non-zero. But when external magnetic fields are present or when there are 'cores', namely points at which the density of particles vanish, we may expect the average vortex strength to be non-zero.

For the sake of completeness, we recall here the well-known argument that leads to quantization of vortices in terms of the winding number. If in a region, the density of particles is non-vanishing, it is easy to see that the circulation of the velocity, namely the vortex strength, is always zero. However, even in the absence of magnetic fields but in the presence of isolated points at which the density becomes zero, we may show that the circulation of the velocity is not zero. To see this we first note that in the absence of magnetic fields, the current and number density are given by,

$$\mathbf{J}(\mathbf{r}) = \frac{1}{2i}(\psi^*(\mathbf{r})(\nabla\psi(\mathbf{r})) - (\nabla\psi^*(\mathbf{r}))\psi(\mathbf{r}))\; ; \; \rho(\mathbf{r}) = \psi^*(\mathbf{r})\psi(\mathbf{r}). \quad (12.17)$$

The velocity is nothing but the ratio of the current density and number density $\mathbf{v}(\mathbf{r}) = \frac{\mathbf{J}(\mathbf{r})}{\rho(\mathbf{r})}$ and $\phi_\Gamma = \oint_\Gamma d\mathbf{r} \cdot \mathbf{v}(\mathbf{r})$ is the vortex strength. Thus,

$$\phi_\Gamma = \frac{1}{2i}\oint_\Gamma d\mathbf{r} \cdot \frac{(\psi^*(\mathbf{r})(\nabla\psi(\mathbf{r})) - (\nabla\psi^*(\mathbf{r}))\psi(\mathbf{r}))}{\psi^*(\mathbf{r})\psi(\mathbf{r})}$$

$$= \frac{1}{2i}\oint_{\Gamma} d\mathbf{r}\cdot\left(\frac{\nabla\psi(\mathbf{r})}{\psi(\mathbf{r})} - \frac{\nabla\psi^*(\mathbf{r})}{\psi^*(\mathbf{r})}\right). \tag{12.18}$$

If the density (and the field) does not vanish either in the region bounded by $\Gamma$ or on the boundary (or boundaries for non-simply connected regions) and the gradient of the field is finite everywhere in the region and on the boundary, then we may use the Stokes theorem to rewrite the line integral in terms of the surface integral of the curl.

$$\phi_{\Gamma} = \frac{1}{2i}\oint_{\Gamma} d\mathbf{r}\cdot\left(\frac{\nabla\psi(\mathbf{r})}{\psi(\mathbf{r})} - \frac{\nabla\psi^*(\mathbf{r})}{\psi^*(\mathbf{r})}\right) = \frac{1}{2i}\oint_{\Gamma} d\mathbf{r}\cdot\left(\frac{\nabla\psi(\mathbf{r})}{\psi(\mathbf{r})} - \frac{\nabla\psi^*(\mathbf{r})}{\psi^*(\mathbf{r})}\right)$$

$$= \frac{1}{2i}\int d\mathbf{a}\cdot\nabla\times\left(\frac{\nabla\psi(\mathbf{r})}{\psi(\mathbf{r})} - \frac{\nabla\psi^*(\mathbf{r})}{\psi^*(\mathbf{r})}\right) = 0 \tag{12.19}$$

The last result is because the integrand itself vanishes identically. Now imagine that the region contains a point which for convenience we choose as the origin at which the density (and therefore, the field) vanishes. In this case we may separate the region bounded by $\Gamma$ into two regions. One is a tiny circular region containing the origin at the center and the other, the rest. Outside the tiny circle, the density is finite and therefore from the earlier argument, Stokes theorem tells us that the circulation around $\Gamma$ is the same as the circulation around the tiny circle of radius $\varepsilon$ (see figure).

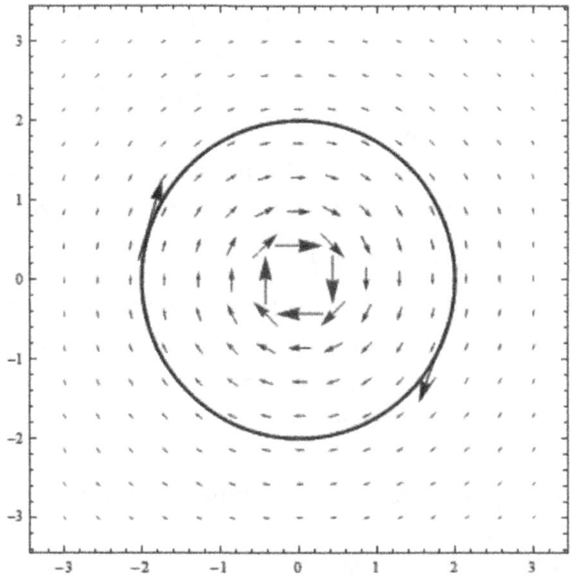


Figure 12.1: Circulation of a velocity field around a loop enclosing a region where the density vanishes, leads to a quantized vortex strength (the region around the center has roundoff errors, which is why it looks like a square).

We may evaluate the integral over the tiny circle as follows (we assume that the loop of integration winds around $n$ times over the tiny circle, for the sake of generality)

$$\phi_\Gamma = \frac{1}{2i}\oint_\Gamma d\mathbf{r}\cdot\left(\frac{\nabla\psi(\mathbf{r})}{\psi(\mathbf{r})} - \frac{\nabla\psi^*(\mathbf{r})}{\psi^*(\mathbf{r})}\right) = \frac{1}{2i}\oint_{\varepsilon-loop}\left(\frac{d\psi}{\psi}\right) + c.c., \qquad (12.20)$$

since $d\mathbf{r}\cdot\nabla \equiv d$. If we think of the last expression as an integral over a complex variable $\psi$, we may see that there is a pole $\psi = 0$ at the center of the $\varepsilon$-circle. Hence this evaluates by residue theorem as $\pi n + \pi n = 2\pi n$ (including the $c.c.$ contribution). Thus the flux threading through a vortex core is quantised. This quantisation is due to topology rather than quantum mechanics. So far we have encountered one kind of non-locality related to vortices and now we go on to describe fermions using non-local operators.

## 12.2 Nonlocal Particle Hole Creation Operators

In this section, we describe fermions using operators that correspond to creation of particle hole pairs across a filled Fermi sea. These operators must be such that the kinetic energy of fermions is purely diagonal in these operators, as was the case in the Fermi picture. These particle-hole-like creation operators come out as nonlocal operators in a sense to be made precise soon.

Consider a spinless Fermi system with annihilation and creation operators $c_{\mathbf{p}}, c^\dagger_{\mathbf{p}}$. We have encountered such operators earlier. Thus $\{c_{\mathbf{p}}, c_{\mathbf{p}'}\} = 0$ and $\{c_{\mathbf{p}}, c^\dagger_{\mathbf{p}'}\} = \delta_{\mathbf{p},\mathbf{p}'}$. We define $|FS>$ to be the filled Fermi sea of the noninteracting system with $N^0$ number of fermions. Thus $n_F(\mathbf{k}) = \theta(k_F - |\mathbf{k}|)$ is the momentum distribution of the noninteracting theory and $N^0 = \sum_{\mathbf{k}} n_F(\mathbf{k})$. We define $c_{\mathbf{p},<} = n_F(\mathbf{p})\, c_{\mathbf{p}}$ and $c_{\mathbf{p},>} = (1 - n_F(\mathbf{p}))c_{\mathbf{p}}$.

$$A_{\mathbf{k}}(\mathbf{q}) = c^\dagger_{\mathbf{k}-\mathbf{q}/2,<}\frac{1}{\sqrt{N_>}}c_{\mathbf{k}+\mathbf{q}/2,>}\ ;\ A^\dagger_{\mathbf{k}}(\mathbf{q}) = c^\dagger_{\mathbf{k}+\mathbf{q}/2,>}\frac{1}{\sqrt{N_>}}c_{\mathbf{k}-\mathbf{q}/2,<} \qquad (12.21)$$

where,

$$N_> = \sum_{\mathbf{k}} c_{\mathbf{k},<}c^\dagger_{\mathbf{k},<} = N^0 - \sum_{\mathbf{k}} c^\dagger_{\mathbf{k},<}c_{\mathbf{k},<}. \qquad (12.22)$$

Here $N_>$ measures the number of particle-hole pairs in the state it acts on. It is important to bear in mind that $[c_{\mathbf{p},>}, N^0] = 0$, hence $[c_{\mathbf{p},>}, N_>] = 0$. In other words, $N_> \neq \sum_{\mathbf{p}} c^\dagger_{\mathbf{p},>}c_{\mathbf{p},>}$ but $N_> = \sum_{\mathbf{p}} c^\dagger_{\mathbf{p},>}c_{\mathbf{p},>} - \hat{N} + N^0$. The definition for $A_{\mathbf{k}}(\mathbf{q})$ is unambiguous except when it acts on the filled Fermi sea (where the number of particle-hole pairs is zero). We postulate that in this case,

$$A_{\mathbf{k}}(\mathbf{q})|FS>= 0. \qquad (12.23)$$

With these caveats in mind, it is easy to verify that following are operator identities on the $N^0$ particle subspace of the Fock space of fermions.

$$c^{\dagger}_{\mathbf{k}+\mathbf{q}/2,>}c_{\mathbf{k}-\mathbf{q}/2,>} = \sum_{\mathbf{q}_1} A^{\dagger}_{\mathbf{k}+\mathbf{q}/2-\mathbf{q}_1/2}(\mathbf{q}_1)A_{\mathbf{k}-\mathbf{q}_1/2}(-\mathbf{q}+\mathbf{q}_1) \tag{12.24}$$

$$c^{\dagger}_{\mathbf{k}+\mathbf{q}/2,<}c_{\mathbf{k}-\mathbf{q}/2,<} = n_F(\mathbf{k})\,\delta_{\mathbf{q},0} - \sum_{\mathbf{q}_1} A^{\dagger}_{\mathbf{k}-\mathbf{q}/2+\mathbf{q}_1/2}(\mathbf{q}_1)A_{\mathbf{k}+\mathbf{q}_1/2}(-\mathbf{q}+\mathbf{q}_1) \tag{12.25}$$

The definition is sufficient to ensure idempotence of the number operator since it follows from the definition of the sea-displacement annihilation operator that,

$$A_{\mathbf{p}}(\mathbf{q})A_{\mathbf{p}+\mathbf{Q}/2}(\mathbf{q}+\mathbf{Q}) = A_{\mathbf{p}}(\mathbf{q})A_{\mathbf{p}+\mathbf{Q}/2}(\mathbf{q}-\mathbf{Q}) = 0. \tag{12.26}$$

While it is easy to see that $[A_{\mathbf{k}}(\mathbf{q}),A_{\mathbf{k}'}(\mathbf{q}')] = 0$, the commutator $[A_{\mathbf{k}}(\mathbf{q}),A^{\dagger}_{\mathbf{k}'}(\mathbf{q}')]$ is not a c-number and is quite complicated. Hence these objects are not bosons at all, even though for deriving many properties we may get away with treating them as though they were. If we wish to set our eyes on what the commutator may look like, we may write (keeping in mind that $A_{\mathbf{k}}(\mathbf{q})f(N_>) = f(N_> + 1)A_{\mathbf{k}}(\mathbf{q})$, $f(N_>)A^{\dagger}_{\mathbf{k}}(\mathbf{q}) = A^{\dagger}_{\mathbf{k}}(\mathbf{q})f(N_> + 1)$ ).

$$\begin{aligned}
&(N_> + 1)A_{\mathbf{k}}(\mathbf{q})\,A^{\dagger}_{\mathbf{k}'}(\mathbf{q}') - N_>\,A^{\dagger}_{\mathbf{k}'}(\mathbf{q}')\,A_{\mathbf{k}}(\mathbf{q}) \\
&= n_F(\mathbf{k}-\mathbf{q}/2)(1-n_F(\mathbf{k}+\mathbf{q}/2))\,\delta_{\mathbf{q},\mathbf{q}'}\,\delta_{\mathbf{k},\mathbf{k}'} \\
&-\sum_{\mathbf{q}_1} A^{\dagger}_{\mathbf{k}+\mathbf{q}/2-\mathbf{q}'+\mathbf{q}_1/2}(\mathbf{q}_1)A_{\mathbf{k}-\mathbf{q}'/2+\mathbf{q}_1/2}(\mathbf{q}-\mathbf{q}'+\mathbf{q}_1) \\
&\qquad (1-n_F(\mathbf{k}+\mathbf{q}/2))\,\delta_{\mathbf{k}+\mathbf{q}/2,\mathbf{k}'+\mathbf{q}'/2} \\
&-\sum_{\mathbf{q}_1} A^{\dagger}_{\mathbf{k}-\mathbf{q}/2+\mathbf{q}'-\mathbf{q}_1/2}(\mathbf{q}_1)A_{\mathbf{k}+\mathbf{q}'/2-\mathbf{q}_1/2}(\mathbf{q}-\mathbf{q}'+\mathbf{q}_1) \\
&\qquad n_F(\mathbf{k}-\mathbf{q}/2)\delta_{\mathbf{k}-\mathbf{q}/2,\mathbf{k}'-\mathbf{q}'/2}
\end{aligned} \tag{12.27}$$

In fact, this commutator is not simple even when $(\mathbf{k},\mathbf{q})$ and $(\mathbf{k}',\mathbf{q}')$ are completely unrelated. For example, if $\mathbf{k}+\mathbf{q}/2 \neq \mathbf{k}'+\mathbf{q}'/2$ and $\mathbf{k}-\mathbf{q}/2 \neq \mathbf{k}'-\mathbf{q}'/2$, even then we find the nontrivial commutation rule,

$$(N_> + 1)A_{\mathbf{k}}(\mathbf{q})\,A^{\dagger}_{\mathbf{k}'}(\mathbf{q}') - N_>\,A^{\dagger}_{\mathbf{k}'}(\mathbf{q}')\,A_{\mathbf{k}}(\mathbf{q}) = 0 \tag{12.28}$$

rather than the naive expectation $[A_{\mathbf{k}}(\mathbf{q}),A^{\dagger}_{\mathbf{k}'}(\mathbf{q}')] = 0$. This is particularly important since some have argued that this naive rule holds whereas the kinetic energy operator continues to be $K = const. + \sum_{\mathbf{k},\mathbf{q}} \frac{\mathbf{k}.\mathbf{q}}{m}A^{\dagger}_{\mathbf{k}}(\mathbf{q})A_{\mathbf{k}}(\mathbf{q})$ as it is in the present approach. This is clearly untenable. Even if we claim that maybe we have to restrict the Hilbert space to contain only states with a large number of particle-holepairs

($N_>$ large) so that the naive expectation is realized, this spoils the commutator $[A_{\mathbf{k}}(\mathbf{q}), A_{\mathbf{k}}^\dagger(\mathbf{q})]$, which is not of order unity, indeed it is not even a c-number. The repeated appearance of the object $N_>$ points to the importance of fluctuations in the number of particle-hole pairs. Thus, if we wish to rigorously justify using the kinetic energy operator that is a simple quadratic diagonal form in the A-operators, then we are stuck with complicated commutation rules for the A-operators, the simplest of which is the random phase approximation (RPA) obtained by making the axiom that we shall set $c^\dagger_{\mathbf{k}+\mathbf{q}/2,<}c_{\mathbf{k}-\mathbf{q}/2,<} = 0$ and $c^\dagger_{\mathbf{k}+\mathbf{q}/2,>}c_{\mathbf{k}-\mathbf{q}/2,>} = 0$ if $\mathbf{q} \neq 0$. Further, if $a_{\mathbf{k}}(\mathbf{q}) = c^\dagger_{\mathbf{k}-\mathbf{q}/2,<}c_{\mathbf{k}+\mathbf{q}/2,>}$, then we could postulate that the commutator $[a_{\mathbf{k}}(\mathbf{q}), a^\dagger_{\mathbf{k}'}(\mathbf{q}')]$ is a c-number. This is the usual RPA. A possible proof of the controlled nature of the random phase approximation comes from the Hubbard-Stratanovich approach to bosonization explained by some authors cited. There it is argued that RPA becomes exact in the limit $k_F, m \to \infty$ such that $k_F/m = v_F < \infty$ (flat Fermi surface). This fact is also well known by the experts of 1d bosonization such as Giamarchi (see his textbook). We shall not dwell on this question, but simply assert that in the limit that the curvature of the Fermi surface disappears, our results are asymptotically exact. In practice, comparison with Bethe ansatz results and so on, shows this to be a reasonable assertion for weakly coupled systems with short range interactions.

### 12.2.1 A Systematic Approach for Going beyond RPA

Here we wish to discuss a systematic approach for going beyond the random phase approximation. We shall not employ this method, but propose this approach as the natural technique for systematically approaching the many-body problem. We make no claims about our approach being controlled in any sense of the term. It is merely a systematic and natural way of going beyond the random phase approximation. The RPA itself has never been shown to be a 'controlled' approximation in the literature (in whatever sense of the term), hence any approach that uses the RPA as the starting point is equally suspect. For example, in the excellent and authoritative text, *Quantum Physics in One Dimension*, Giamarchi defends his dropping of the $g_1$ term (backward scattering term) in the geology language thereby making the RPA a valid starting point, using phrases such as 'it is a pain in the neck'. Clearly then, the RPA is not a controlled approximation. This criticism that approximations may not be 'controlled' applies to conventional bosonization approaches, certainly to conventional perturbation theory (the coupling may be small, but the integrals that appear as coefficients of these couplings may diverge), to variational approaches, even dynamical mean field theory is not above such criticism (it is exact only in infinite dimensions, there is no proof that a systematic perturbation series in inverse powers of the dimension exists or is mathematically meaningful). Indeed, the biggest uncontrolled approximation routinely used in physics is the independent particle approximation. All these approaches are nevertheless quite successful in describing the real world, albeit in limited domains. The same point of view is adopted with regard to approximations. Now we wish to describe the approach we recommend. Consider the operator $a_{\mathbf{k}}(\mathbf{q}) = c^{\dagger}_{\mathbf{k}-\mathbf{q}/2,<}c_{\mathbf{k}+\mathbf{q}/2,>}$ rather than $A_{\mathbf{k}}(\mathbf{q})$. The reason is because the number-conserving product of two Fermi fields is related to this object in a simple way in the RPA sense. We now wish to ascertain then in what sense RPA has to be invoked. At the outset we have $[a_{\mathbf{k}}(\mathbf{q}), a_{\mathbf{k}'}(\mathbf{q}')] = 0$. However, one finds that a natural way of thinking about the other commutator is,

$$
\begin{aligned}
[a_{\mathbf{k}}(\mathbf{q}), a^{\dagger}_{\mathbf{k}'}(\mathbf{q}')] &= (\hat{n}_{\mathbf{k}-\mathbf{q}/2} - \hat{n}_{\mathbf{k}+\mathbf{q}/2}) \\
& n_F(\mathbf{k}-\mathbf{q}/2)(1-n_F(\mathbf{k}+\mathbf{q}/2))\ \delta_{\mathbf{q},\mathbf{q}'}\ \delta_{\mathbf{k},\mathbf{k}'} \\
-\varepsilon \sum_{\mathbf{q}_1} \frac{1}{N_>} & a^{\dagger}_{\mathbf{k}+\mathbf{q}/2-\mathbf{q}'+\mathbf{q}_1/2}(\mathbf{q}_1) a_{\mathbf{k}-\mathbf{q}'/2+\mathbf{q}_1/2}(\mathbf{q}-\mathbf{q}'+\mathbf{q}_1) \\
& (1-n_F(\mathbf{k}+\mathbf{q}/2))\ \delta_{\mathbf{k}+\mathbf{q}/2,\mathbf{k}'+\mathbf{q}'/2}(1-\delta_{\mathbf{q},\mathbf{q}'}) \\
-\varepsilon \sum_{\mathbf{q}_1} \frac{1}{N_>} & a^{\dagger}_{\mathbf{k}-\mathbf{q}/2+\mathbf{q}'-\mathbf{q}_1/2}(\mathbf{q}_1) a_{\mathbf{k}+\mathbf{q}'/2-\mathbf{q}_1/2}(\mathbf{q}-\mathbf{q}'+\mathbf{q}_1) \\
& n_F(\mathbf{k}-\mathbf{q}/2)\delta_{\mathbf{k}-\mathbf{q}/2,\mathbf{k}'-\mathbf{q}'/2}(1-\delta_{\mathbf{q},\mathbf{q}'}).
\end{aligned}
\tag{12.29}
$$

The number-conserving product of two Fermi fields is given in terms of these operators as,

$$
\begin{aligned}
c^{\dagger}_{\mathbf{k}+\mathbf{q}/2}c_{\mathbf{k}-\mathbf{q}/2} &= \hat{n}_{\mathbf{k}}\ \delta_{\mathbf{q},0} + a_{\mathbf{k}}(-\mathbf{q}) + a^{\dagger}_{\mathbf{k}}(\mathbf{q}) \\
+\varepsilon\ (1-\delta_{\mathbf{q},0}) & \sum_{\mathbf{q}_1} \frac{1}{N_>}\ a^{\dagger}_{\mathbf{k}+\mathbf{q}/2-\mathbf{q}_1/2}(\mathbf{q}_1) a_{\mathbf{k}-\mathbf{q}_1/2}(-\mathbf{q}+\mathbf{q}_1) \\
-\varepsilon\ (1-\delta_{\mathbf{q},0}) & \sum_{\mathbf{q}_1} \frac{1}{N_>}\ a^{\dagger}_{\mathbf{k}-\mathbf{q}/2+\mathbf{q}_1/2}(\mathbf{q}_1) a_{\mathbf{k}+\mathbf{q}_1/2}(-\mathbf{q}+\mathbf{q}_1).
\end{aligned}
\tag{12.30}
$$

Here,

$$
\begin{aligned}
\hat{n}_{\mathbf{k}} = n_F(\mathbf{k}) &+ \sum_{\mathbf{q}_1} \frac{1}{N_>}\ a^{\dagger}_{\mathbf{k}-\mathbf{q}_1/2}(\mathbf{q}_1) a_{\mathbf{k}-\mathbf{q}_1/2}(\mathbf{q}_1) \\
&- \sum_{\mathbf{q}_1} \frac{1}{N_>}\ a^{\dagger}_{\mathbf{k}+\mathbf{q}_1/2}(\mathbf{q}_1) a_{\mathbf{k}+\mathbf{q}_1/2}(\mathbf{q}_1).
\end{aligned}
\tag{12.31}
$$

The implication is that we expand in powers of $\varepsilon$ and at the end of a calculation set $\varepsilon = 1$. Here we have used $[N_>, a_{\mathbf{k}}(\mathbf{q})] = -a_{\mathbf{k}}(\mathbf{q})$ and $[N_>, a^{\dagger}_{\mathbf{k}}(\mathbf{q})] = a^{\dagger}_{\mathbf{k}}(\mathbf{q})$, therefore, $[N_>, a^{\dagger}_{\mathbf{k}}(\mathbf{q})a_{\mathbf{k}'}(\mathbf{q}')] = 0$. In general $f(N_>)a_{\mathbf{k}}(\mathbf{q}) = a_{\mathbf{k}}(\mathbf{q})f(N_> - 1)$ and $f(N_>)a^{\dagger}_{\mathbf{k}}(\mathbf{q}) = a^{\dagger}_{\mathbf{k}}(\mathbf{q})f(N_> + 1)$. Finally, $N^2_> = \sum_{\mathbf{k},\mathbf{q}} a^{\dagger}_{\mathbf{k}}(\mathbf{q})a_{\mathbf{k}}(\mathbf{q})$. So far the scheme has been perfectly general. Now we wish to make the approximation that would amount to setting $\varepsilon = 0$.

**The (Generalized) RPA** (setting $\varepsilon = 0$):

$$[a_{\mathbf{k}}(\mathbf{q}), a^{\dagger}_{\mathbf{k}'}(\mathbf{q}')] \approx (\hat{n}_{\mathbf{k}-\mathbf{q}/2} - \hat{n}_{\mathbf{k}+\mathbf{q}/2})\, n_F(\mathbf{k}-\mathbf{q}/2)(1-n_F(\mathbf{k}+\mathbf{q}/2))\, \delta_{\mathbf{q},\mathbf{q}'}\, \delta_{\mathbf{k},\mathbf{k}'} \tag{12.32}$$

$$[a_{\mathbf{k}}(\mathbf{q}), a_{\mathbf{k}'}(\mathbf{q}')] = 0\, ;\ [a_{\mathbf{k}}(\mathbf{q}), \hat{n}_{\mathbf{p}}] = (\delta_{\mathbf{p},\mathbf{k}+\mathbf{q}/2} - \delta_{\mathbf{p},\mathbf{k}-\mathbf{q}/2})\, a_{\mathbf{k}}(\mathbf{q}) \tag{12.33}$$

$$c^{\dagger}_{\mathbf{k}+\mathbf{q}/2} c_{\mathbf{k}-\mathbf{q}/2} \approx \hat{n}_{\mathbf{k}}\, \delta_{\mathbf{q},0} + a_{\mathbf{k}}(-\mathbf{q}) + a^{\dagger}_{\mathbf{k}}(\mathbf{q}) \tag{12.34}$$

$$c^{\dagger}_{\mathbf{k}} c_{\mathbf{k}} \equiv \hat{n}_{\mathbf{k}} = n_F(\mathbf{k}) + \sum_{\mathbf{q}_1} \frac{1}{N_>}\, a^{\dagger}_{\mathbf{k}-\mathbf{q}_1/2}(\mathbf{q}_1) a_{\mathbf{k}-\mathbf{q}_1/2}(\mathbf{q}_1) - \sum_{\mathbf{q}_1} \frac{1}{N_>}\, a^{\dagger}_{\mathbf{k}+\mathbf{q}_1/2}(\mathbf{q}_1) a_{\mathbf{k}+\mathbf{q}_1/2}(\mathbf{q}_1). \tag{12.35}$$

The RPA is consistent since the following simplified commutation rules are obeyed. If $\mathbf{q}, \mathbf{q}' \neq 0$,

$$[c^{\dagger}_{\mathbf{k}+\mathbf{q}/2} c_{\mathbf{k}-\mathbf{q}/2}, c^{\dagger}_{\mathbf{k}'+\mathbf{q}'/2} c_{\mathbf{k}'-\mathbf{q}'/2}] = \delta_{\mathbf{k},\mathbf{k}'} \delta_{\mathbf{q},-\mathbf{q}'}\, (\hat{n}_{\mathbf{k}+\mathbf{q}/2} - \hat{n}_{\mathbf{k}-\mathbf{q}/2}) D(\mathbf{k},\mathbf{q}) \tag{12.36}$$

and,

$$[c^{\dagger}_{\mathbf{k}+\mathbf{q}/2} c_{\mathbf{k}-\mathbf{q}/2}, c^{\dagger}_{\mathbf{k}'} c_{\mathbf{k}'}] = c^{\dagger}_{\mathbf{k}+\mathbf{q}/2} c_{\mathbf{k}-\mathbf{q}/2}\, (\delta_{\mathbf{k}',\mathbf{k}-\mathbf{q}/2} - \delta_{\mathbf{k}',\mathbf{k}+\mathbf{q}/2}). \tag{12.37}$$

Finally,

$$[c^{\dagger}_{\mathbf{k}} c_{\mathbf{k}}, c^{\dagger}_{\mathbf{k}'} c_{\mathbf{k}'}] = 0, \tag{12.38}$$

where $D(\mathbf{k},\mathbf{q}) = n_F(\mathbf{k}-\mathbf{q}/2)(1-n_F(\mathbf{k}+\mathbf{q}/2)) + n_F(\mathbf{k}+\mathbf{q}/2)(1-n_F(\mathbf{k}-\mathbf{q}/2))$. We now wish to show that the RPA is a conserving approximation. If we define $\rho_{\mathbf{q}} = \sum_{\mathbf{k}} c^{\dagger}_{\mathbf{k}+\mathbf{q}/2} c_{\mathbf{k}-\mathbf{q}/2}$ and $\mathbf{j}_{\mathbf{q}} = \sum_{\mathbf{k}} \mathbf{k}\, c^{\dagger}_{\mathbf{k}+\mathbf{q}/2} c_{\mathbf{k}-\mathbf{q}/2}$ and set the Hamiltonian to be $H = \sum_{\mathbf{k}} \frac{k^2}{2m} c^{\dagger}_{\mathbf{k}} c_{\mathbf{k}}$, the RPA ensures that $i\frac{\partial}{\partial t}\rho_{\mathbf{q}} = -\frac{\mathbf{q}}{m} \cdot \mathbf{j}_{\mathbf{q}}$, which is nothing but the equation of continuity. Thus the RPA is a conserving approximation in the sense of Kadanoff and Baym. Now we wish to perform some concrete computations using this approximation scheme. In particular, it would be desirable to calculate the momentum distribution of the free Fermi theory at finite temperature and see if it agrees with the one obtain from Fermi algebra. The reason why this is interesting is because the finite temperature calculation of the free theory is the simplest way of considering excited states, in other words, fluctuations in the number of particle-hole pairs. This exercise also teaches us that the simple RPA, where the right-hand side of Eq. (12.32) is replaced by a c-number, is not sufficient. The reason for this is that the fluctuations in the number of particle-hole pairs have to be taken into account self-consistently in order to reproduce the right finite temperature momentum distribution. To compute the momentum distribution at finite temperature, it is better to calculate the following finite temperature correlation function:

$$G(\mathbf{k},\mathbf{q};\lambda) = < e^{-\lambda N_>} a^{\dagger}_{\mathbf{k}}(\mathbf{q}) a_{\mathbf{k}}(\mathbf{q}) > \equiv \frac{Tr\left(e^{-\beta(H-\mu N)}\, e^{-\lambda N_>} a^{\dagger}_{\mathbf{k}}(\mathbf{q}) a_{\mathbf{k}}(\mathbf{q})\right)}{Tr\left(e^{-\beta(H-\mu N)}\right)}. \quad (12.39)$$

Therefore,

$$\int_{\infty}^{\lambda} d\lambda' \; G(\mathbf{k},\mathbf{q};\lambda') = - < e^{-\lambda N_>} \frac{1}{N_>} a^{\dagger}_{\mathbf{k}}(\mathbf{q}) a_{\mathbf{k}}(\mathbf{q}) > \quad (12.40)$$

and,

$$< \hat{n}_{\mathbf{k},\lambda} > = < e^{-\lambda N_>} c^{\dagger}_{\mathbf{k}} c_{\mathbf{k}} > = n_F(\mathbf{k}) < e^{-\lambda N_>} >$$

$$- \sum_{\mathbf{q}} \int_{\infty}^{\lambda} d\lambda' \; G(\mathbf{k}-\mathbf{q}/2,\mathbf{q};\lambda') + \sum_{\mathbf{q}} \int_{\infty}^{\lambda} d\lambda' G(\mathbf{k}+\mathbf{q}/2,\mathbf{q};\lambda'). \quad (12.41)$$

Using the cyclic permutation property of the trace and the RPA algebra, we obtain the following expression for $G$.

$$G(\mathbf{k},\mathbf{q};\lambda) = \frac{e^{-\lambda}\, e^{-\beta \frac{\mathbf{k}.\mathbf{q}}{m}}}{(1 - e^{-\lambda}\, e^{-\beta \frac{\mathbf{k}.\mathbf{q}}{m}})} (< \hat{n}_{\mathbf{k}-\mathbf{q}/2,\lambda} > - < \hat{n}_{\mathbf{k}+\mathbf{q}/2,\lambda} >)$$

$$n_F(\mathbf{k}-\mathbf{q}/2)(1 - n_F(\mathbf{k}+\mathbf{q}/2)) \quad (12.42)$$

Let $D(\varepsilon)$ be the density of states of the free theory. Thus, $D(\varepsilon)d\varepsilon = \frac{V}{(2\pi)^d}\Omega_d k^{(d-1)}dk$. Note that $D(\varepsilon)$ is an extensive quantity as is the summation $\sum_{\mathbf{q}}$. Thus we have to ensure that the dependence of $n(\lambda,\varepsilon)$ on $\lambda$ is such that when integrated over $\lambda$, leads to an extensive quantity in the denominator. This matter may be made more explicit by differentiating with respect to $\lambda$.

$$\frac{d}{d\lambda} n_<(\lambda,\varepsilon) = -\theta(\varepsilon_F - \varepsilon)u(\lambda)$$

$$+ \int_{\varepsilon_F}^{\infty} d\varepsilon' D(\varepsilon') \frac{1}{(e^{\lambda}\, e^{\beta(\varepsilon'-\varepsilon)} - 1)} (n_<(\lambda,\varepsilon) - n_>(\lambda,\varepsilon'))\; \theta(\varepsilon_F - \varepsilon) \quad (12.43)$$

$$\frac{d}{d\lambda} n_>(\lambda,\varepsilon) =$$

$$- \int_{0}^{\varepsilon_F} d\varepsilon' D(\varepsilon') \frac{1}{(e^{\lambda}\, e^{\beta(\varepsilon-\varepsilon')} - 1)} (n_<(\lambda,\varepsilon') - n_>(\lambda,\varepsilon))\; \theta(\varepsilon - \varepsilon_F) \quad (12.44)$$

$$u(\lambda) = \left\langle e^{-\lambda N_>} N_> \right\rangle = N^0 \left\langle e^{-\lambda N_>} \right\rangle - \int_{0}^{\varepsilon_F} d\varepsilon\; D(\varepsilon) n_<(\lambda,\varepsilon)$$

$$= \int_{\varepsilon_F}^{\infty} d\varepsilon\, D(\varepsilon) n_{>}(\lambda,\varepsilon) \tag{12.45}$$

We may suspect, that these equations can be solved by the following ansatz.[1]

$$n_{>,<}(\lambda,\varepsilon) = \tilde{n}_{>,<}(\lambda,\varepsilon)\, e^{I(\lambda)} \tag{12.47}$$

where the function $I(\lambda)$ is extensive and is independent of the energy variable $\varepsilon$, whereas $\tilde{n}$ is intensive and depends on both the variables in general. Substituting this ansatz into the equations we find,

$$I^{'}(\lambda)\tilde{n}_{<}(\lambda,\varepsilon) + \frac{d}{d\lambda}\tilde{n}_{<}(\lambda,\varepsilon) = -\theta(\varepsilon_F - \varepsilon)\tilde{u}(\lambda)$$

$$+\int_{\varepsilon_F}^{\infty} d\varepsilon^{'} D(\varepsilon^{'}) \frac{1}{(e^{\lambda}\, e^{\beta(\varepsilon^{'}-\varepsilon)} - 1)} (\tilde{n}_{<}(\lambda,\varepsilon) - \tilde{n}_{>}(\lambda,\varepsilon^{'}))\, \theta(\varepsilon_F - \varepsilon) \tag{12.48}$$

$$I^{'}(\lambda)\tilde{n}_{>}(\lambda,\varepsilon) + \frac{d}{d\lambda}\tilde{n}_{>}(\lambda,\varepsilon) =$$

$$-\int_{0}^{\varepsilon_F} d\varepsilon^{'} D(\varepsilon^{'})\, \frac{1}{(e^{\lambda}\, e^{\beta(\varepsilon-\varepsilon^{'})} - 1)} (\tilde{n}_{<}(\lambda,\varepsilon^{'}) - \tilde{n}_{>}(\lambda,\varepsilon))\, \theta(\varepsilon - \varepsilon_F) \tag{12.49}$$

$$\tilde{u}(\lambda) = \int_{\varepsilon_F}^{\infty} D(\varepsilon)\, d\varepsilon\, \tilde{n}_{>}(\lambda,\varepsilon). \tag{12.50}$$

Since $I^{'}(\lambda), \tilde{u}$ and $D(\varepsilon)$ are extensive and $\tilde{n}$ is intensive, we may write after setting $\lambda = 0$,

$$I^{'}(0)\tilde{n}_{<}(0,\varepsilon) = -\theta(\varepsilon_F - \varepsilon)\tilde{u}(0)$$

$$+\int_{\varepsilon_F}^{\infty} d\varepsilon^{'} D(\varepsilon^{'}) \frac{1}{(e^{\beta(\varepsilon^{'}-\varepsilon)} - 1)} (\tilde{n}_{<}(0,\varepsilon) - \tilde{n}_{>}(0,\varepsilon^{'}))\, \theta(\varepsilon_F - \varepsilon) \tag{12.51}$$

$$I^{'}(0)\tilde{n}_{>}(0,\varepsilon) =$$

$$-\int_{0}^{\varepsilon_F} d\varepsilon^{'} D(\varepsilon^{'})\, \frac{1}{(e^{\beta(\varepsilon-\varepsilon^{'})} - 1)} (\tilde{n}_{<}(0,\varepsilon^{'}) - \tilde{n}_{>}(0,\varepsilon))\, \theta(\varepsilon - \varepsilon_F). \tag{12.52}$$

Dividing both sides of Eq. (12.52) by $\tilde{n}_{>}(0,\varepsilon)$ allows us to suspect that it should be possible to write

$$\left(\frac{\tilde{n}_{<}(0,\varepsilon^{'})}{\tilde{n}_{>}(0,\varepsilon)} - 1\right) = (e^{\beta(\varepsilon-\varepsilon^{'})} - 1)\, h(\varepsilon^{'}), \tag{12.53}$$

[1]As the reader may have suspected, the authors thought of this by inspecting the exact result obtained by elementary means.

$$n(\lambda,\varepsilon) = \frac{1}{(e^{\beta(\varepsilon-\mu)-\lambda\theta(\varepsilon_F-\varepsilon)} + 1)} e^{-\lambda N^0} Exp\left(\int_0^{\varepsilon_F} D(\varepsilon^{'})\, d\varepsilon^{'}\, Log\left(\frac{\left(1+e^{-\beta(\varepsilon^{'}-\mu)+\lambda}\right)}{\left(1+e^{-\beta(\varepsilon^{'}-\mu)}\right)}\right)\right) \tag{12.46}$$

so that for some $h$ we have,

$$I'(0) = -\int_0^{\varepsilon_F} d\varepsilon' D(\varepsilon')\, h(\varepsilon'). \tag{12.54}$$

Interchanging $\varepsilon$ and $\varepsilon'$ in Eq. (12.53) and substituting into Eq. (12.51) we obtain

$$I'(0)\tilde{n}_<(0,\varepsilon) = -\theta(\varepsilon_F - \varepsilon)\tilde{u}(0) + h(\varepsilon)\, \tilde{u}(0)\, \theta(\varepsilon_F - \varepsilon). \tag{12.55}$$

We now multiply by the density of states and integrate to obtain, $I'(0)N^{0,<} = -N^0\, N^{0,>} - I'(0)\, N^{0,>}$, where the notation is self-explanatory. Thus $I'(0) = -N^{0,>} = -\tilde{u}(0)$. In other words, $\tilde{n}_<(0,\varepsilon) = 1 - h(\varepsilon)$. Hence we find,

$$\tilde{n}_>(0,\varepsilon) = \frac{1}{1 + \frac{h(\varepsilon')}{1-h(\varepsilon')} e^{\beta(\varepsilon-\varepsilon')}}. \tag{12.56}$$

Therefore, we may conclude that there exists a constant $\mu$ such that, $\frac{h(\varepsilon')}{1-h(\varepsilon')} = e^{\beta(\varepsilon'-\mu)}$, or $h(\varepsilon') = \frac{1}{e^{-\beta(\varepsilon'-\mu)}+1}$. Thus $\tilde{n}_>(0,\varepsilon) = \tilde{n}_<(0,\varepsilon) = \frac{1}{e^{\beta(\varepsilon-\mu)}+1}$. It is remarkable indeed that the Fermi-Dirac distribution emerges from a theory that is bosonic in character. However, it is important to impress upon the reader that it is the *generalized* RPA that takes into account fluctuations in the number of particle-hole pairs in a self-consistent manner that leads to the Fermi-Dirac distribution, whereas the simple-minded RPA fails to do so. This latter fact is easily seen by replacing the commutator $[a_{\mathbf{k}}(\mathbf{q}), a^{\dagger}_{\mathbf{k}}(\mathbf{q})] = n_F(\mathbf{k} - \mathbf{q}/2)(1 - n_F(\mathbf{k} + \mathbf{q}/2))$, which is nothing but the simple-minded RPA. This choice will lead to a result for the momentum distribution at finite temperature as follows.

$$n(0,\varepsilon) = \theta(\varepsilon_F - \varepsilon) - \int_0^{\varepsilon_F} d\varepsilon' D(\varepsilon') \int_{\infty}^{0} d\lambda'$$

$$\frac{1}{(e^{\lambda'}\, e^{\beta(\varepsilon-\varepsilon')} - 1)} (\theta(\varepsilon_F - \varepsilon') - \theta(\varepsilon_F - \varepsilon))\, \theta(\varepsilon - \varepsilon_F)$$

$$+ \int_{\varepsilon_F}^{\infty} d\varepsilon' D(\varepsilon') \int_{\infty}^{0} d\lambda' \frac{1}{(e^{\lambda'}\, e^{\beta(\varepsilon'-\varepsilon)} - 1)} (\theta(\varepsilon_F - \varepsilon) - \theta(\varepsilon_F - \varepsilon'))\, \theta(\varepsilon_F - \varepsilon) \tag{12.57}$$

This is clearly wrong since $D(\varepsilon)$ is an extensive quantity whereas the momentum distribution is not. Thus we have to solve for the momentum distribution self-consistently using the generalized RPA, which ensures that the integral over $\lambda'$ brings to the denominator an extensive quantity making the momentum distribution intensive as it should be.

One may also try to see if the GRPA reproduces the correct finite temperature correlation functions of the a-operators themselves. If we set $\lambda = 0$ in the above discussion, we obtain,

$$< a^{\dagger}_{\mathbf{k}}(\mathbf{q})a_{\mathbf{k}}(\mathbf{q}) >_{Bose-Lang} = \frac{1}{(e^{\beta\frac{\mathbf{k}.\mathbf{q}}{m}} - 1)}(< n_{\mathbf{k}-\mathbf{q}/2} > - < n_{\mathbf{k}+\mathbf{q}/2} >)$$

$$n_F(\mathbf{k}-\mathbf{q}/2)(1 - n_F(\mathbf{k}+\mathbf{q}/2)), \tag{12.58}$$

whereas, in the Fermi language it is

$$< a^{\dagger}_{\mathbf{k}}(\mathbf{q})a_{\mathbf{k}}(\mathbf{q}) >_{Fermi-Lang}$$

$$= < n_{\mathbf{k}+\mathbf{q}/2} > (1 - < n_{\mathbf{k}-\mathbf{q}/2} >)\ n_F(\mathbf{k}-\mathbf{q}/2)(1 - n_F(\mathbf{k}+\mathbf{q}/2)). \tag{12.59}$$

We may easily see that these two are the same formulas since $e^{\beta\frac{\mathbf{k}.\mathbf{q}}{m}} = e^{\beta\epsilon_{\mathbf{k}+\mathbf{q}/2}}e^{-\beta\epsilon_{\mathbf{k}-\mathbf{q}/2}} = \frac{(\frac{1}{<n_{\mathbf{k}+\mathbf{q}/2}>} - 1)}{(\frac{1}{<n_{\mathbf{k}-\mathbf{q}/2}>} - 1)}$. Thus the important point is that we have to invoke the GRPA that retains the number operator on the right-hand side rather than the simple-minded RPA that replaces the right-hand side by the zero temperature c-number functions.

## 12.3 The Sea-Boson

Now we introduce the concept of the 'sea-boson'. We regard it as merely a convenient way of parameterizing the properties of fermions using bosonic operators. Its validity is checked by comparing with limiting cases and other exactly solved models. This approach is common in many areas of physics, where proofs, when they are hard to come by, are replaced by plausibility arguments. In this sea-boson theory, we make the following prescriptions (that are justified a posteriori). Consider as before, a spinless Fermi system with annihilation and creation operators $c_{\mathbf{p}}, c^{\dagger}_{\mathbf{p}}$. Thus $\{c_{\mathbf{p}}, c_{\mathbf{p}^{'}}\} = 0$ and $\{c_{\mathbf{p}}, c^{\dagger}_{\mathbf{p}^{'}}\} = \delta_{\mathbf{p},\mathbf{p}^{'}}$. We define $|FS>$ to be the filled Fermi sea of the noninteracting system with $N^0$ number of fermions. Thus $n_F(\mathbf{k}) = \theta(k_F - |\mathbf{k}|)$ is the momentum distribution of the noninteracting theory and $N^0 = \sum_{\mathbf{k}} n_F(\mathbf{k})$. We define $c_{\mathbf{p},<} = n_F(\mathbf{p})\ c_{\mathbf{p}}$ and $c_{\mathbf{p},>} = (1 - n_F(\mathbf{p}))c_{\mathbf{p}}$.

In the sea-boson framework, the operator $c^{\dagger}_{\mathbf{k}-\mathbf{q}/2,<}c_{\mathbf{k}+\mathbf{q}/2,>} \rightarrow a_{\mathbf{k}}(\mathbf{q})$ is identified with a canonical boson obeying the commutation rules,

$$[a_{\mathbf{k}}(\mathbf{q}), a^{\dagger}_{\mathbf{k}^{'}}(\mathbf{q}^{'})] = n_F(\mathbf{k}-\mathbf{q}/2)(1 - n_F(\mathbf{k}+\mathbf{q}/2))\delta_{\mathbf{k},\mathbf{k}^{'}}\delta_{\mathbf{q},\mathbf{q}^{'}} \tag{12.60}$$

and

$$[a_{\mathbf{k}}(\mathbf{q}), a_{\mathbf{k}^{'}}(\mathbf{q}^{'})] = 0. \tag{12.61}$$

The kinetic energy of electrons is identified with,

$$K \to \sum_{\mathbf{k},\mathbf{q}} \frac{\mathbf{k}.\mathbf{q}}{m_e} a^{\dagger}_{\mathbf{k}}(\mathbf{q}) a_{\mathbf{k}}(\mathbf{q}) + E_0 \tag{12.62}$$

where $E_0$ is the kinetic energy c-number of the noninteracting electrons in the ground state. For the purposes of this article, we shall be needing one more prescription, namely that the momentum distribution of electrons is obtained by computing the average $< a^{\dagger}_{\mathbf{k}}(\mathbf{q}) a_{\mathbf{k}}(\mathbf{q}) >$. Now we take the point of view that either one uses the exact nonlocal operator with fluctuation number of particle-hole pairs discussed earlier or tries to make-do with the present ad hoc approach by somehow guessing a formula that takes into account these effects albeit in an indirect way. Consider the following claim. The average momentum distribution is given by,

$$< n_{\mathbf{k}} > = \frac{1}{2}\left(1 + e^{-2\sum_{\mathbf{q}} < a^{\dagger}_{\mathbf{k}+\mathbf{q}/2}(\mathbf{q}) a_{\mathbf{k}+\mathbf{q}/2}(\mathbf{q}) >}\right) n_F(\mathbf{k})$$

$$+ \frac{1}{2}\left(1 - e^{-2\sum_{\mathbf{q}} < a^{\dagger}_{\mathbf{k}-\mathbf{q}/2}(\mathbf{q}) a_{\mathbf{k}-\mathbf{q}/2}(\mathbf{q}) >}\right)(1 - n_F(\mathbf{k})). \tag{12.63}$$

The above peculiar form with operators in the exponent ensures that divergences at small $q$ are 'tamed' by the above procedure. This appears to be the price one has to pay to make do with simple bosons when in fact the real operators are nonlocal and highly sensitive to fluctuations in the number of particle-hole pairs as we have seen. For higher dimensions when there are no small $q$ divergences (infrared divergences), we may write,

$$< n_{\mathbf{k}} > \approx n_F(\mathbf{k}) + \sum_{\mathbf{q}} < a^{\dagger}_{\mathbf{k}-\mathbf{q}/2}(\mathbf{q}) a_{\mathbf{k}-\mathbf{q}/2}(\mathbf{q}) > - \sum_{\mathbf{q}} < a^{\dagger}_{\mathbf{k}+\mathbf{q}/2}(\mathbf{q}) a_{\mathbf{k}+\mathbf{q}/2}(\mathbf{q}) > . \tag{12.64}$$

This 'theory' works only at zero temperature, unlike the rigorous approach which gives the Fermi Dirac distribution with great difficulty but is otherwise nearly useless even though it is rigorous. We now wish to reproduce Galitskii's perturbative results for the quasiparticle residue in three dimensions. Application to Luttinger liquids in 1D and 2D, 1D Wigner crystal, and so on, are left to the exercises. One of the main purposes of this article is to show how to do this, which requires taking into account particle-hole modes carefully. To do this we have to first set some ground rules. We have to use a model of interacting fermions that becomes soluble when recast in this sea-boson language. We have shown repeatedly in earlier works that this corresponds to interactions that scatter pairs of fermions across the Fermi surface with momentum transfer that is small compared to the Fermi momentum (forward scattering only). Thus we wish to analyze the following Hamiltonian in the Fermi language.

$$H = \sum_{\mathbf{k}} \varepsilon_{\mathbf{k}} c^{\dagger}_{\mathbf{k}} c_{\mathbf{k}}$$

$$+\sum_{\mathbf{q}\neq 0}\frac{v_{\mathbf{q}}}{2V}\sum_{\mathbf{k}\neq\mathbf{k}'}(c^{\dagger}_{\mathbf{k}+\mathbf{q}/2,<}c_{\mathbf{k}-\mathbf{q}/2,>}+c^{\dagger}_{\mathbf{k}+\mathbf{q}/2,>}c_{\mathbf{k}-\mathbf{q}/2,<})$$

$$(c^{\dagger}_{\mathbf{k}'-\mathbf{q}/2,<}c_{\mathbf{k}'+\mathbf{q}/2,>}+c^{\dagger}_{\mathbf{k}'-\mathbf{q}/2,>}c_{\mathbf{k}'+\mathbf{q}/2,<}) \tag{12.65}$$

Here $v_{\mathbf{q}}=0$ if $|\mathbf{q}|>\Lambda\ll k_F$. The additional restriction that the scattering be across the Fermi surface makes the vectors $\mathbf{k},\mathbf{k}'$ lie close to the Fermi surface. This may be regarded as the RPA version of the usual jellium Hamiltonian. As we have included only processes that scatter across the Fermi surface, this is equivalent to setting $c^{\dagger}_{\mathbf{k}+\mathbf{q}/2,<}c_{\mathbf{k}-\mathbf{q}/2,<}\approx 0$ and $c^{\dagger}_{\mathbf{k}+\mathbf{q}/2,>}c_{\mathbf{k}-\mathbf{q}/2,>}\approx 0$ for $\mathbf{q}\neq 0$. In the sea-boson language, this is simply given by,

$$H=\sum_{\mathbf{k},\mathbf{q}}\frac{\mathbf{k}.\mathbf{q}}{m}a^{\dagger}_{\mathbf{k}}(\mathbf{q})a_{\mathbf{k}}(\mathbf{q})$$

$$+\sum_{\mathbf{q}\neq 0}\frac{v_{\mathbf{q}}}{2V}(a(-\mathbf{q})+a^{\dagger}(\mathbf{q}))(a(\mathbf{q})+a^{\dagger}(-\mathbf{q})) \tag{12.66}$$

where $a(\mathbf{q})=\sum_{\mathbf{k}}a_{\mathbf{k}}(\mathbf{q})$. We shall not repeat the details here since we have shown several times that the average in question may be easily calculated as

$$<a^{\dagger}_{\mathbf{k}}(\mathbf{q})a_{\mathbf{k}}(\mathbf{q})>=\frac{v_{\mathbf{q}}}{V}\sum_{i=modes}\frac{1}{|\epsilon'(\mathbf{q},\omega_i(\mathbf{q}))|}\frac{n_F(\mathbf{k}-\mathbf{q}/2)(1-n_F(\mathbf{k}+\mathbf{q}/2))}{(\omega_i(\mathbf{q})-\frac{\mathbf{k}.\mathbf{q}}{m_e})^2}, \tag{12.67}$$

where $\omega_i(\mathbf{q})$ are all allowed zeros of the RPA dielectric function,

$$\epsilon(\mathbf{q},\omega)=1-\frac{v_{\mathbf{q}}}{V}\sum_{\mathbf{k}}\frac{n_F(\mathbf{k}-\mathbf{q}/2)-n_F(\mathbf{k}+\mathbf{q}/2)}{\omega-\frac{\mathbf{k}.\mathbf{q}}{m_e}}. \tag{12.68}$$

In one spatial dimension we could afford to be casual about the interpretation of the sum over modes since eventually only the collective modes survive. However, it is crucial to reinterpret this sum in more than one dimension in order to recover the perturbation theory result. But we have to do it in a way that does not disturb the 1D result that was already coming out right. To do this, let us first examine the perturbation theory result. One way would be to expand the ground state in a perturbation series to second order and compute the expectation value of $n_{\mathbf{k}}$. The new ground state with interactions has a formal expression,

$$|G>=|FS>+\sum_{|k>\neq|FS>}\frac{|k><k|V|FS>}{E_0-E_k}$$

$$+\sum_{|k>\neq|FS>,|l>\neq|FS>}|k>\frac{<k|V|l><l|V|FS>}{(E_0-E_k)(E_0-E_l)}$$

$$- \sum_{|k>\neq|FS>} |k> \frac{<FS|V|FS><k|V|FS>}{(E_0-E_k)^2}$$
$$-\frac{1}{2}\sum_{|k>\neq|FS>} |FS> \frac{<FS|V|k><k|V|FS>}{(E_k-E_0)^2}. \tag{12.69}$$

Therefore,

$$<G|n_{\mathbf{k}}|G>=n_F(\mathbf{k}) - \sum_{|j>\neq|FS>} n_F(\mathbf{k})\frac{<FS|V|j><j|V|FS>}{(E_j-E_0)^2}$$
$$+ \sum_{|j>\neq|FS>} \frac{<FS|V|j><j|n_{\mathbf{k}}|j><j|V|FS>}{(E_0-E_j)^2}. \tag{12.70}$$

It is clear that we have to choose excited states $|j>$ that have two particle-hole pairs in them.

We choose $|j>=c^{\dagger}_{\mathbf{P}-\mathbf{Q}/2,>}c_{\mathbf{P}+\mathbf{Q}/2,<}c^{\dagger}_{\mathbf{P}'+\mathbf{Q}/2,>}c_{\mathbf{P}'-\mathbf{Q}/2,<}|FS>$. As usual we are going to insist that $\mathbf{P}\neq\mathbf{P}'$ and $\mathbf{Q}$ is small compared to $k_F$. Also, interchanging $\mathbf{P},\mathbf{P}'$ and changing the sign of $\mathbf{Q}$ is not going to yield anything different. So we shall remember to divide by a factor of two while summing over $|j>$ to compensate for this. The matrix element of the interaction is,

$$<FS|V|j>=\frac{v_{\mathbf{Q}}}{V}n_F(\mathbf{P}'-\mathbf{Q}/2)(1-n_F(\mathbf{P}'+\mathbf{Q}/2))$$
$$n_F(\mathbf{P}+\mathbf{Q}/2)(1-n_F(\mathbf{P}-\mathbf{Q}/2)) \tag{12.71}$$
$$<j|n_{\mathbf{k}}|j>=n_F(\mathbf{k})-\delta_{\mathbf{k},\mathbf{P}'-\mathbf{Q}/2}-\delta_{\mathbf{k},\mathbf{P}+\mathbf{Q}/2}+\delta_{\mathbf{k},\mathbf{P}-\mathbf{Q}/2}+\delta_{\mathbf{k},\mathbf{P}'+\mathbf{Q}/2} \tag{12.72}$$
$$E_j=E_0-\varepsilon_{\mathbf{P}'-\mathbf{Q}/2}-\varepsilon_{\mathbf{P}+\mathbf{Q}/2}+\varepsilon_{\mathbf{P}-\mathbf{Q}/2}+\varepsilon_{\mathbf{P}'+\mathbf{Q}/2}=E_0+\frac{\mathbf{P}'\cdot\mathbf{Q}}{m_e}-\frac{\mathbf{P}\cdot\mathbf{Q}}{m_e}. \tag{12.73}$$

We note that each of the particle-hole pairs contributes a positive energy to the excited state as the cutoff functions ensure that $\frac{\mathbf{P}'\cdot\mathbf{Q}}{m_e}>0$ and $-\frac{\mathbf{P}\cdot\mathbf{Q}}{m_e}>0$. This will be important in helping us interpret the sum over modes in the sea-boson formalism. After some simplification, we arrive at the following formula, which is what simple perturbation theory in the coupling in the Fermi operator language tells us.

$$<G|n_{\mathbf{k}}|G>\approx n_F(\mathbf{k})$$
$$-\sum_{\mathbf{k}',\mathbf{q}}\left(\frac{v_{\mathbf{q}}}{V}\right)^2\frac{n_F(\mathbf{k}'+\mathbf{q}/2)(1-n_F(\mathbf{k}'-\mathbf{q}/2))n_F(\mathbf{k})(1-n_F(\mathbf{k}+\mathbf{q}))}{(\frac{\mathbf{k}'\cdot\mathbf{q}}{m_e}-\frac{\mathbf{k}\cdot\mathbf{q}}{m_e}-\varepsilon_q)^2}$$
$$+\sum_{\mathbf{k}',\mathbf{q}}\left(\frac{v_{\mathbf{q}}}{V}\right)^2\frac{n_F(\mathbf{k}'+\mathbf{q}/2)(1-n_F(\mathbf{k}'-\mathbf{q}/2))n_F(\mathbf{k}-\mathbf{q})(1-n_F(\mathbf{k}))}{(\frac{\mathbf{k}'\cdot\mathbf{q}}{m_e}-\frac{\mathbf{k}\cdot\mathbf{q}}{m_e}+\varepsilon_q)^2} \tag{12.74}$$

Now we wish to see if the above formula can be recovered by the sea-boson approach. For this we have to use the simplified version of Eq. (12.63) obtained by expanding in powers of $< a^{\dagger}a >$ and retaining the leading term as this quantity is itself of order $v_{\mathbf{q}}^2$. This leads to the following expression,

$$< n_{\mathbf{k}} >= n_F(\mathbf{k}) + \sum_{\mathbf{q}} < a^{\dagger}_{\mathbf{k}-\mathbf{q}/2}(\mathbf{q}) a_{\mathbf{k}-\mathbf{q}/2}(\mathbf{q}) >$$

$$- \sum_{\mathbf{q}} < a^{\dagger}_{\mathbf{k}+\mathbf{q}/2}(\mathbf{q}) a_{\mathbf{k}+\mathbf{q}/2}(\mathbf{q}) > . \tag{12.75}$$

By using Eq. (12.67) we may rewrite this as,

$$< n_{\mathbf{k}} >= n_F(\mathbf{k}) + \sum_{\mathbf{q}} \frac{v_{\mathbf{q}}}{V} \sum_{i=modes} \frac{1}{|\varepsilon'(\mathbf{q},\omega_i(\mathbf{q}))|} \frac{n_F(\mathbf{k}-\mathbf{q})(1-n_F(\mathbf{k}))}{(\omega_i(\mathbf{q}) - \frac{\mathbf{k}.\mathbf{q}}{m_e} + \varepsilon_q)^2}$$

$$- \sum_{\mathbf{q}} \frac{v_{\mathbf{q}}}{V} \sum_{i=modes} \frac{1}{|\varepsilon'(\mathbf{q},\omega_i(\mathbf{q}))|} \frac{n_F(\mathbf{k})(1-n_F(\mathbf{k}+\mathbf{q}))}{(\omega_i(\mathbf{q}) - \frac{\mathbf{k}.\mathbf{q}}{m_e} - \varepsilon_q)^2} . \tag{12.76}$$

Comparing the denominators of the two formulas Eq. (12.76) and Eq. (12.74), we conclude that $\omega_i(\mathbf{q}) \rightarrow \frac{\mathbf{k}'.\mathbf{q}}{m_e} < 0$. This energy has to be negative in order for the two formulas to be consistent. This is not impossible since the zeros of the dielectric function admit both signs. In the noninteracting theory the energies are always positive, but with interactions it can be of either sign. We may conclude that in this case, only negative energies will be allowed in the ground state to minimize energy. In fact, we can go even further and make sure that the two formulas are in fact identical. We rewrite the condition for the zero of the dielectric function as follows.

$$0 = 1 - \frac{v_{\mathbf{q}}}{V} \sum_{\mathbf{k} \neq \mathbf{k}_i} \frac{n_F(\mathbf{k}-\mathbf{q}/2) - n_F(\mathbf{k}+\mathbf{q}/2)}{\omega_i(\mathbf{q}) - \frac{\mathbf{k}.\mathbf{q}}{m_e}}$$

$$- \left(\frac{v_{\mathbf{q}}}{V}\right) \frac{n_F(\mathbf{k}_i-\mathbf{q}/2) - n_F(\mathbf{k}_i+\mathbf{q}/2)}{\omega_i(\mathbf{q}) - \frac{\mathbf{k}_i.\mathbf{q}}{m_e}} \tag{12.77}$$

We are going to choose only the negative zeros. This means $n_F(\mathbf{k}_i - \mathbf{q}/2) = 0$ and $n_F(\mathbf{k}_i + \mathbf{q}/2) = 1$. Thus,

$$\omega_i(\mathbf{q}) = \frac{\mathbf{k}_i.\mathbf{q}}{m_e} - \frac{\left(\frac{v_{\mathbf{q}}}{V}\right)}{\left(1 - \frac{v_{\mathbf{q}}}{V} \sum_{\mathbf{k} \neq \mathbf{k}_i} \frac{n_F(\mathbf{k}-\mathbf{q}/2) - n_F(\mathbf{k}+\mathbf{q}/2)}{\omega_i(\mathbf{q}) - \frac{\mathbf{k}.\mathbf{q}}{m_e}}\right)} \approx \frac{\mathbf{k}_i.\mathbf{q}}{m_e} - \left(\frac{v_{\mathbf{q}}}{V}\right) + \ldots \tag{12.78}$$

Also,

$$\varepsilon'(\mathbf{q},\omega_i) = \frac{v_{\mathbf{q}}}{V} \frac{n_F(\mathbf{k}_i - \mathbf{q}/2) - n_F(\mathbf{k}_i + \mathbf{q}/2)}{(\omega_i(\mathbf{q}) - \frac{\mathbf{k}_i.\mathbf{q}}{m_e})^2}$$

$$+\frac{v_{\mathbf{q}}}{V}\sum_{\mathbf{k}\neq\mathbf{k}_i}\frac{n_F(\mathbf{k}-\mathbf{q}/2)-n_F(\mathbf{k}+\mathbf{q}/2)}{(\omega_i(\mathbf{q})-\frac{\mathbf{k}.\mathbf{q}}{m_e})^2}\approx-\left(\frac{v_{\mathbf{q}}}{V}\right)^{-1}+\ldots \tag{12.79}$$

Substituting these into Eq. (12.76) we find the same formula as Eq. (12.74). The sum over modes may be elegantly redefined by summing over all negative energies weighted by the spectral function. Hence we choose to adopt the following prescription,

$$< a^{\dagger}_{\mathbf{k}}(\mathbf{q})a_{\mathbf{k}}(\mathbf{q}) >= \frac{v_{\mathbf{q}}}{V}\int_{-\infty}^{0} d\omega\,\frac{-\varepsilon_i(\mathbf{q},\omega)/\pi}{\varepsilon_r^2(\mathbf{q},\omega)+\varepsilon_i^2(\mathbf{q},\omega)}$$
$$\frac{n_F(\mathbf{k}-\mathbf{q}/2)(1-n_F(\mathbf{k}+\mathbf{q}/2))}{(\omega-\frac{\mathbf{k}.\mathbf{q}}{m_e})^2}. \tag{12.80}$$

Here $\varepsilon_i$ and $\varepsilon_r$ are the imaginary and real parts of the dielectric function, respectively. Thus,

$$\varepsilon_i(\mathbf{q},\omega)=-\pi\frac{v_{\mathbf{q}}}{V}\sum_{\mathbf{k}'}(n_F(\mathbf{k}'+\mathbf{q}/2)-n_F(\mathbf{k}'-\mathbf{q}/2))\delta(\omega-\frac{\mathbf{k}'.\mathbf{q}}{m_e}) \tag{12.81}$$

and $\varepsilon_r - 1$ is related to $\varepsilon_i$ by the Kramers-Kronig relations.

$$\varepsilon_r(\mathbf{q},\omega)-1=\frac{2}{\pi}\int_0^{\infty}\frac{\omega'\,\varepsilon_i(\mathbf{q},\omega')}{\omega'^2-\omega^2}\,d\omega' \tag{12.82}$$

In order to recover naive perturbation theory from Eq. (12.80) we realize that as $v_{\mathbf{q}}\to 0$, $\varepsilon_r\to 1$ and $\varepsilon_i\sim O(v_{\mathbf{q}})$. Hence we obtain,

$$< a^{\dagger}_{\mathbf{k}}(\mathbf{q})a_{\mathbf{k}}(\mathbf{q}) >\approx \frac{v_{\mathbf{q}}}{V}\int_{-\infty}^{0} d\omega\,\frac{v_{\mathbf{q}}}{V}$$
$$\sum_{\mathbf{k}'}(n_F(\mathbf{k}'+\mathbf{q}/2)-n_F(\mathbf{k}'-\mathbf{q}/2))\delta(\omega-\frac{\mathbf{k}'.\mathbf{q}}{m_e})$$
$$\frac{n_F(\mathbf{k}-\mathbf{q}/2)(1-n_F(\mathbf{k}+\mathbf{q}/2))}{(\omega-\frac{\mathbf{k}.\mathbf{q}}{m_e})^2}$$
$$=\left(\frac{v_{\mathbf{q}}}{V}\right)^2\sum_{\mathbf{k}'}\frac{n_F(\mathbf{k}-\mathbf{q}/2)(1-n_F(\mathbf{k}+\mathbf{q}/2))}{(\frac{\mathbf{k}'.\mathbf{q}}{m_e}-\frac{\mathbf{k}.\mathbf{q}}{m_e})^2}$$
$$n_F(\mathbf{k}'+\mathbf{q}/2))(1-n_F(\mathbf{k}'-\mathbf{q}/2)). \tag{12.83}$$

This, when substituted into Eq. (12.75), yields the right series for $< n_{\mathbf{k}} >$ (Eq. (12.74)). The only thing that remains now is to ascertain whether the right exponents of the Luttinger model are recovered when one specializes this to one dimension and uses Eq. (12.63) to compute $< n_{\mathbf{k}} >$. If we choose $v_q = v_0$, from Eq. (12.83) we get,

$$< a_k^\dagger(q)a_k(q) >\approx \left(\frac{v_0^2}{L}\right)\frac{n_F(k-q/2)(1-n_F(k+q/2))}{(2v_Fq)^2}\frac{|q|}{2\pi}. \tag{12.84}$$

If we use this in Eq. (12.63) we find,

$$< n_k >= \frac{1}{2} + const.\, sgn(|k|-k_F)\, ||k|-k_F|^\beta \tag{12.85}$$

where the exponent $\beta = \frac{v_0^2}{8\pi^2 v_F^2}$. Now we wish to see if this is the same as that predicted by standard bosonization methods. There we find $\beta = \frac{(1-K)^2}{2K}$ and for the Hamiltonian in consideration in this article, $K = \left(\frac{v_F}{(v_F+v_0/\pi)}\right)^{\frac{1}{2}}$. Expanding in powers of $v_0$, one finds $\beta$ that is same as that predicted by our approach. This is somewhat surprising since conventional wisdom tells us that in one dimension, only the collective mode contributes to the summation over the modes, but in the above discussion we have completely ignored this. This issue can be clarified by studying the general case. Now we wish to go back and do this more completely by not expanding in powers of $v_0$. Let us now try to do the integral over $\omega$ in Eq. (12.80) without expanding in powers of $v_0$. In one dimension, the imaginary part of the dielectric function evaluates as,

$$\varepsilon_i(q,\omega) = -\frac{m_e v_0}{2|q|}\left(\theta(k_F - |\frac{m_e\omega}{q} + q/2|) - \theta(k_F - |\frac{m_e\omega}{q} - q/2|)\right). \tag{12.86}$$

From this we may obtain the real part by Kramers-Kronig relations.

$$\varepsilon_r(q,\omega) = 1 + \frac{m_e v_0}{2\pi|q|}\, Log\left|\frac{\omega^2-(v_F|q|+\varepsilon_q)^2}{\omega^2-(v_F|q|-\varepsilon_q)^2}\right| \tag{12.87}$$

In the RPA limit we have,

$$\varepsilon_r(q,\omega) \approx 1 - \frac{v_F v_0}{\pi}\frac{q^2}{\omega^2-(v_F|q|)^2}. \tag{12.88}$$

The region of negative $\omega$ for which $\varepsilon_i \neq 0$ is $-v_F|q| - \varepsilon_q < \omega < -v_F|q| + \varepsilon_q$. Away from this region, only the collective mode contributes since the Lorentzian becomes a delta function. Thus we have,

$$< a_k^\dagger(q)a_k(q) >\approx \frac{v_0}{L}\varepsilon_q\int_{-1}^{1} dx\, \frac{\frac{m_e v_0}{2\pi|q|}}{\left(1+\frac{m_e v_0}{2\pi|q|}\, Log\left|\frac{(x+1)}{(x-1)}\right|\right)^2 + \left(\frac{m_e v_0}{2|q|}\right)^2}$$

$$\frac{n_F(k-q/2)(1-n_F(k+q/2))}{(2v_F|q|)^2}$$

$$+\frac{2\pi}{L}\frac{(v-v_F)^2}{4vv_F|q|}n_F(k-q/2)(1-n_F(k+q/2)). \tag{12.89}$$

The integral over $x$ is the particle-hole mode contribution and the other term is the collective mode contribution. Here we have used the RPA approximation to the real part of the dielectric function to simplify the collective mode contribution. Also $v=\sqrt{v_F^2+\frac{v_0 v_F}{\pi}}$. From the above expression it is easy to see that in the RPA limit, the particle-hole mode drops out completely and we are left with only the collective mode. This is in contrast to what happens in higher dimensions, where the collective mode leads to an exponentially suppressed contribution for weak coupling, but unlike in one dimension where the imaginary part of the dielectric function is nonzero in a rather tiny region, in higher dimensions this region is quite large, meaning that we expect that particle-hole mode contributions will dominate in higher dimensions. These expectations will be verified explicitly later on. We wish to complete the calculation of the momentum distribution. Using Eq. (12.63) we find the same form as Eq. (12.85) but with a new $\beta$ given by $\beta=\frac{(v-v_F)^2}{2vv_F}$. This is the same as the exact result given by conventional bosonization methods since $K=\frac{v_F}{v}$. Now we wish to continue in this vein and compute the dynamical density–density correlation function and see if it matches with the standard results of conventional bosonization/RPA. By now the reader should be convinced that it will. For later use we calculate the general density–density correlation function without committing to any specific dimension (this is for $|\mathbf{q}| \ll k_F$).

$$<T\,\rho(-\mathbf{q},t)\rho(\mathbf{q},t^{'})>=\frac{i}{-i\beta}\sum_{\mathbf{k},n}e^{w_n(t-t^{'})}\frac{n_F(\mathbf{k}-\mathbf{q}/2)-n_F(\mathbf{k}+\mathbf{q}/2)}{\varepsilon(\mathbf{q},iw_n)(iw_n-\frac{\mathbf{k}.\mathbf{q}}{m_e})} \tag{12.90}$$

Specializing to the case when $t^{'}>t$ in the imaginary time sense we obtain, according to our agreement of the interpretation of the summation over modes (for $v_{\mathbf{q}}\neq 0$, $iw_n=\frac{\mathbf{k}.\mathbf{q}}{m_e}$ is not a pole),

$$<\rho(\mathbf{q},t^{'})\rho(-\mathbf{q},t)>=\sum_{\mathbf{k}}\int_{-\infty}^{0}d\omega\, e^{-i\omega(t-t^{'})}\frac{-\varepsilon_i(\mathbf{q},\omega)/\pi}{\varepsilon_r^2(\mathbf{q},\omega)+\varepsilon_i^2(\mathbf{q},\omega)}$$

$$\frac{n_F(\mathbf{k}-\mathbf{q}/2)-n_F(\mathbf{k}+\mathbf{q}/2)}{(\omega-\frac{\mathbf{k}.\mathbf{q}}{m_e})}. \tag{12.91}$$

The integral over $\omega$ is in the sense of a principal value. Let us now evaluate this in one dimension. In the RPA limit, only the collective mode survives at $\omega=-v|q|$. In this case we obtain the familiar result,

$$<\rho(q,t^{'})\rho(-q,t)>=e^{iv|q|(t-t^{'})}\frac{v_F}{v}NS(q). \tag{12.92}$$

In higher dimensions, we may check the validity of this formula by comparing with standard perturbation theory. For this we again rewrite this as sum over modes. Basically the formula in Eq. (12.91) is what we get when we take the thermodynamic limit. If we do not wish to take the thermodynamic limit, then we must use the formula below.

$$< \rho(\mathbf{q},t')\rho(-\mathbf{q},t) >= \sum_{\mathbf{k},i,\omega_i(\mathbf{q})<0} e^{-i\omega_i(\mathbf{q})(t-t')}$$

$$\frac{1}{|\varepsilon'(\mathbf{q},\omega_i(\mathbf{q}))|} \frac{n_F(\mathbf{k}-\mathbf{q}/2) - n_F(\mathbf{k}+\mathbf{q}/2)}{(\omega_i(\mathbf{q}) - \frac{\mathbf{k}.\mathbf{q}}{m_e})} \tag{12.93}$$

We use the results obtained earlier, namely, $\omega_i(\mathbf{q}) \approx \frac{\mathbf{k}_i\cdot\mathbf{q}}{m_e} - \frac{v_\mathbf{q}}{V}$, $|\varepsilon'(\mathbf{q},\omega_i)| \approx (\frac{v_\mathbf{q}}{V})^{-1}$ where $\frac{\mathbf{k}_i\cdot\mathbf{q}}{m_e} < 0$. Substituting these into the formula Eq. (12.93) we obtain,

$$< \rho(\mathbf{q},t')\rho(-\mathbf{q},t) >= \sum_{\mathbf{k},\mathbf{k}_i} e^{-i\frac{\mathbf{k}_i\cdot\mathbf{q}}{me}(t-t')} \left(\frac{v_\mathbf{q}}{V}\right)$$

$$\frac{-n_F(\mathbf{k}+\mathbf{q}/2)(1-n_F(\mathbf{k}-\mathbf{q}/2))}{(\frac{\mathbf{k}_i\cdot\mathbf{q}}{m_e} - \frac{v_\mathbf{q}}{V} - \frac{\mathbf{k}.\mathbf{q}}{m_e})}$$

$$= \sum_{\mathbf{k}} e^{-i\frac{\mathbf{k}\cdot\mathbf{q}}{me}(t-t')} \; n_F(\mathbf{k}+\mathbf{q}/2)(1-n_F(\mathbf{k}-\mathbf{q}/2)) + ..., \tag{12.94}$$

which is as it should be. Lastly, we wish to evaluate the momentum distribution perturbatively in three dimensions for short-range interactions and see if we can reproduce Galitskii's well-known result for the quasiparticle residue. One must exercise caution, however, since the model we have considered is not exactly the one consider by Galitskii. We have only included forward-scattering terms and scattering across the Fermi surface only. Thus we do not expect all the features to be identical. However, we have already demonstrated quite generally that our theory is consistent with conventional perturbation theory. We start with Eq. (12.83). To evaluate the sum over $\mathbf{k}'$, we treat $\mathbf{k},\mathbf{k}'$ as close to the Fermi surface and $\mathbf{q}$ small compared to the Fermi momentum. This is the only limit in which our theory works in the way in which we have presented it. However, Galitskii deals with the more general case in his paper. Hence we should not be surprised if we find a discrepancy in the final results. After performing the summation over $|\mathbf{q}| < \Lambda \ll k_F$ our formulas yield,

$$\sum_{\mathbf{q}} < a^\dagger_{\mathbf{k}_F-\mathbf{q}/2}(\mathbf{q}) a_{\mathbf{k}_F-\mathbf{q}/2}(\mathbf{q}) >$$

$$= \sum_{\mathbf{q}} < a^\dagger_{\mathbf{k}_F+\mathbf{q}/2}(\mathbf{q}) a_{\mathbf{k}_F+\mathbf{q}/2}(\mathbf{q}) > \approx \left(\frac{U_0^2}{(2\pi)^2}\right) \frac{k_F^2}{(2\pi)^2} \frac{1}{v_F^2} \frac{\Lambda^2}{2} \, Log[2]. \tag{12.95}$$

Now we insert the above formula into Eq. (12.64) to arrive at a formula for the quasiparticle residue (jump across the Fermi surface). The quasiparticle residue is given by $Z_F = 1 - \frac{(\Lambda a)^2}{\pi^2} Log[2]$ taking into account the definition of the scattering length $a = \frac{m_e U_0}{4\pi}$, whereas Galitskii obtains $Z_F = 1 - \frac{8}{\pi^2} Log[2](k_F a)^2$ (this result is also reproduced in a more accessible textbook on statistical physics by Landau and Lifshitz).

### 12.3.1 Asymptotically Exact Momentum Distribution for Forward Scattering Interactions

Assuming that the reader is convinced that the momentum distribution provided by sea-boson approach is asymptotically exact for forward scattering, we may proceed to derive explicit expressions for this in two and three spatial dimensions for potentials of the form, $v_q = g\,\frac{v_F}{k_F^{(d-1)}}\,(q/\Lambda)^{\eta}\,\theta(\Lambda - |q|)$. The reason for this particular form is because we want the dielectric function to be meaningful in the RPA limit $k_F, m_e \to \infty$. If we do not choose $k_F^{(d-1)}$ in the denominator, then we get a divergent result for the dielectric function in the RPA limit. We examine the momentum distribution as $g$ is varied and look for signatures of quantum phase transition. First we need the imaginary and real parts of the dielectric function. In the RPA limit (long-wavelength low-energy limit) we may write,

$$\varepsilon_i^{3D}(\mathbf{q},\omega) = \frac{m_e^2}{q}\frac{v_{\mathbf{q}}}{(2\pi)}\,\omega\,\theta(v_F q - |\omega|)$$

$$;\ \varepsilon_i^{2D}(\mathbf{q},\omega) = \frac{v_{\mathbf{q}}}{(2\pi)}\frac{\omega}{v_F}k_F\,\frac{\theta(v_F q - |\omega|)}{\sqrt{(v_F q)^2 - \omega^2}} \tag{12.96}$$

$$\varepsilon_r^{3D}(\mathbf{q},\omega) - 1 = \frac{m_e^2\omega^2}{q}\frac{v_{\mathbf{q}}}{\pi^2}\frac{1}{2\omega}Log\left|\frac{v_F|q| - \omega}{v_F|q| + \omega}\right| + \frac{m_e^2}{q}\frac{v_{\mathbf{q}}}{\pi^2}(v_F q) \tag{12.97}$$

$$\varepsilon_r^{2D}(\mathbf{q},\omega) - 1 = -\frac{v_{\mathbf{q}}}{(2\pi)}\frac{k_F}{v_F}\left(-1 + \theta(\omega^2 - (v_F|q|)^2)\sqrt{\frac{\omega^2}{\omega^2 - (v_F|q|)^2}}\right). \tag{12.98}$$

### 12.3.2 Short-Range Interactions

First we consider short-range interactions in two and three dimensions: $v_q^{2D} = g\frac{v_F}{k_F}\theta(\Lambda - |q|)$, $v_q^{3D} = g\frac{v_F}{k_F^2}\theta(\Lambda - |q|)$. In this case the collective modes are sound waves (zero sound) with dispersion, $\omega_c(q) = \pm v_c|q|$ where in three dimensions the velocity is given as the solution to

$$\frac{v_c}{v_F} \, ArcTanh\left(\frac{v_F}{v_c}\right) = \left(\frac{\pi^2}{g} + 1\right), \tag{12.99}$$

whereas in two dimensions it is simpler $v_c = v_F\sqrt{\frac{(\frac{2\pi}{g}+1)^2}{(\frac{2\pi}{g}+1)^2-1}}$. From this we may evaluate the sea-boson occupation in two and three dimensions. We consider two extreme limits $g \to 0^+$ and $g \to \infty$. In three dimensions, the velocity collective mode is exponentially close to the Fermi velocity. This leads to the collective mode to be exponentially suppressed. Thus we find in three dimensions as, $g \to 0^+$, $v_c \approx v_F(1+2e^{-\frac{2\pi^2}{g}})$, the following formula for the sea-boson occupation.

$$< a^{\dagger}_{\mathbf{k}}(\mathbf{q})a_{\mathbf{k}}(\mathbf{q}) >_{3D} \approx \frac{(4\pi^2)v_F^2|q|}{Vk_F^2} e^{-\frac{2\pi^2}{g}} \frac{n_F(\mathbf{k}-\mathbf{q}/2)(1-n_F(\mathbf{k}+\mathbf{q}/2))}{(-v_F|q| - \frac{\mathbf{k.q}}{m_e})^2}$$

$$-\frac{1}{V}\frac{v_F}{k_F^2}\frac{g^2}{(2\pi^2)}(v_F|q|)\int_{-1}^{0} ds\, s\frac{n_F(\mathbf{k}-\mathbf{q}/2)(1-n_F(\mathbf{k}+\mathbf{q}/2))}{(v_F|q|s - \frac{\mathbf{k.q}}{m_e})^2} \tag{12.100}$$

The first term with $e^{-\frac{2\pi^2}{g}}$ is due to the collective mode, the second is due to the particle-hole mode. We can see that for weak coupling in three dimensions the collective mode contribution to the momentum distribution is exponentially suppressed. The main contribution therefore is from the particle-hole mode, which is proportional to the square of the coupling. Substituting this into Eq. (12.63) we find the following expression for the quasiparticle residue

$$Z_F = 1 - Log[2]\,\frac{2g^2}{(2\pi)^4}\frac{\Lambda^2}{k_F^2}, \tag{12.101}$$

which as we have seen earlier is consistent with Galitskii's result. The opposite limit $g \to \infty$ is more interesting. But in three dimensions, for $g \to \infty$ we get $v_c = \sqrt{\frac{gv_F^2}{3\pi^2}}$. The formula for the boson occupation in this case is,

$$< a^{\dagger}_{\mathbf{k}}(\mathbf{q})a_{\mathbf{k}}(\mathbf{q}) >_{3D} =$$

$$\frac{\sqrt{3\pi^2 g}}{2Vk_F^2}\,\frac{n_F(\mathbf{k}-\mathbf{q}/2)(1-n_F(\mathbf{k}+\mathbf{q}/2))}{|q|}$$

$$+\frac{v_F}{Vk_F^2}v_F|q|\int_{-1}^{0} ds\,\frac{-\frac{g^2}{(2\pi^2)}s}{[1+\frac{g}{\pi^2}(1-\frac{s}{2}Log|\frac{s+1}{s-1}|)]^2+[\frac{g}{(2\pi)}s]^2}$$

$$\frac{n_F(\mathbf{k}-\mathbf{q}/2)(1-n_F(\mathbf{k}+\mathbf{q}/2))}{(v_F q s - \frac{\mathbf{k}.\mathbf{q}}{m_e})^2}. \tag{12.102}$$

We may see from this formula that there are two contributions. The first due to the collective mode diverges as $g \to \infty$ (albeit rather slowly as $\sqrt{g}$), however, the second term, which is due to particle-hole excitations, saturates. We may see from this formula that this is true only if we ignore the collective mode. In this article, we make the sea-boson formula Eq. (12.63) plausible by comparing with both trivial (read free theory), semi-trivial (read leading order perturbation theory), and non-trivial (read Luttinger exponents in 1d) limiting cases. We may proceed to evaluate the quasiparticle residue.

$$Z_F \approx e^{-2\sum_{\mathbf{q}} < a^{\dagger}_{\mathbf{k}_F-\mathbf{q}/2}(\mathbf{q}) a_{\mathbf{k}_F-\mathbf{q}/2}(\mathbf{q}) >} \approx e^{-\frac{\sqrt{3\pi^2 g}}{(2\pi)^2}\frac{\Lambda^2}{2k_F^2}} \tag{12.103}$$

It appears that this quantity shrinks to zero as $g \to \infty$. However, one must bear in mind that the theory presented is valid only in the RPA limit where $k_F, m_e \to \infty$ but $k_F/m_e = v_F < \infty$. The random phase approximation continues to be valid for $\Lambda \ll k_F$, thus we may fix $\Lambda/k_F = r_c \ll 1$. In this case the quasiparticle residue shrinks to zero as the coupling tends to infinity. In two spatial dimensions, we may see that the collective mode and particle-hole modes contribute nearly equally even for weak coupling. Thus we expect something more interesting for strong coupling. For weak coupling in two spatial dimensions we have,

$$< a^{\dagger}_{\mathbf{k}}(\mathbf{q}) a_{\mathbf{k}}(\mathbf{q}) >_{2D} =$$

$$\frac{1}{A} \frac{2\pi|q| v_F^3 \frac{g^3}{(2\pi)^3}}{k_F v_F} \frac{n_F(\mathbf{k}-\mathbf{q}/2)(1-n_F(\mathbf{k}+\mathbf{q}/2))}{(-v_F|q| - \frac{\mathbf{k}.\mathbf{q}}{m_e})^2}$$

$$-\frac{g^2 \frac{v_F}{k_F}}{(2\pi^2)A} v_F|q| \int_{-1}^{0} \frac{ds\, s}{\sqrt{1-s^2}} \frac{n_F(\mathbf{k}-\mathbf{q}/2)(1-n_F(\mathbf{k}+\mathbf{q}/2))}{(v_F|q|s - \frac{\mathbf{k}.\mathbf{q}}{m_e})^2}. \tag{12.104}$$

Here again the first term is the collective mode and the second is the particle-hole mode. One sees that their contribution is comparable. While the collective mode is of order $g^3$, the particle-hole mode is of order $g^2$. This is in contrast to the case in three dimensions where the collective mode is exponentially suppressed. One may evaluate the quasiparticle residue for weak coupling to get,

$$Z_F \approx 1 - \frac{\Lambda}{k_F}\frac{g^2}{(2\pi^4)}(0.9) + ... \tag{12.105}$$

The numerical constant appears in an integral that we have evaluated numerically. For strong coupling, again, the particle-hole mode saturates but the collective mode grows so that we obtain,

$$< a^{\dagger}_{\mathbf{k}}(\mathbf{q})a_{\mathbf{k}}(\mathbf{q}) >_{2D} \approx \frac{(\pi g)^{\frac{1}{2}}}{A}\ \frac{n_F(\mathbf{k}-\mathbf{q}/2)(1-n_F(\mathbf{k}+\mathbf{q}/2))}{(k_F|q|)}. \qquad (12.106)$$

Thus the momentum distribution is given by the formula,

$$< n_{\mathbf{k}} > \approx \frac{1}{2}\left(1+e^{-\frac{\sqrt{(\pi g)}}{2\pi}\frac{\Lambda}{k_F}}\right)n_F(\mathbf{k}) + \frac{1}{2}\left(1-e^{-\frac{\sqrt{(\pi g)}}{2\pi}\frac{\Lambda}{k_F}}\right)(1-n_F(\mathbf{k})). \quad (12.107)$$

### 12.3.3 Long-Range Interactions

Now we consider long-range interaction of the gauge potential type. In two dimensions we choose $v^{2D}_q = g\ \frac{v_F}{k_F}\ (q/\Lambda)^{-2}\ \theta(\Lambda-|q|)$ and in three dimensions we choose $v_q = g\ \frac{v_F}{k_F^2}\ (q/\Lambda)^{-2}\ \theta(\Lambda-|q|)$. The former is a $Log(r)$ type of potential in real space, whereas the latter is the usual Coulomb potential. The real part of the dielectric functions have the following zeros: $\omega_{c,2D} \approx \sqrt{\frac{g}{(4\pi)}}(v_F\Lambda)$ , $\omega_{c,3D} \approx \sqrt{\frac{g}{3\pi^2}}(v_F\Lambda)$. Thus the collective excitations are of the plasma type. After performing the computations we find the following expressions for the boson occupations. In two spatial dimensions, we obtain,

$$< a^{\dagger}_{\mathbf{k}}(\mathbf{q})a_{\mathbf{k}}(\mathbf{q}) >=$$

$$-\frac{\frac{v_F}{k_F}\ \theta(\Lambda-|q|)}{A}(v_F|q|)^{-1}\int_{-1}^{0} dx\ 2x\ \sqrt{1-x^2}\frac{\theta(\hat{k}\cdot\hat{q})}{(x-\hat{k}\cdot\hat{q})^2}$$

$$+\theta(\omega_c - v_F|q|)\frac{g\ \frac{v_F}{k_F}\ (q/\Lambda)^{-2}}{A}\ \frac{2\pi\omega_c}{g(v_F\Lambda)^2}\,n_F(\mathbf{k}-\mathbf{q}/2)(1-n_F(\mathbf{k}+\mathbf{q}/2))\ \theta(\Lambda-|q|). \qquad (12.108)$$

Here we find that as $q \to 0$ the most singular contribution is coming from the collective mode. Hence the Luttinger behavior of the system is due to the collective mode and not the particle-hole mode. In three dimensions, our formalism yields the following expression.

$$< a^{\dagger}_{\mathbf{k}}(\mathbf{q})a_{\mathbf{k}}(\mathbf{q}) >=$$

$$\frac{g\ \frac{v_F}{k_F^2}\ (q/\Lambda)^{-2}\ \theta(\Lambda-|q|)}{V}\int_{-1}^{0}(v_F|q|)\ dx$$

$$\frac{-\frac{1}{\pi}\frac{g\ (\Lambda/q)^2}{(2\pi)}x}{(1+x\frac{g}{\pi^2}\frac{\Lambda^2}{2q^2}Log\left|\frac{1-x}{1+x}\right|\ +\frac{g\ \Lambda^2}{\pi^2 q^2})^2+(\frac{g^2\ (\Lambda/q)^4}{(2\pi)^2}x^2)}$$

$$\frac{n_F(\mathbf{k}-\mathbf{q}/2)(1-n_F(\mathbf{k}+\mathbf{q}/2))}{(xv_F|q| - \frac{\mathbf{k.q}}{m_e})^2}$$

$$+\frac{\theta(\omega_c - v_F|q|)}{V}\frac{\pi^2}{(k_F q)^2}\left(\frac{3\omega_c}{2v_F}\right)n_F(\mathbf{k}-\mathbf{q}/2)(1-n_F(\mathbf{k}+\mathbf{q}/2)). \qquad (12.109)$$

The importance of the two contributions may be made clear by examining the occupation number summed over the momentum label $\mathbf{q}$.

$$\sum_{\mathbf{q}} < a^{\dagger}_{\mathbf{k}_F-\mathbf{q}/2}(\mathbf{q})a_{\mathbf{k}_F-\mathbf{q}/2}(\mathbf{q}) >=$$

$$-\frac{1}{(2\pi)^2}g\frac{\Lambda^2}{k_F^2}\int_{-1}^{0}\frac{dx}{(1-x)}\frac{1}{2\pi x}$$

$$\left(ArcTan\left(\frac{\pi x}{2+\frac{2\pi^2}{g}+xLog\left(\frac{1-x}{1+x}\right)}\right)-ArcTan\left(\frac{\pi x}{2+xLog\left(\frac{1-x}{1+x}\right)}\right)\right)$$

$$+\frac{1}{(2\pi)^2}min(\Lambda,\frac{\omega_c}{v_F})\frac{\pi^2}{k_F^2}\left(\frac{3\omega_c}{2v_F}\right) \qquad (12.110)$$

As $g \to 0$, we obtain the following expression for the momentum distribution.

$$< n_{\mathbf{k}} >\approx (1-\frac{g}{(2\pi)^2}\frac{\Lambda^2}{k_F^2}(0.64))\, n_F(\mathbf{k})$$

$$+\frac{g}{(2\pi)^2}\frac{\Lambda^2}{k_F^2}(0.64)\,(1-n_F(\mathbf{k})) \qquad (12.111)$$

As $g \to \infty$, the particle-hole mode contribution saturates whereas the collective mode contribution grows, albeit slowly, to give

$$< n_{\mathbf{k}} >=\frac{1}{2}\left(1+e^{-\frac{\Lambda^2}{4k_F^2}\sqrt{\frac{3g}{\pi^2}}}\right)n_F(\mathbf{k})+\frac{1}{2}\left(1-e^{-\frac{\Lambda^2}{4k_F^2}\sqrt{\frac{3g}{\pi^2}}}\right)(1-n_F(\mathbf{k})). \qquad (12.112)$$

In the exercises, the reader is invited to study the situations we have left out, viz. long-range interactions in one and two dimensions, etc.

## 12.4 Exercises

**Q.1** Verify that Eq. (12.5) obeys the relations $\rho(x) \equiv \psi^{\dagger}(x)\psi(x)$ and $\rho(x)\mathbf{v}(x) \equiv Im[\psi^{\dagger}(x)\nabla\psi(\mathbf{x})]$.

**Q.2** Add a source term $\int_0^{-i\beta\hbar} dt \int d^3x\, \mathbf{W}(x)\cdot\mathbf{v}(x)$ to Eq. (12.10) and evaluate the generating function,

$$Z[W] \equiv \int D[\mathbf{v}]e^{iS_{eff}}e^{\int_0^{-i\beta\hbar}dt\int d^3x\,\mathbf{W}(x)\cdot\mathbf{v}(x)}. \qquad (12.113)$$

By differentiating appropriately with respect to $W$, derive Eq. (12.11).

**Q.3** Perform the summation over **k** directly in the equal-time version of Eq.(12.12) and then take the large $\omega_0$ limit to verify that the final answers are the same as what has been described in the text. Hint: You should get a $K_1$ Bessel function.

**Q.4** What's wrong if we simply write $\sqrt{\omega_0^2 + c^2k^2} \approx \omega_0$ in Eq. (12.12) and arrive at,

$$< \phi_\Gamma(t)\phi_{\Gamma'}(t') > \approx \frac{4\pi q^2}{V} \sum_{\mathbf{k}} e^{-i\omega_0(t-t')} \frac{\oint d\mathbf{x} \cdot \oint d\mathbf{x}' e^{i\mathbf{k}\cdot(\mathbf{x}-\mathbf{x}')}}{(2\hbar\omega_0)}$$

$$= 4\pi q^2 \, e^{-i\omega_0(t-t')} \frac{\oint d\mathbf{x} \cdot \oint d\mathbf{x}' \delta^3(\mathbf{x}-\mathbf{x}')}{(2\hbar\omega_0)} ? \tag{12.114}$$

Hint: It is an indeterminate form.

**Q.5** Calculate the unequal time version of Eq. (12.12) by writing,

$$e^{-i\sqrt{\omega_0^2+c^2k^2}(t-t')} \approx e^{-i\omega_0(t-t')} e^{-i\frac{c^2k^2}{2\omega_0}(t-t')}. \tag{12.115}$$

**Q.6** Using the sea-boson formula Eq. (12.63), evaluate the momentum distribution for fermions in one dimension with forward-scattering potentials (i) $v_q = v_0 = const.$ and (ii) $v_q = q_0^2/q^2$ (this is called a Wigner crystal). What is the nature of the momentum distribution if the same expressions are used in two dimensions? To answer this use the Hamiltonian,

$$H = \sum_{\mathbf{k},\mathbf{q}} \frac{\mathbf{k}.\mathbf{q}}{m} a^\dagger_{\mathbf{k}}(\mathbf{q}) a_{\mathbf{k}}(\mathbf{q}) + \sum_{\mathbf{k},\mathbf{k}',\mathbf{q}} \frac{v_q}{2\Omega} (a_{\mathbf{k}}(-\mathbf{q}) + a^\dagger_{\mathbf{k}}(\mathbf{q}))(a_{\mathbf{k}'}(\mathbf{q}) + a^\dagger_{\mathbf{k}'}(-\mathbf{q})) \tag{12.116}$$

together with the commutation rules $[a_{\mathbf{k}}(\mathbf{q}), a^\dagger_{\mathbf{k}'}(\mathbf{q}')] = \delta_{\mathbf{k},\mathbf{k}'}\delta_{\mathbf{q},\mathbf{q}'} n_F(\mathbf{k}-\mathbf{q}/2)(1 - n_F(\mathbf{k}+\mathbf{q}/2))$ and $[a_{\mathbf{k}}(\mathbf{q}), a_{\mathbf{k}'}(\mathbf{q}')] = 0$ evaluate the zero temperature sea-boson occupation $< a^\dagger_{\mathbf{k}}(\mathbf{q}) a_{\mathbf{k}}(\mathbf{q}) >$.

The next two are research-level questions.

**Q.7** How would you evaluate vortex–vortex correlations for a Fermi fluid?

**Q.8** How would you reconcile Eq. (12.63) using the concepts of nonlocal particle-hole creation discussed in the beginning? Is a derivation of Eq. (12.63) possible using those more rigorous concepts?

# Chapter 13

# Non-chiral Bosonization of Fermions in One Dimension

In this chapter, an alternative to the conventional approach to bosonization in one dimension that invokes the Dirac equation in 1+1 dimension with chiral 'right-movers' and 'left-movers' is proposed (the conventional approach is discussed, e.g., in the text book by Giamarchi, *Quantum Physics in One Dimension*, Oxford). This technique allows us to use a basis different from the plane wave basis that makes this non-chiral approach ideally suited to study Luttinger liquids that have a boundary(s) or impurities that break translational symmetry. We provide a simple solution to the electron Green function for the problem of a Luttinger liquid (LL) with a boundary and also for a LL with a single impurity. The present method is significantly easier than the g-ology based standard bosonization and other methods that require a combination of RG (renormalization group) along with bosonization/refermionization techniques. This method is superior because it reproduces, as a limiting case, the exact closed formula for the (random phase approximation) RPA Green function of free fermions with the single impurity without mutual interactions obtainable trivially using Fermi algebra, whereas the conventional bosonization/RG approaches are not able to do so.

The basic idea of bosonization is the inversion of the bilinear relation between local currents and densities $J(x) = Im[\psi^{\dagger}(x)\partial_x\psi(x)]$ and $\rho(x) = \psi^{\dagger}(x)\psi(x)$—performing a kind of square root operation—so that the Fermi fields themselves are expressible in terms of currents and densities so that $\psi(x) = F(\rho, J; x)$. The goal of this approach is to determine to what extent this is possible and meaningful and how one should use this idea in practical calculations. One may use a simile to describe this process. Imagine trying to understand the nature of ripples on the surface of water. Perhaps one is interested in the velocity of the wave, the energy versus wave vector relation, and so on. These may be determined in principle, with the knowledge of the microscopic constituents of the material. In other words, one may derive the bulk properties of the system from a knowledge of the microscopic constituents and their interactions. This situation would be analogous in the present instance to finding $J(x), \rho(x)$—the ripples—from a knowledge of the constituents viz. $\psi(x), \psi^{\dagger}(x)$. This is appropriate since the continuum fluid description is indeed described by currents and densities as we have seen in earlier chapters. One could now become inquisitive and ask if the reverse is possible. Could one infer the nature of the atomic constituents of a fluid by examining in detail the nature of the ripples that appear in the medium? Bosonization would mean exactly this. Inferring the correlation function of a few particles with a knowledge of multi-particle correlations. Of course, at this stage, there is no reason why one approach should be practically superior to the other. There is a special situation where the currents and densities and their mutual correlations are simple, but the correlations between single particles are complicated. This is the typical situation when one attempts to study the asymptotic properties of the Green functions of fermion systems as we have already alluded to in the introduction to the earlier chapter. Therefore, it is worthwhile studying this method in detail. The rest of this chapter is written in a somewhat terse and non-pedagogical style. In places, we are also combative while trying to point out what we believe are flaws in the conventional approaches. The reader is encouraged to consult those ideas first (Giamarchi's text, e.g.) and the present chapter and judge for oneself which approach makes more sense.

## 13.1 A Critique of Conventional Chiral Bosonization

The subject of what is currently referred to as bosonization started with the works of Coleman and also independently by Luther. While Coleman showed that the fermion Green function massive of the Thirring model has an independent description in terms of bosonic variables of an equivalent model involving commuting variables, namely the so-called Sine-Gordon theory, other authors such as Luther and Mandelstam took this to mean that the Fermi field operator itself has an expression in terms of bosonic variables. This latter assertion is much stronger and is stated without proof in those articles, making them subject to criticism. The stronger assertion has been used by later researchers in condensed matter physics to generate Hamiltonians of fermions in one dimension that go under the name 'g-ology'. Here we argue that the stronger assertion is in fact false, making the g-ology program in condensed matter physics of questionable validity. This approach has been used in highly cited works of Kane and Fisher on impurities in Luttinger liquids and nearly everywhere in the textbook by Giamarchi. An alternative is proposed, mainly for condensed matter problems, involving an action for fermions in terms of hydrodynamic variables and a prescription for generating the propagator of fermions (that potentially also allows one to go beyond the linear dispersion approximation and also the random phase approximation). This action approach for fermions in one dimension is not new, however. The particle-hole excitations of the Fermi system that make the kinetic energy diagonal in these operators are not bosons even 'approximately' making the term 'bosonization' a misnomer. (The belief in conventional bosonization that the kinetic energy $K = \sum_k \varepsilon_k c_k^\dagger c_k = \sum_{p>0} v_F p \, b_{p,R}^\dagger b_{p,R} + \sum_{p>0} v_F p \, b_{p,L}^\dagger b_{p,L} + const$ is an operator identity is false, [1] it is merely a mnemonic for generating the correlation functions—and rather poor one at that, judging by the content of the rest of this chapter. Coleman's assertion that $\bar{\psi}\gamma^\mu\psi \equiv -\frac{\beta}{2\pi}\varepsilon^{\mu\nu}\partial_\nu\varphi$ is a metaphor only, for the left-hand side is a Grassmann number and the right-hand side is a real number—and Grassmann number can never be equal to a real number though it can be equivalent to a real number). We have some serious issues with the conventional approach to bosonization. The purpose of this section is to motivate the rest of the chapter by these criticisms. In his otherwise excellent textbook, based on pioneering works of Haldane and others, Giamarchi proposes the following formula for the field operator (Luther-Haldane construction),

$$\psi_r^\dagger(x) = Lim_{\varepsilon\to 0}\, \psi_r^\dagger(x;\varepsilon) = Lim_{\varepsilon\to 0} \frac{1}{\sqrt{2\varepsilon L}} e^{-ir(k_F - \frac{\pi}{L})x} e^{i\phi_r^\dagger(x,\varepsilon) + i\phi_r(x,\varepsilon)} U_r^\dagger, \quad (13.1)$$

where $r = \pm$ corresponds to right and left movers and,

$$\phi_r(x,\varepsilon) = -\frac{\pi r x}{L} N_r + i \sum_{p\neq 0} \left(\frac{2\pi}{L|p|}\right)^{\frac{1}{2}} e^{-\frac{L\varepsilon|p|}{2\pi}} Y(rp) b_p e^{ipx}. \quad (13.2)$$

The main point we are making is that Eq. (13.1) is true only in a weak sense to be made precise soon. In particular, it is not true that matrix elements of the right-hand side of Eq. (13.1) are the matrix elements of the field operator (at least no such proof is forthcoming). Some authors recommend that we postpone taking the $\varepsilon \to 0$ limit until the end of the 'calculation'. Until the end of what calculation? What if I don't want to do any calculation? What if I just want to stare at this operator itself? If one insists on a calculation, how about calculating the matrix elements of the field operator? All these rhetorical questions lead to one conclusion - that is, all the $\varepsilon$'s have be the same for all the field operators in the N-point function calculation and the fermion commutation rules are recovered only at the level of correlation functions and not at the level of operators. This is in fact clear if one consults Coleman's pioneering paper on the equivalence of the massive Thirring model and Sine-Gordon equation, He only shows that the Green functions come out right in both the languages; it is never shown that the matrix elements of the

[1]The actual operators involved are nonlocal particle-hole creation operators; see earlier chapter.

field operator come out right. Thus the g-ology program which involves a literal interpretation of the Luther construction is on shaky ground.

In the literature, the operator description is sometimes replaced by a path integral version based on Hubbard-Stratanovich transformation making these ideas appear more legitimate. However, both these approaches are flawed for the same reason—they manipulate infinities under the euphemism known as 'normal ordering'. Next, we provide some technical details justifying the reservations we have expressed. This is important if we are to convince those who are already familiar with the conventional approach (instructors of courses who may be using this book, for example). Others who are new to the whole idea are strongly urged to skip to the parts where we discuss the 'non-chiral' approach.

### 13.1.1 Chiral Fermions on a Ring

The central formalism-focused pioneering papers on this subject are by (i) Haldane and (ii) Heidenreich et al. (see bibliography). Neither address the crucial issues we are about to address here. We start with the Luther-Haldane construction,

$$\psi_r^\dagger(x) = Lim_{\varepsilon\to 0}\ \psi_r^\dagger(x;\varepsilon);\quad \psi_r(x) = Lim_{\varepsilon\to 0}\ \psi_r(x;\varepsilon). \tag{13.3}$$

One may use the normal ordered form,

$$\psi_r^\dagger(x;\varepsilon) = \frac{1}{\sqrt{L}} e^{-ik_F rx} e^{i\phi_r^\dagger(x,\varepsilon)} U_r^\dagger e^{i\phi_r(x,\varepsilon)} \tag{13.4}$$

or the unitary form,

$$\psi_r^\dagger(x;\varepsilon) = \frac{1}{\sqrt{2\varepsilon L}} e^{-ik_F rx} e^{i\phi_r^\dagger(x,\varepsilon)+i\phi_r(x,\varepsilon)} U_r^\dagger \tag{13.5}$$

$$U_r^\dagger = Lim_{\delta\to 0}\frac{1}{\sqrt{L}}\int_0^L dy\ e^{irk_F y}\ e^{-i\phi_r^\dagger(y,\delta)}\Psi_r^\dagger(y;?)e^{-i\phi_r(y,\delta)} \tag{13.6}$$

where,

$$\Psi_r^\dagger(x;\delta) = \frac{1}{\sqrt{L}} e^{-irk_F x}\sum_p e^{-ipx} c_{r,p}^\dagger e^{-\delta\frac{L|p|}{2\pi}} \tag{13.7}$$

where $\delta > 0$ and $\Psi_r(x) = Lim_{\delta\to 0^+}\Psi_r(x;\delta)$. The formula for $U_r^\dagger$ is seldom given properly in these supposedly rigorous works (should I use $\Psi_r^\dagger(y)$ or $\psi_r^\dagger(y)$ or $\Psi_r^\dagger(y;\delta)$ on the right-hand side of Eq. (13.6)?, hence the question mark). We have inferred the above version from those works where $r = \pm$ corresponds to right and left movers and,

$$\phi_r(x,\varepsilon) = -\frac{\pi rx}{L}N_r + i\sum_{p\neq 0}\left(\frac{2\pi}{L|p|}\right)^{\frac{1}{2}} e^{-\frac{L\varepsilon|p|}{2\pi}} Y(rp) b_p e^{ipx} \tag{13.8}$$

$$b_p^{\dagger} = \left(\frac{2\pi}{L|p|}\right)^{\frac{1}{2}} \sum_{r'=\pm} Y(r'p)\rho_{r'}^{\dagger}(p) \tag{13.9}$$

$$b_p = \left(\frac{2\pi}{L|p|}\right)^{\frac{1}{2}} \sum_{r'=\pm} Y(r'p)\rho_{r'}^{\dagger}(-p) \tag{13.10}$$

and $Y(x) = 1$; $x > 0$ and $Y(x) = 0$; $x \leq 0$ is Heaviside's step function, and

$$\rho_{r'}^{\dagger}(p \neq 0) = \sum_k c_{r',k+p}^{\dagger} c_{r',k}, \tag{13.11}$$

and $\{c_{r,k}, c_{r',k'}\} = 0$, and $\{c_{r,k}, c_{r',k'}^{\dagger}\} = \delta_{r,r'}\delta_{k,k'}$ are the usual chiral fermion operators. Also $U_r$ is a unitary operator independent of $\varepsilon$ and the $b_q$'s. Another useful form is,

$$\psi_r^{\dagger}(x;\varepsilon) = \frac{1}{\sqrt{2\varepsilon L}} e^{-ik_F rx} e^{-\sum_{p\neq 0}\left(\frac{2\pi r}{Lp}\right) e^{-\frac{L\varepsilon|p|}{2\pi}} \rho_r^{\dagger}(-p) e^{ipx}} U_r^{\dagger}. \tag{13.12}$$

A relatively minor issue involves taking the thermodynamic limit. We implicitly imagine $L$ in these formulas to be the perimeter of a circle and the momenta to be discrete $p = \frac{n}{R}$ where $L = 2\pi R$ so that $\psi_r(x+L) = \psi_r(x)$. The main point we are making is that Eq. (13.3) is not an operator identity but rather a mnemonic for generating the correlation functions. *The practical consequence of these seemingly pedantic distinctions is that an operator identity is universal and absolute but a mnemonic can be model dependent and ad hoc. We wish to establish here that for the formulas of chiral bosonization, it is the latter that is true.* We also go on to show how these formulas should be modified under different circumstances.

We now point out some other difficulties with the well-known notion of the Schwinger anomaly that is central to the chiral bosonization program. We wish to critically examine the assumptions underlying the algebra in conventional bosonization approaches and see to what extent they are conscious of the importance of taking into account fluctuations in the momentum distribution. In the earlier chapter, we have seen the important role played by fluctuating momentum distributions in taming infrared divergences and so on. In the conventional approach, one postulates two species of fermions (chiralities) $c_{R,L}(p)$ that obey canonical fermion commutation rules. Consider the conventional right-mover $\rho_R(q) = \sum_p c_R^{\dagger}(p+q)c_R(p)$. If the Hilbert space consists of states that have an unbounded number of momentum states occupied (as it is in the case of the Luttinger liquid), clearly this operator is mathematically meaningless (for example, for $q = 0$ we get infinity). Thus one is forced to implicitly assume a cutoff around the Fermi momentum that is set to infinity at the end of the calculation. But this cutoff has to be much smaller than the Fermi momentum in order for the Luttinger model with linear energy dispersion to mimic the non-relativistic spectrum close to the Fermi momentum. Therefore, the Fermi momentum should run off to infinity much before the cutoff does. But this makes the excitation energies infinite unless the mass also becomes infinite such that the Fermi velocity if finite. This forces the theory to become equivalent to the RPA. If one redefines $\rho_R(q) = \sum_p c_R^\dagger(p+q)c_R(p)\Gamma(p,q)$ where $\Gamma(p,q) = \theta(\Lambda - |p+q-k_F|)\theta(\Lambda - |p-k_F|)$, remembering to set $k_F \sim m \gg \Lambda \to \infty$, then,

$$[\rho_R(q), \rho_R(q^{'})] = \sum_p c_R^\dagger(p+q^{'}+q)c_R(p)$$

$$\times(\theta(\Lambda - |k_F - p - q^{'}|) - \theta(\Lambda - |k_F - p - q|))\theta(\Lambda - |k_F - p - q^{'} - q|)\theta(\Lambda - |k_F - p|). \tag{13.13}$$

Here $|p-k_F| < \Lambda$ and $|q|, |q^{'}| < 2\Lambda$. At this stage it is by no means obvious that the right-hand side of the above commutator is a c-number in the said limit. Giamarchi argues that we may define the normal ordered quantities : $c_R^\dagger(p+q^{'}+q)c_R(p) :\equiv c_R^\dagger(p+q^{'}+q)c_R(p) - \delta_{q+q^{'},0}\theta(k_F - p)$ and simply asserts without proof that the term involving the normal ordered quantities may be dropped. The reason why this is not acceptable to us is that there is no reason why, for example, we should choose the noninteracting ground state distribution $\theta(k_F - p)$. Indeed, in the spirit of the generalized RPA that we have been discussing, perhaps it is more appropriate to assume that the right side is zero unless $q + q^{'} = 0$ (which appears to be an uncontrolled approximation but may be justified in the limit $\Lambda \to \infty$ by taking matrix elements with respect to the states of the noninteracting system) but the non-vanishing term on the right is related to the number operator rather than a c-number,

$$[\rho_R(q), \rho_R(q^{'})] = \delta_{q+q^{'},0} sgn(q) \sum_{-|q|<p<0} c_R^\dagger(k_F + p + \Lambda)c_R(k_F + p + \Lambda)$$

$$- sgn(q)\delta_{q+q^{'},0} \sum_{|q|>p>0} c_R^\dagger(k_F + p - \Lambda)c_R(k_F + p - \Lambda). \tag{13.14}$$

The important point here is that $|q|$ is fixed while $k_F \gg \Lambda \to \infty$. This means that the first term is the number operator at momenta much above the Fermi surface and the second is the operator for momenta much below the Fermi surface. We assume that the interactions (or temperatures) are such that they do not affect the states in these regions. Thus the number operator is purely classical in these regimes with a value given by the noninteracting Fermi distribution. This is the real reason why the commutator $[\rho_R(q), \rho_R(-q)]$ is a c-number with the value $-\frac{qL}{(2\pi)}$, rather than the unconvincing reason involving a casual manipulation of infinite series by adding and subtracting infinite quantities (in fact, no explanation is given as to why the momentum distribution of the *noninteracting* ground state is used to 'regularize' rather than the interacting quantities), given by Haldane, which is reproduced almost verbatim in the text by Giamarchi.

Also for the same reason, Haldane's expression that harmonically analyzes the density fluctuation that includes terms that oscillate rapidly with wave numbers that are even multiples of $k_F$ is equally meaningless since these expressions are formally divergent and cannot be regarded as operator identities. They are indeed regarded as such in the g-ology program; for instance in the much studied work of Kane and Fisher on impurities in Luttinger liquids. A better way of relating the higher harmonics to the lower ones is to adopt Haldane's idea in a different way. We note that in one dimension, the number of particles in any interval $[x,x']$ is an integer. This means $\int_{x'}^{x} dy\, \rho(y)$ is an integer where $\rho(x) \equiv \psi^{\dagger}(x)\psi(x)$. Or, $e^{2\pi i \int_{x'}^{x} dy\, \rho(y)} = 1$. We write,

$$\rho(y) = \rho_0 + \frac{1}{L} \sum_{0<|q|\ll k_F} e^{-iqy}\rho_{q,s} + \frac{1}{L} \sum_{0<|q|\ll k_F, n=\pm1,\pm2,...} e^{-iqy}e^{-2ik_F n y}\rho_{q,2n}. \tag{13.15}$$

Thus,

$$e^{2ik_F x + \frac{2\pi i}{L}\sum_{q\neq 0}\frac{e^{-iqx}}{-iq}\rho_{q,s} + \frac{2\pi i}{L}\sum_{q\neq 0, n=\pm1,\pm2,...}\frac{e^{-iqx}e^{-2ik_F n x}}{-2ik_F n}\rho_{q,2n}} = const. \tag{13.16}$$

The following ansatz appears to fit the purpose.

$$\frac{1}{L}\sum_{q\neq 0} e^{-iqx}\,\rho_{q,2n} = \rho_0\, e^{-\frac{2n\pi i}{L}\sum_{q\neq 0}\frac{e^{-iqx}}{-iq}\rho_{q,s}} \tag{13.17}$$

Substituting this back into Eq. (13.16) we get,

$$e^{2ik_F x + \frac{2\pi i}{L}\sum_{q\neq 0}\frac{e^{-iqx}}{-iq}\rho_{q,s} - a\sum_{n=\pm1,\pm2,...}\frac{z^n}{n}} = \left(e^{2ik_F x + \frac{2\pi i}{L}\sum_{q\neq 0}\frac{e^{-iqx}}{-iq}\rho_{q,s}}\right)\frac{1-z}{1-\frac{1}{z}} = const. \tag{13.18}$$

where,

$$z = e^{-2ik_F x}e^{-\frac{2\pi i}{L}\sum_{q\neq 0}\frac{e^{-iqx}}{-iq}\rho_{q,s}}, \tag{13.19}$$

thereby verifying the claim. The above assertions are also merely mnemonics and not operator identities since $\rho_0 \to \infty$ is essential for the algebra to make sense and in this limit none of the operators are meaningful. Thus Luther's construction, Haldane's harmonic analysis of the field operator etc., are at best only mnemonics for the correlation functions, not to be taken literally and used as an operator to construct Hamiltonians. The harmonic analysis presented here is only valid for translationally invariant systems. For systems with impurities, a different kind of harmonic analysis is needed that reproduces the correct density–density correlations of otherwise free fermions but with the impurity. Therefore, the following set of judgmental characterizations are in order.

(a) *Preposterous*: The Fermi field operator has an expression in terms of bosons constructed out of Fermi bilinears and other objects like Klein factors. No such claim has ever been proven in the literature. In other words, no proof exists that all matrix elements, or indeed any matrix element of the nonlocal combination of bosons is equal to the corresponding matrix elements of the field operator.

(b) *Plausible but still untrue*: Equal-time number-conserving products of Fermi fields are expressible in terms of Fermi bilinears that are bosonic in character. This has also never been proven; just showing that the propagator comes out right is not enough. All matrix elements have to come out right, showing the finite temperature case is also not enough—that is just the diagonal matrix elements.

(c) *Possible Fact*: N-point functions have a nonlocal integral representation involving commuting variables that may be simply related to Fermi bilinears such as current and densities.

For the chiral bosonization scheme to be meaningful, the following conjecture should be proven.

*The bosonization conjecture*:

$$\Psi_r^\dagger(x) \equiv \psi_r^\dagger(x) \tag{13.20}$$

Eq. (13.20) means, among other things:

*Conjecture A*:

$$\{\Psi_r^\dagger(x), c_{r,p}\} \equiv \frac{1}{\sqrt{L}} e^{-irk_F x} e^{-ipx} = Lim_{\varepsilon \to 0}\{\psi_r^\dagger(x;\varepsilon), c_{r,p}\}. \tag{13.21}$$

This Eq. (13.21) is both necessary and sufficient to verify that matrix elements of $\psi_r^\dagger(x)$ and $\Psi_r^\dagger(x)$ are identical, for we may imagine states $|P;n-1>= c^\dagger_{p_1}c^\dagger_{p_2}..c^\dagger_{p_{n-1}}|G>$ and $|P';n>= c^\dagger_{p'_1}c^\dagger_{p'_2}..c^\dagger_{p'_n}|G>$, where $|G>$ is the ground state of the noninteracting system (say), and evaluate $<P',n|\psi_r^\dagger(x)|P;n-1>$ by passing the field across others. For this to come out right, it is necessary and sufficient for Eq. (13.21) to be valid. We now have to insert the expression for $U_r^\dagger$.

$$U_r^\dagger = Lim_{\delta \to 0}\frac{1}{\sqrt{L}}\int_0^L dy\; e^{irk_F y}\; e^{-i\phi_r^\dagger(y,\delta)}\Psi_r^\dagger(y;?)e^{-i\phi_r(y,\delta)} \tag{13.22}$$

where we use $\Psi_r^\dagger(y;?)$ to denote the ambiguity in the existing literature associated with this choice as we have alluded to already. With some effort it can be shown that for the advertised properties of $U_r$ to come out right (viz. $U_rU_r^\dagger = U_r^\dagger U_r$, etc.) the choice $\Psi_r^\dagger(y;?) \equiv \Psi_r^\dagger(y)$ must be made. It is important that $U_r^\dagger$ be independent of 'regularization' parameters like $\varepsilon, \delta$, and so on (these limits have to be taken and $U_r$ should be free from these parameters), since otherwise

$Lim_{\varepsilon\to 0^+}\{\psi_r(x;\varepsilon),\psi_r^{\dagger}(x';\varepsilon)\}$ will not come out right[2] (see later). Now we insert the definition of $U_r$ to get,

$$\psi_r^{\dagger}(x) \equiv Lim_{\varepsilon,\delta\to 0^+}\frac{1}{L}\int_0^L dy\, e^{-ik_Fr(x-y)}e^{i\phi_r^{\dagger}(x,\varepsilon)}\,e^{-i\phi_r^{\dagger}(y,\delta)}\psi_r^{\dagger}(y)e^{-i\phi_r(y,\delta)}e^{i\phi_r(x,\varepsilon)} = \Psi_r^{\dagger}(x). \tag{13.23}$$

For chiral bosonization to be meaningful, it is important that the above correspondence be literally true, i.e., an operator identity. It is easy to verify that this is not the case. For this we rewrite the above relation using the identities,

$$(e^{-ia\phi_r^{\dagger}(y',\delta')}\psi_r^{\dagger}(y)e^{ia\phi_r^{\dagger}(y',\delta')}) = e^{a(i\frac{\pi ry'}{L}+Log(1-e^{\frac{i(y-y')}{R}-\delta'}))}\,\psi_r^{\dagger}(y). \tag{13.24}$$

They lead to

$$Lim_{\varepsilon,\delta\to 0^+}\frac{1}{L}\int_0^L dy\, e^{-i(k_F+\frac{\pi}{L})r(x-y)}\frac{(1-e^{-\delta})}{(1-e^{\frac{i(y-x)}{R}-\varepsilon})}\,\psi_r^{\dagger}(y)\,e^{i\phi_r^{\dagger}(x,\varepsilon)-i\phi_r^{\dagger}(y,\delta)}e^{-i\phi_r(y,\delta)+i\phi_r(x,\varepsilon)}$$
$$= \Psi_r^{\dagger}(x). \tag{13.25}$$

This is an operator identity only if,

$$Lim_{\varepsilon,\delta\to 0^+}\frac{1}{L}e^{-i(k_F+\frac{\pi}{L})r(x-y)}\frac{(1-e^{-\delta})}{(1-e^{\frac{i(y-x)}{R}-\varepsilon})}\,e^{i\phi_r^{\dagger}(x,\varepsilon)-i\phi_r^{\dagger}(y,\delta)}e^{-i\phi_r(y,\delta)+i\phi_r(x,\varepsilon)} = \delta(x-y). \tag{13.26}$$

The delta function in this context means,

$$\delta(x-y) = Lim_{\delta\to 0^+}\frac{1}{L}\sum_p e^{ip(x-y)}e^{-\frac{L|p|}{2\pi}\delta} = \frac{1}{L}Lim_{\delta\to 0^+}\frac{sinh(\delta)}{cosh(\delta)-cos(\frac{(x-y)}{R})}. \tag{13.27}$$

Thus, if the bosonization formulas are operator identities as their proponents believe, Eq. (13.26) should be literally true. It is easy to verify that this is not true. For example, we may evaluate the expectation value of Eq. (13.26) in the non-interacting ground state ($\phi, b$ etc. annihilate the ground state) to obtain,

$$Lim_{\varepsilon,\delta\to 0^+}\frac{1}{L}e^{-i(k_F+\frac{\pi}{L})r(x-y)}\frac{(1-e^{-\delta})}{(1-e^{\frac{i(y-x)}{R}-\varepsilon})} =_? \delta(x-y) \tag{13.28}$$

where $=_?$ means - a conjecture. If $x\neq y$, it is fine but for $x-y\to 0$,

$$LHS = Lim_{\varepsilon,\delta\to 0^+}\frac{1}{L}\frac{\delta}{\varepsilon} = ? \tag{13.29}$$

[2]It must be stressed that it is not $\{\psi_r(x),\psi_r^{\dagger}(x')\} \equiv Lim_{\varepsilon,\varepsilon'\to 0^+}\{\psi_r(x;\varepsilon),\psi_r^{\dagger}(x';\varepsilon')\}$ that comes out right, but the weaker version with the two $\varepsilon$'s the same.

This is telling us that ε should go to zero faster than δ. In this limit we may approximate Eq. (13.28) as,

$$Lim_{\varepsilon,\delta\to 0^+} \frac{(-i(x-y)+R\varepsilon)\ \delta}{2R\varepsilon} \frac{R\varepsilon/\pi}{(x-y)^2+R^2\varepsilon^2} =_? \delta(x-y). \tag{13.30}$$

Since ε should go to zero before δ we get,

$$LHS = Lim_{\varepsilon,\delta\to 0^+} \frac{(-i(x-y)+R\varepsilon)\ \delta}{2R\varepsilon}\ \delta(x-y) =_? \delta(x-y). \tag{13.31}$$

This is clearly not possible as we have reached a contradiction. A quicker way of seeing this contradiction is to note that in the integral in Eq. (13.25), the integrand is sharply peaked at $x=y$ due to the factor $(1-e^{\frac{i(y-x)}{R}-\varepsilon})^{-1}$. This means in all other places we are entitled to replace $y$ with $x$ and take it outside the integral (mean value theorem). The resulting expression will not involve ε but will involve δ as a proportionality constant which vanishes in the end so that the left-hand side becomes zero but the right-hand side remains nonzero. It is possible that our appraisal of the situation is too simplistic, but we believe that the general idea that the chiral bosonization relations are not operator identities but merely mnemonics which are model dependent will hold up to more rigorous scrutiny by mathematicians.

*Conjecture B*: Fermion commutation rules between $\psi_r(x)$ and $\psi_r^\dagger(x')$ come out right.

This would be a kind of 'saving grace', but alas, this is untrue as well (true only in a weak sense). To test this, we first prove that $U_r$ is independent of the bosons (see exercises and Giarmarchi's text). It may be shown that $U_rU_r^\dagger = U_r^\dagger U_r$. This will be useful subsequently. Now we note that,

$$[\phi_r(x,\varepsilon),\phi_r^\dagger(x',\varepsilon')] = -Log(1-e^{\frac{ir}{R}(x-x')-\varepsilon-\varepsilon'}). \tag{13.32}$$

The Dirac delta is given by Eq. (13.27). The step function is related to this through the correspondence

$$\frac{d}{dx}\theta(x-x') = \delta(x-x'), \tag{13.33}$$

so that $(x\neq x')$,

$$\theta(x-x') = \frac{1}{2} + Lim_{\delta\to 0}\sum_n \frac{sin(\frac{n}{R}r(x-x'))}{2\pi nr}e^{-|n|\delta}$$

$$= \frac{1}{2} + \frac{1}{2\pi ir}(Log(e^{\frac{ir(x-x')}{R}}-1) - Log(1-e^{\frac{ir(x-x')}{R}})). \tag{13.34}$$

Now consider,

$$\psi_r^{\dagger}(x';\varepsilon') = \frac{1}{\sqrt{2\varepsilon' L}} e^{-ik_F r x'} e^{-\sum_{p\neq 0}\left(\frac{2\pi r}{Lp}\right) e^{-\frac{L\varepsilon'|p|}{2\pi}} \rho_r^{\dagger}(-p) e^{ipx'}} U_r^{\dagger} \tag{13.35}$$

$$\psi_r(x;\varepsilon) = \frac{1}{\sqrt{2\varepsilon L}} e^{ik_F r x} e^{-\sum_{p\neq 0}\left(\frac{2\pi r}{Lp}\right) e^{-\frac{L\varepsilon|p|}{2\pi}} \rho_r^{\dagger}(p) e^{-ipx}} U_r. \tag{13.36}$$

The fermion commutator is,

$$\{\psi_r(x), \psi_r^{\dagger}(x')\} = Lim_{\varepsilon,\varepsilon'\to 0} \frac{1}{\sqrt{2\varepsilon L}\sqrt{2\varepsilon' L}} e^{ik_F r(x-x')} U_r^{\dagger} U_r$$

$$\{e^{-\sum_{p\neq 0}\left(\frac{2\pi r}{Lp}\right) e^{-\frac{L\varepsilon|p|}{2\pi}} \rho_r^{\dagger}(p) e^{-ipx}}, e^{-\sum_{p\neq 0}\left(\frac{2\pi r}{Lp}\right) e^{-\frac{L\varepsilon'|p|}{2\pi}} \rho_r^{\dagger}(-p) e^{ipx'}}\}. \tag{13.37}$$

Since

$$[\rho_r^{\dagger}(p), \rho_{r'}^{\dagger}(-p')] = -\delta_{r,r'}\delta_{p,p'}\left(\frac{rpL}{2\pi}\right), \tag{13.38}$$

this means

$$\{\psi_r(x), \psi_r^{\dagger}(x')\} = Lim_{\varepsilon,\varepsilon'\to 0} \frac{1}{\sqrt{2\varepsilon L}\sqrt{2\varepsilon' L}} e^{ik_F r(x-x')} U_r^{\dagger} U_r$$

$$e^{-\sum_{p\neq 0}\left(\frac{2\pi r}{Lp}\right) e^{-\frac{L\varepsilon|p|}{2\pi}} \rho_r^{\dagger}(p) e^{-ipx} - \sum_{p\neq 0}\left(\frac{2\pi r}{Lp}\right) e^{-\frac{L\varepsilon'|p|}{2\pi}} \rho_r^{\dagger}(-p) e^{ipx'}}$$

$$e^{-\frac{1}{2}\sum_{p\neq 0}\left(\frac{2\pi r}{Lp}\right) e^{-\frac{L(\varepsilon+\varepsilon')|p|}{2\pi}} e^{-ip(x-x')}} e^{-\frac{1}{2}\sum_{p\neq 0}\left(\frac{2\pi r}{Lp}\right) e^{-\frac{L(\varepsilon+\varepsilon')|p|}{2\pi}} e^{-ip(x-x')}}$$

$$(1 + e^{\sum_{p\neq 0}\left(\frac{2\pi r}{Lp}\right) e^{-\frac{L(\varepsilon+\varepsilon')|p|}{2\pi}} e^{-ip(x-x')}}). \tag{13.39}$$

When $x - x' \neq 0$ we may write,

$$S = \sum_{p\neq 0}\left(\frac{2\pi r}{Lp}\right) e^{-\frac{L(\varepsilon+\varepsilon')|p|}{2\pi}} e^{-ip(x-x')}$$

$$= -i\sum_{n\neq 0} e^{-(\varepsilon+\varepsilon')|n|} \frac{sin(\frac{n}{R} r(x-x'))}{n} = -2\pi i r(\theta(x-x') - \frac{1}{2}). \tag{13.40}$$

Therefore when $x - x' \neq 0$,

$$\{\psi_r(x), \psi_r^{\dagger}(x')\} = Lim_{\varepsilon,\varepsilon'\to 0} \frac{1}{\sqrt{2\varepsilon L}\sqrt{2\varepsilon' L}} e^{ik_F r(x-x')} U_r^{\dagger} U_r$$

$$e^{-\sum_{p\neq 0}\left(\frac{2\pi r}{Lp}\right) e^{-\frac{L\varepsilon|p|}{2\pi}} \rho_r^{\dagger}(p) e^{-ipx} - \sum_{p\neq 0}\left(\frac{2\pi r}{Lp}\right) e^{-\frac{L\varepsilon'|p|}{2\pi}} \rho_r^{\dagger}(-p) e^{ipx'}}$$

$$e^{-\frac{1}{2}\sum_{p\neq 0}\left(\frac{2\pi r}{Lp}\right)e^{-\frac{L(\varepsilon+\varepsilon')|p|}{2\pi}}e^{-ip(x-x')}}(1+e^{-2\pi i r(\theta(x-x')-\frac{1}{2})})=0. \tag{13.41}$$

Thus, fermion commutation rules come out correctly when $x \neq x'$ but for the situation when $x-x' \to 0$, it does not come out correctly *unless* $\varepsilon=\varepsilon'$ in the above formulas. That is we have *contracted* the uv regulators into one. Thus it is valid in a weaker sense and the formulas cannot be considered operator identities. This kind of contraction of a similar parameter is needed also in our non-chiral bosonization technique (one meant to supplant the chiral approach). To see these issues more clearly, set $x-x'=X \to 0$ (means take this limit first) and $x+x'=2R$.

$$\{\psi_r(x),\psi_r^{\dagger}(x')\} = Lim_{\varepsilon,\varepsilon'\to 0}\frac{1}{\sqrt{2\varepsilon L}\sqrt{2\varepsilon' L}}e^{ik_F rX}U_r^{\dagger}U_r$$

$$e^{-\sum_{p\neq 0}\left(\frac{2\pi r}{Lp}\right)e^{-\frac{L\varepsilon|p|}{2\pi}}\rho_r^{\dagger}(p)e^{-ip(R+\frac{X}{2})}-\sum_{p\neq 0}\left(\frac{2\pi r}{Lp}\right)e^{-\frac{L\varepsilon'|p|}{2\pi}}\rho_r^{\dagger}(-p)e^{ip(R-\frac{X}{2})}}$$

$$(e^{-\frac{1}{2}\sum_{p\neq 0}\left(\frac{2\pi r}{Lp}\right)e^{-\frac{L(\varepsilon+\varepsilon')|p|}{2\pi}}e^{-ipX}}+e^{\frac{1}{2}\sum_{p\neq 0}\left(\frac{2\pi r}{Lp}\right)e^{-\frac{L(\varepsilon+\varepsilon')|p|}{2\pi}}e^{-ipX}}) \tag{13.42}$$

In general,

$$(e^{-\frac{1}{2}\sum_{p\neq 0}\left(\frac{2\pi r}{Lp}\right)e^{-\frac{L(\varepsilon+\varepsilon')|p|}{2\pi}}e^{-ipX}}+e^{\frac{1}{2}\sum_{p\neq 0}\left(\frac{2\pi r}{Lp}\right)e^{-\frac{L(\varepsilon+\varepsilon')|p|}{2\pi}}e^{-ipX}})=$$

$$2Cosh\left(\frac{r}{2}\left(\frac{iX}{R}+Log(1-e^{-i\frac{X}{R}-\delta})-Log(1-e^{i\frac{X}{R}-\delta})\right)\right). \tag{13.43}$$

We wish to investigate the small $X$ behavior of this function. Anticipating that this has the appearance of a Lorentzian, we Taylor expand the reciprocal of the above expression in powers of $X$ and retain the leading terms. This means,

$$(e^{-\frac{1}{2}\sum_{p\neq 0}\left(\frac{2\pi r}{Lp}\right)e^{-\frac{L(\varepsilon+\varepsilon')|p|}{2\pi}}e^{-ipX}}+e^{\frac{1}{2}\sum_{p\neq 0}\left(\frac{2\pi r}{Lp}\right)e^{-\frac{L(\varepsilon+\varepsilon')|p|}{2\pi}}e^{-ipX}})\approx$$

$$\frac{4R^2(\varepsilon+\varepsilon')^2}{X^2+2R^2(\varepsilon+\varepsilon')^2} \tag{13.44}$$

(since $0<\varepsilon+\varepsilon' \ll 1$)

$$\{\psi_r(x),\psi_r^{\dagger}(x')\} = Lim_{\varepsilon,\varepsilon'\to 0}\left(\frac{(\varepsilon+\varepsilon')}{\sqrt{2\varepsilon\varepsilon'}}\frac{\frac{\sqrt{2}R(\varepsilon+\varepsilon')}{\pi}}{X^2+2R^2(\varepsilon+\varepsilon')^2}\right)e^{ik_F rX}U_r^{\dagger}U_r$$

$$e^{-\sum_{p\neq 0}\left(\frac{2\pi r}{Lp}\right)e^{-\frac{L\varepsilon|p|}{2\pi}}\rho_r^{\dagger}(p)e^{-ip(R+\frac{X}{2})}-\sum_{p\neq 0}\left(\frac{2\pi r}{Lp}\right)e^{-\frac{L\varepsilon'|p|}{2\pi}}\rho_r^{\dagger}(-p)e^{ip(R-\frac{X}{2})}}. \tag{13.45}$$

From the above expression it is really clear that unless we set $\varepsilon = \varepsilon'$, the above expression will not be the Dirac delta function (in fact it diverges if they are independent). A similar observation also holds for four-point functions. While the quantity below is meaningless,

$$Lim_{\varepsilon_i \to 0^+} < \psi^\dagger(x_1, \varepsilon_1)\psi^\dagger(x_2, \varepsilon_2)\psi(x_3, \varepsilon_3)\psi(x_4, \varepsilon_4) >= \tag{13.46}$$

it is meaningful to write,

$$Lim_{\varepsilon_1, \varepsilon_2 \to 0^+} < \psi^\dagger(x_1, \varepsilon_1)\psi^\dagger(x_2, \varepsilon_2)\psi(x_3, \varepsilon_1)\psi(x_4, \varepsilon_2) > \tag{13.47}$$

or,

$$Lim_{\varepsilon_1, \varepsilon_2 \to 0^+} < \psi^\dagger(x_1, \varepsilon_1)\psi^\dagger(x_2, \varepsilon_2)\psi(x_3, \varepsilon_2)\psi(x_4, \varepsilon_1) > . \tag{13.48}$$

This distinction holds also for the 'point-splitting' definition of the density (fluctuation) of right movers. The density fluctuation of right movers is *not*,

$$\rho_r(x) \neq Lim_{a \to 0}(\psi_r^\dagger(x)\psi_r(x+a) - \frac{1}{2\pi i a})$$

$$\equiv Lim_{a \to 0} Lim_{\varepsilon, \varepsilon' \to 0}(\psi_r^\dagger(x;\varepsilon)\psi_r(x+a;\varepsilon') - \frac{1}{2\pi i a}) \tag{13.49}$$

but rather the contracted version,

$$\rho_r(x) \equiv Lim_{a \to 0} Lim_{\varepsilon \to 0}(\psi_r^\dagger(x;\varepsilon)\psi_r(x+a;\varepsilon) - \frac{1}{2\pi i a}). \tag{13.50}$$

These nuances may seem like pedantry gone haywire, but these have important ramifications to the problem of finding the solution to Luttinger liquids with impurities or boundaries where the contraction of similar objects plays a crucial role in generating the right Green functions. The point repeatedly being made is that the bosonization formulas are not operator identities but mnemonics for generating the n-point functions. Unlike operator identities, mnemonics are not mandated to be unique, universal, and model independent. The clinching evidence comes when applying standard chiral bosonization to free electrons plus a single delta function impurity. This model is solved in undergraduate textbooks and the asymptotic correlation function all have trivial exponents (identical to free fermions). However, evaluating the density–density correlations using Haldane's harmonic analysis shows a nontrivial exponent. This clearly demonstrates that chiral bosonization is wrong when applied to translationally non-invariant systems.

Our approach differs from chiral bosonization in several respects. The propagator is reduced to a closed form that is shown to be exact in the RPA (asymptotic) limit. It is shown to reproduce the expected results when the impurity is turned off and the two-body potential is on and vice versa (there are some interesting exceptions) when applied to the Kane and Fisher problem of a Luttinger liquid in the presence of a delta function potential. These features are unlike the 'standard' approach where some sort of renormalization group analysis is also needed. As far as we are concerned, the litmus test is to ask which of the two approaches is able to reproduce the Green function of free fermions together with the impurity in the RPA limit—a result trivially deduced using Fermi algebra. Our approach is able to do so, whereas the competing ones are not. Besides, no mathematically questionable manipulations such as 'normal ordering' are made, nor do we find the need to map the Fermi model to a bogus and ill-defined bosonic theory. Now we go on to discuss our approach.

## 13.2 General Considerations

Here, we explain how to write down explicit formulas for the field operator in terms of currents and densities. The general idea is the following. Given that $\rho(x) = \psi^\dagger(x)\psi(x)$ and $J(x) = Im[\psi^\dagger(x)\partial_x\psi(x)]$ are bilinears in $\psi$, one would like to invert these relations—take a sort of 'square root' of these two defining relations and express $\psi(x)$ alone in terms of $J$ and $\rho$. This enables the computation of single-particle properties in terms of the correlations between currents and densities, which in turn may be evaluated independently using some controlled approximation under some restrictive conditions. For now we shall treat this activity of expressing $\psi$ in terms of $\rho$ and $J$ as merely an interesting mathematical question. One issue immediately presents itself. A relation of the form $\psi(x) = F(\rho, J; x)$ interpreted as operator identity with no operators other than $\rho$ and $J$ on the right-hand side would not be correct since the field operator annihilates a particle whereas the right-hand side conserves the number of particles. Hence, one is forced to do one of two things. Either we insist that this should be an operator identity and introduce further variables which take care of adding and subtracting particles known as Klein factors well-known in the traditional bosonizing community, or advocate the point of view of the present article, namely think of the relation not as an operator identity but as a mnemonic for the n-point functions it generates. The n-point function being number-conserving quantities pose no such difficulties. The inversion in question is accomplished by introducing the conjugate to the density through the relation $J(x) = -\rho(x)\partial_x\pi(x)$. More explicitly, $\pi(x) = -\int_{-\infty}^{x} dy \frac{J(y)}{\rho(y)}$. Of course one is immediately faced with the problem of having to define the inverse of the density operator; this is particularly acute in places where it vanishes. Practically, we shall always assume that it is legitimate to write $\rho(x) = \rho_0 + \tilde{\rho}(x)$ where $\rho_0 \neq 0$ is the uniform average density and expand in powers of $\tilde{\rho}(x)$. The claim is that the field $\psi(x)$ may be written as,

$$\psi(x) = U([\rho]; x)e^{-i\pi(x)} \sqrt{\rho(x)}. \tag{13.51}$$

The current and density variables obey a closed algebra that is insensitive to the statistics of the underlying particles—they are the same for both bosons and fermions. Inverting the correspondence between current/density and the field means the expression for the field should contain some additional ingredient that addresses the issue of the nature of the underlying particles, because the currents and densities certainly do not. The object $U$ that appears in the above inverted correspondence (Eq. (13.51)) is precisely such an ingredient. In our approach, Eq. (13.51) is not an operator identity but a shorthand way of writing down correlation functions (n-point functions) using this relation. This means there is no guarantee that $U$ is going to be system independent. For now, let us try and determine as many properties as possible about this object $U$. Since $\pi(x)$ and $\rho(x)$ are Hermitian and $\psi^{\dagger}(x)\psi(x) \equiv \rho(x)$, it follows that $U$ has to be unitary.

$$U^{\dagger}([\rho];x)U([\rho];x) = 1 \tag{13.52}$$

Furthermore, using Eq. (13.51) we conclude that $Im[\psi^{\dagger}(x)\partial_x\psi(x)] \equiv -\rho(x)\partial_x\pi(x)$ if an only if,

$$U^{\dagger}([\rho];x)(\partial_x U([\rho];x)) \equiv 0. \tag{13.53}$$

Strictly speaking, the imaginary part of the above equation has to be zero. The real part of the left-hand side is also zero since $U$ is unitary. This appears to imply that $U$ is independent of $x$ since $U^{\dagger}$ is the inverse of $U$. This conclusion, however, is inconsistent with the requirement that $\psi(x)$ is a Fermi field. The fallacy here is that we are interpreting Eq. (13.51) as an operator identity. The difficulty with writing the Fermi field in this manner may be overcome if one is content with a weaker assertion, namely, the ability to generate all the n-point functions, as opposed to the matrix elements of the fields themselves. This is probably why this line of thought was abandoned in the seventies by those working in the field of current algebra. To see these difficulties more clearly, let us for the moment go along with the premise that we are dealing with operator identities. Now impose fermion commutation rules.

$$\begin{aligned}\psi(x)\psi(x') &= U([\rho];x)e^{-i\pi(x)}\sqrt{\rho(x)}U([\rho];x')e^{-i\pi(x')}\sqrt{\rho(x')}\\ &= U([\rho];x)U([\{\rho(y)+\delta(y-x)\}];x')\\ &\times e^{-i\pi(x)}e^{-i\pi(x')}\sqrt{\rho(x)-\delta(x-x')}\sqrt{\rho(x')}\end{aligned} \tag{13.54}$$

In the above we have used the identity $e^{-i\pi(x)}\rho(y)e^{i\pi(x)} \equiv \rho(y)+\delta(y-x)$. Imposition of fermion commutation rules $\psi(x)\psi(x') = -\psi(x')\psi(x)$ tells us that we have to impose,

$$U([\rho];x)U([\{\rho(y)+\delta(y-x)\}];x') = -U([\rho];x')U([\{\rho(y)+\delta(y-x')\}];x). \tag{13.55}$$

It is immediately obvious upon setting $x = x'$ in the first constraint (viz. Eq. (13.55)) $U \equiv 0$, if these are interpreted as operator identities. This is because, if these are operator identities, $U$ is invertible from the unitarity condition, and one may multiply by its inverse and come to this conclusion. Hence, these should not be thought of as operator identities. Whereas the second fermion commutation rule gives,

$$
\begin{aligned}
\delta(x-x') = \\
U([\{\rho(y)-\delta(y-x)\}];x)\ \sqrt{\rho(x)}\sqrt{\rho(x')}\ U^{\dagger}([\{\rho(y)-\delta(y-x')\}];x') \\
+\sqrt{(\rho(x')-\delta(x-x'))}\sqrt{(\rho(x)-\delta(x-x'))} \\
\times U([\{\rho(y)-\delta(y-x')-\delta(y-x)\}];x)U^{\dagger}([\{\rho(y)-\delta(y-x)-\delta(y-x')\}];x').
\end{aligned}
\tag{13.56}
$$

Set,

$$
U([\{\rho(y)-\delta(y-x)\}];x)\ \sqrt{\rho(x)} \equiv W([\rho];x). \tag{13.57}
$$

This means,

$$
\begin{aligned}
\delta(x-x') = W([\rho];x)W^{\dagger}([\rho];x') \\
+W([\{\rho(y)-\delta(y-x')\}];x)W^{\dagger}([\{\rho(y)-\delta(y-x)\}];x').
\end{aligned}
\tag{13.58}
$$

There are many solutions of Eq. (13.55) and Eq. (13.58). We have to use some information about the model we are studying in order make further progress. If we are going to interpret these assertions as operator identities (which they are not) we may go quite far in pinning down the nature of the functional $U([\rho];x)$. This is accomplished by examining the commutators between the field and the densities and currents. To this end, consider,

$$
[\psi(x),\rho(x')] = \delta(x-x')\psi(x) \tag{13.59}
$$

and

$$
[\psi(x),j(x')] = -i\delta(x-x')\nabla\psi(x) + \frac{1}{2i}(\nabla\delta(x-x'))\psi(x). \tag{13.60}
$$

The relation Eq. (13.59) is automatically obeyed by Eq. (13.51). But for Eq. (13.60) to be obeyed, a constraint on $U([\rho];x)$ has to be imposed. The constraint equation for $U(x)$ is,

$$
\rho(x')[U([\rho];x),\nabla'\pi(x')] = i\delta(x-x')(\nabla U([\rho];x)). \tag{13.61}
$$

To obtain this we have had to insert Eq. (13.51) into both sides of Eq. (13.60) and use the formula for current $j(x') = -\rho(x')\partial_{x'}\pi(x')$. But crucial use has been made of the idea that it is permissible to divide by the density $\rho(x)$. This is clearly not valid when the system has a rigid boundary or other translationally non-invariant systems, for at those points, the density vanishes and this constraint is not valid.

Thus, strictly speaking, it is valid for translationally invariant systems. One can see that Eq. (13.61) has many solutions, for if $U(x)$ is a solution, so is,

$$\tilde{U}([\rho];x) = U([\rho];x)\, e^{i\lambda \int_{-\infty}^{\infty} dy\, \theta(x-y)\, \rho(y)}. \tag{13.62}$$

The constraint equation Eq. (13.61) is a linear first-order functional "ODE" in $\rho$. This means that the most general solution will have to be a linear combination of the above solutions.

$$U([\rho];x) = \int_{-\infty}^{\infty} d\lambda\, D(\lambda)\, e^{i\lambda \int_{-\infty}^{\infty} dy\, \theta(x-y)\, \rho(y)} \tag{13.63}$$

Imposing fermion commutation rules on the form Eq. (13.51) leads to an additional constraint that $\lambda = n\pi$ where $n$ is an odd integer.

$$U([\rho];x) = \sum_{n=odd} D_n\, e^{in\pi \int_{-\infty}^{\infty} dy\, \theta(x-y)\, \rho(y)} \tag{13.64}$$

For translationally noninvariant systems, this constraint is violated and we will have to deal with them on a case-by-case basis. In earlier works (also see later), we have shown that the following choice for the field works for homogeneous Fermi systems.

$$\psi(x) = \frac{1}{\sqrt{N^0}} \sum_k n_F(k)\, e^{i\pi r(k)} e^{i\pi sgn(k) \int_{-\infty}^{x} \rho(y) dy} e^{-i\pi(x)} \sqrt{\rho(x)} \tag{13.65}$$

Here $r(k)$ are independent random (commuting with everything) real numbers for each $k$, taking on values in the interval $[-1,1]$ with a uniform probability distribution, so that $< f >= \frac{1}{2}\int_{-1}^{1} f(r)dr$. Also $n_F(k) = \theta(k_F - |k|)$ is the Fermi distribution and $N^0 = \sum_k n_F(k)$. This choice makes everything work out fine. As the reader may verify, the following conditions are obeyed.

$$< \psi(x)\hat{O} >= 0 \tag{13.66}$$

$$< \{\psi(x), \psi(x^{'})\}\hat{O} >= 0; \;\; < \{\psi(x), \psi^{\dagger}(x^{'})\}\hat{O} >= \delta(x - x^{'})\hat{O} \tag{13.67}$$

Here the average is over the random numbers and $\hat{O}$ is any function of $\rho$ and $J$, or anything else not involving the random numbers. Thus the fermion commutation rules are obeyed in this weaker sense of average over the random numbers. Similarly, the following is also true,

$$< \psi^{\dagger}(x)\psi(x)\hat{O} >= \rho(x)\hat{O}; \;\; < Im[\psi^{\dagger}(x)\partial_x\psi(x)]\hat{O} >$$
$$= -\rho(x)\partial_x\pi(x)\hat{O} = J(x)\hat{O}. \tag{13.68}$$

One may see that since

$$< e^{i\pi r(k)} e^{i\pi r(k^{'})} >= 0. \tag{13.69}$$

(even when $k = k'$) it follows that

$$< \psi(x)\psi(x') >= 0. \tag{13.70}$$

It is important to note that when more than two fields are involved, things are not so simple. The reason is because we have to take into account the observation that while

$$< e^{i\pi r(k)} e^{-i\pi r(k')} >= \delta_{k,k'}, \tag{13.71}$$

the product of four such random phasors (complex number of unit modulus) is,

$$< e^{i\pi r(k)} e^{-i\pi r(k')} e^{i\pi r(p)} e^{-i\pi r(p')} >= \delta_{k,k'}\delta_{p,p'} + \delta_{k,p'}\delta_{p,k'}. \tag{13.72}$$

The above assertion is incorrect when $k = k' = p = p'$, but this choice contributes negligibly to the correlation functions since the $k$'s take on continuous values. As a result, the two-point function may be written as,

$$< \psi^{\dagger}(x',t')\psi(x,t) >= \frac{1}{N^0}\sum_{k,k'} n_F(k')n_F(k)\ \delta_{k,k'}$$

$$\times < \sqrt{\rho(x',t')} e^{i\pi(x',t')} e^{-i\pi sgn(k')\int_{-\infty}^{x'}\rho(y',t')dy'}$$

$$\times e^{i\pi sgn(k)\int_{-\infty}^{x}\rho(y,t)dy} e^{-i\pi(x,t)} \sqrt{\rho(x,t)} >. \tag{13.73}$$

For the four-point functions we have similarly,

$$< \psi^{\dagger}(x',t')\psi(x,t)\psi^{\dagger}(y',s')\psi(y,s) >=$$

$$\frac{1}{(N^0)^2}\sum_{k,k',p,p'} (\delta_{k,k'}\delta_{p,p'} + \delta_{k,p'}\delta_{p,k'})\ n_F(k)n_F(k')n_F(p)n_F(p')$$

$$\times < \sqrt{\rho(x',t')} e^{i\pi(x',t')} e^{-i\pi sgn(k')\int_{-\infty}^{x'}\rho(z'',t')dz''}$$

$$\times e^{i\pi sgn(k)\int_{-\infty}^{x}\rho(z'',t)dz''} e^{-i\pi(x,t)} \sqrt{\rho(x,t)}$$

$$\sqrt{\rho(y',s')} e^{i\pi(y',s')} e^{-i\pi sgn(p')\int_{-\infty}^{y'}\rho(z',s')dz'}$$

$$\times e^{i\pi sgn(p)\int_{-\infty}^{y}\rho(z,s)dz} e^{-i\pi(y,s)} \sqrt{\rho(y,s)} >. \tag{13.74}$$

It has been shown in earlier works that when specialized to the noninteracting theory, these give the right answers. It involves writing $\rho(x,t) = \rho_0 + \tilde{\rho}(x,t)$ and writing the correlations between $\tilde{\rho}$ and $\pi$ using the random phase approximation. Thus, the results are exact only in the asymptotic sense (the separation between spatial locations tends to infinity and/or the separation between times also diverges). At this stage the reader may feel quite satisfied that everything has been given 'explicitly'. This is an illusion, however. There is nothing unique about the 'explicit construction' in Eq. (13.65). The device of introducing the random phasors was nothing more than a mnemonic for generating the above correlation functions (Eq. (13.73) and Eq. (13.74)). The non-uniqueness of this construction may be seen by noting that a choice,

$$\psi(x) = \frac{1}{\sqrt{N^0}} \sum_k n_F(k)\, e^{i\pi r(k)} e^{i\pi sgn(k) \int_{-\infty}^{x} \rho(y)dy}$$

$$\times e^{2i\pi sgn(k) \int_{-\infty}^{x} \rho(-y)dy} e^{-i\pi(x)} \sqrt{\rho(x)}, \tag{13.75}$$

also reproduces the expressions for current and density variables and also fermion commutation rules. The only thing it does not do, is reproduce the correct correlation functions of the noninteracting theory. Hence they cannot possibly be operator identities.

### 13.2.1 Free Propagator and Luttinger Liquid

Here, we explicitly evaluate the propagator of the noninteracting system using the prescription in Eq. (13.73). Later on we evaluate the propagator of fermions mutually interacting via short-range interactions, restricting our attention to only forward-scattering terms (this means the momentum transfer is (vanishingly) small compared to the Fermi momentum in the RPA limit). Even though the claim is that Eq. (13.73) is generally valid, practically, we have to restrict ourselves to the asymptotic limit. This means we look at intervals such that $|x - x'| \to \infty$ and/or $|t - t'| \to \infty$. This limit may be investigated elegantly by taking the RPA limit at the outset viz. the Fermi momentum and the mass both diverge, but their ratio—the Fermi velocity—is held fixed ($k_F, m \to \infty$ and $v_F = k_F/m < \infty$). In this limit, it can be shown that higher moments such as $< \tilde{\rho}(x,t)\tilde{\rho}(x',t')\tilde{\rho}(x'',t'') >$, $< \pi(x,t)\tilde{\rho}(x',t')\tilde{\rho}(x'',t'') >$,... all vanish. This means that the only non-zero correlations are $< \pi(x,t)\tilde{\rho}(x',t') >$, $< \tilde{\rho}(x,t)\tilde{\rho}(x',t') >$, and $< \pi(x,t)\pi(x',t') >$ (here $\tilde{\rho}(x,t) = \rho(x,t) - \rho_0$ is the density fluctuation from the uniform average). Thus Eq. (13.73) may be significantly simplified, first by replacing $\sqrt{\rho}$ by the constant $\sqrt{\rho_0}$ and using (a version of) the cumulant expansion.

$$< e^A e^B e^C e^D > = e^{\frac{1}{2}<A^2>} e^{\frac{1}{2}<B^2>} e^{\frac{1}{2}<C^2>} e^{\frac{1}{2}<D^2>} e^{<AB>} e^{<AC>} e^{<AD>} e^{<BC>} e^{<BD>} e^{<CD>} \tag{13.76}$$

The notion of the cumulant owes its origin to statistics. Let $X$ be a random variable. Define $Exp[g(t)] =< e^{tX} >$ where $t$ is some parameter. The idea is to relate the Taylor expansion of $g(t)$ in terms of averages $< X^n >$. This is accomplished as follows.

$$g(t) = -\sum_{n=1}^{\infty}\frac{1}{n}\left(-\sum_{m=1}^{\infty}L_m\frac{t^m}{m!}\right)^n = L_1 t + (L_2 - L_1^2)\frac{t^2}{2!}$$
$$+(L_3 - 3L_2L_1 + 2L_1^3)\frac{t^3}{3!} + ... \qquad (13.77)$$

where $L_n \equiv < X^n >$.

The current is given by $J(x) = -\rho(x)\partial_x\pi(x)$. The idea now is that one is interested in the RPA limit. In this limit, the fields are peaked in momentum or wave number space at $k = \pm k_F$ corresponding to right and left movers as we have explained already in earlier chapters. Therefore, it is legitimate to write $\psi(x) = e^{ik_Fx}\psi_R(x) + e^{-ik_Fx}\psi_L(x)$. This means the density is of the form

$$\rho(x) = \rho_0 + \rho_R(x) + \rho_L(x) + e^{2ik_Fx}\rho_f(x) + e^{-2ik_Fx}\rho_f^*(x) \qquad (13.78)$$

and same with current,

$$J(x) = J_R(x) + J_L(x) + e^{2ik_Fx}J_f(x) + e^{-2ik_Fx}J_f^*(x). \qquad (13.79)$$

Here $\rho_R(x)$, $\rho_L(x)$ and $\rho_f(x)$ are slowly varying on a scale of $k_F^{-1}$ and are designated 'slow variables'. But since the current also has similar terms and no higher ones, it follows that the velocity potential $\pi(x)$ is always going to be a 'slow' variable. In the RPA limit we may write,

$$J_s(x) \equiv J_R(x) + J_L(x) \approx -\rho_0\partial_x\pi(x) \qquad (13.80)$$

$$J_f(x) \approx -\rho_f(x)\partial_x\pi(x). \qquad (13.81)$$

The velocity potential may be related to the time derivatives of the slow part of the density using the equation of continuity.

$$\frac{1}{m}\partial_x J_s(x,t) + \partial_t\rho_s(x,t) = 0 \qquad (13.82)$$

or,

$$-\frac{\rho_0}{m}\partial_x^2\pi(x) + \partial_t\rho_s(x,t) = 0 \qquad (13.83)$$

where $\frac{\rho_0}{m} = \frac{v_F}{\pi}$. Thus we are able to relate $\pi(x)$ to the time derivative of $\rho_s(x)$. Therefore, while evaluating the Green function using Eq. (13.73) we have to evaluate the correlation between the densities and their time derivatives. The crucial point about bosonization is to relate the rapidly varying $\rho_f(x)$ to the slowly varying $\rho_s(x)$. For translationally invariant systems, this is well known (see Giamarchi, Haldane, etc.) but for the Kane and Fisher problem it is new and will be discussed in this chapter. We may now substitute Eq. (13.17) into Eq. (13.15) to obtain,

$$\rho(y) = \rho_0 + \frac{1}{L}\sum_{0<|q|\ll k_F} e^{-iqy}\rho_{q,s} + \sum_{n=\pm 1} e^{-2ik_F ny}\,\rho_0\, e^{-\frac{2n\pi i}{L}\sum_{q\neq 0}\frac{e^{-iqx}}{-iq}\rho_{q,s}}. \tag{13.84}$$

Substituting these into Eq. (13.51) e.g., enables us to read off mnemonics for the right and left movers ($\nu = +1(R), -1(L)$).

$$\psi_\nu(x,t) \sim e^{i\pi\nu\int_{-\infty}^{x}\rho_s(y,t)dy}e^{-i\pi(x,t)}\sqrt{\rho_0} \tag{13.85}$$

The propagators may be written as,

$$<\psi_\nu(x,t)\psi_\nu^\dagger(x',t')> \sim < e^{i\pi\nu\int_{-\infty}^{x}\rho_s(y,t)dy}e^{-i\pi(x,t)}e^{i\pi(x',t')}e^{-i\pi\nu\int_{-\infty}^{x'}\rho_s(y',t')dy'} > \rho_0. \tag{13.86}$$

For translationally invariant systems, it is convenient to work in momentum space where we write,

$$\rho_s(x,t) = \frac{1}{L}\sum_{0<|q|<\Lambda\ll k_F} e^{-iqy}\rho_{q,s}(t). \tag{13.87}$$

Similarly, we may write,

$$\pi(x,t) = \sum_{0<|q|<\Lambda\ll k_F} e^{iqy}X_q(t). \tag{13.88}$$

Substituting these into Eq. (13.83) we get,

$$\frac{Nq^2}{m}X_q(t) + \partial_t\rho_{-q,s}(t) = 0. \tag{13.89}$$

We may consolidate these results to write,

$$\begin{aligned}<\psi_\nu(x,t)\psi_\nu^\dagger(x',t')> \;\sim\; &< e^{\sum_q(i\pi\nu\frac{1}{-iqL}+i(\frac{m}{Nq^2})\partial_t)e^{-iqx}\rho_{q,s}(t)} \\ &\times e^{\sum_q(-i\pi\nu\frac{1}{iqL}-i(\frac{m}{Nq^2})\partial_{t'})e^{iqx'}\rho_{-q,s}(t')} > \rho_0.\end{aligned} \tag{13.90}$$

Now we have to use the cumulant expansion to evaluate this. In order to do this, we have to know how to compute the density–density correlations. There is an elegant and well-known method using functional bosonization (see the paper by D. K. K. Lee and Y. Chen and also by Yurkevich). The procedure is as follows. We formally write the action of free fermions as $S_0[\rho,\pi]$. If we now consider mutual interaction between particles due to a potential $v_q$, we may write (in the Matsubara formalism),

$$S_{full}[\rho,\pi] = S_0[\rho,\pi] - \int_0^{-i\beta} dt \sum_q \frac{v_q}{2L}\rho_q(t)\rho_{-q}(t)$$

$$= S_0[\rho,\pi] + i\beta \sum_{q,n} \frac{v_q}{2L}\rho_{q,n}\rho_{-q,-n}. \tag{13.91}$$

Here we set $\sum_n e^{-w_n t}\rho_{q,n} \equiv \rho_q(t)$ where $w_n = \frac{2\pi n}{\beta}$ is the Matsubara frequency. This is nothing but $S_{full} = \int dt L = \int dt(T-V)$. Now we define the generating function of density correlations as,

$$Z[U] \equiv \left\langle e^{\sum_{q,n} U_{q,n}\rho_{q,n}} \right\rangle = \int D[\rho] e^{\sum_{q,n} U_{q,n}\rho_{q,n}} \int D[\pi]\ e^{iS_{full}[\rho,\pi]}. \tag{13.92}$$

Here $Z[U]$ is such that,

$$\frac{\delta^2}{\delta U_{q,n}\delta U_{-q,-n}} Z[U] \equiv < \rho_{q,n}\rho_{-q,-n} > \tag{13.93}$$

and so on. This notation simply means,

$$< T\ \rho_q(t)\rho_{-q}(t') > = \sum_n e^{-w_n(t-t')} < \rho_{q,n}\rho_{-q,-n} > . \tag{13.94}$$

Hence the name generating function. We may also define the generating function for the noninteracting system.

$$Z_0[U] \equiv \left\langle e^{\sum_{q,n} U_{q,n}\rho_{q,n}} \right\rangle_0 = \int D[\rho] e^{\sum_{q,n} U_{q,n}\rho_{q,n}} \int D[\pi]\ e^{iS_0[\rho,\pi]} \tag{13.95}$$

Since the density correlations of the noninteracting system are simple and known beforehand, we regard $Z_0[U]$ as fully known. The idea is to eliminate the unknown viz. $S_0$ and express the important quantity $Z[U]$ directly in terms of $Z_0[U]$. To do this, we may assume $U_{q,n}$ is purely imaginary and write,

$$\int D[U']\ Z_0[U']\ e^{-\sum_{q,n} U'_{q,n}\rho_{q,n}} =$$

$$\int D[\rho'] \int D[\pi'] \int D[U']\ e^{\sum_{q,n} U'_{q,n}(\rho'_{q,n}-\rho_{q,n})}\ e^{iS_0[\rho',\pi']}$$

$$= \int D[\rho'] \int D[\pi'] \left( \prod_{q,n} \delta(\rho'_{q,n} - \rho_{q,n}) \right) e^{iS_0[\rho',\pi']} = \int D[\pi']\ e^{iS_0[\rho,\pi']}. \tag{13.96}$$

Therefore,

$$\int D[\pi]\ e^{iS_0[\rho,\pi]} = \int D[U']\ Z_0[U']\ e^{-\sum_{q,n} U'_{q,n}\rho_{q,n}}. \tag{13.97}$$

We insert the above relation into Eq. (13.92) to get,

$$Z[U] = \int D[U'] \, Z_0[U'] \int D[\rho] \, e^{\sum_{q,n}(U_{q,n}-U'_{q,n})\rho_{q,n}} e^{-\beta\sum_{q,n}\frac{v_q}{2L}\rho_{q,n}\rho_{-q,-n}}$$

$$= \int D[U'] \, Z_0[U'] \, e^{\sum_{q,n}\frac{L}{2\beta v_q}(U_{q,n}-U'_{q,n})(U_{-q,-n}-U'_{-q,-n})}. \tag{13.98}$$

Therefore, we have the final important relation,

$$Z[U] = \int D[U'] \, Z_0[U'] \, e^{\sum_{q,n}\frac{L}{2\beta v_q}(U_{q,n}-U'_{q,n})(U_{-q,-n}-U'_{-q,-n})}. \tag{13.99}$$

Using cumulant expansion, we may write (we assume that $q \neq 0$),

$$\begin{aligned} Z[U] &= \left\langle e^{\sum_{q,n}U_{q,n}\rho_{q,n}} \right\rangle \\ &= e^{\frac{1}{2!}\sum_{q,n}U_{q,n}U_{-q,-n}<\rho_{q,n}\rho_{-q,-n}>} e^{\frac{1}{3!}\sum_{q,n}U_{q,n}U_{q',n'}<\rho_{q,n}\rho_{q',n'}\rho_{-q-q',-n-n'}>} \ldots \end{aligned} \tag{13.100}$$

similarly for $Z_0$,

$$\begin{aligned} Z_0[U] &= \left\langle e^{\sum_{q,n}U_{q,n}\rho_{q,n}} \right\rangle \\ &= e^{\frac{1}{2!}\sum_{q,n}U_{q,n}U_{-q,-n}<\rho_{q,n}\rho_{-q,-n}>_0} e^{\frac{1}{3!}\sum_{q,n}U_{q,n}U_{q',n'}<\rho_{q,n}\rho_{q',n'}\rho_{-q-q',-n-n'}>_0} \ldots \end{aligned} \tag{13.101}$$

The claim now is that in the RPA limit, and for forward scattering only $v_q = 0$ for $|q| > \Lambda \ll k_F$. This means from Eq. (13.99) $U'_{q,n} \equiv U_{q,n}$ whenever $|q| > \Lambda$. But we are only interested in the correlation functions for momenta $|q| < \Lambda$. This means we may restrict our attention to integrating over $U'_{q,n}$ only for $|q| < \Lambda$. It can be shown that if all momenta are much smaller than the Fermi momentum, then only the second moment $< \rho_{q,n}\rho_{-q,-n} >$ survives and all higher ones such as $< \rho_{q,n}\rho_{q',n'}\rho_{-q-q',-n-n'} >$ vanish. This makes the theory Gaussian and we obtain,

$$Z[U] = \int D[U'] \, e^{\frac{1}{2!}\sum_{q,n}U'_{q,n}U'_{-q,-n}<\rho_{q,n}\rho_{-q,-n}>_0} \, e^{\sum_{q,n}\frac{L}{2\beta v_q}(U_{q,n}-U'_{q,n})(U_{-q,-n}-U'_{-q,-n})}$$

$$= e^{\frac{1}{2!}\sum_{q,n}\frac{<\rho_{q,n}\rho_{-q,-n}>_0}{1+\frac{\beta v_q}{L}<\rho_{q,n}\rho_{-q,-n}>_0}U_{q,n}U_{-q,-n}}. \tag{13.102}$$

Therefore in the RPA sense we get,

$$< \rho_{q,n}\rho_{-q,-n} >= \frac{< \rho_{q,n}\rho_{-q,-n} >_0}{1+\frac{\beta v_q}{L} < \rho_{q,n}\rho_{-q,-n} >_0}. \tag{13.103}$$

For the free theory we may write,

$$< \rho_{q,n}\rho_{-q,-n} >_0 = \frac{L|q|}{\pi\beta}\frac{v_F|q|}{w_n^2+(v_Fq)^2}. \tag{13.104}$$

This means in the time domain,

$$< \rho_q(t)\rho_{-q}(t') > = N\frac{|q|}{2mv}e^{-iv|q|(t-t')} \tag{13.105}$$

where $v = ((v_F)^2 + \frac{v_q}{\pi}v_F)^{\frac{1}{2}}$ is the Fermi velocity modified by interactions (assuming $v_q$ is independent of $q$). We may now use the cumulant expansion on Eq. (13.90) and insert the form of the density–density correlations to obtain the following expressions for the propagators of a Luttinger liquid (see exercises).

$$< T\ \Psi_R(x,t)\Psi_R^{\dagger}(x',t') > =$$

$$\frac{e^{ik_F(x-x')}}{((x-x')-v(t-t'))}\left(\frac{i}{2\pi}\right)e^{\left(-\frac{v_F}{4v}+\frac{1}{2}-\frac{v}{4v_F}\right)\ Log(|(x-x')^2-v^2(t-t')^2|)} \tag{13.106}$$

and,

$$< T\ \Psi_L(x,t)\Psi_L^{\dagger}(x',t') > =$$

$$-\frac{e^{-ik_F(x-x')}}{((x-x')+v(t-t'))}\left(\frac{i}{2\pi}\right)e^{\left(-\frac{v_F}{4v}-\frac{v}{4v_F}+\frac{1}{2}\right)\ Log(|(x-x')^2-v^2(t-t')^2|)} \tag{13.107}$$

## 13.3 Impurity in a Luttinger Liquid

First, we consider free spinless electrons in an impurity potential $V(x) = V_0\delta(x)$, and list the following expressions for the RPA limit of the propagator, which is nothing but the sum of these four pieces shown below (here $\theta(z)$ is the Heaviside unit step function). But first we have to define the right and left moving fields. Let $\psi(x)$ be the fermion field obeying $\{\psi(x),\psi(x')\} = 0$ and $\{\psi(x),\psi^{\dagger}(x')\} = \delta(x-x')$. Then we define ($\lambda_F = \frac{2\pi}{mv_F}$) figuratively speaking,

$$\Psi_R(x) = e^{ik_Fx}\ Lim_{m\to\infty}\frac{1}{2\lambda_F}\int_{x-\lambda_F}^{x+\lambda_F} dx'\ \psi(x')\ e^{-imv_Fx'} \tag{13.108}$$

$$\Psi_L(x) = e^{-ik_Fx}\ Lim_{m\to\infty}\frac{1}{2\lambda_F}\int_{x-\lambda_F}^{x+\lambda_F} dx'\ \psi(x')\ e^{imv_Fx'}. \tag{13.109}$$

Literally, however, the assertions are weaker. They really mean,

$$< T\ \Psi_a(x,t)\Psi_b^{\dagger}(x',t') > = e^{iak_Fx}e^{-ibk_Fx'}\ Lim_{m\to\infty}\frac{1}{2\lambda_F}\int_{x-\lambda_F}^{x+\lambda_F} d\xi\frac{1}{2\lambda_F}$$

$$\int_{x'-\lambda_F}^{x'+\lambda_F} d\xi' \; e^{imv_F(b\xi'-a\xi)} < T\,\psi(\xi,t)\,\psi^{\dagger}(\xi',t') > . \tag{13.110}$$

The full field in the RPA limit would then be $\Psi(x) = \Psi_R(x) + \Psi_L(x)$. In this case we may write (at zero temperature, see exercises of Chapter 5),

$$< T\,\Psi_R(x,t)\Psi_R^{\dagger}(x',t') >= e^{ik_F(x-x')}\frac{1}{(x-x')-v_F(t-t')}$$
$$\times\left(\frac{i}{2\pi} - \frac{V_0}{2\pi v_F}\left(\frac{\theta(x')\theta(-x)}{\left(1-V_0\frac{i}{v_F}\right)} - \frac{\theta(x)\theta(-x')}{\left(1+V_0\frac{i}{v_F}\right)}\right)\right)$$

$$< T\,\Psi_L(x,t)\Psi_L^{\dagger}(x',t') >= e^{-ik_F(x-x')}\frac{-1}{(x-x')+v_F(t-t')}$$
$$\times\left(\frac{i}{2\pi} - \frac{V_0}{2\pi v_F}\left(\frac{\theta(-x')\theta(x)}{\left(1-V_0\frac{i}{v_F}\right)} - \frac{\theta(-x)\theta(x')}{\left(1+V_0\frac{i}{v_F}\right)}\right)\right)$$

$$< T\,\Psi_R(x,t)\Psi_L^{\dagger}(x',t') > \;=\; e^{ik_F(x+x')}\frac{V_0}{2\pi v_F}\left(\frac{\theta(-x')\theta(-x)}{\left(1-V_0\frac{i}{v_F}\right)} - \frac{\theta(x)\theta(x')}{\left(1+V_0\frac{i}{v_F}\right)}\right)$$
$$\times\frac{1}{(-x'-x)-v_F(t'-t)}$$

$$< T\,\Psi_L(x,t)\Psi_R^{\dagger}(x',t') > = e^{-ik_F(x+x')}\frac{V_0}{2\pi v_F}\left(\frac{\theta(x')\theta(x)}{\left(1-V_0\frac{i}{v_F}\right)} - \frac{\theta(-x)\theta(-x')}{\left(1+V_0\frac{i}{v_F}\right)}\right)$$
$$\times\frac{1}{(x'+x)-v_F(t'-t)}. \tag{13.111}$$

The average density and density correlation may be written as follows.

$$< \rho(x,t) > -\rho_0 = -\frac{V_0}{\pi^2}\frac{mv_F^2}{(V_0^2+v_F^2)}\int_0^{\infty}\frac{dq}{q}\,cos(qx)\,Log\left[\frac{2k_F+q}{|2k_F-q|}\right] \tag{13.112}$$

The density–density correlation function is (note that at this stage, we are dealing with external fields, only not mutual interactions, hence Wick's theorem applies)

$$< T\,\rho(x,t)\rho(x',t') > - < T\,\rho(x,t) >< T\,\rho(x',t') >$$
$$= - < T\,\Psi(x,t)\Psi^{\dagger}(x',t') >< T\,\Psi(x',t')\Psi^{\dagger}(x,t^{+}) > . \tag{13.113}$$

If $\rho_s(x,t)$ is the slowly varying part of the density,

$$< T\ \rho_s(x,t)\rho_s(x',t') >=$$

$$-\frac{V_0^2}{(2\pi)^2}\frac{\theta(xx')}{(v_F^2+V_0^2)}\left(\frac{1}{((x'+x)+v_F(t'-t))^2}+\frac{1}{((x+x')+v_F(t-t'))^2}\right)$$
$$+\frac{1}{((x-x')+v_F(t-t'))^2}\left(-\frac{\theta(xx')}{(2\pi)^2}-\frac{\theta(-xx')}{(2\pi)^2}\frac{v_F^2}{(v_F^2+V_0^2)}\right)$$
$$+\frac{1}{((x-x')-v_F(t-t'))^2}\left(-\frac{\theta(xx')}{(2\pi)^2}-\frac{\theta(-xx')}{(2\pi)^2}\frac{v_F^2}{(v_F^2+V_0^2)}\right). \quad (13.114)$$

## 13.3.1 Anomalous Scaling

We now prove the remarks made earlier, namely that the g-ology-based chiral bosonization gives the wrong answer when applied to translationally noninvariant systems such as the one considered here. In that approach, Haldane's harmonic analysis of the field operator leads to the following expression for the density fluctuation including the rapidly varying parts.

$$\rho(x,t)=\rho_0+\tilde{\rho}_s(x,t)+\rho_0\ e^{2ik_Fx}\ e^{2\pi i\int_{-\infty}^{x}dy\ \tilde{\rho}_s(y,t)}+\rho_0\ e^{-2ik_Fx}\ e^{-2\pi i\ \int_{-\infty}^{x}dy\ \tilde{\rho}_s(y,t)} \quad (13.115)$$

Here $\tilde{\rho}_s(x,t)$ is the slowly varying part (on a scale of $k_F^{-1}$) of the density fluctuation. This follows apparently from the simple observation that $\int_{x'}^{x}dy\ \rho(y,t)$ is an integer since it counts the number of particles in the interval $(x,x')$. This then means $e^{2\pi i\int_{x'}^{x}dy\rho(y,t)}=1$. The crucial next step is to write $\rho(y,t)=\rho_0+\tilde{\rho}(y,t)+e^{2ik_Fy}\ \rho_f(y,t)+e^{-2ik_Fy}\ \rho_f^*(y,t)$ and match the powers of $e^{\pm ik_Fx}$ on both sides to recover the result. Consider the part of the density–density correlation that oscillates as $e^{2ik_F(x-x')}$. Using the standard approach we are forced to conclude,

$$< \rho_f(x,t)\rho_f^*(x',t') >_{chiral}= e^{2ik_F(x-x')}\left\langle e^{2\pi i\int_{-\infty}^{x}dy\ \tilde{\rho}_s(y,t)}e^{-2\pi i\int_{-\infty}^{x'}dy'\ \tilde{\rho}_s(y',t')}\right\rangle. \quad (13.116)$$

This is easily evaluated using Eq. (13.114) to yield, for $xx'>0$,

$$< T\ \rho_f(x,t)\rho_f^*(x',t') >_{chiral}= e^{2ik_F(x-x')}$$

$$\times\, e^{\frac{v_0^2}{(v_F^2+V_0^2)}Log((x+x')^2-v_F^2(t-t')^2)-Log((x-x')^2-v_F^2(t-t')^2)}\ e^{-\frac{V_0^2}{(v_F^2+V_0^2)}Log(4xx')}. \quad (13.117)$$

This prediction of the standard bosonization method is wrong, for this quantity is trivially evaluated using Fermi algebra and Wick's theorem.

$$< T\ \rho_f(x,t)\rho_f^*(x',t') >_{correct} = -< T\ \psi_L(x',t')\psi_L^{\dagger}(x,t) >< T\ \psi_R(x,t)\psi_R^{\dagger}(x',t') > \tag{13.118}$$

We now substitute the values of the right and left moving Green function from the exact solution trivially obtained using Fermi algebra (Eq. (13.111)) to obtain,

$$< T\ \rho_f(x,t)\rho_f^*(x',t') >_{correct} = -e^{2ik_F(x-x')} \frac{1}{(2\pi)^2((x-x')^2 - v_F^2(t-t')^2)}. \tag{13.119}$$

Basically, this is saying that for a system of fermions with no mutual interactions, there should not be an anomalous exponent—all exponents are trivial. This is in contrast to the incorrect result inferred using Haldane's harmonic analysis (Eq. (13.117)), which shows anomalous scaling when applied to this model of free fermions plus a delta function potential (the correct solution and the chiral one agree only when $V_0 = 0$).

We now proceed to outline a formalism that does get this result right. We assert that the correct harmonic analysis for this problem is *model dependent* and *non-universal*. Besides, statements such as harmonic analysis of the density fluctuation are *not operator identities* but mnemonics for generating the correlation functions, as we have been repeatedly stressing. The correct way of doing harmonic analysis for the case of fermions with a single delta-function impurity is model dependent and is given by,

$$\rho_f(x,t) = e^{2ik_F x}\ r_2(x)\ e^{2\pi i \int_{sgn(x)\infty}^{x} dy\ (\tilde{\rho}_s(y,t) + \tilde{\rho}_s(-y,t))}. \tag{13.120}$$

In the next section (see also, exercises), we see that this choice does indeed correctly reproduce the correlation functions of free fermions plus impurity. Inserting Eq. (13.120) into the model-independent field operator prescription [3] in Eq. (13.65) allows us to write,

$$\psi(x) = \frac{1}{\sqrt{N^0}} \sum_k n_F(k)\ e^{i\pi r(k)}$$

$$e^{i\pi sgn(k)(\int_{-\infty}^{x} \rho_s(y)dy + e^{2ik_F x}\ \frac{r_2(x)}{2ik_F}\ e^{2\pi i \int_{sgn(x)\infty}^{x} dy\ (\tilde{\rho}_s(y,t)+\tilde{\rho}_s(-y,t))} + e^{-2ik_F x}\ \frac{r_2^*(x)}{-2ik_F}\ e^{-2\pi i \int_{sgn(x)\infty}^{x} dy\ (\tilde{\rho}_s(y,t)+\tilde{\rho}_s(-y,t))})}$$

$$e^{-i\pi(x)}\ \sqrt{\rho(x)}. \tag{13.121}$$

Upon expanding the exponent and matching the harmonics, we may read off a formula for the right mover (e.g.) in case of this system with a single impurity.

$$\psi_R(x,t) = C_1(x)\ e^{i\pi \int_{-\infty}^{x} \rho_s(y,t)dy} + C_2(x)\ e^{i\pi \int_{-\infty}^{x} \rho_s(y,t)dy} e^{2i\pi \int_{\infty sgn(x)}^{x} \rho_s(-y,t)dy} \tag{13.122}$$

[3]This is not strictly correct, that is only for translationally invariant systems, but it is a formula more readily accepted by the reader.

We have therefore found a derivation of the form of the field operator we are going to use for this problem. We can be confident that when mutual interactions between fermions are included, the results are also going to be right.

### 13.3.2 The Field 'Operator' Construction

The preceding discussion makes the need for an alternative to standard chiral bosonization amply clear. Its predictions are simply wrong when applied to a system of free fermions plus a delta function potential. We wish to reproduce the correct correlation functions using a version of the field operator obtained by exponentiating the commutators of the field with currents and densities. This idea has a long history, as can be seen by consulting the bibliography of our earlier works on the subject. For now, inspired by our earlier works, we make the following assertions that we justify a posteriori. Set $\theta_R(x,t) = \pi \int_{sgn(x)\infty}^{x} dy\, \tilde{\rho}_s(y,t) + \int_{\infty\, sgn(x)}^{x} dy\, \nu(y,t)$ and $\theta_L(x,t) = -\pi \int_{sgn(x)\infty}^{x} dy\, \tilde{\rho}_s(y,t) + \int_{\infty\, sgn(x)}^{x} dy\, \nu(y,t)$. Then we may write down the following expression for the field operator (this a mnemonic for the correlation functions it generates, it is not an operator identity),

$$\Psi(x,t) = C_{R,1}(x)\, e^{ik_F x}\, e^{i\theta_R(x,t)} + C_{R,2}(x)\, e^{ik_F x}\, e^{i\theta_R(x,t)+2\pi i \int_{sgn(x)\infty}^{x} dy\, \tilde{\rho}_s(-y,t)}$$

$$+C_{L,1}(x)\, e^{-ik_F x}\, e^{i\theta_L(x,t)} + C_{L,2}(x)\, e^{-ik_F x}\, e^{i\,\theta_L(x,t)-2\pi i \int_{sgn(x)\infty}^{x} dy\, \tilde{\rho}_s(-y,t)}. \quad (13.123)$$

The n-point functions deduced from this expression will involve products of several of the C-functions. Our earlier works have explained how to write down explicit formulas in case of a homogeneous system for these C-functions in terms of what we have called 'singular complex numbers'. It must be stressed that these objects commute with everything unlike the Klein factors of conventional bosonization. Fermion commutation rules and so on do come out correctly. These C-functions have to be independently fixed by making contact with the corresponding expressions obtained from Fermi algebra. This model-dependent field may be derived from a *model-independent* universal Fermi field (Eq. (13.65)) together with the *model dependent* harmonic analysis of the density fluctuation (Eq. (13.120)) by inserting the latter into the former. The *model dependence* of the harmonic analysis stems from a need to ensure that all the exponents related to the model of free fermions plus an external potential are trivial.

We now wish to write down expressions for the Green function of the system at hand, namely spinless fermions in the presence of a single impurity. One last point before we do this. Fermion commutation rules inferred from Eq. (13.123) do come out as expected namely $\Psi(x,t)\Psi(x',t) = -\Psi(x',t)\Psi(x,t)$ and $\Psi(x,t)\Psi^{\dagger}(x',t) = -\Psi^{\dagger}(x',t)\Psi(x,t)$ for $x \neq x'$ as can be easily verified. The additional term $Exp[-2\pi i \int_{sgn(x)\infty}^{x} dy\, \tilde{\rho}_s(-y,t)]$ is just $Exp[-2\pi i \times (integer)]$, which does not change the statistics of the field, which is determined by the term $e^{i\pi\int_{sgn(x)\infty}^{x} dy\ \tilde{\rho}_s(y,t)} = (-1)^{(integer)}$.

Some experimentation with Eq. (13.123) shows that the only combination of these terms that reproduces the correct propagator of free fermions plus impurity for both $xx' > 0$ and $xx' < 0$ are given below. The situation when $x$ and $x'$ are on different sides of the impurity ($xx' < 0$) is particularly tricky and more will be said about this case later. One may include all possible combinations from Eq. (13.123) and successively rule out all but the ones given below.

If $xx' > 0$:

$$< T\ \Psi_R(x,t)\Psi_R^\dagger(x',t') >= e^{ik_F(x-x')}\left(\frac{1}{2\pi}\right)\ < e^{i\theta_R(x,t)}e^{-i\theta_R(x',t')} > \quad (13.124)$$

$$< T\ \Psi_L(x,t)\Psi_L^\dagger(x',t') >= e^{-ik_F(x-x')}\left(\frac{1}{2\pi}\right)\ < e^{i\theta_L(x,t)}e^{-i\theta_L(x',t')} > \quad (13.125)$$

$$\begin{aligned}
&< T\ \Psi_R(x,t)\Psi_L^\dagger(x',t') >= \\
&e^{ik_F(x+x')}\frac{V_0}{4\pi v_F}\left(\frac{\theta(-x')\theta(-x)}{\left(1-V_0\frac{i}{v_F}\right)} - \frac{\theta(x)\theta(x')}{\left(1+V_0\frac{i}{v_F}\right)}\right)e^{\frac{2V_0^2+v_F^2}{2(V_0^2+v_F^2)}Log(2x)} \\
&\times\left\langle e^{(i\theta_R(x,t)+2\pi i\int_{sgn(x)\infty}^{x} dy\ \tilde{\rho}_s(-y,t))}e^{-i\theta_L(x',t')}\right\rangle \\
&+e^{ik_F(x+x')}\frac{V_0}{4\pi v_F}\left(\frac{\theta(x')\theta(x)}{\left(1+V_0\frac{i}{v_F}\right)} - \frac{\theta(-x)\theta(-x')}{\left(1-V_0\frac{i}{v_F}\right)}\right)e^{\frac{2V_0^2+v_F^2}{2(V_0^2+v_F^2)}Log(2x')} \\
&\times\left\langle e^{i\theta_R(x,t)}e^{(-i\theta_L(x',t')+2\pi i\int_{sgn(x')\infty}^{x'} dz'\ \tilde{\rho}_s(-z',t'))}\right\rangle
\end{aligned} \quad (13.126)$$

$$\begin{aligned}
&< T\ \Psi_L(x,t)\Psi_R^\dagger(x',t') >= \\
&e^{-ik_F(x+x')}\frac{V_0}{4\pi v_F}\left(\frac{\theta(x')\theta(x)}{\left(1-V_0\frac{i}{v_F}\right)} - \frac{\theta(-x)\theta(-x')}{\left(1+V_0\frac{i}{v_F}\right)}\right)e^{\frac{2V_0^2+v_F^2}{2(V_0^2+v_F^2)}Log(2x)} \\
&\times\left\langle e^{(i\theta_L(x,t)-2\pi i\int_{sgn(x)\infty}^{x} dy\ \tilde{\rho}_s(-y,t))}e^{-i\theta_R(x',t')}\right\rangle \\
&+e^{-ik_F(x+x')}\frac{V_0}{4\pi v_F}\left(\frac{\theta(-x')\theta(-x)}{\left(1+V_0\frac{i}{v_F}\right)} - \frac{\theta(x)\theta(x')}{\left(1-V_0\frac{i}{v_F}\right)}\right)e^{\frac{2V_0^2+v_F^2}{2(V_0^2+v_F^2)}Log(2x')}
\end{aligned}$$

$$\times \left\langle e^{i\theta_L(x,t)} e^{(-i\theta_R(x',t') - 2\pi i \int_{sgn(x')\infty}^{x'} dz' \, \tilde{\rho}_s(-z',t'))} \right\rangle . \quad (13.127)$$

If $xx' < 0$:

$$< T \, \Psi_R(x,t)\Psi_R^{\dagger}(x',t') >$$

$$= e^{ik_F(x-x')} \, e^{Log(2x)} \, [\frac{1}{4(x+x')\pi} - \frac{V_0}{4i(x+x')\pi v_F} \left( \frac{\theta(x')\theta(-x)}{\left(1 - V_0 \frac{i}{v_F}\right)} - \frac{\theta(x)\theta(-x')}{\left(1 + V_0 \frac{i}{v_F}\right)} \right)]$$

$$< e^{(i\theta_R(x,t) + 2i\pi \int_{sgn(x)\infty}^{x} dy \, \tilde{\rho}_s(-y,t))} e^{-i\theta_R(x',t')} >$$

$$+ e^{ik_F(x-x')} \, e^{Log(2x')} \, [\frac{1}{4(x+x')\pi} - \frac{V_0}{4i(x+x')\pi v_F} \left( \frac{\theta(x')\theta(-x)}{\left(1 - V_0 \frac{i}{v_F}\right)} - \frac{\theta(x)\theta(-x')}{\left(1 + V_0 \frac{i}{v_F}\right)} \right)]$$

$$< e^{i\theta_R(x,t)} e^{-i\theta_R(x',t') - 2i\pi \int_{sgn(x')\infty}^{x'} dy' \, \tilde{\rho}_s(-y',t')} > \quad (13.128)$$

$$< T \, \Psi_L(x,t)\Psi_L^{\dagger}(x',t') >$$

$$= e^{-ik_F(x-x')} \, [\frac{1}{4(x+x')\pi} - \frac{V_0}{4i(x+x')\pi v_F} ( \frac{\theta(-x')\theta(x)}{\left(1 - V_0 \frac{i}{v_F}\right)} - \frac{\theta(-x)\theta(x')}{\left(1 + V_0 \frac{i}{v_F}\right)} )]$$

$$e^{Log(2x)} \; < e^{(i\theta_L(x,t) - 2i\pi \int_{sgn(x)\infty}^{x} dy \, \tilde{\rho}_s(-y,t))} e^{-i\theta_L(x',t')} >$$

$$+ e^{-ik_F(x-x')} \, [\frac{1}{4(x+x')\pi} - \frac{V_0}{4i(x+x')\pi v_F} ( \frac{\theta(-x')\theta(x)}{\left(1 - V_0 \frac{i}{v_F}\right)} - \frac{\theta(-x)\theta(x')}{\left(1 + V_0 \frac{i}{v_F}\right)} )]$$

$$e^{Log(2x')} \; < e^{i\theta_L(x,t)} e^{(-i\theta_L(x',t') + 2i\pi \int_{sgn(x')\infty}^{x'} dy' \, \tilde{\rho}_s(-y',t'))} > \quad (13.129)$$

$$< T \, \Psi_L(x,t)\Psi_R^{\dagger}(x',t') > = < T \, \Psi_R(x,t)\Psi_L^{\dagger}(x',t') > = 0. \quad (13.130)$$

We have zeroed in on the above expressions after much trial and error. It is quite hard to think of any other expression that does these things at the same time—(i) reproduce Eq. (13.111) when the mutual interactions are turned off. This is especially tricky when $xx' < 0$. (ii) Formal expressions that make it clear how the propagators obey fermion commutation rules. This means that no operators in the exponent are allowed that contain system-dependent multiplicative factors.

The long wavelength part of the density–density correlation in the RPA limit with mutual interaction $v_q$ (forward scattering only) and the impurity potential $V_0$ is given by,

$$< \rho(x,t)\rho(x',t') > - < \rho(x,t) >< \rho(x',t') > =$$

$$\frac{\theta(xx')}{2\pi}\left(\frac{1}{((x+x')+v'(t-t'))^2}\frac{v_F^3}{2\pi v'(V_0^2+v_F^2)}-\frac{1}{((x+x')+v(t-t'))^2}\frac{v_F}{2\pi v}\right)$$

$$+\frac{1}{2\pi}\left(-\theta(-xx')\frac{1}{((x-x')+v'(t-t'))^2}\frac{v_F^3}{2\pi v'(V_0^2+v_F^2)}-\theta(xx')\frac{1}{((x-x')+v(t-t'))^2}\frac{v_F}{2\pi v}\right)$$

$$+\frac{\theta(xx')}{2\pi}\left(\frac{1}{((x+x')-v'(t-t'))^2}\frac{v_F^3}{2\pi v'(V_0^2+v_F^2)}-\frac{1}{((x+x')-v(t-t'))^2}\frac{v_F}{2\pi v}\right)$$

$$+\frac{1}{2\pi}\left(-\theta(-xx')\frac{1}{((x-x')-v'(t-t'))^2}\frac{v_F^3}{2\pi v'(V_0^2+v_F^2)}-\theta(xx')\frac{1}{((x-x')-v(t-t'))^2}\frac{v_F}{2\pi v}\right) \tag{13.131}$$

where,

$$v^2 = v_F^2 + \frac{v_F v_q}{\pi} \tag{13.132}$$

$$v'^2 = v_F^2\frac{(v^2+V_0^2)}{(V_0^2+v_F^2)}. \tag{13.133}$$

We may see that the equation Eq. (13.131) reduces to the correct expression when mutual interaction between fermions is absent ($v = v' = v_F$, $V_0 \neq 0$). Conversely, when the impurity is absent but mutual interactions are present ( $V_0 = 0, v' = v \neq v_F$) then,

$$(<\rho(x,t)\rho(x',t')> - <\rho(x,t)><\rho(x',t')>)_{V_0=0} =$$

$$-\frac{v_F}{v}\frac{1}{(2\pi)^2}\left(\frac{1}{((x-x')+v(t-t'))^2}+\frac{1}{((x-x')-v(t-t'))^2}\right) \tag{13.134}$$

as it should be. Lastly, when $V_0 = \infty$ we expect the results to coincide with those of a Luttinger liquid with a boundary obtained in an earlier section. The velocity and density are related in the RPA limit as follows.

$$v(x,t) = -\pi\,\partial_{v_F t}\int_{sgn(x)\infty}^{x}\tilde{\rho}(y',t)\,dy' \tag{13.135}$$

Using the form of the density–density correlation in the long-wavelength limit, and the relation between velocity and density and the Baker-Hausdorff theorem, we may evaluate the single particle propagator. The propagator for right movers may be evaluated to yield the following:

For $xx' > 0$

$$<T\Psi_R(x,t)\Psi_R^{\dagger}(x',t')> = e^{ik_F(x-x')}\, e^{-[\frac{v_F^2}{(V_0^2+v_F^2)}(-\frac{v_F}{4v'}+\frac{v'}{4v_F})+(\frac{v_F}{4v}-\frac{v}{4v_F})]Log(4xx')}$$

$$e^{\frac{v_F^2}{(V_0^2+v_F^2)}(-\frac{v_F}{4v'}+\frac{v'}{4v_F})Log((x+x')^2-v'^2(t-t')^2)}$$

$$e^{(\frac{v_F}{4v}-\frac{v}{4v_F})Log((x+x^{'})^2-v^2(t-t^{'})^2)}e^{(\frac{1}{2}-\frac{v_F}{4v}-\frac{v}{4v_F})Log((x-x^{'})+v(t-t^{'}))}$$
$$e^{(-\frac{v_F}{4v}-\frac{v}{4v_F}-\frac{1}{2})Log((x-x^{'})-v(t-t^{'}))}. \tag{13.136}$$

It is easy to see that this also has all the right limits. For instance, when the impurity is absent, $V_0 = 0$, $v = v^{'} \neq v_F$, and Eq. (13.136) reduces to,

$$< \Psi_R(x,t)\Psi_R^{\dagger}(x^{'},t^{'}) >_{V_0=0}= e^{ik_F(x-x^{'})}$$
$$\times e^{(\frac{1}{2}-\frac{v_F}{4v}-\frac{v}{4v_F})Log((x-x^{'})+v(t-t^{'}))}e^{(-\frac{v_F}{4v}-\frac{v}{4v_F}-\frac{1}{2})Log((x-x^{'})-v(t-t^{'}))} \tag{13.137}$$

as it should. Conversely, if mutual interactions between fermions are absent but the impurity is present, then $V_0 \neq 0$ but $v = v^{'} = v_F$. In this case,

$$< \Psi_R(x,t)\Psi_R^{\dagger}(x^{'},t^{'}) >_{v=v^{'}=v_F}= e^{-Log((x-x^{'})-v_F(t-t^{'}))} \tag{13.138}$$

again as it should be. Lastly, if $V_0 = \infty$ (this also means $v = v^{'}$) we expect the results to coincide with those of a LL with a boundary. In this case Eq. (13.136) becomes,

$$< \Psi_R(x,t)\Psi_R^{\dagger}(x^{'},t^{'}) >_{V_0=\infty}= e^{ik_F(x-x^{'})}$$
$$\times e^{-(\frac{v_F}{4v}-\frac{v}{4v_F})\ Log(4xx^{'})}e^{(\frac{v_F}{4v}-\frac{v}{4v_F})Log((x+x^{'})^2-v^2(t-t^{'})^2)}$$
$$e^{(\frac{1}{2}-\frac{v_F}{4v}-\frac{v}{4v_F})Log((x-x^{'})+v(t-t^{'}))}e^{(-\frac{v_F}{4v}-\frac{v}{4v_F}-\frac{1}{2})Log((x-x^{'})-v(t-t^{'}))}. \tag{13.139}$$

Now we may extract the dynamical density of states for right movers. For this we examine the equal space and unequal time part of the Green function. First we verify that far away from the impurity we get what we expect.

$$< \Psi_R(x \to \infty,t)\Psi_R^{\dagger}(x \to \infty,t^{'}) > \ \sim e^{(-\frac{v_F}{2v}-\frac{v}{2v_F})Log(v(t-t^{'}))} \tag{13.140}$$

So that far away from the impurity $D(\omega) \sim |\omega|^{\delta}$ where $\delta = \frac{v}{2v_F} + \frac{v_F}{2v} - 1$. At the impurity we expect the exponent to be very different.

$$< \Psi_R(x=0,t)\Psi_R^{\dagger}(x=0,t^{'}) >= e^{[\frac{v_F^2}{(V_0^2+v_F^2)}(-\frac{v_F}{2v}+\frac{v^{'}}{2v_F})-\frac{v}{v_F}]\ Log(t-t^{'})} \tag{13.141}$$

From this we may conclude that at the impurity, the density of states is $D(\omega) = |\omega|^{\delta^{'}}$ where $\delta^{'} = \frac{v_F^2}{(V_0^2+v_F^2)}(\frac{v_F}{2v^{'}} - \frac{v^{'}}{2v_F}) + \frac{v}{v_F} - 1$ . A plot in Fig.13.1 of $\delta^{'}$ and $\delta$ versus $v$ indicates that $\delta^{'} > \delta$ for repulsive interactions ($v > v_F$) and $\delta^{'} < \delta$ for attractive interactions ($v < v_F$). This plot is a concrete realization of Kane and Fisher's metaphor of the impurity 'cutting the chain' when the interaction between the particles is repulsive and 'healing the chain' when the mutual interaction is attractive. This is

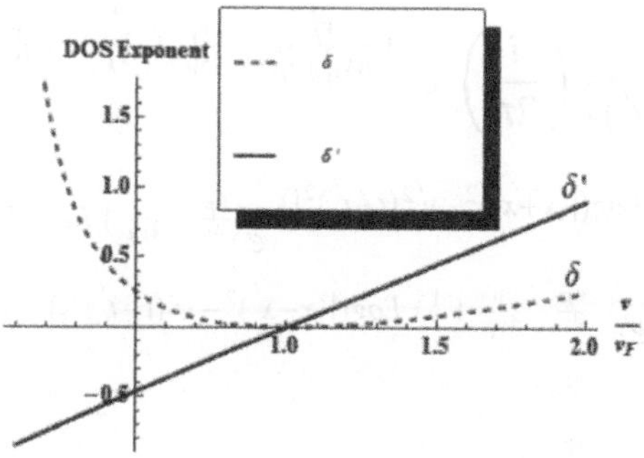


Figure 13.1: This is a plot of the density of state exponents at the location of the impurity when the impurity strength is zero ($\delta$) and when it is not ($\delta'$).

because the exponent associated with the density of states is negative in the attractive case and positive in the repulsive case only when the impurity is present. When the impurity is absent, however, this exponent is always positive. This state of affairs may be contrasted with the large number of results obtained by various groups cited in the references that fall well short of providing a simple answer to the question of what is the exponent associated with the dynamical density of states at the location of the impurity.

## 13.4 Closed Form for Full Propagator

Now we write down the closed expression for the full propagator including all the terms. Set $v^2 = v_F^2 + \frac{v_F v_q}{\pi}$ where $v_q = const$ mutual interaction forward scattering only and $v'^2 = v_F^2 \frac{(v^2+V_0^2)}{(V_0^2+v_F^2)}$ where $V_0$ is impurity strength. In general, we may write for the full particle propagator,

$xx' > 0$:

$$< T\, \Psi_R(x,t)\Psi_R^{\dagger}(x',t') >=$$

$$\frac{e^{ik_F(x-x')}}{((x-x')-v(t-t'))}\left(\frac{i}{2\pi}\right) e^{-(\frac{v_F^2}{2(V_0^2+v_F^2)}(-\frac{v_F}{2v'}+\frac{v'}{2v_F})+\frac{v_F}{4v}-\frac{v}{4v_F})\, Log(|4xx'|)}$$
$$e^{\frac{v_F^2}{2(V_0^2+v_F^2)}(-\frac{v_F}{2v'}+\frac{v'}{2v_F})\, Log(|(x+x')^2-v'^2(t-t')^2|)}\; e^{(\frac{v_F}{4v}-\frac{v}{4v_F})\, Log(|(x+x')^2-v^2(t-t')^2|)}$$
$$e^{(-\frac{v_F}{4v}+\frac{1}{2}-\frac{v}{4v_F})\, Log(|(x-x')^2-v^2(t-t')^2|)} \qquad (13.142)$$

and,

$$< T\, \Psi_L(x,t)\Psi_L^{\dagger}(x',t') >=$$

$$-\frac{e^{-ik_F(x-x')}}{((x-x')+v(t-t'))}\left(\frac{i}{2\pi}\right) e^{-(\frac{v_F^2}{2(V_0^2+v_F^2)}(-\frac{v_F}{2v'}+\frac{v'}{2v_F})+(\frac{v_F}{4v}-\frac{v}{4v_F}))\,Log(|4xx'|)}$$
$$e^{\frac{v_F^2}{2(V_0^2+v_F^2)}(-\frac{v_F}{2v'}+\frac{v'}{2v_F})\,Log(|(x+x')^2-v'^2(t-t')^2|)}\; e^{(\frac{v_F}{4v}-\frac{v}{4v_F})\,Log(|(x+x')^2-v^2(t-t')^2|)}$$
$$e^{(-\frac{v_F}{4v}-\frac{v}{4v_F}+\frac{1}{2})\,Log(|(x-x')^2-v^2(t-t')^2|)} \tag{13.143}$$

and,

$$< T\ \Psi_R(x,t)\Psi_L^\dagger(x',t') >= \frac{e^{ik_F(x+x')}}{((x+x')-v(t-t'))}\frac{V_0}{4\pi v_F}\left(\frac{\theta(x)\theta(x')}{\left(1+V_0\frac{i}{v_F}\right)}-\frac{\theta(-x')\theta(-x)}{\left(1-V_0\frac{i}{v_F}\right)}\right)$$
$$\times e^{\frac{v_F^2}{2(V_0^2+v_F^2)}(-\frac{v_F}{2v'}+\frac{v'}{2v_F})\,Log(|(x+x')^2-v'^2(t-t')^2|)}\; e^{(-\frac{v_F}{4v}-\frac{v}{4v_F}+\frac{1}{2})\,Log(|(x+x')^2-v^2(t-t')^2|)}$$
$$\times e^{(\frac{v_F}{4v}-\frac{v}{4v_F})\,Log(|(x-x')^2-v^2(t-t')^2|)}(e^{\gamma_1\,Log(|2x|)}$$
$$\times e^{\gamma_3\,Log(|2x'|)}+e^{\gamma_2 Log(|2x'|)}e^{\gamma_3\,Log(|2x|)}) \tag{13.144}$$

and,

$$\begin{aligned} < T\ \Psi_L(x,t)\Psi_R^\dagger(x',t') > &= \frac{e^{-ik_F(x+x')}}{((x+x')+v(t-t'))}\frac{V_0}{4\pi v_F} \\ &\times\left(\frac{\theta(x')\theta(x)}{\left(1-V_0\frac{i}{v_F}\right)}-\frac{\theta(-x)\theta(-x')}{\left(1+V_0\frac{i}{v_F}\right)}\right)\end{aligned}$$
$$e^{\frac{v_F^2}{2(V_0^2+v_F^2)}(-\frac{v_F}{2v'}+\frac{v'}{2v_F})\,Log((x+x')^2-v'^2(t-t')^2)}\; e^{(-\frac{v_F}{4v}-\frac{v}{4v_F}+\frac{1}{2})\,Log((x+x')+v(t-t'))}$$
$$e^{(-\frac{v_F}{4v}-\frac{v}{4v_F}+\frac{1}{2})\,Log((x+x')-v(t-t'))}\; e^{(\frac{v_F}{4v}-\frac{v}{4v_F})\,Log((x-x')^2-v^2(t-t')^2)}$$
$$(e^{\gamma_2 Log(2x)}e^{\gamma_3 Log(2x')}+e^{\gamma_1 Log(2x')}e^{\gamma_3\,Log(2x)}) \tag{13.145}$$

where,

$$\gamma_1 = -\frac{v_F^3}{2v'(V_0^2+v_F^2)}(\frac{1}{2}-2+\frac{1}{2}\frac{v'^2}{v_F^2})-(\frac{v_F}{4v}-\frac{v}{4v_F}+\frac{v_F}{v})+\frac{2V_0^2+v_F^2}{2(V_0^2+v_F^2)} \tag{13.146}$$

$$\gamma_3 = -\frac{v_F^3}{2v'(V_0^2+v_F^2)}(-\frac{1}{2}+\frac{1}{2}\frac{v'^2}{v_F^2})-(\frac{v_F}{4v}-\frac{v}{4v_F}) \tag{13.147}$$

$$\gamma_2 = -\frac{v_F^3}{2v'(V_0^2+v_F^2)}(-\frac{1}{2}-1+\frac{1}{2}\frac{v'^2}{v_F^2})-(\frac{v_F}{4v}-\frac{v}{4v_F}+\frac{v_F}{v})+\frac{2V_0^2+v_F^2}{2(V_0^2+v_F^2)}. \tag{13.148}$$

The really interesting case is the one given below. In this expression, the propagator is derived when $x$ and $x^{'}$ are on different sides of the impurity. These formulae are not guaranteed to converge to the expression for the propagator of LL minus the impurity, as the impurity strength is made smaller and smaller unless the mutual interaction has already been turned off. The reason for this has already been given by Kane and Fisher. The impurity 'cuts' the system into two, making it qualitatively different from the ordinary LL. However, when both the position variables are on the same side, then the system does behave as an ordinary LL when the impurity goes to zero. More mathematically, the Green function is a discontinuous function of the impurity strength for points on either side of the impurity.

$xx^{'} < 0:$

$$< T\, \Psi_R(x,t)\Psi_R^{\dagger}(x^{'},t^{'}) >$$

$$= \frac{e^{ik_F(x-x^{'})}}{(x-x^{'})-v(t-t^{'})}\,[\frac{i}{4(x+x^{'})\pi} - \frac{V_0}{4(x+x^{'})\pi v_F}\left(\frac{\theta(x^{'})\theta(-x)}{\left(1-V_0\frac{i}{v_F}\right)} - \frac{\theta(x)\theta(-x^{'})}{\left(1+V_0\frac{i}{v_F}\right)}\right)]$$

$$e^{(-\frac{v^{'}}{2v_F}+\frac{v_F}{2v^{'}})\ Log(|(x-x^{'})^2-v^{'2}(t-t^{'})^2|)\frac{v_F^2}{2(V_0^2+v_F^2)}}$$

$$e^{(-\frac{v_F}{2v}+\frac{1}{2})\ Log(|(x-x^{'})^2-v^2(t-t^{'})^2|)}\ e^{(-\frac{1}{2}+\frac{v_F}{2v})\ Log(|(x+x^{'})^2-v^2(t-t^{'})^2|)}$$

$$e^{-\frac{1}{2}Log(4xx^{'})\ ((\frac{v^{'}}{v_F}-\frac{v_F}{v^{'}})\frac{v_F^2}{2(V_0^2+v_F^2)}+(\frac{v_F}{2v}-\frac{v}{2v_F}))}$$

$$\left(((x+x^{'})+v(t-t^{'}))e^{Log(2x)\ (1-\frac{v_F}{v})}+((x+x^{'})-v(t-t^{'}))e^{Log(2x^{'})\ (1-\frac{v_F}{v})}\right) \tag{13.149}$$

$$< T\, \Psi_L(x,t)\Psi_L^{\dagger}(x^{'},t^{'}) >$$

$$= -\frac{e^{-ik_F(x-x^{'})}}{(x-x^{'})+v(t-t^{'})}\,[\frac{i}{4(x+x^{'})\pi} - \frac{V_0}{4(x+x^{'})\pi v_F}(\frac{\theta(-x^{'})\theta(x)}{\left(1-V_0\frac{i}{v_F}\right)} - \frac{\theta(-x)\theta(x^{'})}{\left(1+V_0\frac{i}{v_F}\right)})]$$

$$e^{(-\frac{v^{'}}{2v_F}+\frac{v_F}{2v^{'}})\ Log(|(x-x^{'})^2-v^{'2}(t-t^{'})^2|)\frac{v_F^2}{2(V_0^2+v_F^2)}}$$

$$e^{(-\frac{v_F}{2v}+\frac{1}{2})\ Log(|(x-x^{'})^2-v^2(t-t^{'})^2|)}\ e^{(\frac{v_F}{2v}-\frac{1}{2})\ Log(|(x+x^{'})^2-v^2(t-t^{'})^2|)}$$

$$e^{-\frac{1}{2}Log(4xx^{'})\ ((\frac{v^{'}}{v_F}-\frac{v_F}{v^{'}})\frac{v_F^2}{2(V_0^2+v_F^2)}+(\frac{v_F}{2v}-\frac{v}{2v_F}))}$$

$$\left(((x+x^{'})+v(t-t^{'}))\ e^{Log(2x^{'})\ (1-\frac{v_F}{v})}+((x+x^{'})-v(t-t^{'}))\ e^{Log(2x)\ (1-\frac{v_F}{v})}\right) \tag{13.150}$$

and,

$$< T\, \Psi_L(x,t)\Psi_R^{\dagger}(x^{'},t^{'}) >=< T\, \Psi_R(x,t)\Psi_L^{\dagger}(x^{'},t^{'}) >= 0. \tag{13.151}$$

Several observations are in order. First, when the mutual interaction between particles is absent, the above expression reduces to the simple expression from Eq. (13.111). However, when the impurity strength is made to tend to zero, the above expressions *do not* reduce to the well-known Luttinger expressions unless the mutual interactions are also absent. This does not mean the above expression is wrong. It simply means the Green function is a discontinuous function of the impurity strength for the Green function evaluated at points on either side of the impurity. We conclude by looking at transport across the impurity.

### 13.4.1 Transport across the Impurity

We wish to evaluate the amplitude for the electron to be transported across the impurity. In particular we wish to extract the exponent associated with the conductance. The idea is, we are going to evaluate the propagator when $xx^{'} < 0$ at fixed locations and set $t - t^{'} \to \infty$. Then upon replacing $t - t^{'} \sim \frac{1}{\omega}$ we obtain the transmission amplitude, which we take to be proportional to the conductance.

$$T(\omega) \sim \int_{0}^{\infty} e^{i\omega t}\, dt\ < T\ \Psi_L(x,t)\Psi^{\dagger}_L(x^{'},0) >= \omega^{\alpha} \tag{13.152}$$

where $\alpha = (\frac{v^{'}}{2v_F} - \frac{v_F}{2v^{'}})\frac{v_F^2}{(V_0^2+v_F^2)} - 1$. To conclude, our results represent a concrete realization of Kane and Fisher's cutting of the chain and healing of the chain metaphor. To us though, the more interesting outcomes are the closed expressions for the Green functions especially when $xx^{'} < 0$, where the answer depends discontinuously on the impurity strength.

## 13.5 Exercises

**Q.1** Show that $U_r^{\dagger}U_r = U_rU_r^{\dagger}$ and $[U_r, b_q] = 0$ provided we make the choice $\Psi(y;?) \equiv \Psi(y)$.

**Q.2** Show that if $|q|, |q^{'}| \ll k_F$ (held fixed), the three density correlation $< \rho_q\rho_{q^{'}}\rho_{-q-q^{'}} >$ vanishes in the RPA limit $k_F, m \to \infty, k_F/m = v_F < \infty$.

**Q.3** Derive Eq. (13.106) and Eq. (13.107).

**Q.4** Evaluate $< \rho_f(x,t)\rho_f(x^{'},t^{'}) >$ and $< \rho_f(x,t)\rho^{*}_f(x^{'},t^{'}) >$ using Eq. (13.120) for free fermions with a single impurity and verify that the resulting exponents are trivial. How should Eq. (13.120) be interpreted so that the resulting formulas for the above correlation functions are identical to what may be obtained from Wick's theorem?

**Q.5** What would the density–density correlation function look like when there are two impurities, or three?

# Bibliography

[1] H. Goldstein, *Classical Mechanics* (3rd ed.), Addison-Wesley Publishing Company (Massachusetts) (2001).

[2] L. D. Landau and E. M. Lifshitz, *Mechanics: A Course of Theoretical Physics*, vol.1 (3rd edn.) , Pergamon Press (1976).

[3] C. Lanczos, *The Variational Principles of Mechanics* (4th ed.), (Reprint of University of Toronto), Courier Dover (1970).

[4] J. R. Taylor, *Classical Mechanics*, University Science Books (2005).

[5] L. N. Hand and J. D. Finch, *Analytical Mechanics*, Cambridge University Press (1998).

[6] D. A. Wells, *Lagrangian Dynamics*, Schaum's Outlines (1967).

[7] C. L. Lewis, Explicit gauge covariant Euler-Lagrange equation, *American Journal of Physics* 77, 839 (2009).

[8] J. Casey, Geometrical derivation of Lagrange's equations for a system of particles, *American Journal of Physics* 62, 836 (1994).

[9] M. C. Whatley, The role of kinematics in Hamiltonian dynamics: Momentum as a Lagrange multiplier, *American Journal of Physics* 58, 1006 (1990).

[10] D. H. Kobe, G. Reali and S. Sieniutycz, Lagrangian for dissipative systems, *American Journal of Physics* 54, 997 (1986).

[11] E. A. Desloge and E. Eriksen, Lagrange's equations of motion for a relativistic particle, *American Journal of Physics* 53, 83 (1985).

[12] N. A. Lemos, Note on the Lagrangian description of dissipative systems, *American Journal of Physics* 49, 1181 (1981).

[13] J. Hanc, S. Tuleja and M. Hancova, Simple derivation of Newtonian mechanics from the principle of least action, *American Journal of Physics* 71, 386 (2003).

[14] Francis E. Low, *Classical Field Theory: Electromagnetism and Gravitation*, John Wiley and Sons (1997).

[15] E.C.G. Sudarshan and N. Mukunda, *Classical Dynamics: A Modern Perspective*, John Wiley and Sons (1974).

[16] H. A. Al-Kuwari and M. O. Taha, Noether's theorem and local gauge invariance, *American Journal Physics* 59, 85 (1991).

[17] N. B-. Ares, Noether's theorem in discrete classical mechanics, *American Journal of Physics* 56, 174 (1988).

[18] L. Y. Bahar and H. G. Kwatny, Generalized Lagrangian and conservation law for the damped harmonic oscillator, *American Journal of Physics* 49, 1062 (1981).

[19] W. J. Thompson, *Angular Momentum: An Illustrated Guide to Rotational Symmetries for Physical Systems*, Wiley-VCH Verlag GmbH & Co. (Weinheim) (2004).

[20] E. Noether, Invariante Variationsprobleme, Nachr. v. d. Ges. d. Wiss. zu Gottingern, *Math-phys. Klasse*, 235 (1918); M. A. Tavel, Invariant variation problem, *Transport Theory Stat. Mech.* 1, 183 (1971) (English translation).

[21] N. Byers, E. Noether's discovery of the deep connection between symmetries and conservation laws, *Isr. Math. Conf. Proc.* 12, 67 (1999).

[22] L. D. Landau and E. M. Lifshitz, *Classical Theory of Fields: A Course of Theoretical Physics*, vol.2 (4th edn.), Butterworth Heinemann (1975).

[23] P. Havas and J. Stachel, Invariances of approximately relativistic Lagrangians and the center of mass theorem. I, *Physical Review* 185, 1636 (1969).

[24] N. C. Bobillo-Ares, Noether's theorem in discrete classical mechanics, *American Journal of Physics* 56, 174 (1988).

[25] C. M. Giordano and A. R. Plastino, Noether's theorem, rotating potentials, Jacobi's integral of motion, *American Journal of Physics* 66, 989 (1998).

[26] E. Candotti, C. Palmieri and B. Vitale, On the inversion of Noether's theorem in the Lagrangian formalism, *Il Nuovo Cimento* A 70, 233 (1970).

[27] U. E. Schrder, Neother's theorem and the conservation laws in classical field theories, *Fortschritte der Physik* 16, 357 (1968).

[28] T. W. B. Kibble, Conservation laws for free fields, *Journal of Mathematical Physics* 6, 1952 (1965).

[29] D. M. Fradkin, Conserved Quantities Associated with Symmetry Transformations of Relativistic Free-Particle Equations of Motion, Journal of Mathematical Physics 6, 879 (1965).

[30] D. B. Fairlie, Conservation laws and invariance principles, Il Nuovo Cimento 37, 897 (1965).

[31] R. F. O'Connell and D. R. Tompkins, Physical interpretation of generalized conservation laws, Il Nuovo Cimento A 63, 391 (1965).

[32] H. R. Brown and P. Holland, Dynamical versus variational symmetries: understanding Noether's first theorem, Molecular Physics 102, 1133 (2004).

[33] D. M. Lipkin, Existence of a New Conservation Law in Electromagnetic Theory, Journal of Mathematical Physics 5, 696 (1964).

[34] N. H. Ibragimov, Symmetries, Lagrangian and Conservation Laws for the Maxwell's Equations, Acta Applicandea Mathematicae 105, 157 (2009).

[35] I. Y. Krivskii and V. M. Simulik, Noether analysis of zilch conservation laws and their generalization for the electromagnetic field. I. Use of different formulations of the principle of least action, Theoretical and Mathematical Physics 80, 864 (1990).

[36] D. J. Candlin, Analysis of the new conservation law in electromagnetic theory, Il Nuovo Cimento 37, 1390 (1965).

[37] T. A. Morgan and D. W. Joseph, Tensor Lagrangians and generalized conservation laws for free fields, *Il Nuovo Cimento* 39, 494 (1965).

[38] N. Anderson and A. M. Arthurs, A variational principle for Maxwell's equations, *International Journal of Electronics* 45, 333 (1978).

[39] A. Gersten, Tensor Lagrangians: Lagrangians equivalent to the Hamilton-Jacobi equations and relativistic dynamics, *Foundations of Physics* 41, 88 (2011).

[40] V. M. Simulik, Connection between the symmetry properties of the Dirac and Maxwell equations. Conservation laws, *Theoretical and Mathematical Physics*, 386 (1991).

[41] W. I. Fushchich, I. Y. Krivsky and V. M. Simulik, On vector and pseudovector Lagrangians for electromagnetic field, *Il Nuovo Cimento B* 103, 423 (1989).

[42] J. Rosen, Redundancy and superfluity for electromagnetic fields and potentials, *American Journal of Physics* 48, 1071 (1980).

[43] A. Sudbery, A vector Lagrangian for the electromagnetic field, *Journal of Physics A: Mathematical and General* 19, L33 (1986).

[44] M. Born and E. Wolf, Principles of Optics: Electromagnetic Theory of Propagation, Interference and Diffraction of Light, (7th ed.) Cambridge University Press (1999).

[45] A. Ghatak, Optics (4th edn.) Tata McGraw Hill (2009).

[46] R. M. Fano, L. J. Chu and R. B. Adler, *Electromagnetic Fields, Energy and Forces*, Wiley (New York) (1960).

[47] W. Israel, The dynamics of polarization, *General Relativity and Gravitation* 9, 451 (1978).

[48] G. Gyrgyi, Elementary considerations on the dynamics of light waves, *American Journal of Physics* 28, 85 (1960).

[49] R. A. Grot and A. C. Eringen, Relativistic continuum mechanics: Part II—electromagnetic interactions with matter, *International Journal of Engineering Science* 4, 639 (1966).

[50] J. P. Gordon, Radiation forces and momenta in dielectric media, *Physical Review A* 8, 14 (1973).

[51] V. L. Ginzburg, The laws of conservation of energy and momentum in emission of electromagnetic waves (photons) in a medium and the energy-momentum tensor in macroscopic electrodynamics, *Soviet Physics Uspekhi* 16, 434 (1973).

[52] S. R. De Groot and L. G. Suttorp, The relativistic energy-momentum tensor in polarized media: I-VII, *Physica* 37, 284 (1967); *ibid* 37, 297 (1967); *ibid* 39, 28 (1968); *ibid* 39, 41 (1968); *ibid* 39, 61 (1968); *ibid* 77, 28 (1968); *ibid* 39, 84 (1968).

[53] J. A. Arnaud, Momentum of photons, *American Journal of Physics* 42, 71 (1974).

[54] N. L. Balazs, The energy momentum tensor of the electromagnetic field inside matter, *Physical Review* 91, 408 (1953).

[55] I. Brevik, Experiments in phenomenological electrodynamics and the electromagnetic energy momentum tensor, *Physics Reports* 52, 133 (1979).

[56] R. N. C. Pfeifer, R. A. Nieminen, N. R. Heckenberg and H. R. Dunlop, Colloquium: Momentum of an electromagnetic wave in dielectric media, *Review of Modern Physics* 79, 1197 (2007).

[57] C. L. Tang and J. Meixner, Relativistic theory of the propagation of plane electromagnetic waves, *Physics of Fluids* 4, 148 (1961).

[58] D. V. Skobel'tsyn, The momentum-energy tensory of the electromagnetic field, *Soviet Physics Uspekhi* 16, 381 (1973).

[59] S. Schwarz, Electromagnetic forces and the energy-momentum tensor in the presence of electric polarization and magnetization, *International Journal of Engineering Science* 30, 963 (1992).

[60] S. Stallinga, Energy and momentum of light in dielectric media, *Physical Review E* 73, 026606 (2006).

[61] F. N. H. Robinson, Electromagnetic stress and momentum in matter, *Physics Reports* 16, 313 (1975).

[62] Y. N. Obukhov and F. W. Hehl, Electromagnetic energy-momentum and forces in matter, *Physics Letters* A 311, 277 (2003).

[63] D. F. Nelson, Momentum, pseudomomentum, and wave momentum: Toward resolving the Minkowski-Abraham controversy, *Physical Review A* 44, 3985 (1991).

[64] S. Mikura, Variational formulation of the electrodynamics of fluids and its application to the radiation pressure problem, *Physical Review A* 13, 2265 (1976).

[65] H. M. Lai, Electromagnetic momentum in static fields and the Abraham-Minkowski controversy, *American Journal of Physics* 48, 658 (1980).

[66] Miroslav Krany, The Minkowski and Abraham tensors, and the non-uniqueness of non-closed systems resolution of the controversy, *International Journal of Engineering Science* 20, 1193 (1982).

[67] C. W. Kisner, K. S. Thorne and J. A. Wheeler, *Gravitation*, W. H. Freeman and Company, San Francisco (1973).

[68] R. A. d'Invero, *Introducing Einstein's Relativity*, Oxford University Press, New York (1992).

[69] S. P. Timoshenko and J. N. Goodier, *Theory of Elasticity* (3rd edn.) McGraw Hill (1970).

[70] A. A. Shabana, *Computational Continuum Mechanics*, Cambridge University Press (2008).

[71] T. J. Chung, *General Continuum Mechanics*, 2nd edn., New Jersey: Prentice Hall (1962).

[72] A. J. M. Spencer, *Continuum Mechanics*, Longman Group Limited (London) (1980).

[73] R. C. Batra, *Elements of Continuum Mechanics*, American Institute of Aeronautics and Astronautics, Reston, VA (2006).

[74] Eringen, A.C., *Mechanics of Continua*. Robert E. Krieger Publishing Company, Huntington, New York (1980).

[75] Y. C. Fung, *A First Course in Continuum Mechanics*, 2nd edn., Prentice-Hall, Inc. (1977).

[76] M. E. Gurtin, *An Introduction to Continuum Mechanics*, Academic Press, New York (1981).

[77] W. M. Lai, D. Rubin and E. Krempl, *Introduction to Continuum Mechanics*, 3rd edn., Elsevier (1996).

[78] G. E. Mase, Schaum's outline series, *Continuum Mechanics: Theory and Problems of Continuum Mechanics*, McGraw-Hill, New York (1970).

[79] L. E. Malvem, *Introduction to the Mechanics of a Continuous Medium*, Prentice-Hall (1969).

[80] F. Irgens, *Continuum Mechanics*, Springer, Berlin (2008).

[81] A. L. Besse, *Einstein Manifolds*, Springer (1987).

[82] P. Constantin, Euler and Navier-Stokes equations, *Publicacions Mathemtiques* 52, 235 (2008).

[83] S. B. Pope, *Turbulent Flows*, Cambridge University Press (2000).

[84] M. Potter, D. C. Wiggert, *Fluid Mechanics*, Schaum's Outlines, McGraw-Hill (USA) (2008).

[85] R. Aris, Vectors, *Tensors and the Basic Equations of Fluid Mechanics*, Dover Publications (1989).

[86] G. Emanuel, *Analytical Fluid Dynamics* (2nd edn.) CRC Press (2001).

[87] R. Temam, *Navier-Stokes Equations: Theory and Numerical Analysis*, AMS Chelsea, Providence RI (2001).

[88] C. Y. Wang, Exact solutions of the steady-state Navier-Stokes equations, *Annual Review of Fluid Mechanics* 23, 159 (1991).

[89] C. R. Ethier and D. A. Steinman, Exact fully 3D Navier-Stokes solutions for benchmarking, *International Journal for Numerical Methods in Fluids* 19, 369 (1994).

[90] G. K. Batchelor, *An Introduction to Fluid Dynamics*, Cambridge University Press (1967).

[91] L. D. Landau and E. M. Lifshitz, *Fluid Mechanics: A Course of Theoretical Physics*, vol.6 (2nd edn.), Pergamon Press (1987).

[92] J. Kay, *Fluid Mechanics & Heat Transfer* (2nd ed.), Cambridge University Press (1962).

[93] T. Kato and G. Ponce, Commutator estimates and the Euler and Navier-Stokes equations, *Communications on Pure and Applied Mathematics* 41, 891 (1988).

[94] J. Carlson, A. Jaffe and A. Wiles, *The Millennium Prize Problems*, Library of Congress Cataloging-in-Publication Data (2006).

[95] A. de Masi, R. Esposito and J. L. Lebowitz, Incompressible Navier-Stokes and Euler limits of the Boltzmann equation, *Communications in Pure and Applied Mathematics* 42, 1189 (1989).

[96] J. E. Marsden and S. Shkoller, The anisotropic Lagrangian averaged Euler and Navier-Stokes equations, *Archive for Rational Mechanics and Analysis* 166, 27 (2003).

[97] M. Nelkin, Resource Letter TF-1: Turbulence in fluids, *American Journal of Physics* 68, 310 (2000).

[98] R. J. Becker, Lagrangian/Hamiltonian formalism for description of Navier Stokes fluids, *Physical Review Letters* 58, 1419 (1987).

[99] G. Rosen, Restricted invariance of the Navier-Stokes equation, *Physical Review A* 22, 313 (1980).

[100] T. D. Lee, Difference between Turbulence in a two-dimensional fluid and in a three-dimensional fluid, *Journal of Applied Physics* 22,524 (1951).

[101] G. Rosen, Navier-Stokes initial value problem for boundary free incompressible fluid flow, *Physics of Fluids* 13, 2891 (1970).

[102] F. M. White, *Fluid mechanics* (7th edn.), McGraw-Hill Higher Education (2011).

[103] J. Gleick, *Chaos: Making a New Science*, Penguin Books (1988).

[104] T. M. Atanackovic, A. Guran, *Theory Elasticity for Scientists and Engineers*, Springer (2000).

[105] J. Chakrabarty, *Theory of Plasticity* (3rd edn.), Elsevier Butterworth-Heinemann, Burlington, MA (2006).

[106] W-F. Chen and D-J. Han, Plasticity for Structural Engineers, J. Ross Publishing (2007).

[107] K. D. Hjelmstand, *Fundamentals of Structural Mechanics* (2nd edn.), Springer Science, New York, NY (2005).

[108] J. Lubliner, *Plasticity Theory*, Dover Publications (2008).

[109] H-C. Wu, *Continuum Mechanics and Plasticity*, CRC Press (2005).

[110] W. A. Backofen, *Deformation Processing*, Addison-Wesley (1972).

[111] E. W. Billington, *Introduction to the Mechanics and Physics of Solids*, Adam Hilger Bristol and Boston (1986).

[112] E. W. Billington and A. Tate, *The Physics of Deformation and Flow*, McGraw-Hill (1981).

[113] A. P. Boresi and O. M. Sidebottom, *Advanced Mechanics of Materials*, Wiley and Sons (1985).

[114] L. R. Calcote, *Introduction to Continuum Mechanics*, van Nostrand (1968).

[115] H. Ford and J. M. Alexander, *Advanced Mechanics of Materials*, Longman (1963).

[116] D. W. A. Rees, *Mechanics of Solids and Structures*, IC Press, World Scientific (2000).

[117] R. Zwanzig and R. D. Mountain, High-frequency elastic moduli of simple fluids, *Journal of Chemical Physics* 43, 4464 (1965); R. Zwanzig, Frequency-dependent transport coefficients in fluid mechanics, *Journal of Chemical Physics* 43, 714 (1965).

[118] K. F. Herzfeld, Bulk viscosity and shear viscosity in fluids according to the theory of irreversible processes, *Journal of Chemical Physics* 28, 595 (1958).

[119] J. A. McLennan, Statistical mechanics of transport in fluids, *Physics of Fluids* 3, 493 (1960).

[120] S. Y. Auyang, *How Is Quantum Field Theory Possible?*, Oxford University Press (1995).

[121] N. N. Bogoliubov, A. A. Logunov and I. T. Todorov, *Introduction to Axiomatic Quantum Field Theory*.[14] Reading, Mass.: W. A. Benjamin, Advanced Book Program (1975).

[122] H. R. Brown and R. Harr, *Philosophical Foundations of Quantum Field Theory*, Oxford:Clarendon Press (1988).

[123] D. Buchholz, Current trends in axiomatic quantum field theory, *Lecture Notes in Physics*, vol. 558, Springer (2000).

[124] T. Y. Cao, *Conceptual Foundations of Quantum Field Theories*, Cambridge Univesity Press (Cambridge) (1999).

[125] O. Darrigol, The origin of quantized matter waves, *Historical Studies in the Physical and Biological Sciences* 16, 197 (1986).

[126] Dieks, D. (in Kuhlmann, M., Lyre, H., Wayne, A. (eds.)). Events and covariance in the interpretation of quantum field theory, *Ontological Aspects of Quantum Field Theory*, pp 215–234, World Scientific Publishing Company, Singapore (2002).

[127] R. Haag, *Local Quantum Physics: Fields, Particles, Algebras* (2nd edn.), Texts and monographs in physics, Springer (1996).

[128] R. Haag and D. Kastler, An algebraic approach to quantum field theory, *Journal of Mathematical Physics* 5, 848 (1964).

[129] M. E. Peskin and D. V. Schroeder, *An Introduction to Quantum Field Theory*, Perseus Books, Cambridge, Mass. (1995).

[130] H. R. Brown and R. Harr, *Philosophical Foundations of Quantum Field Theory*, Clarendon Press (1988).

[131] M. Kuhlmann, *The Ultimate Constituents of the Material World: In Search of an Ontology for Fundamental Physics*, Volume 37 of *Philosophische Analyse* (2002); see esp. M. L. G. Redhead, The interpretation of gauge symmetry, pp. 281–301 of this work.

[132] I. Duck and E. C. G. Sudarshan, Towards an understanding of the spin-statistics theorem, *Am. J. Phys.* 66, 284 (1998).

[133] W. Pauli, The connection between spin and statistics, Phys. Rev. 58, 716 (1940).

[134] W. Pauli, On the connection between spin and statistics, *Prog. Theor. Phys.* 5 (4) 526 (1950).

[135] R. F. Streater and A. S. Wightman, *PCT Spin Statistics and All That*, Benjamin, New York (1964).

[136] L. H. Ryder, *Quantum Field Theory*, 2nd edn., Cambridge University Press, Cambridge (1996).

[137] H. R. Brown and S. Saunders, *The Philosophy of Vacuum*, Oxford: Clarendon Press (1991).

[138] P. Teller, What the quantum field is not, *Philosophical Topics* 18, 175 (1990).

[139] P. Teller, *An Interpretive Introduction to Quantum Field Theory*, Princeton: Princeton University Press (1995).

[140] S. Weinberg, *The Quantum Theory of Fields: Foundations*, Volume 1, Cambridge: Cambridge University Press (1995).

[141] N. N. Bogoliubov, D. V. Shirkov, *Quantum Fields*, Benjamin-Cummings Pub. Co. (1982).

[142] P. H. Frampton, Gauge Field Theories, Frontiers in Physics Series, 2nd edn., Wiley (2000).

[143] C. Itzykson and J. -B. Zuber, *Quantum Field Theory*, McGraw-Hill (1980).

[144] A. Zee, *Quantum Field Theory in a Nutshell*, Princeton University Press (2003).

[145] G. 't Hooft, The conceptual basis of quantum field theory, in J. Butterfield and J. Earman, eds., *Philosophy of Physics*, Part A. Elsevier 661 (2007).

[146] F. Wilczek, *Quantum field theory*, Reviews of Modern Physics 71, S83 (1999).

[147] F. Mandl and G. Shaw, *Quantum Field Theory*, John Wiley & Sons Ltd (England) (2002).

[148] A. Das, *Field Theory: A Path Integral Approach*, World Scientific (1993).

[149] S. Minwalla, M. V. Raamsdonk and N. Seiberg, Noncommutative pertubative dynamics, *Journal of High-Energy Physics* JHEP02, 020 (2000).

[150] R. J. Szabo, Quantum field theory on noncommutative spaces, *Physics Reports* 378, 207 (2003).

[151] B. Dragovich and Z. Rakic, Path integrals in noncommutative quantum mechanics, *Theoretical Mathematical Physics* 140, 1299 (2004).

[152] J. R. Klauder, The action option and a Feynman quantization of spinor fields in terms of ordinary c-numbers, *Annals of Physics* 11, 123 (1960).

[153] C. N. Yang and D. Feldman, The S-Matrix in the Heisenberg Representation, *Physical Review* 79, 972 (1950).

[154] F. J. Dyson, The radiation theories of Tomonaga, Schwinger and Feynman, *Physical Review* 75, 486 (1949); F. J. Dyson, The S-Matrix in quantum electrodynamics, *Physical Review* 75, 1736 (1949).

[155] S. Hori, On the well-ordered S-Matrix, *Progress in Theoretical Physics* 7, 578 (1952).

[156] P. J. Daniell, Integrals in an infinite number of dimensions, *The Annals of Mathematics*. Second Series 20, 281 (1919).

[157] O. G. Smolyanov and E. T. Shavgulidze, *Continual Integrals*, Moscow State University Press (Moscow) (1990).

[158] M. Reed & B. Simon, *Methods of Modern Mathematical Physics*, vols. I & II, Academic Press (1980).

[159] Z. G. Arenas and D. G. Barci, Functional integral approach for multiplicative stochastic processes, *Physical Review E* 81, 051113 (2010).

[160] D. Kochan, Functional integral for non-Lagrangian systems, *Physical Review A* 81, 022112 (2010).

[161] V. D. Rushai and Y. Y. Lobanov, Studying open quantum systems by means of deterministic approach to approximate functional integration, *Physical Review E* 71, 066708 (2005).

[162] J. E. Moyal, Quantum mechanics as a statistical theory, *Mathematical Proceedings of the Cambridge Philosophical Society* 45, 99 (1949).

[163] B. Jouvet and R. Phythian, Quantum aspects of classical and statistical fields, *Physical Review A* 19, 1350 (1979).

[164] R. Phythian, The functional formalism of classical statistical dynamics, *Journal of Physics A: Mathematical and General* 10, 777 (1977).

[165] P. -L. Chow, Applications of function space integrals to problems in wave propagation in random media, *Journal of Mathematical Physics* 13, 1224 (1972).

[166] I. Hosokawa, A functional treatise on statistical hydromechanics with randon force action, *Journal of Physical Society of Japan* 25, 271 (1968).

[167] G. Rosen, Functional calculus theory for incompressible fluid turbulence, *Journal of Mathematical Physics* 12, 812 (1971).

[168] A. Siegel and T. Burke, Approximate functional integral methods in statistical mechanics. I. Moment expansions, *Journal of Mathematical Physics* 13, 1681 (1972).

[169] I. M. Gel'fand and A. M. Yaglom, Integration in functional spaces and its applications in quantum physics, *Journal of Mathematical Physics* 1, 48 (1960).

[170] S. G. Brush, Functional integrals and statistical physics, *Review of Modern Physics* 33, 79 (1961).

[171] S. Deser, *Functional integrals and adiabatic limits in field theory*, Physical Review 99, 325 (1955).

[172] M. Carreau, E. Farhi and S. Gtmann, Functional integral for a free particle in a box, *Physical Review D* 42, 1194 (1990).

[173] X.X. Dai and W. E. Evenson, Functional integral approach: A third formulation of quantum statistical mechanics, *Physical Review E* 65, 026118 (2002).

[174] I. W. Mayes and J. S. Dowker, Hamiltonian orderings and functional integrals, *Journal of Mathematical Physics* 14, 434 (1973).

[175] P. Ramond, *Field Theory: A Modern Primer* (2nd edn.), Addison Wesley (1989).

[176] E. C. Kemble, *The Fundamental Principles of Quantum Theory*, McGraw-Hill Book Company, Inc., New York, (1937).

[177] G. Wentzel, *Quantum Theory of Fields*, Interscience Publishers, Inc., New York, (1949).

[178] C. P. Enz and L. Garrido, Perturbation theory for classical thermodynamics Green's functions, *Physical Review A* 14, 1258 (1976).

[179] D. W. McLaughlin, Path integrals, asymptotics and singular perturbations, *Journal of Mathematical Physics* 13, 784 (1972).

[180] A. L. Fetter and J. D. Walecka, *Quantum Theory of Many-Particle Systems*, McGraw-Hill, New York (1971).

[181] C. P. Enz, *The Many-Body Problem*, edited by A. Cruz and T. W. Preist, Plenum Press, New York (1969).

[182] F. Langouche, D. Roekaerts and E. Tirapegui, Perturbation expansions through functional integrals for nonlinear systems, *Physical Review D* 20, 433 (1979).

[183] H. Leschke, A. C. Hirshfeld and T. Suzuki, Canonical perturbation theory for nonlinear systems, *Physical Review D* 18, 2834 (1978).

[184] A. Salam and J. Strathdee, Equivalent formulations of massive vector field theories, *Physical Review D* 2, 2869 (1970).

[185] M. C. Bergre and Y. -M. P. Lam, Equivalence theorem and Faddeev-Popov ghosts, *Physical Review* D 13, 3247 (1976).

[186] J. Honerkamp and K. Meetz, Chiral-invariant perturbation theory, *Physical Review D* 3, 1996 (1971).

[187] S. Schweber, Perturbation theory and configuration space methods in field theory, *Physical Review* 84, 1259 (1951).

[188] S. -S. Wu, A remark on the conventional perturbation theory, *Physical Review* 83, 730 (1951).

[189] E. Feenberg, A note on perturbation theory, *Physical Review* 74, 206 (1948).

[190] E. L. Hill, Perturbation theory and electron scattering, *Physical Review* 43, 303 (1933).

[191] P. S. Epstein, Problems of quantum theory in the light of the theory of perturbations, *Physical Review* 19, 578 (1922).

[192] K. Nishijima, Asymptotic conditions and perturbation theory, *Physical Review* 119, 485 (1960).

[193] S. Mandelstam, Analytic properties of transition amplitudes in perturbation theory, *Physical Review* 115, 1741 (1959).

[194] R. Karplus, C. M. Sommerfield and E. H. Wichmann, Spectral representations in perturbation theory, I. Vertex function & II. Two-particle scattering, *Physical Review* 111, 1187 (1958); *ibid* 114, 376 (1959).

[195] J. E. Young, Perturbation expansions in the formal theory of scattering, *Physical Review* 109, 2141 (1958).

[196] W. H. Young and N. H. March, Perturbation theory in wave mechanics, *Physical Review* 109, 1854 (1958).

[197] W. Tobocman, Many-body perturbation theory, *Physical Review* 107, 203 (1957).

[198] A. J. Kromminga and M. Bolsterli, Perturbation theory of many-boson systems, *Physical Review* 128, 2887 (1962).

[199] M. Chaichian, A. P. Demichev, *Path Integrals in Physics* Volume 1: *Stochastic Process & Quantum Mechanics*, Taylor & Francis (2001).

[200] P. A. M. Dirac, The Lagrangian in quantum mechanics, *Physikalische Zeitschrift der Sowjetunion* 3, 64 (1933).

[201] H. Kleinert, *Gauge Fields in Condensed Matter*, Vol. 1, World Scientific, Singapore (1989).

[202] R. P. Feynman, Space-time approach to nonrelativistic quantum mechanics, *Review of Modern Physics* 20, 367 (1948).

[203] H. Duru and H. Kleinert, Solution of the path integral for the H-atom, *Physics Letters* 84B, 11 (2007).

[204] R. P. Feynman and A. R. Hibbs, *Quantum Mechanics and Path Integrals*, McGraw-Hill, New York (1965).

[205] H. Ikemori, Path-integral representation of the wave function: The relativistic particle, *Physical Review* D 40, 3512 (1989).

[206] P. A. M. Dirac, On the analogy between classical and quantum mechanics, *Review of Modern Physics* 17, 195 (1945).

[207] H. Kleinert, *Path Integrals in Quantum Mechanics, Statistics, Polymer Physics and Financial Markets* (4th edn.), Work Scientific, Singapore (2004).

[208] J. J. Zinn, *Path Integrals in Quantum Mechanics*, Oxford University Press (2004).

[209] L. S. Schulman, *Techniques & Applications of Path Integration*, John Wiley & Sons, New York (1981).

[210] C. Grosche and F. Steiner, *Handbook of Feynman Path Integrals*, Springer Tracts in Modern Physics 145, Springer-Verlag (1998).

[211] R. J. Rivers, *Path Integrals Methods in Quantum Field Theory*, Cambridge University Press (1987).

[212] S. Albeverio and R. H-. Krohn, *Mathematical Theory of Feynman Path Integral*, Lecture Notes in Mathematics 523, Springer-Verlag (1976).

[213] L. Chetouani, L. Guechi, T. F. Hammann and M. Letlout, Path integral for the damped harmonic oscillator coupled to its dual, *Journal of Mathematical Physics* 35, 1185 (1994).

[214] G. Ghosh and R. W. Hasse, Coherent state and the damped harmonic oscillator, *Physical Review* A 24, 1621 (1981)90.

[215] P. Cartier, D. -M. Cécile, A new perspective on functional integration, *Journal of Mathematical Physics* 36, 2137 (1995).

[216] B. K. Cheng, Exact evaluation of the propagator for the damped harmonic oscillator, *Journal of Physics A: Mathematical and General* 17, 2475 (1984).

[217] M. G. G. Laidlaw and C. M. Dewitt, Feynman functional integrals for systems of indistinguishable particles, *Physical Review D* 3, 1375 (1971).

[218] D. -M. Cécile, Feynman's path integral: Definition without limiting procedure, *Communication in Mathematical Physics* 28, 47 (1972).

[219] L. Cohen, Hamiltonian operators via Feynman path integrals, *Journal of Mathematical Physics* 11, 3296 (1970).

[220] J. S. Dowker, Path Integrals and ordering rules, *Journal of Mathematical Physics* 17, 1873 (1976).

[221] L. Cohen, Correspondence rules and path integrals, *Journal of Mathematical Physics* 17, 597 (1976).

[222] C. Acatrinei, Path integral formulation of noncommutative quantum mechanics, *Journal of High-Energy physics*. JHEP 09, 007 (2001).

[223] H. S. Snyder, Quantized spacetime, *Physical Review* 71, 38 (1947)

[224] J. Glimm and A. Jaffe, *Quantum Physics: A Functional Integral Point of View*, Springer-Verlag, New York (1981).

[225] C. Grosche and F. Steiner, *Handbook of Feynman Path Integrals*, Springer Tracts in Modern Physics, Vol. 145, Springer, Berlin (1998).

[226] R. Mackenzie, *Path Integral Methods and Applications*, Lectures given at Rencotres du Vietnam: VIth Vietnam School of Physics, Vung Tau, Vietnam, 27 December 1999–8 January 2000.

[227] V. Fock, Konfigurationsraum und zweite Quantulung, *Z. Phys.* 75, 622 (1932).

[228] E. Zeidler, *Quantum Field Theory II: Quantum Electrodynamics*, Springer, Berlin (2000).

[229] M. Danos and V. Gillet,*Angular Momentum Calculus in Quantum Physics*, World Scientific (1990).

[230] P. Grassberger and M. Scheunert, Fock-Space methods for identical classical objects, *Progress of Physics* 28, 547 (1980).

[231] R. Becker and G. Leibfried, On the method of second quantization, *Physical Review* 69, 34 (1946).

[232] E. M. Corson, Second quantization and representation theory, *Physical Review* 70, 728 (1946).

[233] R. K. Osborn, The second quantized theory of Spin-1/2 particles in the nonrelativistic limit, *Physical Review* 86, 340 (1952).

[234] T. J. Nelson and Jr. R. H. Good, Second quantization process for particles with any spin and with internal symmetry, *Review of Modern Physics* 40, 508 (1968).

[235] C. L. Hammer and Jr. R. H. Good, Quantization process for massless particles, *Physical Review* 111, 342 (1958).

[236] L. L. Foldy, Synthesis of covariant particle equations, *Physical Review* 102, 568 (1956).

[237] D. H. Kobe, Second quantization in nonrelativistic quantum mechanics, *American Journal of Physics* 51, 312 (1983).

[238] B. R. Judd and J. C. Slater, Second quantization and atomic spectroscopy, *American Journal of Physics* 36, 69 (1968).

[239] M A. L. Fetter and J. D. Walecka, *Quantum Theory of Many-Particle Systems*, McGraw-Hill, San Francisco (1971).

[240] G. D. Mahan, *Many-Particle Physics: Physics of Solids and Liquids*, 2nd edn., Plenum Press, New York (1990).

[241] R. Mills, *Propagators for Many-Particle Systems: An Elementary Treatment*, Gordon and Breach, New York (1969).

[242] A. Altland and B. Simons, *Condensed Matter Field Theory*, Cambridge University Press (2009).

[243] B. Leuenberger, Second quantization as a tool for the calculation of effective Hamiltonians, *Molecular Physics: An International Journal at the Interface between Chemistry and Physics* 59, 241 (1986).

[244] R. H. Landau, *Quantum Mechanics II: A Second Course in Quantum Theory*, Wiley-VCH Verlag GmbH & Co. (2004).

[245] K. G. Wilson, Model Hamiltonians for local quantum field theory, *Physical Review* 140, B445 (1965).

[246] E. M. Henley and W. Thirring, *Elementary Quantum Field Theory*, McGraw-Hill Book Company, Inc., New York (1962).

[247] P. S. Epstein, Problems of quantum theory in the light of the theory of perturbations, *Physical Review* 19, 578 (1922).

[248] A. M. Astashov and A. M. Vinogradov, On the structure of Hamiltonian operators in the field theory, *Journal of Geometry and Physics* 3, 263 (1986).

[249] D. B. Fairlie and J. Nuyts, Fock space representations for non-Hermitian

Hamiltonians, *Journal of Physics A: Mathematical and General* 38, 3611 (2005).

[250] J. Hubbard, Electron correlations in narrow energy bands, *Proceedings of the Royal Society London A* 276, 238 (1963).

[251] I. Montvay and G. Mnster, *Quantum Fields on a Lattice*, Cambridge Monographs on Mathematical Physics, Cambridge (1994).

[252] B. Simon, *The Statistical Mechanics of Lattice Gases*, vol.1, Princeton U. Press (1993).

[253] J. R. Schrieffer and P. A. Wolff, Relation between Anderson and Kondo Hamiltonians, *Phys. Rev.* 149, 491 (1966).

[254] P. Yu and M. Cardona, *Fundamentals of Semiconductors*, (4th edn.) Springer (2010).

[255] L. P. Kadanoff and G. Baym, *Quantum Statistical Mechanics: Green's Function Methods in Equilibrium and Non-Equilibrium Problems*, W. A. Benjamin, New York (1962).

[256] V. L. B-. Bruevich and S. V. Tyablikov, *The Green Function Method in Statistical Mechanics*, North Holland Publishing Co. (1962).

[257] A. A. Abrikosov, L. P. Gorkov and I. E. Dzyaloshinski, *Methods of Quantum Field Theory in Statistical Physics*, Englewood Cliffs: Prentice-Hall (1963).

[258] W. C. Schieve and L.P. Horwitz, *Quantum Statistical Mechanics*, Cambridge University Press (2009).

[259] R. D. Mattuck, *A Guide to Feyman Diagrams in the Many-Body Problem*, Dover Publications (1992).

[260] S. Fujita and R. Hirota, Quantum statistics of interacting particles: Thermodynamic quantities and pair distribution function, *Physical Review* 118, 6 (1960).

[261] P. C. Martin and J. Schwinger, Theory of many-particle systems I, *Physical Review* 115, 1342 (1959).

[262] T. Matsubara, A new approach to quantum statistical mechanics, *Progress of Theoretical Physics* 14, 351 (1955).

[263] P. Nozires, *Theory of Interacting Fermi Systems*, Addison-Wesley (1997).

[264] D. J. Thouless, *The Quantum Mechanics of Many-Body Systems*, Academic Press, New York (1972).

[265] S. Fujita, Resolution of the hierarchy of Green's functions for fermions, *Physical Review A* 4, 1114 (1971).

[266] A. G. Hall, Non equilibrium Green's functions: Generalized Wick's theorem and diagrammatic perturbation theory with initial correlations, *Journal of Physics A: Mathematical and General* 8, 214 (1975).

[267] D. J. Thouless, Use of field theory techniques in quantum statistical mechanics, *Physical Review* 107, 1162 (1957).

[268] M. Balzer and M. Potthoff, Nonequilibrium cluster perturbation theory, *Physical Review B* 83, 195132 (2011).

[269] M. P. Krl, Nonequilibrium linked cluster expansion for steady state quantum transport, *Physical Review B* 56, 7293 (1997).

[270] M. Wagner, Expansions of nonequilibrium Green's functions, *Physical Review B* 44, 6104 (1991).

[271] G. Baym and L. P. Kadanoff, Conservation laws and correlation functions, *Physical Review* 124, 287 (1961).

[272] P. Danielewicz, Quantum theory of nonequilibrium processes I, *Annals of Physics* 152, 239 (1984).

[273] J. Rammer and H. Smith, Quantum field theoretical methods in transport theory of metals, *Review of Modern Physics* 58, 232 (1986).

[274] A. G. Hall, Nonequilibrium Green functions: Generalized Wick's theorem and diagrammatic perturbation with initial correlations, *Journal of Physics A: Mathematical and General* 8, 214 (1975).

[275] H. Bruus and K. Flensberg, *Many-Body Quantum Theory in Condensed Matter Physics*, Oxford University Press (2004).

[276] W. R. Bandy and A. J. Glick, Tight binding Green's function calculation of electron tunneling. I & II, One-dimensional two band model, *Physical Review B* 13, 3368 (1976), *ibid* 16, 2346 (1977).

[277] L. Allen and J.H. Eberly, *Optical Resonance and Two-Level Atoms*, John Wiley and Sons (1975).

[278] L. Mandel and E. Wolf, *Optical Coherence and Quantum Optics*, Cambridge University Press (1995).

[279] J. Rammer, *Quantum Field Theory of Nonequilibrium States*, Cambridge University Press (2007).

[280] G. S. Setlur, and Y.-C. Chang, Role of non-equilibrium dynamical screening in carrier thermalization, *Phys. Rev. B* 55, 1517 (1997).

[281] P.M. Bakshi and K.T. Mahantappa, Expectation value formalism in quantum field theory, *J. Math. Phys.* 4, 1 (1963); 4, 12 (1963).

[282] L.V. Keldysh, Diagram technique for nonequilibrium processes, *Z. Eksp. Teor. Fiz.* 47, 1515 (1964) [*Sov. Phys. JETP* 20, 4 (1965)].

[283] J. R. Klauder and B-S. Skagerstam (editors), *Coherent States: Applications in Physics and Mathematical Physics*, World Scientific, Singapore (1985).

[284] L. S. Schulman, *Techniques and Applications of Path Integration*, Wiley, New York (1981).

[285] T. Kashiwas and Y. Ohnuki and M. Suzuki, *Path Integral Methods*, Oxford University Press, New York (1997).

[286] J. Zak, Finite Translations in solid state physics, *Physical Review Letters* 19, 1385 (1967).

[287] L. S. Schulman, A path integral for spin, *Physical Review* 176, 1558 (1968).

[288] L. S. Schulman, Green's function for an electron in a lattice, *Physical Review* 188, 1139 (1969).

[289] J. Shibata and S. Takagi, A note on (spin) coherent state path integral, *International Journal of Modern Physics B* 13, 107 (1999).

[290] H. G. Solari, Semiclassical treatment of spin system by means of coherent states, *Journal of Mathematical Physics* 28, 1097 (1987).

[291] J. R. Klauder, Path integrals and stationary phase approximations, *Physical Review D* 19, 2349 (1979).

[292] H. Kuratsuji, Classical quantization of time dependent Hartree Fock solutions by coherent state path integrals, *Physics Letters B* 103, 79 (1981).

[293] R. J. Glauber, Coherent and incoherent states of the radiation field, *Physical Review* 131, 2766 (1963).

[294] Y. Weissman, Semiclassical approximation in the coherent states representation, *Journal of Chemical Physics* 76, 4067 (1982).

[295] Y. Weissman, On the stationary phase evaluation of path integrals in teh coherent states representation, *Journal of Physics A: Mathematical and General* 16, 2693 (1983).

[296] T. -K. Ng, *Introduction to Classical and Quantum Field Theory*, Wiley-VCH Verlag GmbH & Co. (2009).

[297] J. -P. Gazeau, *Coherent States in Quantum Physics*, Wiley-VCH Verlag GmbH & Co. (2009).

[298] A. Altland and B. Simons, *Condensed Matter Field Theory*, Cambridge University Press (2009).

[299] H. S. Tan, A coherent state based path integral for quantum mechanics on the Moyal plane, *Journal of Physics A: Mathematical and General* 39, 15299 (2006).

[300] A. M. Perelomov, *Generalized Coherent States and Their Applications*, Springer (Berlin) (1986).

[301] Jean Zinn-Justin, *Quantum Field Theory and Critical Phenomena*, Oxford University Press, 4th edn. (2002).

[302] A. A. Grib and Jr. W. A. Rodrigues, *Nonlocality in Quantum Physics*, Springer (1999).

[303] M S. Popescu and D. Rohrlich, Quantum nonlocality as an axiom, *Foundations of Physics* 24, 379 (1994).

[304] H. Yukawa, Quantum theory of nonlocal fields. Part I. Free fields, *Physical Review* 77, 219 (1950); H. Yukawa, Quantum Theory of non-local fields. Part II. Irreducible Fields and their interaction, *Physical Review* 80, 1047 (1950).

[305] G. V. Efimov, Nonlocal quantum theory of the scalar field, *Communications in Mathematical Physics* 5, 42 (1967).

[306] D. Lee and H. Rabitz, Scaling of nonlocal operators, *Physical Review A* 32, 877 (1985).

[307] H. Hafermann, S. Brener, A. N. Rubtsov, M. I. Katsnelson and A. I. Lichtenstein, Cluster dual fermion approach to nonlocal correlations, *JETP Letters* 86, 677 (2007).

[308] A. D. Mironov and A. V. Zabrodin, Finite size effects in conformal field theories and nonlocal operators in one-dimensional quantum systems, *Journal of Physics A: Mathematical and General* 23, L493 (1990).

[309] D. Feldman, On realistic field theories and the polarization of the vacuum, *Physical Review* 76, 1369 (1949).

[310] M. A. L. Capri, V. E. R. Lemes, R. F. Sobreiro, S. P. Sorella and R. Thibes, Local renormalizable gauge theories from nonlocal operators, *Annals of Physics* 323, 752 (2008).

[311] M. A. L. Capri, V. E. R. Lemes, R. F. Sobreiro, S. P. Sorella, R. Thibes, Local renormalizable gauge theories from nonlocal operators, *Annals of Physics* 323 (2007).

[312] A. D. Mironov and A. V. Zabrodin, Finite size effects in conformal field theories and nonlocal operators in one dimensional quantum systems, *Journal of Physics A: Mathematical and General* 23, L493 (1990).

[313] D. R. Yennie, Some remarks on nonlocal field theory, *Physical Review* 80, 1053 (1950).

[314] G. Wataghin, On a nonlocal field theory, *Il Nuovo Cimento* 10, 1602 (1953).

[315] W. Güttinger, Non-local structure of quantized field theories of the second kind, *Nuclear Physics* 9, 429 (1958).

[316] L. D. Landau and E. M. Lifshitz, *Statistical Physics, Part 2: Theory of the Condensed State, A Course of Theoretical Physics*, vol. 9, Butterworth Heinemann (2002).

[317] S. Coleman, The quantum Sine-Gordon equation as the massive Thirring model, *Phys. Rev. D* 11, 2088 (1975).

[318] A. Luther, Tomonaga fermions and the Dirac equation in three dimensions, *Phys. Rev. B* 19, 320 (1979).

[319] S. Mandelstam, Soliton operators for the quantized Sine-Gordon equation, *Phys. Rev. D* 11, 3026 (1975).

[320] F.D.M. Haldane, 'Luttinger liquid theory' of one-dimensional quantum fluids, *J. Phys. C* 14, 2585 (1981); *Phys. Lett. A* 93, 464(1983); *Phys. Rev. Lett.* 50, 1153(1983).

[321] R. Heidenreich, R. Seiler and D.A. Uhlenbrock, The Luttinger model, *J. Stat. Phys.* 22, 27 (1980).

[322] F. D. M. Haldane, *Helv. Phys. Acta.* 65, 152 (1992); Perspectives in many-particle physics, *Proceedings of the International School of Physics 'Enrico Fermi' Course CXXI*, Varenna, 1992 edited by R. Schreiffer and R. A. Broglia (North-Holland, New York, 1994), also available as cond-mat/0505529.

[323] A. H. Castro Neto and E. H. Fradkin, Bosonization of quantum fluids, *Phys. Rev. Lett.* 72, 1393 (1994); *Phys. Rev. B* 49, 10877 (1994); 51, 4084 (1995).

[324] A. Houghton and J. B. Marston, Multidimensional Bosonization, *Phys. Rev. B* 48, 7790 (1993); A. Houghton, H. J. Kwon and J. B. Marston, *Phys. Rev.*

*B* 50, 1351 (1994); A. Houghton, H. J. Kwon, J. B. Marston and R. Shankar, *J. Phys. C* 6, 4909 (1994).

[325] P. Kopietz, J. Hermisson and K. Schonhammer, Bosonization of interacting fermions in arbitrary dimensions, *Phys. Rev. B* 52, 10877 (1995); L. Bartosch and P. Kopietz, *Phys. Rev. B* 59, 5377 (1999); P. Kopietz and K. Schonhammer, *Z. Phys. B* 100, 259 (1996).

[326] P. Kopietz in *Proceedings of the Raymond L. Orbach Symposium* edited by D. Hone (World Scientific, Singapore, 1996) pp. 101–119; P. Kopietz and G. E. Castilla, *Phys. Rev. Lett.* 76, 4777 (1996); P. Kopietz, *Bosonization of Interacting Fermions in Arbitrary Dimensions* (Springer Verlag, Berlin, 1997).

[327] P.-A. Bares and X.-G. Wen, Breakdown of Fermi liquid due to long-range interactions, *Phys. Rev. B* 48, 8636 (1993).

[328] L. Bartosch and P. Kopietz, Correlation functions of higher dimensional Luttinger liquids, *Phys. Rev. B* 59, 5377 (1999).

[329] G. S. Setlur and V. Meera, A general approach to higher dimensional bosonization, *Pramana J. Phys.*, 69, 639(2007); G. S. Setlur, *Pramana J. Phys.*, 66, 575(2006); 62, 115(2004); 62, 101(2004); G. S. Setlur and D. S. Citrin, *Phys. Rev. B* 65, 165111(2002); G. S. Setlur and Y. C. Chang, *Phys. Rev. B* 57, 15144(1998); See also the comment and reply, L. C. Cune and M. Apostol, *Phys. Rev. B* 60, 8388 (1999); G. S. Setlur and Y. C. Chang, *Phys. Rev. B* 60, 8390(1999).

[330] F. Schutz, L. Bartosch, P. Kopietz, Collective fields in the functional renormalization group for fermions, Ward identities, and the exact solution of the Tomonaga-Luttinger model, *Phys. Rev. B* 72, 035107 (2005); I. V. Yurkevich, in *Bosonization as the Hubbard Stratanovich Transformation*, edited by I. V. Lerner et al. (Kluwer Academic Dordrecht, 2002) vol. 69. (see also I. Yurkevich cond-mat/0112270).

[331] T. Giamarchi, *Quantum Physics in One Dimension*, International Series of Monographs on Physics 121, Oxford University Press (2004).

[332] D. K. K. Lee and Y. Chen, Functional bosonisation of the Tomonaga-Luttinger model, *Journal of Physics A: Mathematical and General* 21, 4155 (1988).

# Index

For Product Safety Concerns and Information please contact our EU
representative GPSR@taylorandfrancis.com Taylor & Francis Verlag GmbH,
Kaufingerstraße 24, 80331 München, Germany

Printed and bound by CPI Group (UK) Ltd, Croydon, CR0 4YY
01/05/2025
01858546-0008